Forum for Interdisciplinary Mathematics

The *Forum for Interdisciplinary Mathematics* is a Scopus-indexed book series. It publishes high-quality textbooks, monographs, contributed volumes and lecture notes in mathematics and interdisciplinary areas where mathematics plays a fundamental role, such as statistics, operations research, computer science, financial mathematics, industrial mathematics, and bio-mathematics. It reflects the increasing demand of researchers working at the interface between mathematics and other scientific disciplines.

Mohamed Kharrat · Nouressadat Touafek ·
Moez Krichen
Editors

Modeling of Discrete and Continuous Systems

Ordinary, Partial and Fractional Derivatives

Springer

Editors
Mohamed Kharrat
Department of Mathematics
Jouf University
Sakakah, Saudi Arabia

Nouressadat Touafek
Department of Mathematics
University of Jijel
Jijel, Algeria

Moez Krichen
Department of Information Technology
Al-Baha University
Alaqiq, Saudi Arabia

ISSN 2364-6748 ISSN 2364-6756 (electronic)
Forum for Interdisciplinary Mathematics
ISBN 978-981-97-8714-2 ISBN 978-981-97-8715-9 (eBook)
https://doi.org/10.1007/978-981-97-8715-9

Mathematics Subject Classification: 60Hxx, 34-XX, 60Gxx, 26A33, 35-XX, 37-XX, 39-XX

This Springer imprint is published by the registered company Springer Nature Singapore Pte Ltd.
The registered company address is: 152 Beach Road, #21-01/04 Gateway East, Singapore 189721, Singapore

Preface

The exploration of mathematical modeling, particularly in the context of discrete and continuous systems governed by ordinary, partial and fractional derivatives, stands as a testament to the human endeavor to comprehend and harness the complexities of the natural and engineered world. This book is a curated collection of 17 chapters, each of which sheds light on different facets of this multifaceted field.

The chapters in this volume cover a broad spectrum of topics, reflecting the diverse applications and theoretical underpinnings of mathematical modeling. For instance, the book opens with a chapter on the application of fractional differential equations in engineering sciences, specifically in modeling the quasi-static mechanical behavior of polyurethane foam. This is followed by a discussion on a new operator approach for solving time-fractional nonlinear Burgess equations, showcasing the evolution of methodologies in tackling complex mathematical problems.

Further enriching the content, the book delves into explicit formulas for the solutions of higher-order systems of difference equations, stochastic integrodifferential equations with deviating arguments driven by Poisson jumps, and a comprehensive review of fractional calculus in the modeling of cancer. These chapters not only highlight the theoretical advancements in the field but also underscore the relevance of these mathematical tools in addressing real-world challenges.

The volume also includes chapters that focus on the development of conformable fractional Taylor series algorithms for solving temporal fractional reaction-diffusion-convection equations, the analysis of Alzheimer's disease using nonlocal operators, and the evaluation of European options under different time-fractional models. These contributions demonstrate the interdisciplinary nature of mathematical modeling, bridging the gap between mathematics, biology, finance and other domains.

The volume editors have meticulously assembled this collection to provide a comprehensive overview of the field, ensuring that each chapter contributes to a holistic understanding of the subject. Their diverse backgrounds in mathematics, computer science and engineering add a rich multidisciplinary flavor to the book, making it accessible to a wide audience.

The book is aimed at a broad spectrum of readers, including researchers, academics, graduate students and professionals who are engaged in the study, teaching, or application of mathematical modeling. It is our hope that this volume will serve as a valuable resource for those seeking to deepen their understanding of discrete and continuous systems and their derivatives, as well as inspire further research and innovation in this dynamic field.

In closing, we express our heartfelt gratitude to all the contributors for their dedication and insightful contributions to this volume. Their collective efforts have resulted in a book that not only serves as a repository of knowledge but also as a beacon for future explorations in the fascinating world of mathematical modeling.

Sakakah, Saudi Arabia — Mohamed Kharrat
Jijel, Algeria — Nouressadat Touafek
Alaqiq, Saudi Arabia — Moez Krichen

Contents

Fractional Differential Equations in Engineering and for Polyurethane Foam Modelling

Hamdi Jmal and Raphaël Dupuis

Abstract Differential equations form the basis of consecutive equations in the field of engineering sciences, in particular in mechanics and material sciences. In fact, the use of rheological constitutive equations incorporating fractional derivatives has garnered significant attention over the past fifty years. Fractional derivative equations have demonstrated consistency with molecular theories [1]. In terms of thermodynamic compatibility, rheological models incorporating fractional derivatives offer a more precise fit with laboratory measurement results [2, 3]. Therefore, in this chapter, we demonstrate in a first part (Sect. 1) the interest of differential equations with fractional order in the description of the mechanical behaviour of materials. Due to its ability to absorb deformation energy, flexible polyurethane (PU) foam finds widespread usage in various comfort applications, including automotive seat cushions and mattresses. Designing a mechanical model to characterize its behaviour under different test conditions holds considerable interest for optimizing comfort of the personalized seating occupant system under static, quasi-static, and dynamic conditions. When it is subject to a significant compression deformation, PU exhibits highly nonlinear elasticity and viscoelastic behaviour. The theory of viscoelasticity holds significant importance in describing materials characterized by time-dependent mechanical behaviour. Constitutive models for small one-dimensional deformations in viscoelastic materials can be expressed in either integer or fractional differential equations. In the second part (Sect. 2) of this chapter, we focus on modelling the quasi-static behaviour of polyurethane foam using a memory integer and fractional models. The parameters of two memory models (fractional memory model and integer memory model) were determined through original minimization and identification methods. The identified parameters meet the thermodynamic conditions. The results show good agreement between experimental behaviour of polyurethane

H. Jmal (✉)
ICube, Université de Strasbourg, CNRS UMR 7357, Manufacture des Tabacs, 1 Cr des Cigarières, 67000 Strasbourg, France
e-mail: jmal@unistra.fr

R. Dupuis
IRIMAS, ENSISA, Université de Haute Alsace, Mulhouse, 12 rue des frères Lumières, 68093 Mulhouse, France
e-mail: raphael.dupuis@uha.fr

M. Kharrat et al. (eds.), *Modeling of Discrete and Continuous Systems*, Forum for Interdisciplinary Mathematics, https://doi.org/10.1007/978-981-97-8715-9_1

foam and those predicted by two memory models. Finally, a comparison between these models was conducted.

Acronyms

Symbol	Units	Description
t	$[s]$	Time
T	$[s]$	Cycle period
T_e	$[s]$	Sampling period
x	$[m]$	Displacement
(L_0, b_0, h_0)	(m, m, m)	Initial dimensions of polyurethane foam samples
ϵ	$[mm.mm^{-1}]$	Strain
$\dot{\epsilon}$	$[s^{-1}]$	Strain rate
ϵ_{max}	$[mm.mm^{-1}]$	Maximum strain
ϵ_0	$[mm.mm^{-1}]$	Initial strain
$(q_i)_{i=1..g}$		Coefficient related to the $i-th$-order derivative of the function $\sigma(t)$
$(p_i)_{i=1..m}$		Coefficient related to the $i-th$-order derivative of the function $\epsilon(t)$
s		Laplace operator
E	$[N.m^{-2}]$	Elastic modulus
μ	$[N.m^{-2}.s]$	Damping coefficient
β		Fractional derivation order
η	$[N.m^{-2}.s^{\beta}]$	Fractional dashpot coefficient
Γ		Gamma function
F	$[N]$	Foam force response
F_e	$[N]$	Viscoelastic force
F_{ve}	$[N]$	Elastic force (spring force)
K_i	$[N.m^i]$	Spring stiffness of order i
k_i	$[N.m^i]$	Global elastic stiffness of order i
K_{ve}	$[N.m^1]$	Viscoelastic stiffness
N_W		Integer memory viscoelastic order
Re		Real part of a complex value
Im		Imaginary part of a complex value
α_l	$[s^{-1}]$	Complex number representing the lth viscoelastic mode in the integer memory model
Re α_l	$[s^{-1}]$	Real part of lth viscoelastic mode representing the lth time relaxation in the integer memory model
Im α_l	$[s^{-1}]$	Imaginary part of lth viscoelastic mode representing the lth frequency in the integer memory model
a_l	$[N.m^{-2}.s^{-1}]$	Complex number indicating the lth viscoelastic residue in the integer memory model
N_{de}		Fractional memory viscoelastic order
M		Order of elastic polynomial
N		Number of cycles in quasi-static tests
$\Gamma(g, f)$		Incomplete Gamma function defined by: $\Gamma(g, f) = \int_0^f \tau^{g-1} e^{-\tau} d\tau; g > 0$ and $f \in \Re$
$\Gamma(g)$		Gamma function defined by $\Gamma(g) = \int_0^{+\infty} \tau^{g-1} e^{-\tau} d\tau; g > 0$

1 Part 1- Fractional Differential Equations in Engineering Sciences

1.1 General Context

The use of rheological constitutive equations incorporating fractional derivatives has garnered significant attention over the past fifty years [4–9]. Bagley and Torvik1 demonstrated the consistency of constitutive models employing fractional derivatives with molecular theories. Some researchers have explored thermodynamic compatibility in the context of fractional constitutive models [10–14]. Paola et al. [2] proposed that a fractional rheological element provides a more accurate fit to observed results from laboratory measurements. In this part of the chapter, we demonstrate the importance of the fractional derivation in engineering sciences and in the description of the viscoelastic behaviour of materials. The mathematically proven equation is also described in this part.

1.2 Importance of Fractional Derivation for Viscoelastic Material and Mathematical Equations

Partial or total derivative equations are widely used in the field of engineering sciences to describe thermodynamic, electrical, mechanical solid or fluid systems, chemical systems, etc. Those equations are established whenever it is necessary to study the variation of a physical quantity or a variable with respect to another. In the most common and simplest cases, the relationship between causes and effects, regardless of the nature of the physical system and the operating conditions, is defined by an equation of the following form:

$$p_0\sigma(t) + p_1\frac{d\sigma(t)}{dt} + \ldots + p_m\frac{d^m\sigma(t)}{dt^m} = q_0\varepsilon(t) + q_1\frac{d\varepsilon(t)}{dt} + \ldots + q_g\frac{d^g\varepsilon(t)}{dt^g} \tag{1}$$

where (g,m) are two arbitrary integers and $(q_i)_{i=1..g}$ (resp. $(p_i)_{i=0..m}$) being characteristic coefficient related to the ith-order derivative of the cause function $\sigma(t)$ (resp. the effect function $\varepsilon(t)$). From a mathematical point of view, this description is valid only if the functions $\sigma(t)$ and $\varepsilon(t)$ are of class $C^{(m)}$ and $C^{(g)}$, respectively. The cause-and-effect functions can be scalar, vector, or more generally tensor functions. The characteristic coefficients can be constants or functions of the dependent variables or functions of their derivatives with respect to the independent variable. Based on the types of coefficients, various families of integer derivative equations are distinguished. When the characteristic coefficients of Eq. 1 are constants, it is called a linear differential equation for which general methods of analytical solution may exist. First-order linear equations given generally by Eq. 2 and second-order linear equations given generally by Eq. 3 are frequently encountered in engineering sciences. A significant applied mathematics lecture, especially in engineering, focuses on these types of equations.

$$q_0\varepsilon(t) + q_1\frac{d\varepsilon(t)}{dt} = \sigma(t) \tag{2}$$

$$q_0\varepsilon(t) + q_1\frac{d\varepsilon(t)}{dt} + q_2\frac{d^2\varepsilon(t)}{dt^2} = \sigma(t) \tag{3}$$

For example, to study the uniform rectilinear motion of a damped and forced mechanical system, the relationship between the cause, which is the force $F(t)$, and the effect which is the displacement $x(t)$ of the system, is given by Eq. 4. When the system undergoes accelerated or decelerated motion, the relationship is described by Eq. 5.

$$kx(t) + c\frac{dx(t)}{dt} = F(t) \tag{4}$$

$$kx(t) + c\frac{dx(t)}{dt} + M\frac{d^2x(t)}{dt^2} = F(t) \tag{5}$$

The analytical solution $\varepsilon(t)$ of a first-order or second-order linear differential equations is composed of a homogeneous solution $\epsilon_h(t)$, determined by assuming a null effect function $\sigma(t) = 0$, and a particular function $\epsilon_p(t)$ that has the same form as the effect function $\sigma(t)$. The analytic solutions of a first-order or second-order linear differential equations are widely detailed and it will not be the subject of this chapter.

For the case of a more general linear equation respecting the form of Eq. 1, using a Laplace transform is the basis of the method for analytical solution by incorporating the initial conditions of the cause-and-effect functions.

When the initial conditions are zero, applying a Laplace transform establishes the following relationship:

$$\boldsymbol{H_{ve}}^*(s) = \frac{\boldsymbol{\sigma}^*(s)}{\boldsymbol{\epsilon}^*(s)} \Rightarrow \boldsymbol{\sigma}^*(s) = \boldsymbol{H_{ve}}^*(s)\boldsymbol{\epsilon} * (s) = \frac{\sum_{l=0}^{m} q_l s^l}{\sum_{l=0}^{w} p_l s^l}\boldsymbol{\epsilon}^*(s) \tag{6}$$

Here, **s** is the Laplace operator and ($\boldsymbol{\sigma}^*$, $\boldsymbol{\epsilon}^*$, $\mathbf{H_{ve}}^*$) are representing the Laplace domain representations of the cause function $\sigma(t)$, the effect function $\varepsilon(t)$, and the transfer function $H_{ve}(t)$, respectively.

The transformation back to the time domain allows us to express the function $\sigma(t)$ in terms of $\varepsilon(t)$ (Eq. 7).

$$\sigma(t) = \int_0^t H_{ve}(t-\tau)\varepsilon(\tau)d\tau \tag{7}$$

As seen with Eq. 7, this modelling takes into account the memory effect through the convolution product. This description is useful for description of shape memory material in which the response at time t depends on the material's history.

The derivation used in Eq. 1 is called natural or ordinary because it is total with an integer order. This implies that fractional derivation with non-integer order is non-natural derivation. However, fractional derivation is also natural as integer derivation. To demonstrate this, let us revisit the phenomenological elements commonly used in describing the mechanical behaviour of materials.

A material exhibits a linear elastic behaviour when the relationship between the cause function, the stress $\sigma(t)$, is proportional to the strain $\varepsilon(t)$:

$$\sigma(t) = E\varepsilon(t) \tag{8}$$

Equation 8 is a specific case of Eq. 1 with $m = 0$, $g = 0$, $q_0 = 1$, and $p_0 = E$, where E represents the elastic modulus, which indicates the material's rigidity. Higher the modulus, more the material is stiffer. An elastic material is described by a spring element (Fig. 1). Note that the strain $\varepsilon(t)$ is mathematically the zero's order derivation of the strain function.

A material exhibits a linear viscous behaviour if the stress $\sigma(t)$ is proportional to its strain rate $\dot{epsilon} = \frac{d\epsilon(t)}{dt}$.

$$\sigma(t) = \mu \frac{d\varepsilon(t)}{dt} \tag{9}$$

Equation 9 is a specific case of Eq. 1 with $m = 0$, $g = 0$, $q_0 = 1$, $p_0 = 0$, and $p_1 = \mu$, where μ represents the damping coefficient which characterizes the capacity of material to absorb energy. A viscous material is described by a dashpot element (Fig. 1).

A material that exhibits viscoelastic behaviour is a material that is both viscous and elastic at the same time. From this perspective, a combination of a spring and a damper in series or parallel (Fig. 2) can be used to model viscoelastic behaviour. Here, we can distinguish different models such as the Kelvin-Voigt model (Eq. 10), Maxwell model (Eq. 11), Zener model (Eq. 12), and among others to describe viscoelastic material.

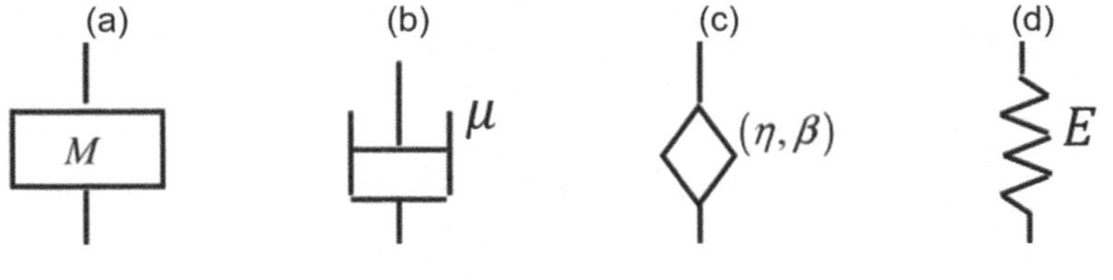

Fig. 1 Basic rheologic mechanical elements: **a** Mass element, **b** Dashpot element, **c** Fractional dashpot element, and **d** Spring element

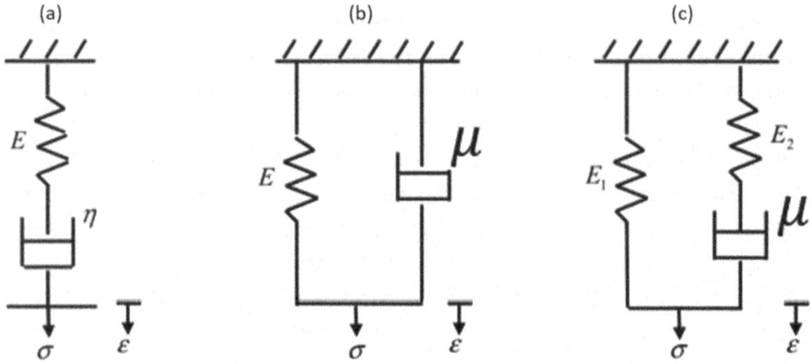

Fig. 2 Viscoelastic rheologic model: **a** Maxwell model, **b** Kelvin-Voigt model, and **c** Zener model

$$\frac{d\varepsilon(t)}{dt} = \frac{1}{E}\frac{d\sigma(t)}{dt} + \frac{1}{\mu}\sigma(t) \tag{10}$$

$$\phi(t) = E\varepsilon(t) + \mu\frac{d\varepsilon(t)}{dt} \tag{11}$$

$$\frac{E_1 + E_2}{E_1}\frac{d\varepsilon(t)}{dt} + \frac{E_2}{\mu}\varepsilon = \frac{1}{E_1}\frac{d\sigma(t)}{dt} + \frac{1}{\mu}\sigma(t) \tag{12}$$

From another point of view, viscoelastic behaviour is neither purely elastic nor purely viscous but rather intermediate between the two. In the case of purely viscous behaviour, the stress is proportional to the first-order derivative of deformation, while in the case of purely elastic behaviour, the stress is proportional to the zero-order derivative of deformation. Therefore, behaviour is an intermediate behaviour that is defined by stress proportional to a derivative between 0 and 1 of deformation, meaning it is proportional to a fractional-order derivative (Eq. 13). In this case, the viscoelastic material is described by a fractional dashpot element (Fig. 1).

$$\sigma(t) = \eta\frac{d^{\beta}\varepsilon(t)}{dt^{\beta}} \tag{13}$$

where η is the fractional dashpot coefficient and β a fractional derivation order. To calculate the term $\frac{d^{\beta}\varepsilon(t)}{dt^{\beta}}$, Riemann-Liouville definition (Eq. 14) is commonly used in the literature for fractional differentiation.

$$\frac{d^{\beta}\varepsilon(t)}{dt^{\beta}} = \frac{1}{\Gamma(n-\beta)}\left(\frac{d^n}{dt^n}\left(\int_0^t \varepsilon(\tau)(t-\tau)^{n-\beta-1}d\tau\right)\right) \tag{14}$$

where $n = Int(\beta) + 1$.

$Int(\beta)$ denotes the integer part of β and $\Gamma(t)$is gamma function. In the case of linear evolution of $\epsilon(\tau)$, where $\epsilon(\tau) = z\tau + w$,we propose to consider a general expression of the integral $\int_0^t \varepsilon(\tau)(t-\tau)^{n-\beta-1}d\tau$ given by

$$I\left(t_0, t_1, t_2, z, w\right) = \int_{t_0}^{t_1} (z\tau + w)\left(t_2 - \tau\right)^{n-\beta-1} d\tau \tag{15}$$

Replacing $(n - \beta - 1)$ by γ and using a part integration, the $\int_{t_0}^{t_1} \tau\left(t_2 - \tau\right)^{\gamma} d\tau$ is equal to $\frac{1}{(\gamma+1)(\gamma+2)}\left(\left(t_2 - t_0\right)^{(\gamma+1)}\left((\gamma+1)t_0 + t_2\right) - \left(t_2 - t_1\right)^{(\gamma+1)}\left((\gamma+1)t_1 + t_2\right)\right)$. Thus, the development of $I\left(t_0, t_1, t_2, z, w\right)$is given by

$$\begin{aligned}
I\left(t_0, t_1, t_2, z, w\right) &= \int_{t_0}^{t_1} (z\tau + w)\left(t_2 - \tau\right)^{n-\beta-1} d\tau \\
&= z\int_{t_0}^{t_1} \tau\left(t_2 - \tau\right)^{\gamma} d\tau + w\int_{t_0}^{t_1} \left(t_2 - \tau\right)^{\gamma} d\tau \\
&= \frac{z}{(\gamma+1)(\gamma+2)}\begin{pmatrix} \left(t_2 - t_0\right)^{\gamma+1}\left((\gamma+1)t_0 + t_2\right) \\ -\left(t_2 - t_1\right)^{\gamma+1}\left((\gamma+1)t_1 + t_2\right) \end{pmatrix} + \frac{w}{\gamma+1}\left[\tau^{\gamma+1}\right]_{t_2-t_1}^{t_2-t_0} \\
&= \frac{z}{(n-\beta)(n-\beta+1)}\begin{pmatrix} \left(t_2 - t_0\right)^{n-\beta}\left((n-\beta)t_0 + t_2\right) \\ -\left(t_2 - t_1\right)^{n-\beta}\left((n-\beta)t_1 + t_2\right) \end{pmatrix} + \frac{w}{n-\beta}\begin{pmatrix} \left(t_2 - t_0\right)^{n-\beta} \\ -\left(t_2 - t_1\right)^{n-\beta} \end{pmatrix} \\
&= \frac{z}{(n-\beta)(n-\beta+1)}\begin{pmatrix} D2\left(t_2, t_0, n-\beta\right) \\ -D2\left(t_2, t_1, n-\beta\right) \end{pmatrix} + \frac{w}{n-\beta}\begin{pmatrix} D1\left(t_2, t_0, n-\beta\right) \\ -D1\left(t_2, t_1, n-\beta\right) \end{pmatrix}
\end{aligned} \tag{16}$$

where $D2\left(t, t_x, \lambda\right) = \left(t - t_x\right)^{\lambda}\left(\lambda t_x + t\right)$ and $D1\left(t, t_x, \lambda\right) = \left(t - t_x\right)^{\lambda}$. In the case where $t_2 = t_0$, the integration I is given by Eq. 17.

$$I\left(t_2, t_1, t_2, z, w\right) = \frac{-z}{(n-\beta)(n-\beta+1)} D2\left(t_2, t_1, n-\beta\right) + \frac{-w}{n-\beta} D1\left(t_2, t_1, n-\beta\right) \tag{17}$$

In the particular case where $t_1 = t_2$, the integration I is given by Eq. 18.

$$I\left(t_0, t_2, t_2, z, w\right) = \frac{z}{(n-\beta)(n-\beta+1)} D2\left(t_2, t_0, n-\beta\right) + \frac{w}{n-\beta} D1\left(t_2, t_0, n-\beta\right) \tag{18}$$

Considering $S = t - t_x$, the derivation until the nth order of the functions $D1\left(t, t_x, \lambda\right)$ and $D2\left(t, t_x, \lambda\right)$ according to time t are, respectively, given by Eqs. 19 and 20.

$$\frac{d^n\left(D1\left(t,t_x,\lambda\right)\right)}{dt^n}=\left(\prod_{i=0}^{n-1}(\lambda-i)\right)(t-t_x)^{\lambda-n} \tag{19}$$

$$\frac{d^n\left(D2\left(t,t_x,\lambda\right)\right)}{dt^n}=\begin{pmatrix}(\lambda+1)t_x\left(\prod_{i=0}^{n-1}(\lambda-i)\right)(t-t_x)^{\lambda-n}\\ +\left(\prod_{i=0}^{n-1}(\lambda+1-i)\right)(t-t_x)^{\lambda+1-n}\end{pmatrix} \tag{20}$$

The proof of Eq. 19 is detailed as follows:

$$\begin{cases}\frac{d(D1(t,t_x,\lambda))}{dt}=\frac{d(t-t_x)^{\lambda}}{dt}=\frac{dS^{\lambda}}{dS}\frac{dS}{dt}=\frac{dS^{\lambda}}{dS}=\lambda S^{\lambda-1}\\ \frac{d^{n-1}(D1(t,t_x,\lambda))}{dt^{n-1}}=\frac{d^{n-1}S^{\lambda}}{dS^{n-1}}=\left(\prod_{i=0}^{n-2}(\lambda-i)\right)S^{\lambda-n+1}\\ \frac{d^n(D1(t,t_x,\lambda))}{dt^n}=\frac{d\left(\left(\prod_{i=0}^{n-2}(\lambda-i)\right)S^{\lambda-n+1}\right)}{dS}=\left(\prod_{i=0}^{n-1}(\lambda-i)\right)(t-t_x)^{\lambda-n}\end{cases} \tag{21}$$

The proof of Eq. 20 is detailed in the following procedure:

$$\frac{d^n\left(D2\left(t,t_x,\lambda\right)\right)}{dt^n}=\frac{d^n\left((t-t_x)^{\lambda}\left(\lambda t_x+t\right)\right)}{dt^n}$$

$$=\frac{d^n\left((t-t_x)^{\lambda}\left((\lambda+1)t_x+(t-t_x)\right)\right)}{dt^n}$$

$$=\left((\lambda+1)t_x\frac{d^n\left((t-t_x)^{\lambda}\right)}{dt^n}+\frac{d^n\left((t-t_x)^{\lambda+1}\right)}{dt^n}\right)$$

$$=\left((\lambda+1)t_x\frac{d^n\left(D1\left(t,t_x,\lambda\right)\right)}{dt^n}+\frac{d^n\left(D1\left(t,t_x,\lambda+1\right)\right)}{dt^n}\right)$$

$$=\left((\lambda+1)t_x\left(\prod_{i=0}^{n-1}(\lambda-i)\right)(t-t_x)^{\lambda-n}\right)+\left(\prod_{i=0}^{n-1}(\lambda+1-i)\right)(t-t_x)^{\lambda+1-n}\Bigg) \tag{22}$$

From Eqs. 19 and 20, it becomes possible to calculate the derivation until the nth order of the function $I(t_0, t_1, t_2, z, w)$ corresponding to t_2:

$$\frac{d^n\left(I\left(t_0,t_1,t_2,z,w\right)\right)}{dt_2{}^n}=$$

$$\frac{z}{(n-\beta)(n-\beta+1)}\left(\frac{d^n\left(D2\left(t_2,t_0,n-\beta\right)\right)}{dt_2{}^n}-\frac{d^n\left(D2\left(t_2,t_1,n-\beta\right)\right)}{dt_2{}^n}\right)+\frac{w}{n-\beta}\left(\frac{d^n\left(D1\left(t_2,t_0,n-\beta\right)\right)}{dt_2{}^n}-\frac{d^n\left(D1\left(t_2,t_1,n-\beta\right)\right)}{dt_2{}^n}\right)$$

$$=\frac{z}{(n-\beta)(n-\beta+1)}\left(\begin{array}{l}\left((n-\beta+1)t_0\left(\prod_{i=0}^{n-1}(n-\beta-i)\right)(t_2-t_0)^{-\beta}\right)\\+\left(\prod_{i=0}^{n-1}(n-\beta+1-i)\right)(t_2-t_0)^{-\beta+1}\end{array}\right)$$

$$-\frac{z}{(n-\beta)(n-\beta+1)}\left(\begin{array}{l}\left((n-\beta+1)t_1\left(\prod_{i=0}^{n-1}(n-\beta-i)\right)(t_2-t_1)^{-\beta}\right)\\+\left(\prod_{i=0}^{n-1}(n-\beta+1-i)\right)(t_2-t_1)^{-\beta+1}\end{array}\right)$$

$$+\frac{w}{n-\beta}\left(\left(\prod_{i=0}^{n-1}(n-\beta-i)\right)(t_2-t_0)^{-\beta}-\left(\prod_{i=0}^{n-1}(n-\beta-i)\right)(t_2-t_1)^{-\beta}\right) \tag{23}$$

The derivation until the nth order of the function $I(t_0, t_1, t_2, z, w)$ corresponding to t_2 in the case of $t_2 = t_0$ is obtained from Eq. 17:

$$\frac{d^n\left(I\left(t_2,t_1,t_2,z,w\right)\right)}{dt_2{}^n}=$$

$$\left(\begin{array}{l}\frac{-w}{n-\beta}(\prod_{i=0}^{n-1}(n-\beta-i))(t_2-t_1)^{-\beta}\\ +\frac{-z}{(n-\beta)(n-\beta+1)}\left((n-\beta+1)t_1\left(\prod_{i=0}^{n-1}(n-\beta-i)\right)(t_2-t_1)^{-\beta}\right)\\ +\frac{-z}{(n-\beta)(n-\beta+1)}\left((n-\beta+1)t_1\left(\prod_{i=0}^{n-1}(n-\beta+1-i)\right)(t_2-t_1)^{-\beta+1}\right)\end{array}\right) \tag{24}$$

The derivation until the nth order of the function $I(t_0, t_1, t_2, z, w)$ corresponding to t_2 in the case of $t_2 = t_1$ is obtained from Eq. 18:

$$\frac{d^n\left(I\left(t_0,t_2,t_2,z,w\right)\right)}{dt_2{}^n}=$$

$$\left(\begin{array}{l}\frac{z}{(n-\beta)(n-\beta+1)}\left((n-\beta+1)t_0\left(\prod_{i=0}^{n-1}(n-\beta-i)\right)(t_2-t_0)^{-\beta}\right)\\ +\frac{z}{(n-\beta)(n-\beta+1)}\left(\prod_{i=0}^{n-1}(n-\beta+1-i)\right)(t_2-t_0)^{-\beta+1}\\ +\frac{w}{n-\beta}\left(\left(\prod_{i=0}^{n-1}(n-\beta-i)\right)(t_2-t_0)^{-\beta}\right)\end{array}\right) \tag{25}$$

Based on Eq. 14, the fractional derivation $\frac{d^{\beta}\varepsilon(t)}{dt^{\beta}}$ where $\epsilon(\tau) = z\tau + w$ is given by

$$\frac{d^{\beta}\varepsilon(t)}{dt^{\beta}} = \frac{1}{\Gamma(n-\beta)}\left(\frac{d^n}{dt^n}\left(\int_0^t \varepsilon(\tau)(t-\tau)^{n-\beta-1}d\tau\right)\right)$$

$$= \frac{1}{\Gamma(n-\beta)}\left(\frac{d^n I\,(t_0=0, t_1=t, t_2=t, z, w)}{dt^n}\right)$$

$$= \frac{1}{\Gamma(n-\beta)}\left(\begin{array}{c} \dfrac{z}{(n-\beta)(n-\beta+1)}\left(\displaystyle\prod_{i=0}^{n-1}(n-\beta+1-i)\right)t^{-\beta+1}. \\ \\ +\dfrac{w}{(n-\beta)}\left(\displaystyle\prod_{i=0}^{n-1}(n-\beta-i)\right)t^{-\beta} \end{array}\right) \tag{26}$$

The fractional derivation has been proven for description of the viscoelastic material behaviour and the mathematic development has been detailed.

1.3 Part1-Conclusions

In the first part of this chapter, we have introduced fractional differentiation and we have established the mathematical equations necessary for solving fractional derivatives. We have also highlighted the relevance of the fractional derivation in the description of materials exhibiting viscoelastic behaviour.

Viscoelastic materials find many applications in various engineering fields, including mechanical systems, household appliances, aerospace, automotive, and civil engineering (e.g., bridges). This wide use stems from their ability to effectively absorb shock and dampen vibrations. The use of fractional derivation for viscoelastic material highlights its importance for engineering applications.

In the subsequent section of this chapter, our focus shifts to an application of fractional derivation: specifically, the quasi-static modelling of polyurethane foams, which are extensively employed in comfort-related applications.

2 Part 2- Modelling the Quasi-Static Mechanical Behaviour Using an Integer and Fractional Memory Models

2.1 Mechanical Context

The theory of viscoelasticity holds significant importance in describing materials characterized by time-dependent mechanical behaviour [3]. Viscoelastic materials are widely used for comfort application [15], shock absorption, and vibration damping [16, 17] thanks to their ability to absorb deformation energy.

Polyurethane foam is a cellular material characterized by a range of interesting mechanical properties, including low density, the ability to absorb deformation energy, and low stiffness. It also exhibits properties such as excellent thermal and acoustic insulation, high liquid absorption, and complex light diffusion. This combination of these properties makes polyurethane foam commonly used in various applications related to noise damping, thermal insulation, vibration isolation, shock absorption, and comfort [18–22]. Polyurethane foam is a good choice in several fields such as automobile, construction aeronautics, biomedical, and packaging industries.

To control the vibration transmitted to seat occupants, several automatic regulation and control devices are being developed, such as active and semi-active dampers. The performance of these devices depends on accurately predicting the behaviours of all seat components, particularly the foam. In general, it is crucial to model the complex mechanical behaviour of polyurethane foam and identify its quasi-static and dynamic properties to optimize the design of systems incorporating foam, especially in terms of comfort optimization. With this objective the main goal of this part is to implement reliable mechanical models of polyurethane foam capable of predicting its response under various quasi-static and dynamic conditions. In this chapter, we consider only quasi-static mechanical conditions.

In the literature, various models have been developed, including memory models. Modelling the mechanical behaviour of viscoelastic components is often challenging due to its strong nonlinearity and dependence on factors such as temperature, prestress, frequency, and excitation amplitude. In many instances, linear viscoelasticity fails to provide a comprehensive understanding of these components' behaviour, as it overlooks aspects such as large deformations and nonlinearities observed even at small excitation amplitudes. To overcome the limitation of linear viscoelastic models, there are two strategies. The first is to improve modelling by developing a generalized nonlinear model such as the Maxwell nonlinear generalized model [16, 17, 23–26]. The second strategy very used for modelling polyurethane foams is to dissociate behaviour into two parts: pure nonlinear elastic behaviour and linear viscoelastic behaviour [27–31]. As seen, in the second strategy, the nonlinearity is only associated to elastic behaviour which is simpler to model. Despite that the second strategy associate nonlinearity only to elastic behaviour, which is simplest to describe mathematically, the major drawback of models of polyurethane foams present in the literature is their parameter dependence on testing conditions. This dependence affects the general representativeness of their quasi-static behaviour.

In this chapter, we have considered integer and fractional memory models that, through careful selection of identification methods, ensure a more general representativeness of the quasi-static behaviour of polyurethane foam. Specifically, we have demonstrated that the dimensional parameters of the developed models can be expressed as the product of two independent parts: one related to testing conditions and the other defining the dimensionless and invariant parameters that characterize the material.

Furthermore, a discrimination between fractional and integer models was performed, and the advantages and limitations of each model were discussed.

2.2 Materials and Methods—Mechanical Behaviour of Polyurethane Foam

Polyurethane foam is generally considered homogeneous and isotropic, with a constant cross section. Furthermore, it is often assumed that Poisson's ratio is zero. This hypothesis is shared by researchers who have developed memory models for polyurethane foams [31–33]. Therefore, in unidirectional tests, strain is directly correlated to displacement and stress to force. This allows the force-displacement response to describe the quasi-static behaviour of the polyurethane foam [31–33]. To determine the mechanical behaviour of the polyurethane foam, described in Table 1, a unidirectional compression cycling test is performed on a universal compression/traction device (Instron 33R4240—Fig. 3) and at three different test conditions (Table 2). All specimens considered in this study have the same mechanical and environmental history: they are pristine and derived from the same cutting mattresses with dimensions of 2 m x 1.2 m x 0.075 m. Each specimen is tested only once.

The compression cycling test consists of two phases: the loading phase, during which compression occurs from an initial strain in a set direction until a specified final strain, and the unloading phase, which proceeds in the same direction but in the opposite sense. If the compression cycles more than once, it is termed a multicycle compression test. The mathematical description of displacement (command) under compression cycling test is given by the following equation:

$$x(t) = x_j(t) = \begin{cases} l_0\dot{\varepsilon}_{c,j}\left(t - t_{0,j}\right) + \varepsilon_{0,j} \text{ if } t_{0,j} \leq t \leq t_{\frac{1}{2}cyc,j} \\ \\ l_0\dot{\varepsilon}_{u,j}\left(t_{\frac{1}{2}cyc,j} - t\right) + l_0\varepsilon_{\max,j} \text{ if } t_{\frac{1}{2}cyc,j} \leq t \leq t_{f,j} \end{cases} \quad (27)$$

Table 1 Studied Polyurethane foam

Designation	Foam A
Foam type	Soft polyurethane foam
Iso-cyanate	Toluene of di-iso-cyanate TDI
Polyol	Polyether designed by PA
Expansion agent	CO2
Fabrication process	Open moulding
Density [kgm^{-3}]	28
Porosity means μm	828
Sample shape	Cubic
Dimensions	(0.075.0.075.0.075) m^3
Alveolar type	Open alveolar

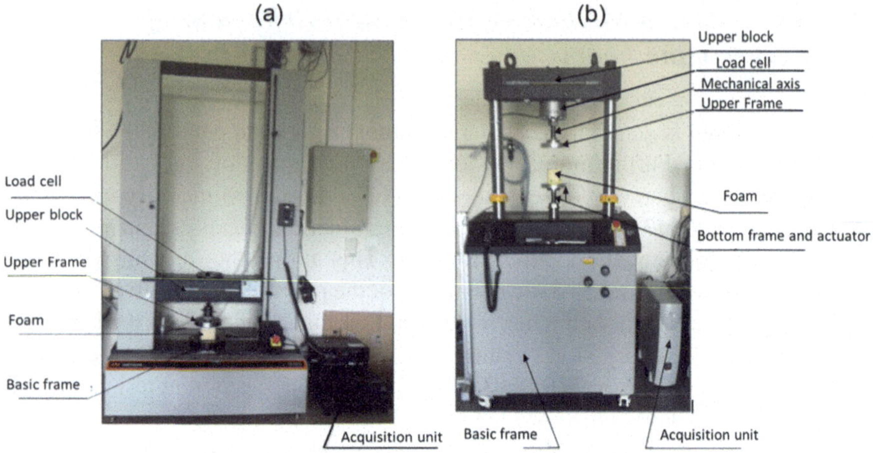

Fig. 3 Unidirectional compression cycling test devices: **a** Quasi-static device Instron 33R4240, **b** dynamic device Instron 8800

Table 2 Condition of quasi-static compression cycling test

Tests	N_{cyc}	$\dot{\varepsilon}[s^{-1}]$	$\epsilon_0[\%]$	$\epsilon_{max}[\%]$	$T[s]$	$T_{ech}[s]$	Specimen number per test
Test 1	1	1.0610^{-2}	0	80	150	0.0625	45
Test 2	1	5.3310^{-3}	0	80	300	0.1250	22
Test 2	1	6.6610^{-4}	0	80	2400	2.0000	24

** N_{cyc} is number of cycles in compression test, $\epsilon$is the test strain rate, ϵ_0 is the initial strain, ϵ_{max} is the final strain, T is cycle duration, and T_{ech} is the sampling time between consecutive saving of time, force, and displacement

In the expression $\left(t_{0,j}, t_{\frac{1}{2}cyc,j}, t_{f,j}\right)$, $t_{0,j}$ is the time in the beginning of the jth cycle, $t_{\frac{1}{2}cyc,j}$ is the time at the end of the loading phase of the jth cycle so at half cycle, and $t_{f,j}$ is the time at the end of unloading phase so at the end of the jth cycle. In the expression$\left(\dot{\varepsilon}_{c,j}, \dot{\varepsilon}_{u,j}\right)$, $\dot{\varepsilon}_{c,j}$ is the strain rate at the loading phase of the j^{th}cycle and $\dot{\varepsilon}_{u,j}$ is the strain rate at the unloading phase at the jth cycle. In the expression $\left(\varepsilon_{0,j}, \varepsilon_{\max,j}\right)$, $\varepsilon_{0,j}$ is the initial strain at the jth cycle and is the specified final strain of the jth cycle. $x_j(t)$ is the displacement at the jth cycle at time t. Note that l_0 is the thickness of polyurethane specimen.

A compression cycling test can be either quasi-static or dynamic, depending on the rate of deformation. Deng R et al [34] state that the strain rate limit for the type of polyurethane foam studied is of the order of $10^{-2}s^{-1}$. All tests with a strain rate below this limit fall within the scope of quasi-static studies. On the other hand, all compression cycling tests with a high strain rate (greater than $10^{-2}s^{-1}$) are part of dynamic studies. The same strain rate limit is obtained for the polyurethane foam

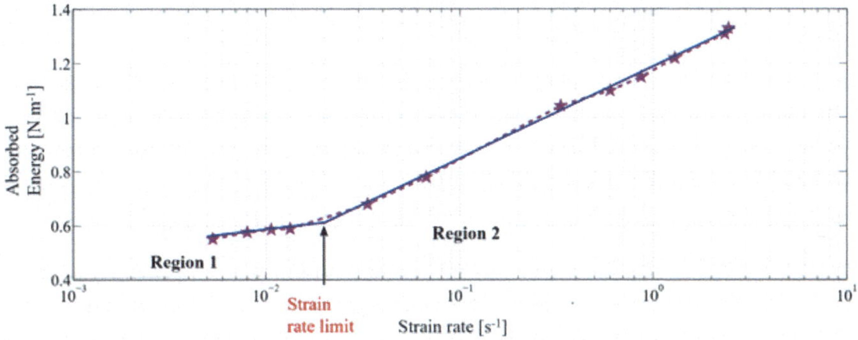

Fig. 4 Strain rate limit of polyurethane foam—evolution of absorbed energy with the strain rate during the 4th cycle of a multicycling compression test

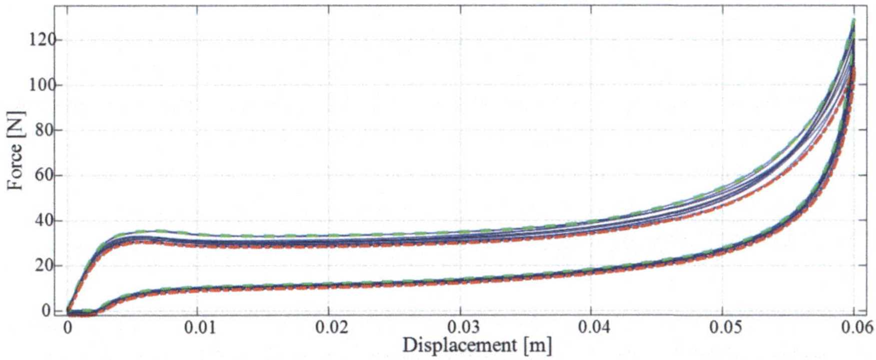

Fig. 5 Force-displacement curves of foam A under quasi-static unidirectional cycling compression test: green curve is the maximum experimental envelop, red curve is the minimum experimental envelop

studied in this work $0.02s^{-1}$ (Fig. 4). Dynamic tests are performed on fatigue device Instron 8800 equipped with dynamic force cell and displacement sensor (Fig. 3).

The experimental force-displacement curves during the first cycle of foam A allow the extraction of the minimum and maximum experimental envelopes (Fig. 5). The force-displacement curves of foams A in the case of test 1 and test 3 are not shown due to their similar shapes to those of test 2.

Figure 5 shows the presence of a hysteresis cycle in the foam's response and a return to a position different from the initial state. This deformation cannot be considered irreversible (plastic) as the polyurethane foam returns to its initial configuration after two days of rest. The hysteresis cycle highlights its viscoelastic behaviour.

The loading phase (Fig. 5) can be divided into three regions: the quasi-linear region, the plateau region, and the densification region [35]. This supports the argument for the nonlinear modelling of the elastic behaviour. It is also observed that the dispersion during the loading phase seems more significant than the dispersion during the unloading phase.

The quasi-static studies carried out by Deng [36], Dupuis et al. [37], and Song et al. [38] show the influence of the strain rate and the final compression level reached on the behaviour of the foam following a compression test. The maximum compression stress and the energy absorbed by the foam increase with the strain rate, with a greater effect when the final compression level and foam density are high [39, 40]. These results demonstrate the viscoelastic behaviour of the polyurethane foam.

The experimental studies of White S [32] and Widdle et al. [41] in quasi-static conditions, on polyurethane foams of the same type as the foam studied in this work, confirm that the quasi-static behaviour of polyurethane foam under large deformation includes three zones: the linear zone (zone 1), the plateau zone (zone 2), and the densification zone (zone 3). These three zones are visible and illustrate the nonlinear elastic behaviour.

At small strain level (zone 1), the force is not very significant. Almost all of the struts of the foam cells are subjected to homogeneous compression; however, only a few struts buckle. For this reason, the first zone is not perfectly linear. Moving on to the plateau zone (zone 2), the behaviour becomes rubbery. The foam starts to collapse. The cell struts have buckled and then bent, which explains the deformation plateau. Following the collapse, the struts on the upper face of the cells contribute to the compression of the struts on the lower face, leading to an increase in stress. This is referred to as foam densification (zone 3).

2.3 *Mathematic Modelling of the Quasi-Static Behaviour of Polyurethane Foam—Fractional and Integer Memory Models*

As shown before, the soft polyurethane foam, under quasi-static compression cycling tests, undergoes both a nonlinear viscoelastic behaviour. To describe the behaviour, we consider the total force exerted by the foam can be modelled as the sum of a nonlinear elastic force and a linear viscoelastic force (Eq. 28):

$$F(t) = F_e(t) + F_{ve}(t) \tag{28}$$

2.3.1 Memory Integer and Fractional Models of Polyurethane Foam—Nonlinear Elastic Behaviour

The elastic part of the foam is typically modelled by a nonlinear spring described by a power function. Some studies [42, 43] introduce a power elastic representation (Eq. 29)) in their models to determine the nonlinear elastic properties of their polymer foams.

$$\sigma(t) = E\left(a\varepsilon^{-p} + b\varepsilon^{-q}\right) \tag{29}$$

$a,p,b,q,$ and$E$ are constants determined from the experimental stress-strain curve.

White [32] described the elastic response using a polynomial of order M_w .The polynomial coefficients $(K_{l,w})_{l=1..M_w}$ constitute the stiffness vector of the foam (Eq. 30). Deng et al. [30, 36] described the elastic response of his foam using a sum of orthogonal polynomials (Eq. 31). The coefficients $(K_{l,de})_{l=1..M_{de}}$ do not constitute the stiffness vector of the foam. This is deduced by expanding the sum of the polynomials. We note $(k_{l,de})_{l=1..M_{de}}$ the stiffness vector for fractional memory model.

$$F_e(t) = \sum_{l=1}^{M_w} K_{l,w} x^l(t) \tag{30}$$

$$F_e(t) = \sum_{l=1}^{M_{de}} K_{i,de} P_l(x(t)) \tag{31}$$

When $M_{de} = 7$, the orthogonal polynomials of Eq. 31 are given by

$$\begin{cases} P_1(x) = x \\ P_2(x) = x^2 - \frac{3}{4} l_0 x \\ P_3(x) = x^3 - \frac{4}{3} l_0 x^2 + \frac{2}{5} {l_0}^2 x \\ P_4(x) = x^4 - \frac{15}{8} l_0 x^3 + \frac{15}{14} {l_0}^2 x^2 - \frac{5}{28} {l_0}^3 x \\ P_5(x) = x^5 - \frac{12}{5} l_0 x^4 + 2{l_0}^2 x^3 - \frac{2}{3} {l_0}^3 x^2 + \frac{1}{14} {l_0}^4 x \\ P_6(x) = x^6 - \frac{35}{12} l_0 x^5 + \frac{35}{11} {l_0}^2 x^4 - \frac{35}{22} {l_0}^3 x^3 + \frac{35}{99} {l_0}^4 x^2 - \frac{7}{264} {l_0}^5 x \\ P_7(x) = x^7 - \frac{24}{7} l_0 x^6 + \frac{60}{13} {l_0}^2 x^5 - \frac{40}{13} {l_0}^3 x^4 + \frac{150}{143} {l_0}^4 x^3 - \frac{24}{143} {l_0}^5 x^2 + \frac{4}{429} {l_0}^6 x \end{cases} \tag{32}$$

Memory models describe the elastic behaviour of the foam through a polynomial function. In this study, we choose, for both integer and fractional memory models, the description given by Eq. 30. However, contrary to literature, the stiffness of order pair is considered zeros to generalize model in the case of tensile loading conditions. In this case, the nonlinear force is given by

$$F_e(t) = \sum_{i=0}^{M} k_{2i+1} x^{2i+1} \tag{33}$$

where $k_{2i+1} = K_{2i+1,w}$ in the case of integer memory model and $k_{2i+1} = k_{2i+1,de}$ in the case of fractional memory model.

Many studies describe the elastic part with a hyperplastic model that supposes that the material exhibits nonlinear elastic behaviour at large deformations [29, 35, 44]. The hyperelastic modelling supposes the existence of strain density energy and that the material doesn't dissipate energy at large deformation.

2.3.2 Memory Integer Model of Polyurethane Foam—Linear Viscoelastic Behaviour

During quasi-static compression test, the strain rate is very low. Thus, purely viscous damping and thermal energy lost are neglected. Therefore, the energy lost due to viscoelasticity is the only mechanism that manifests in this case. If the foam is assumed to be homogeneous, with a constant cross-sectional area, and isotropic behaviour, the viscoelasticity is modelled by a linear description.

When a material dissipates part of the energy received and stores the other part, it is said to be viscoelastic. This definition made it possible to construct from the basic rheological elements, springs (energy-conserving elements) and dampers (viscous energy-dissipating elements), the first models describing viscoelastic behaviour such as the Kelvin-Voigt model (Eq. 10), Maxwell model (Eq. 11), and Zener model (Eq. 12). There are other elementary models of linear viscoelasticity used such as the Dupaix-Boyce model and the Kelvin model. There are other elementary models of linear viscoelasticity used such as the Dupaix-Boyce model, the Generalized Maxwell model [16, 17], Schiessel et Blumen model, and Heymans et Bauwens model [5]. The drawback of these rheological models is that they involve a large number of parameters, which are difficult to identify.

White [32] use for the description of the viscoelastic behaviour of polyurethane foam a more general model which supposes a stochastic combination in series and parallel of basic integer rheological elements. The relationship between stress $\sigma(t)$t and strain $\epsilon(t)$t is given by Eq. 1:

$p_0\sigma(t) + p_1\frac{d\sigma(t)}{dt} + ... + p_m\frac{d^m\sigma(t)}{dt^m} = q_0\varepsilon(t) + q_1\frac{d\varepsilon(t)}{dt} + ... + q_n\frac{d^n\varepsilon(t)}{dt^n}$

The general solution $\sigma(t) = \int_0^t H_{ve}(t-\tau)\epsilon(\tau)d\tau$ of Eq. 1, which is an integer derivation equation, is detailed in the part 1 of this chapter. As seen, this modelling takes into account the memory effect through the convolution product. This description is useful for the description of shape memory material in which the response at time t depends on the material's history. As the cross section S_e of polyurethane foam in our experimentation is constant, the test is unidirectional, and the foam is homogeneous, we can establish a relationship between the viscoelastic force and the displacement (Eq. 34).

$$F_{ve}(t) = \int_0^t G_{ve}(t-\tau)x(\tau)d\tau \tag{34}$$

where $G(t) = \frac{Se}{l_0}H_{ve}(t)$ and l_0 : specimen thickness.

The memory integer description of viscoelastic behaviour (Eq. 35) of the polyurethane foam is obtained when the n Class function order of the strain $\epsilon(t)$in Eq. 1 is more than m Class function order of the stress $\sigma(t)$ and when all pole values of the Laplace transformation of the function $\boldsymbol{H_{ve}}$ are simple and distinct.

$$F_{ve}(t) = \int_0^t \sum_{l=1}^{N_w} a_l e^{-\alpha_l(t-\tau)} x(\tau)d\tau \tag{35}$$

where N_w is the mode number of viscoelastic description or integer memory viscoelastic order,

a_l is a complex value which determines the residue of viscoelastic description associated to l^{th} mode,

α_l is a complex value which determines the pole of viscoelastic description associated to l^{th} mode.

The memory integer description of viscoelastic behaviour (Eq. 35) takes into account the memory effect through the convolution product. The viscoelastic response at time t depends on the history loading on the material. The viscoelastic force is a pondering sum of all responses to previous imposed displacement inputs. The term$\sum_{l=1}^{N_w} a_l e^{-\alpha_l(t-\tau)}$ describes the relaxation effect and shows that the influence of each preceding loading depends on its location in the time scale. The negative sign ($-$) in the exponential (in front α_l) is used to highlight that the history (memory) of the near past has a more significant effect than the effect of the far past on the viscoelastic force at time. Compared to a relaxation test, the real part of defines the relaxation α_l constant. Based on Eq. 27, the displacement function x(t) in the first cycle during a compression cycling test at constant strain rate ($\dot{\varepsilon}_{u,j} = \dot{\varepsilon}_{l,j} = \frac{2\varepsilon_{max}}{T}$) is given by Eq. 36.

$$x(t) = \begin{cases} \frac{2\varepsilon_{\max}}{T} l_0 t & \text{if } 0 \leq t \leq \frac{T}{2} \\ \frac{2\varepsilon_{\max}}{T} l_0 \left(\frac{T}{2} - t\right) + l_0 \varepsilon_{\max} & \text{if } \frac{T}{2} \leq t \leq T \end{cases} \tag{36}$$

If the integer memory viscoelastic order is equal to 2 and the viscoelastic parameters a and α are supposed to be different between load and unload phases, Eq. 35 is equivalent to Eq. 37.

$$F_{ve}(t) = \begin{cases} \int_0^t \sum_{i=1}^{2} a_i^L e^{-\alpha_i^L (t-\tau)} x(\tau) d\tau & \text{if } t \in \left[0, \frac{T}{2}\right] \\ \left(\begin{array}{c} \int_0^{\frac{T}{2}} \sum_{i=1}^{L} a_i^L e^{-\alpha_i^L (t-\tau)} x(\tau) d\tau \\ + \int_{\frac{T}{2}}^{t} \sum_{i=1}^{2} a_i^{UL} e^{-\alpha_i^{UL} (t-\tau)} x(\tau) d\tau \end{array} \right) & \text{if } t \in \left[\frac{T}{2}, T\right] \end{cases} \tag{37}$$

Let's $X = \left[X^L X^{UL}\right]$ the viscoelastic parameters of the integer memory description given by Eq. 37.

$X^L = \left[Re\left(\alpha^L\right), Im\left(\alpha^L\right), Re\left(a^L\right), Im\left(a^L\right)\right] = \left[X^L(1), X^L(2), X^L(3), X^L(4)\right]$is the viscoelastic parameters in loading phase.

$X^{UL} = \left[Re\left(\alpha^{UL}\right), Im\left(\alpha^{UL}\right), Re\left(a^{UL}\right), Im\left(a^{UL}\right)\right] =$
$\left[X^{UL}(1), X^{UL}(2), X^{UL}(3), X^{LU}(4)\right]$ regroups the viscoelastic parameters in loading phase. Note that Re designs the real part of a complex value and Im designs the imaginary part of a complex value. Through replacing the expression of $x(t)$ given by Eqs. 36 into 37, the memory integer description of viscoelastic behaviour is given by

$$F_{ve}(t) = \begin{cases} \frac{4\epsilon_{max} l_0}{T} \left(F_5{}^L(t, A3^L, A4^L) - F_1{}^L(t)\right) & \text{if } t \in \left[0, \frac{T}{2}\right] \\ \frac{4\epsilon_{max} l_0}{T} \left(\begin{array}{c} F_5{}^L(t, A3^L, A4^L) - F_5{}^L((t - \frac{T}{2}), F_1{}^L(\frac{T}{2}), F_2{}^L(\frac{T}{2})) \\ - F_5{}^{UL}((t - \frac{T}{2}), F_3{}^{UL}(\frac{T}{2}), F_4{}^{UL}(\frac{T}{2})) + F_3{}^{UL}((T - t)) \end{array} \right) & \text{if } t \in \left[\frac{T}{2}, T\right] \end{cases} \tag{38}$$

where

$$\left(\begin{array}{l} F_1{}^j(t) = A_3{}^j - tA_1{}^j \quad ; F_2{}^j(t) = A_4{}^j - tA_2{}^j \\ F_1{}^j(t) = A_3{}^j + tA_1{}^j \quad ; F_2{}^j(t) = A_4{}^j + tA_2{}^j \\ F_5{}^j\,(t, S_1, S_2) = e^{-X^j(1)t} \left(S_1 \cos\left(X^j(2)t\right) + S_2 \sin\left(X^j(2)t\right)\right) \end{array} \right)_{j=L \text{ ou } UL},$$

$A1^j = \frac{X^j(1)X^j(3)+X^j(2)X^j(4)}{\left(X^j(1)\right)^2+\left(X^j(2)\right)^2}$,

$$A2^j = \frac{X^j(1)X^j(4) - X^j(2)X^j(3)}{(X^j(1))^2 + (X^j(2))^2}$$

and
$$\begin{cases} A3^j = \dfrac{\left((X^j(1))^2 - (X^j(2))^2\right)X^j(3) + 2X^j(1)X^j(2)X^j(4)}{\left(X^j(1)^2 + X^j(2)^2\right)^2} \\ \\ A4^j = \dfrac{\left((X^j(1))^2 - (X^j(2))^2\right)X^j(4) - 2X^j(1)X^j(2)X^j(3)}{\left(X^j(1)^2 + X^j(2)^2\right)^2} \end{cases}$$

For the next, we suppose that a and α are same in load and unload phases.

2.3.3 Memory Fractional Model of Polyurethane Foam—Linear Viscoelastic Behaviour

Using the same demarche as integer memory description of viscoelastic behaviour, it is possible to suppose a stochastic combination in series and parallel of basic integer and fractional rheological elements to construct a fractional elementary model [45]. The general linear fractional model can be given by the following expression:

$$p_0 \sigma^{\beta_0}(t) + p_1 \frac{d\sigma^{\beta_1}(t)}{dt^{\beta_1}} + \ldots + p_m \frac{d^{\beta_m}\sigma(t)}{dt^{\beta_m}} = q_0 \varepsilon^{\gamma_0}(t) + q_1 \frac{d\varepsilon^{\gamma_1}(t)}{dt^{\gamma_1}} + \ldots + q_n \frac{d^{\gamma_n}\varepsilon(t)}{dt^{\gamma_n}} \quad (39)$$

where $(\beta_i)_{i=1..m}$ and $(\gamma_i)_{i=1..n}$ are fractional values.

Fractional derivation has been proven theoretically and experimentally at different macroscopic and microscopic scales. The fractional derivation constitutes the theoretical basis of work in different physical disciplines, namely the medical field [45], the biological field, and the mechanical field of materials. Fractional derivation is therefore a tool used. However, this derivation requires very complicated mathematical developments (Sect. 1).

Analytical and numerical methods are developed to solve a nonlinear fractional derivation problem such as the Adomian decomposition method [47–49] and the Taylor decomposition [50]. These methods are used to solve the nonlinear version of the Bagley Torvick equation (1984) [51]. The Adomian and Taylor decomposition methods are a systematic approach for solving both linear and nonlinear, deterministic and stochastic operator equations. This method is applicable to a range of equations including ordinary differential equations (ODEs), partial differential equations (PDEs), and integer or fractional differential equations [52, 53]. However, the solution of this method depends on the choice of initial and boundary values. The solution is generally a local solution near a given initial value. In addition, the solution from those methods supposes an infinite series of components and the accuracy of solution depends on truncation order.

No simple technique is able until today to solve analytically the general fractional model given by Eq. 39. Deng R.ref [30] presents a simplest fractional description of the viscoelastic behaviour of polyurethane foam using linear elements verifying Eq. 40.

$$F_{ve}(t) = \sum_{i=1}^{N_{de}} F_{ve,i}(t) \text{ where } \frac{dF_{ve,i}(t)}{dt} + b_i F_{ve,i}(t) = c_{1,i}x(t) + c_{2,i}\frac{d^{\beta_i}x(t)}{dt^{\beta_i}} \quad (40)$$

The parameter b gives a general information associated to the relaxation rate relative to the different levels of compression (load and unload phases). The parameter c_1 is the instantaneous elastic modulus and shows the contribution of the pure elastic part to the viscoelastic behaviour. The parameter c_2 is the loss modulus, which shows the contribution of the fractional component to viscoelastic behaviour.

The second member function of Eq. 40 is given by $f(t) = c_1 x(t) + c_2 \frac{d^{\beta}x(t)}{dt^{\beta}}$. We deduce that Eq. 40 corresponds to a combination of a first-order derivation system with a general expression given by Eq. 41 where the input is the second member function $f(t)$ and the output is the viscoelastic force $F_{ve}(t)$.

$$\frac{dF_{ve}(t)}{dt} + bF_{ve}(t) = f(t) \quad (41)$$

In the case of a quasi-static cyclic compression test, the viscoelastic force $F_{ve}(t)$ and its first derivative $\frac{dF_{ve}(t)}{dt}$ are zero at t=0. Thus, the solution of Eq. 41 is given by a convolution product Eq. 42.

$$F_{ve}(t) = \int_0^t e^{-b(t-\tau)} f(\tau)\, d\tau \quad (42)$$

From Eq. 42, we can deduce that establishing the developed analytical expression of the viscoelastic force during the first cycle requires the calculation of the second member function $f(t)$. The main difficulty lies in calculating the non-integer derivation $\frac{d^{\beta}x(t)}{dt^{\beta}}$.

Using Eqs. 26 and 36, the fractional derivation of displacement in the first cycle of compression cycling test is given by Eq. 43.

$$
\begin{aligned}
\frac{d^{\beta}x(t)}{dt^{\beta}} &= \frac{1}{\Gamma(n-\beta)}\frac{d^n}{dt^n}\begin{cases} (\int_0^t (\frac{2\varepsilon_{max}}{T} l_0 \tau)(t-\tau)^{n-\beta-1} d\tau) & \text{if } t \in [0, \frac{T}{2}] \\ \left(\left\{\begin{array}{l} \int_0^{\frac{T}{2}} (\frac{2\varepsilon_{max} l_0}{T}\tau)(t-\tau)^{n-\beta-1} d\tau \\ + \int_{\frac{T}{2}}^{t} (-\frac{2\varepsilon_{max}}{T} l_0 \tau + 2\varepsilon_{max} l_0)(t-\tau)^{n-\beta-1} d\tau \end{array}\right.\right) & \text{if } t \in [\frac{T}{2}, T] \end{cases} \\
&= \frac{1}{\Gamma(n-\beta)}\begin{cases} \frac{d^n}{dt^n} I(0, t, t, \frac{2\epsilon_{max} l_0}{T}, 0) & \text{if } t \in [0, \frac{T}{2}] \\ \left(\left\{\begin{array}{l} \frac{d^n}{dt^n} I(0, \frac{T}{2}, t, \frac{2\epsilon_{max} l_0}{T}, 0) \\ + \frac{d^n}{dt^n} I(\frac{T}{2}, t, t, -\frac{2\epsilon_{max} l_0}{T}, 2\epsilon_{max} l_0) \end{array}\right.\right) & \text{if } t \in [\frac{T}{2}, T] \end{cases} \\
&= \frac{1}{\Gamma(n-\beta)}\frac{\frac{2\epsilon_{max} l_0}{T}}{(n-\beta)(n-\beta+1)}\begin{cases} (\prod_{i=0}^{n-1}(n-\beta+1-i))t^{-\beta+1} & \text{if } t \in [0, \frac{T}{2}] \\ \left(\left\{\begin{array}{l} (\prod_{i=0}^{n-1}(n-\beta+1-i))t^{-\beta+1} \\ +2(\prod_{i=0}^{n-1}(n-\beta+1-i))(t-\frac{T}{2})^{-\beta+1} \end{array}\right.\right) & \text{if } t \in [\frac{T}{2}, T] \end{cases}
\end{aligned} \tag{43}
$$

If $1 < \beta < 2$ and $n = 2$, Eq. 43 is equivalent to Eq. 44.

$$
\frac{d^{\beta}x(t)}{dt^{\beta}} = \frac{\frac{2\varepsilon_{max} l_0}{T}}{\Gamma(2-\beta)}\begin{cases} t^{-\beta+1} & \text{if } t \in [0, \frac{T}{2}] \\ (t^{-\beta+1} - 2(t-\frac{T}{2})^{-\beta+1}) & \text{if } t \in [\frac{T}{2}, T] \end{cases} \tag{44}
$$

Replacing the expression of the non-integer derivation $\frac{d^{\beta}x(t)}{dt^{\beta}}$, given by Eq. 44, into $f(t) = c_1 x(t) + c_2 \frac{d^{\beta}x(t)}{dt^{\beta}}$, the analytic expression of the second member is equivalent to Eq. 45.

$$
f(t) = \frac{2\varepsilon_{max} l_0}{T}\begin{cases} \left(c_1 t + c_2 \frac{t^{-\beta+1}}{\Gamma(2-\beta)}\right) & \text{if } t \in \left[0, \frac{T}{2}\right] \\ \left(c_1(T-t) + \frac{c_2}{\Gamma(2-\beta)}\left(t^{-\beta+1} - 2\left(t-\frac{T}{2}\right)^{-\beta+1}\right)\right) & \text{if } t \in \left[\frac{T}{2}, T\right] \end{cases} \tag{45}
$$

By replacing the expression of $f(t)$, given by Eq. 45, in Eq. 42, the viscoelastic force $F_{ve}(t)$ derived for fractional derivation have an analytic expression given by Eq. 46.

$$F_{ve}(t) = \frac{2\varepsilon_{max} l_0}{T} \begin{cases} \frac{c_1}{b^2}(bt-1) + A(t) & \text{if } t \in [0, \frac{T}{2}] \\ -\frac{c_1}{b^2}(b(t-T)-1) + A(t) - 2A(t-\frac{T}{2}) & \text{if } t \in [\frac{T}{2}, T] \end{cases} \tag{46}$$

where $A(t) = \left(\frac{c_1}{b^2} + \frac{c_2}{\Gamma(2-\beta)} \frac{\Gamma(2-\beta,-bt)}{(-b)^{2-\beta}}\right) e^{-bt}$. The proof of Eq. 46 is detailed as follows:

$$F_{ve}(t) = \int_0^t e^{-b(t-\tau)} f(\tau) d\tau$$

$$= \begin{cases} F_{ve}^{L}(t) = \frac{2\varepsilon_{max}}{T} l_0 \int_0^t e^{-b(t-\tau)}(c_1\tau + c_2 \frac{\tau^{-\beta+1}}{\Gamma(2-\beta)})d\tau \\ F_{ve}^{UL}(t) = \frac{2\varepsilon_{\max} l_0}{T} \begin{cases} \int_0^{\frac{T}{2}} e^{-b(t-\tau)}(c_1\tau + c_2\frac{\tau^{-\beta+1}}{\Gamma(2-\beta)})d\tau \\ +\int_{\frac{T}{2}}^t e^{-b(t-\tau)} c_1(T-\tau)d\tau \\ +\int_{\frac{T}{2}}^t e^{-b(t-\tau)} \frac{c_2}{\Gamma(2-\beta)}(\tau^{-\beta+1} - 2(\tau-\frac{T}{2})^{-\beta+1})d\tau \end{cases} \end{cases}$$

$$= \begin{cases} F_{ve}^{L}(t) = \frac{2\varepsilon_{\max} l_0}{T}\left(c_1 e^{-bt}\int_0^t \tau e^{b\tau}d\tau + \frac{c_2 e^{-bt}}{\Gamma(2-\beta)}\int_0^t \tau^{-\beta+1}e^{b\tau}d\tau\right) \\ F_{ve}^{UL}(t) = \frac{2\varepsilon_{\max} l_0}{T} \begin{cases} c_1 e^{-bt}\int_0^{\frac{T}{2}} \tau e^{b\tau}d\tau + \frac{c_2 e^{-bt}}{\Gamma(2-\beta)}\int_0^{\frac{T}{2}} \tau^{-\beta+1}e^{b\tau}d\tau \\ +c_1 e^{-bt}\int_{\frac{T}{2}}^t e^{b\tau}(T-\tau)d\tau \\ +\frac{c_2 e^{-bt}}{\Gamma(2-\beta)}\int_{\frac{T}{2}}^t e^{b\tau}\left(\tau^{-\beta+1} - 2\left(\tau-\frac{T}{2}\right)^{-\beta+1}\right)d\tau \end{cases} \end{cases}$$

$$= \begin{cases} F_{ve}^{L}(t) = \frac{2\varepsilon_{\max} l_0}{T} \begin{cases} c_1 e^{-bt}\left(\left[\frac{\tau}{b}e^{bz}\right]_0^t - \left[\frac{1}{b^2}e^{bz}\right]_0^t\right) \\ +\frac{1}{\Gamma(2-\beta)}\frac{c_2 e^{-bt}}{(-b)^{-\beta+2}}\int_0^t (-b\tau)^{-\beta+1}e^{bz}d(-b\tau) \end{cases} \\ F_{ve}^{UL}(t) = \frac{2\varepsilon_{\max} l_0}{T} \begin{cases} c_1 e^{-bt}\left(\left[\frac{\tau}{b}e^{bz}\right]_0^{\frac{T}{2}} - \left[\frac{1}{b^2}e^{bz}\right]_0^{\frac{T}{2}}\right) \\ +\frac{1}{\Gamma(2-\beta)}\frac{c_2 e^{-bt}}{(-b)^{-\beta+2}}\int_0^{\frac{T}{2}} (-b\tau)^{-\beta+1}e^{b\tau}d(-b\tau) \\ -c_1 e^{-b(t-T)}\int_{\frac{T}{2}}^t e^{b(\tau-T)}(\tau-T)d(\tau-T) \\ +\frac{1}{\Gamma(2-\beta)}\frac{c_2 e^{-bt}}{(-b)^{-\beta+2}}\int_{\frac{T}{2}}^t (-b\tau)^{-\beta+1}e^{b\tau}d(-b\tau) \\ -2\frac{1}{\Gamma(2-\beta)}\frac{c_2 e^{-b\left(t-\frac{T}{2}\right)}}{(-b)^{-\beta+2}}\int_{\frac{T}{2}}^T \left(-b\left(\tau-\frac{T}{2}\right)\right)^{-\beta+1} e^{b\left(\tau-\frac{T}{2}\right)}d\left(-b\left(\tau-\frac{T}{2}\right)\right) \end{cases} \end{cases}$$

$$
= \begin{cases}
F_{ve}^{L}(t) = \frac{2\varepsilon_{\max} l_0}{T} \begin{cases}
\frac{c_1}{b^2}((tb-1)+e^{-bt}) \\[2ex]
+\frac{1}{\Gamma(2-\beta)}\frac{c_2}{(-b)^{-\beta+2}} e^{-bt} \int_0^{-bt} (\tau)^{-\beta+1} e^{-\tau} d\tau
\end{cases} \\[4ex]
F_{ve}^{UL}(t) = \frac{2\varepsilon_{\max} l_0}{T} \begin{cases}
\frac{c_1}{b^2}((\frac{T}{2}b-1)e^{-b(t-\frac{T}{2})}+e^{-bt} \\[2ex]
+\frac{1}{\Gamma(2-\beta)}\frac{c_2}{(-b)^{-\beta+2}} e^{-bt} \int_0^{-b\frac{T}{2}} (\tau)^{-\beta+1} e^{-\tau} d\tau \\[2ex]
-c_1 e^{-b(t-T)} \int_{-\frac{T}{2}}^{t-T} \tau e^{b\tau} d\tau \\[2ex]
+\frac{1}{\Gamma(2-\beta)}\frac{c_2}{(-b)^{-\beta+2}} e^{-bt} (-\int_0^{-b\frac{T}{2}} (-\int_0^{-bt} (\tau)^{-\beta+1} e^{-\tau} d\tau) \\[2ex]
+\frac{1}{\Gamma(2-\beta)}\frac{c_2}{(-b)^{-\beta+2}} e^{-bt} (-\int_0^{-b\frac{T}{2}} (\tau)^{-\beta+1} e^{-\tau} d\tau) \\[2ex]
-2\frac{1}{\Gamma(2-\beta)}\frac{c_2 e^{-b\left(t-\frac{T}{2}\right)}}{(-b)^{-\beta+2}} (\int_0^{-b\left(t-\frac{T}{2}\right)} (\tau)^{-\beta+1} e^{-\tau} d\tau) \\[2ex]
-2\frac{1}{\Gamma(2-\beta)}\frac{c_2 e^{-b\left(t-\frac{T}{2}\right)}}{(-b)^{-\beta+2}} (-\int_0^{-b\left(\frac{T}{2}-\frac{T}{2}\right)} (\tau)^{-\beta+1} e^{-\tau} d\tau)
\end{cases}
\end{cases}
$$

$$
= \begin{cases}
F_{ve}^{L}(t) = \frac{2\varepsilon_{\max} l_0}{T} \begin{cases}
\frac{c_1}{b^2}(tb-1) \\[2ex]
+\left(\frac{c_1}{b^2}+\frac{c_2}{(-b)^{-\beta+2}\Gamma(2-\beta)}\Gamma(-\beta+1,-bt)\right) e^{-bt}
\end{cases} \\[4ex]
F_{ve}^{UL}(t) = \frac{2\varepsilon_{\max} l_0}{T} \begin{cases}
\frac{c_1}{b^2}\left(\frac{T}{2}b-1\right) e^{-b\left(t-\frac{T}{2}\right)} \\[2ex]
-c_1 e^{-b(t-T)} \left(\left[\frac{\tau}{b}e^{bz}\right]_{-\frac{T}{2}}^{t-T} - \left[\frac{1}{b^2}e^{bz}\right]_{-\frac{T}{2}}^{t-T}\right) \\[2ex]
+\left(\frac{c_1}{b^2}+\frac{c_2}{(-b)^{-\beta+2}\Gamma(2-\beta)}\Gamma(-\beta+1,-bt)\right) e^{-bt} \\[2ex]
-2\frac{c_2}{(-b)^{-\beta+2}\Gamma(2-\beta)}\Gamma\left(-\beta+1,-b\left(t-\frac{T}{2}\right)\right) e^{-b\left(t-\frac{T}{2}\right)}
\end{cases}
\end{cases}
$$

$$
= \begin{cases}
F^{L}_{ve}(t) = \frac{2\varepsilon_{\max} l_0}{T} \begin{cases} \frac{c_1}{b^2}(tb-1) \\ +\left(\frac{c_1}{b^2} + \frac{c_2}{(-b)^{-\beta+2}\Gamma(2-\beta)}\Gamma(-\beta+1,-bt)\right)e^{-bt} \end{cases} \\
F^{UL}_{ve}(t) = \frac{2\varepsilon_{\max} l_0}{T} \begin{cases} \frac{c_1}{b^2}\left(\frac{T}{2}b-1\right)e^{-b\left(t-\frac{T}{2}\right)} \\ -\frac{c_1}{b^2}e^{-b(t-T)}\left(b(t-T)-1+\left(\frac{T}{2}b+1\right)e^{-b\left(t-\frac{T}{2}\right)}\right) \\ +\left(\frac{c_1}{b^2} + \frac{c_2}{(-b)^{-\beta+2}\Gamma(2-\beta)}\Gamma(-\beta+1,-bt)\right)e^{-bt} \\ -2\frac{c_2}{(-b)^{-\beta+2}\Gamma(2-\beta)}\Gamma\left(-\beta+1,-b\left(t-\frac{T}{2}\right)\right)e^{-b\left(t-\frac{T}{2}\right)} \end{cases}
\end{cases}
$$

$$
= \begin{cases}
F^{L}_{ve}(t) = \frac{2\varepsilon_{\max} l_0}{T} \begin{cases} \frac{c_1}{b^2}(tb-1) \\ +\left(\frac{c_1}{b^2} + \frac{c_2}{(-b)^{-\beta+2}\Gamma(2-\beta)}\Gamma(-\beta+1,-bt)\right)e^{-bt} \end{cases} \\
F^{UL}_{ve}(t) = \frac{2\varepsilon_{\max} l_0}{T} \begin{cases} \left(\frac{c_1}{b^2} + \frac{c_2}{(-b)^{-\beta+2}\Gamma(2-\beta)}\Gamma(-\beta+1,-bt)\right)e^{-bt} \\ -2\left(\frac{c_1}{b^2} + \frac{c_2}{(-b)^{-\beta+2}\Gamma(2-\beta)}\Gamma\left(-\beta+1,-b\left(t-\frac{T}{2}\right)\right)\right)e^{-b\left(t-\frac{T}{2}\right)} \\ -\frac{c_1}{b^2}(b(t-T)-1) \end{cases}
\end{cases}
$$

$$
= \begin{cases}
F^{L}_{ve}(t) = \frac{2\varepsilon_{\max} l_0}{T}\left(\left(\frac{c_1}{b^2}(tb-1) + A(t)\right)\right) \\
F^{UL}_{ve}(t) = \frac{2\varepsilon_{\max} l_0}{T}\left(A(t) - 2A\left(\left(t-\frac{T}{2}\right)\right) - \frac{c_1}{b^2}(b(t-T)-1)\right)
\end{cases}
\tag{47}
$$

where $A(t) = \left(\frac{c_1}{b^2} + \frac{c_2}{\Gamma(2-\beta)}\frac{\Gamma(2-\beta,-bt)}{(-b)^{2-\beta}}\right)e^{-bt}$.

We remember that $F^{L}_{ve}(t)$ is defined for $t \in \left[0, \frac{T}{2}\right]$ and $F^{UL}_{ve}(t)$ is defined for $t \in \left[\frac{T}{2}, T\right]$.

2.3.4 Non-dimensional Parameters and Thermodynamic Conditions

The viscoelastic parameters determined by White [32], Singh et al. [54, 55], Deng [30], and Ippili et al. [33] in quasi-static studies depend on the test conditions (strain rate, final compression level, test period). In our work, in order to make the results independent to the test conditions, the parameters are replaced by non-dimensional parameters (Table 3).

Table 3 Equivalence between dimensional and non-dimensional parameters

Integer memory model		Fractional memory model	
Dimensional	Non-dimensional	Dimensional	Non-dimensional
$X(1) = \mathrm{Re}(\alpha_1)[s^{-1}]$	$Y(1) = X(1)T$	$X(1) = b$	$Y(1) = X(1)T$
$X(2) = \mathrm{Im}(\alpha_1)[s^{-1}]$	$Y(2) = X(2)T$	$X(2) = \beta$	$Y(2) = X(2)$
$X(3) =$ $\mathrm{Re}(a_1)[Nm^{-1}s^{-1}]$	$Y(3) = \frac{X(3)}{F_{max}} l_0 \epsilon_{max} T$	$X(3) = c_1[Nm^{-1}s^{-1}]$	$Y(3) = \frac{X(3)}{F_{\max}} l_0 \epsilon_{max}$
$X(4) =$ $\mathrm{Im}(a_1)[Nm^{-1}s^{-1}]$	$Y(4) = \frac{X(4)}{F_{max}} l_0 \epsilon_{max} T$	$X(4) =$ $c_2[Nm^{-1}s^{\beta-1}]$	$Y(4) =$ $\frac{X(4)}{F_{max}} l_0 \epsilon_{max} T^{1-\beta}$
$k_{i,w}[Nm^{-i}]$	$L_i = \frac{k_{i,w}}{F_{max}} l_0^i$	$k_{i,de}[Nm^{-i}]$	$L_i = \frac{k_{i,de}}{F_{\max}} l_0^i$

$F_{max} = max(F)$ and is determined from experiment

Table 4 Thermodynamic conditions in the case of integer memory model

Viscoelastic parameter to identify: $X = [\mathrm{Re}\,(\alpha), \mathrm{Im}\,(\alpha), \mathrm{Re}\,(a), \mathrm{Im}\,(a)]$	
$k_{1,w} \geq 0$	
$X(3) \leq 0$	
$X(2)X(4) + X(1)X(3) \leq 0$	(Stability of viscoelastic behaviour)
$X(1) \geq 0$	(Stability of viscoelastic behaviour)
$k_{1,w}((X(1)^2 + X(2)^2)) + (X(2)X(4) + X(1)X(3)) \geq 0$	
$X(3)(X(1)^2 + X(2)^2) - 2X(1)(X(2)X(4) + X(1)X(3)) \geq 0$	

The parameters to be identified are the elastic parameters and the viscoelastic parameters (Y vector from which X vector is obtained). To identify valid physical parameters, it is necessary to satisfy the thermodynamic conditions, which are conditions of causality and stability derived from the Clausius inequality [13].

In the quasi-static case, the thermodynamic conditions are determined by checking that the real and imaginary parts of the Fourier transform of the model's response to a Dirac pulse, neglecting the nonlinear terms, are positive, whatever the frequency.

Table 4 summarizes the thermodynamic conditions to be met for the integer memory model and Table 5 for the fractional model. Details of the various inequations are given in previous work [15].

2.4 Identification Process

The identification process is the same for both memory integer model and memory fractional model. In the next section, we present the identification with only the equation of fractional model.

Table 5 Thermodynamic conditions in the case of fractional memory model

Viscoelastic parameter to identify: $X = [b, \beta, c_1, c_2]$
$k_{1,\mathrm{de}} \geq 0$
$k_{1,\mathrm{de}} \geq \frac{-X(3)}{X(1)}$
$1 < X(2) \leq 2$
$X(3) \leq 0$
$X(4) \geq 0$
$X(1) \geq 0$
$X(1)X(4)\omega^{X(2)} \sin\left(\frac{\pi}{2}X(2)\right) - X(3)\omega \geq 0$ where ω is the frequency.

2.4.1 Difference-Forces Method

In the case of quasi-static compression test, the strain rates in load and unload are considered equal and constant. Thus, the displacement curve $x(t)$ is presented by isosceles triangle. The times (t_1, t_2)corresponding to the same position x_0, respectively, in loading and unloading phases are related by

$$t_2 = T - t_1 \tag{48}$$

The approach of the difference-forces method is based on the symmetry of the elastic force between load and unload phases. Therefore, only the viscoelastic parameters are present in the analytical expression of difference forces between load and unload. Using the previous Eq. 46 and Eq. 48, the load time projection of difference forces is given by

$$\delta F_{th}(t_1) = l_0 \dot{\varepsilon}\left(-2\frac{c_1}{b^2} + A(t_1) + 2A\left(\frac{T}{2} - t_1\right) - A(T - t_1)\right) \tag{49}$$

where $A(t) = \left(\frac{c_1}{b^2} + \frac{c_2}{\Gamma(2-\beta)}\frac{\Gamma(2-\beta,-bt)}{(-b)^{2-\beta}}\right)e^{-bt}$.

The difference-forces method consists in establishing, through optimization method, the viscoelastic parameters which allows minimizing the least mean square error between analytical and experimental difference forces (Fig. 6). In the second step, the reconstruction of viscoelastic force in load and unload phases allows the determination of the elastic force. Then, the symmetry condition of elastic force must be verified and the elastic polynomial coefficients are identified using MATLAB.

2.4.2 Optimization Approach

Optimization methods are the basis tools for viscoelastic parameters identification. There are two types of optimization methods namely deterministic methods and random methods. Deterministic methods are effective when the objective function has

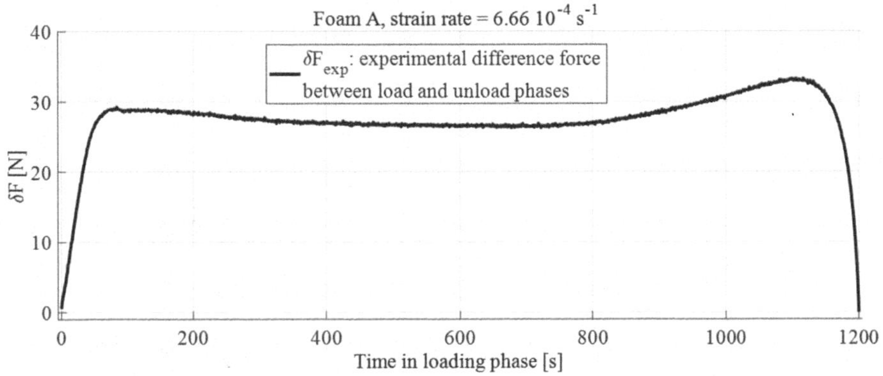

Fig. 6 The projection of difference forces on load time

a known prior form so that it is possible to choose an initialization near the global minimum. However, in the general case, these methods are not able to find the global minimum. The random methods are reliable in the case of non-differentiable functions and noisy functions. In addition, these methods offer a great probability to find the global minimum when the objective function possesses many local minima. However, his time calculation is very important compared with deterministic methods. In order to exploit the advantages of deterministic methods while ensuring a high probability to find global minimum, an approach involving three steps is considered. It consists in using both the genetic algorithm ('gatool' command of MATLAB) and the Trust region reflective method ('lsqnonlin' command). The first step allows having a great chance to find initialization for 'lsqnonlin' solver near global admissible optimum by using the genetic algorithm with gross stopping criteria. In the second step, the search for the best combination of viscoelastic parameters which minimizes the least mean square error between analytical and experimental difference forces is finalized by using 'lsqnonlin' MATLAB solver with tight stopping criteria. Finally, the solution found in the second step is injected into random initial population of genetic algorithm to verify that it is effectively the global admissible minimum. The stopping criterion in this third step is the generation number fixed at 100.

2.4.3 Memory Fractional Model—Incomplete Gamma Challenge

The 'Manual MATLAB Help' states that the value of the incomplete Gamma function $\Gamma(2-\beta, -bt)$ is complex and not accurate when $(-bt) \leq 0$ and $\text{abs}(-bt) < (2-\beta)$. For the first time, those conditions are verified in Eqs. 46 and 49. On other wise, the term $(-b)^{2-\beta}$ is complex. Accordingly, we cannot always have a real value of the function $\frac{\Gamma(2-\beta,-bt)}{(-b)^{2-\beta}}$. The numerical errors translate often complex residue infinitely small. However, as the fractional model is a memory model, the complex residue is accumulated with time and becomes non-negligible. The reduction through

Table 6 Identified parameters for integer memory model

Parameters	Test 1	Test 2	Test 3
$Y(1)$	45.40	46.56	46.22
$Y(2)$	0.2235	0.2275	0.2235
$Y(3)$	−256.19	−257.10	−262.69
$Y(4)$	2.202 10^4	2.187 10^4	2.204 10^4
δF_{fn} and δF_{exp} Error (%)	5.29	4.69	3.84
L_1	9.84	10.08	10.44
L_3	−39.41	−49.82	−63.05
L_5	550.58	710.11	877.41
L_7	−4204.48	−5473.79	−6673.01
L_9	18663.93	24425.20	29494.21
L_{11}	−49362.89	−64852.90	−77708.57
L_{13}	76639.09	101032.12	120218.48
L_{15}	−64435.79	−85210.86	−100703.91
Elastic identification error (%)	0.16	0.30	0.35
Global error (%)	3.99	4.06	4.96

Re $\left(\frac{Gamma(2-\beta,-bt)}{(-b)^{2-\beta})}\right.$ or abs $\left(\frac{\Gamma(2-\beta,-bt)}{(-b)^{2-\beta}}\right)$causes the imprecision of the results optimization and viscoelastic force reconstruction. The solution of this challenge consists in realizing some mathematical modifications in the aim to considerate a numerical integration of real function instead of the function$\frac{\Gamma(2-\beta,-bt)}{(-b)^{2-\beta}}$ (e.g., Riemann integration or Simpson integration).

2.4.4 Results and Validation

To validate the identification results (Tables 6 and 7), it is necessary to examine four criteria: verification of the thermodynamic conditions, statistical quality, estimation accuracy, and reconstruction of force-displacement curve between the minimum and maximum experimental envelopes.

The parameters below verify the thermodynamic conditions presented in Tables 4 and 5. In addition, they are a mean of test specimens (Table 2) which is determined to ensure the statistical quality of all identified parameters. This quality is reviewed through the set at 95% confidence level and the statistical error limit 'SLE' which shall not exceed 10%. For the third criterion, the maximum relative error between the experimental and analytical curves doesn't exceed 7% for both fractional and integer memory models. The two models reconstructed the overall force of type A foam (Figs. 7b and 8b) within the experimental envelopes (fourth criterion) and in good agreement with the experimental overall force curve except at

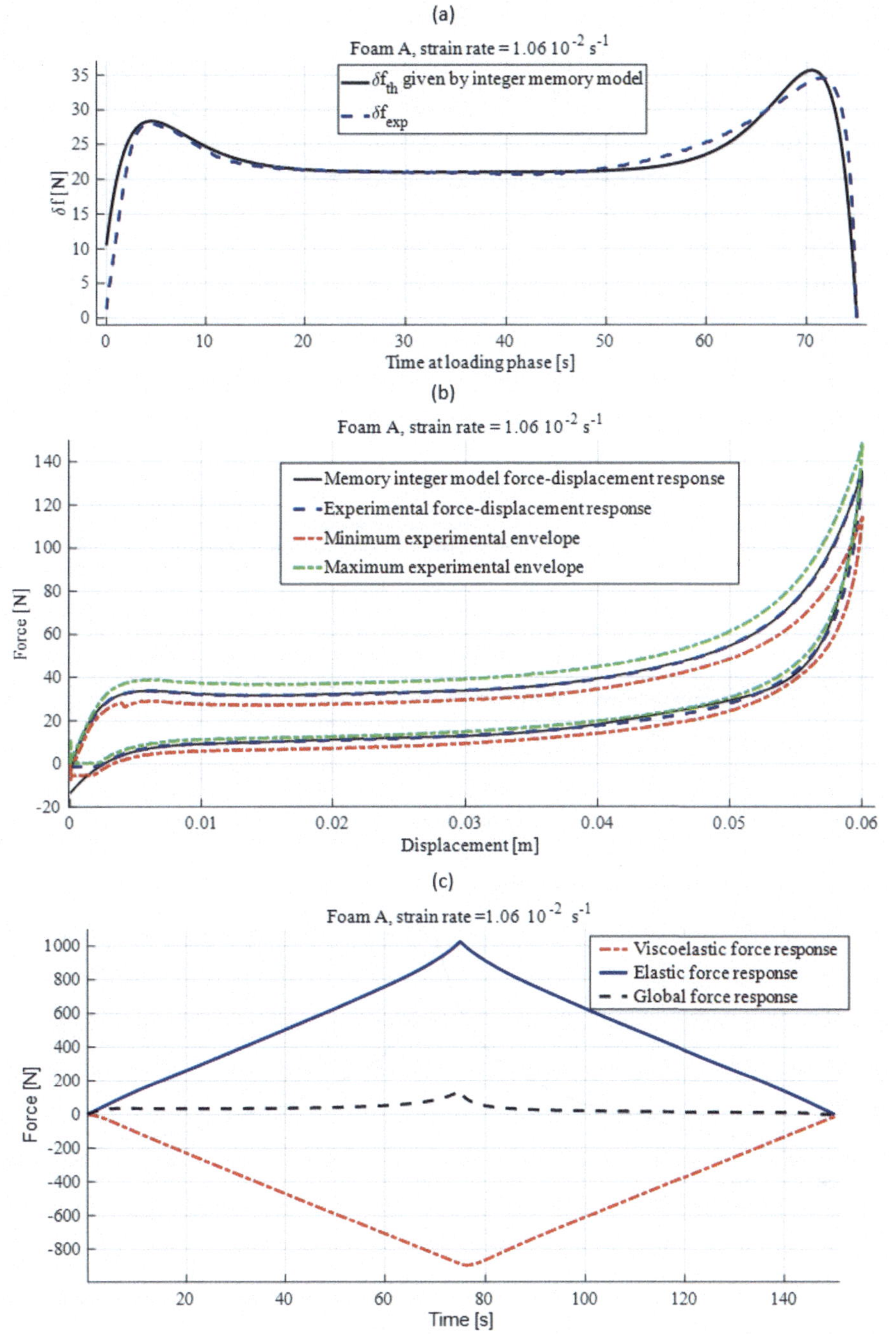

Fig. 7 Integer Memory model: **a** load and unload difference force, **b** global force, **c** elastic and viscoelastic response

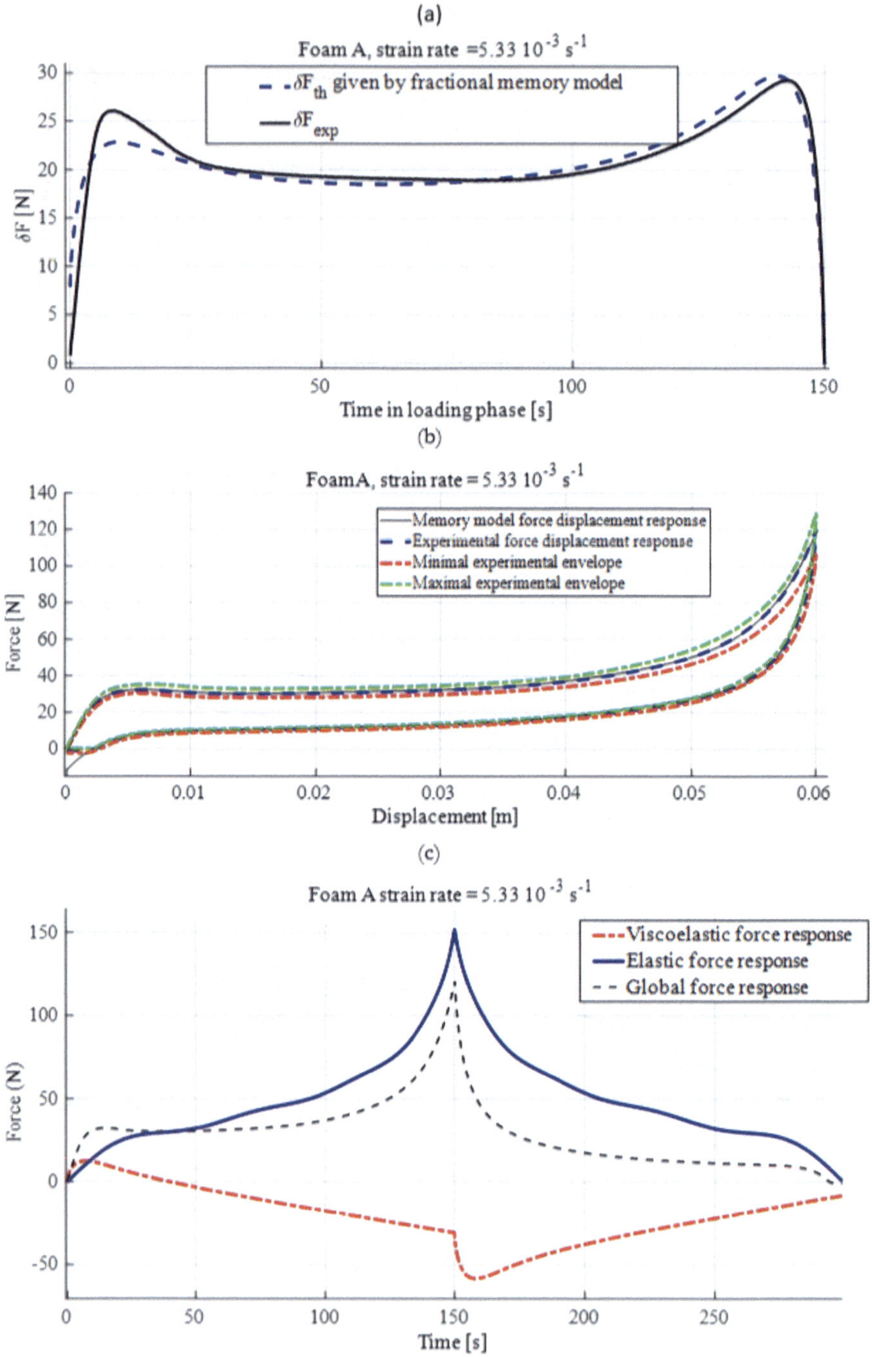

Fig. 8 8 Fractional Memory model: **a** load and unload difference force, **b** global force, **c** elastic and viscoelastic response

Table 7 Identified parameters for fractional memory model

Parameters	Test 1	Test 2	Test 3
$Y(1)$	48.00	48.00	48.95
$Y(2)$	1.355	1.343	1.298
$Y(3)$	−10.947	−14.820	−11.252
$Y(4)$	1.090	1.101	1.435
δF_{fn} and δF_{exp} Error (%)	3.82	4.90	5.51
L_1	1.98	2.38	2.78
L_3	−40.94	−49.33	−51.29
L_5	593.88	737.71	703.65
L_7	−4629.79	−5775.82	−5311.37
L_9	20800.59	26011.71	23360.88
L_{11}	−55438.94	−69475.61	−61318.53
L_{13}	86525.43	108669.80	94577.87
L_{15}	−73015.58	−91911.84	−79030.47
Elastic identification error (%)	1.32	2.43	0.32
Global error (%)	3.84	4.9	5.56

the end of the test where the difference was significant. This difference is explained by the presence of a residual (but not permanent) stress which forces the foam to return to a final position, different from its initial position (viscoelastic effect). Thus, contact between the upper block of the machine and the foam is lost. The experimental global force values are therefore inaccurate at this stage of the test. For this reason, the first part of the difference force between load and unload phases is not considered in the identification of viscoelastic parameters for both integer and fractional memory models. The dimensionless viscoelastic parameters in both integer and fractional memory reconstruct with good accuracy the experimental difference force between load and unload phases (Figs. 7a and 8a).

The dimensionless parameters of the two models are invariant as a function of the strain rate a. It is therefore possible to characterize the viscoelastic behaviour of the material using these parameters. This last result forms the basis for a possible comparison between different types of foam with viscoelastic behaviour. The comparison between the two models shows that the fractional memory model is capable of determining the nonlinearity of the elastic behaviour (Fig. 7c). However, the order value of the global elastic response and of the viscoelastic response in the case of integer memory is very important (Fig. 8c). The global elastic polynomial in the case of integer model is almost linear. This contradicts the studies presented in the introduction and the first assumption: foam under large deformation exhibits a nonlinear elastic behaviour. The linearity of integer memory model is due to the predominance of the global stiffness response of the first order($k_{i,w}x(t) \gg \sum_{i=1}^{M} k_{2i+1,w}x^{2i+1}$). Certain similarities between the two models were identified: the two models have the same relaxation time and the viscoelastic forces are negative almost all the time (the negative sign represents the relaxation phenomenon and explains the presence of a residual stress at the end of the test). The number of parameters for elastic behaviour is equal to 8 which is a very important number. The number of parameters needs to be reduced. Hyper-elastic models are used in different studies to describe the nonlinear elastic behaviour of polyurethane foam. It will be a good alternative. In addition, for both memory models, the stiffness of orders 3, 7,11, and 15 are negative. This supposes the existence of inelastic force. The integration of dry damping to models can involve these inelastic forces.

3 Conclusions and Further Works

This chapter presents an experimental and theoretical investigation into the quasi-static behaviour of polyurethane foam. The experimental study is based on uni-directional quasi-static compression/decompression tests. This study showed that polyurethane foam under large deformations exhibits both viscoelastic behaviour and nonlinear elastic behaviour. The dimensionless parameters of the memory models that we developed in this chapter are almost invariant with respect to the strain rates. These parameters can be considered as characteristic parameters of the material. They therefore allowed us to have a representative description of the quasi-static behaviour of polyurethane foam and to overcome the major drawback of the models presented in the literature: the dependence of the parameters on the test conditions. The memory models that we have developed reconstruct the quasi-static behaviour of the polyurethane foam (viscoelastic and elastic behaviour) in accordance with experimentation.

The 'viscoelastic force' response given by both integer and fractional memory models is generally negative in sign. It models the relaxation effect and explains the presence of a non-permanent residual stress at the end of the tests.

The number of parameters for elastic response of both integer and fractional memory model needs to be reduced. Hyperplastic models are used in different studies to describe the nonlinear elastic behaviour of polyurethane foam. It will be a good alternative.

Finally, it was shown in this chapter that the fractional memory model can describe the nonlinear elastic behaviour, whereas the integer memory model is not capable of doing so. The ability of the fractional memory model to well describe the quasi-static behaviour of different types of foam needs to be verified in further work. In addition, the Mullins effect, which is a particular mechanical response of the foam under multicycle loaded, must be modelled using memory models. Different models are used in literature to describe the dynamic behaviour of polyurethane foam. The fractional derivation can be interesting in this case.

References

1. R.L. Bagley, P.J. Torvik, Fractional calculus—a different approach to the analysis of viscoelastically damped structures. AIAA J. **21** (1983)
2. M. Di Paola, A. Pirrotta, V. Ajm, An easier method for best fitting experimental results, . Visco-elastic behavior through fractional calculus. Mech. Mater. **43**, 799–806 (2011)
3. L.-J. Shen, Fractional derivative models for viscoelastic materials at finite deformations. Int. J. Solids Struct. **190**, 226–237 (2020)
4. R.L. Bagley, P.J. Torvik, Fractional calculus—a different approach to the analysis of viscoelastically damped structures. AIAA J. **21** (1983)
5. N. Heymans, J.-C. Bauwens, Fractal rheological models and fractional differential equations for viscoelastic behavior. Rheol. Acta **33**, 210–219 (1994)
6. W.G. Glöckle, T.F. Nonnenmacher, Fractional relaxation and the time-temperature superposition principle. Rheol. Acta **33**, 337–343 (1994)
7. A. Freed, K. Diethelm, Caputo derivatives in viscoelasticity: a non-linear finite-deformation theory for tissue. Fract. Calc. Appl. Anal. **10**, 219–248 (2007)
8. A.D. Freed, K. Diethelm, Fractional calculus in biomechanics: a 3D viscoelastic model using regularized fractional derivative kernels with application to the human calcaneal fat pad. Biomech. Model Mechanobiol. **5**, 203–215 (2006)
9. A. Arikoglu, A new fractional derivative model for linearly viscoelastic materials and parameter identification via genetic algorithms. Rheol. Acta **53**, 219–233 (2014)
10. R.L. Bagley, P.J. Torvik, On the fractional calculus model of viscoelastic behavior. J. Rheol. (N Y N Y) **30**, 133–155 (1986)
11. S. André, Y. Meshaka, C. Cunat, Rheological constitutive equation of solids: a link between models based on irreversible thermodynamics and on fractional order derivative equations. Rheol. Acta **42**, 500–515 (2003)
12. A. Lion, On the large deformation behaviour of reinforced rubber at different temperatures. J. Mech. Phys. Solids **45**, 1805–1834 (1997)
13. A. Lion, On the thermodynamics of fractional damping elements. Continuum. Mech. Thermodyn. **9**, 83–96 (1997)
14. C.H.R. Friedrich, Relaxation and retardation functions of the Maxwell model with fractional derivatives. Rheol. Acta **30**, 151–158 (1991)
15. H. Jmal, Identification du comportement quasi-statique et dynamique de la mousse de polyuérathane au travers de modèles de mémoire (2012)
16. J.-L. Dion, H. Jrad, F. Renaud, I. Tawfiq, Identification method of Non Linear Generalized Maxwell Model for viscoelastic components
17. H. Jrad, I. H. Tawfiq Mohamed, Etude du comportement dynamique non linéaire des composants viscoélastiques: Caractérisation, modélisation et identification (2014)
18. N.V. Gama, A. Ferreira, A. Barros-Timmons, Polyurethane foams: Past, present, and future. Materials **11**, 1841 (2018)
19. I.K. Harith, Study on polyurethane foamed concrete for use in structural applications. Case Studies Constr. Mater. **8**, 79–86 (2018)
20. Z. Horak et al., Experimental measurements of mechanical properties of pur foam used for testing medical devices and instruments depending on temperature, density and strain rate. Materials **13**, 4560 (2020)
21. S.A. Dabous, T. Ibrahim, S. Shareef, E. Mushtaha, I. Alsyouf, Sustainable façade cladding selection for buildings in hot climates based on thermal performance and energy consumption. Results Eng. **16**, 100643 (2022)
22. S. Etemad, A. Kantzas, S. Bryant, A systematic analysis of foam drainage: Experiment and model. Results Eng. **15**, 100551 (2022)
23. M. Handelshauser, Y.-R. Chiang, M. Marchetti-Deschmann, P.J. Thurner, O.G. Andriotis, Collagen fibril tensile response described by a nonlinear Maxwell model. J. Mech. Behav. Biomed. Mater **145**, 105991 (2023)

24. W.N. Findley, F.A. Davis, Creep and Relaxation of Nonlinear Viscoelastic Materials. (Courier corporation, 2013)
25. R.A. Schapery, Nonlinear viscoelastic solids. Int. J. Solids Struct. **37**, 359–366 (2000)
26. C. Ge, B. Rice, Impact damping ratio of a nonlinear viscoelastic foam. Polym. Test **72**, 187–195 (2018)
27. B. Fazekas, T.J. Goda, Determination of the hyper-viscoelastic model parameters of open-cell polymer foams and rubber-like materials with high accuracy. Mater. Des. **156**, 596–608 (2018)
28. M. Elfarhani, A. Jarraya, S. Abid, M. Haddar, Fractional derivative and hereditary combined model for memory effects on flexible polyurethane foam. Mech. Time Depend Mater. **20**, 197–217 (2016)
29. M.L. Ju, H. Jmal, R. Dupuis, E. Aubry, Visco-hyperelastic constitutive model for modeling the quasi-static behavior of polyurethane foam in large deformation. Polym. Eng. Sci. **55**, 1795–1804 (2015)
30. R. Deng, P. Davies, A.K. Bajaj, A nonlinear fractional derivative model for large uni-axial deformation behaviour of polyurethane foam. Signal Process. **86** (2006)
31. R. Singh, P. Davies, A.K. Bajaj, Identification of nonlinear and viscoelastic properties of flexible polyurethane foam. Nonlinear Dyn. **34** (2003)
32. S.W. White, *Dynamic Modelling and Mesurement of Occupied Car Seats and Seating Foam* (Purdue University, West Lafayette, USA, School of Mechanical Engineering, 1998)
33. R.K. Ippili, R.K., P. Davies, A.K. Bajaj, L. Hagenmeyer, Nonlinear multi-body dynamic modelling of seat-occupant system with polyurethane seat and H-point prediction. Int. J. Ind. Ergon. **38** (2007)
34. R. Deng, Modeling and Characterization of Flexible Polyurethane Foam. School of Mechanical Engineering. (Purdue University, West Lafayette, USA, 2004)
35. A.K. Landauer, X. Li, C. Franck, D.L. Henann, Experimental characterization and hyperelastic constitutive modeling of open-cell elastomeric foams. J. Mech. Phys. Solids **133**, 103701 (2019)
36. R. Deng, Modeling and Characterization of Flexible Polyurethane Foam. School of Mechanical Engineering. (Purdue University, West Lafayette, USA, 2004)
37. R. Dupuis, E. Aubry, *Development and comparison of foam comprehensive law in great deformation* (Proc. of XI International Congers Orlando, Florida, USA, 2008)
38. Y. Song, Z. Wang, L. Zhao, J. Luo, Dynamic crushing behaviour of 3D closed-cell foams based on Voronoi random model. Mater Des 31, (2010)
39. M.C. Saha, Effect of density, microstructure, and strain rate on compression behavior of polymeric foams. Mater. Sci. Eng. A **406**, 328–336 (2005)
40. G. Subhash, Q. Liu, X.-L. Gao, Quasistatic and high strain rate uniaxial compressive response of polymeric structural foams. Int. J. Impact Eng. **32**, 1113–1126 (2006)
41. R.D. Widdle, A.K. Bajaj, P. Davies, Measurement of the Poisson's ratio of flexible polyurethane foam and its influence on a uniaxial compression model. Int. J. Eng. Sci. **46**, 31–49 (2008)
42. K.C. Rusch, Load-compression behavior of flexible foams. J. Appl. Polym. Sci. **13** (1969)
43. W.N. Patten, S. Sha, C. Mo, A vibration model of open celled polyurethane foam automotive seat cushions. J. Sound Vib. **217** (1998)
44. J.D.E. Dalisay, L. Liu, M. Eriten, L.A. Bergman, A.F. Vakakis, Characterization of visco-hyperelastic behavior of open cell polyurethane foam through transient shear testing. Int. J. Solids Struct. **241**, 111482 (2022)
45. R. Lewandowski, B. Chorazyczewski, Identification of parameters of the fractional rheological model of viscoelastic dampers. XXIV Symposium Vibrations in Physical Systems
46. D. Craiem, Fractional Calculus Applied to Model Arterial Viscoelasticity (2008). https://www.researchgate.net/publication/279564053
47. S.S. Ray, R.K. Bera, Analytical solution of the Bagley Torvik equation by Adomian decomposition method. Appl. Math. Comput. **168**, 398–410 (2005)
48. Z.H. Wang, X. Wang, General solution of the Bagley-Torvik equation with fractional-order derivative. Commun. Nonlinear Sci. Numer. Simul. **15**, 1279–1285 (2010)
49. V. Daftardar-Gejji, H. Jafari, Adomian decomposition: a tool for solving a system of fractional differential equations. J. Math. Anal. Appl. **301**, 508–518 (2005)

50. Y. Çenesiz, Y. Keskin, A. Kurnaz, The solution of the Bagley-Torvik equation with the generalized Taylor collocation method. J. Franklin Inst. **347**, 452–466 (2010)
51. P.J. Torvik, R.L. Bagley, On the Appearance of the Fractional Derivative in the Behavior of Real Materials. J. Appl. Mech. **51**, 294–298 (1984)
52. J.-S. Duan, R. Rach, D. Baleanu, A. Wazwaz, A review of the Adomian decomposition method and its applications to fractional differential equations. Commun. Fractional Calculus **3**, 73–99 (2012)
53. A.Ü. Keskin, Adomian Decomposition Method (ADM). in Boundary Value Problems for Engineers: with MATLAB Solutions (ed. Keskin, A. Ü.) 311–359 (Springer International Publishing, Cham, 2019). https://doi.org/10.1007/978-3-030-21080-9_7
54. R. Singh, P. Davies, A.K. Bajaj, Estimation of the dynamical properties of polyurethane foam through use of Prony series. J. Sound Vib. **264** (2003)
55. R. Singh, P. Davies, A.K. Bajaj, Identification of nonlinear and viscoelastic properties of flexible polyurethane foam. Nonlinear Dyn. **34**, 319–346 (2003)

A New Operator Approach for Solving Time-Fractional Nonlinear Burgess Equation

Hassen Arfaoui, Mohamed Kharrat, Hacene Mecheri, and Dumitru Baleanu

Abstract The introduction of fractional derivatives into the cancer treatment model continues to be improved with current cancer treatment. In this work, we define a new time-fractional nonlinear Burgess equation in order to model the brain tumor growth under treatment. The new mathematical model is in agreement with the clinical data proposed by Stupp et al. (2005). The numerical processing of this model is based on the splitting method which makes it possible to avoid and relax several numerical problems. The numerical results obtained are very satisfactory and excellent.

1 Introduction

The glioblastoma multiform is the mainly prevalent and antagonistic brain tumor in humans. It is a fast-growing tumor comprising a mixture of astrocytes that are defective with pleomorphism, necrosis, the proliferation of the vascular, as well as an increased mitosis rate. At any age this can occur; however, it is more common

H. Arfaoui (✉)
Mathematics Department, College of Science, Jouf University, Sakaka, Saudi Arabia

Institut Supérieur des Sciences Appliquées et de Technologie de Mateur, Université de Carthage, Tunis, Tunisia
e-mail: haarfaoui@ju.edu.sa; hassen.arfaoui@enit.rnu.tn

M. Kharrat
Mathematics Department, College of Science, Jouf University, Sakaka, Saudi Arabia
e-mail: mkharrat@ju.edu.sa; mohamed.kharrat@fphm.rnu.tn

Laboratory of Probability and Statistics, LR18ES28 Sfax University, Tunisia

H. Mecheri
Mathematics and Informatics Department, LAMIS Laboratory, Tebessa University, Tebessa, Algeria
e-mail: mecherih2000@yahoo.fr

D. Baleanu
Department of Mathematics, Cankaya University, 06530 Balgat, Ankara, Turkey

Institute of Space Science, Magurele-Bucharest, Romania

M. Kharrat et al. (eds.), *Modeling of Discrete and Continuous Systems*, Forum for Interdisciplinary Mathematics, https://doi.org/10.1007/978-981-97-8715-9_2

in the third and fourth categories of age. In the last few years, some works studied the mathematical modeling of the spread of this tumor. In 1995, Tracqui et al. [1] developed a glioma growth mathematical model that examined the effect of chemical therapy on Spatio-temporal growth. In 1996, the effect of surgical resection has been shown by Woodward et al. [2]. Burgess et al. [3], in 1997, presented a new model which describe glioblastoma growth without any medicinal remedies that might grow without limitations. This model provides information on changes in tumor density anywhere, but does not provide information about the occurrence of some tumor cell erasure. Lastly, in 2015, Gonzlez-Gaxiola and Bernal-Jaquez [4] presented a tumor growth model under medical treatment that would resolve an Adomian decomposition nonlinear partial differential equation.

A natural generalization of the classical integer calculus is the calculus of fractional order. It deals with derivatives and integrals of a real or complex arbitrary order. The fractional calculus is invested in several research axes [5–12]. For instance, fractional derivation models have shown a very important skill to describe character memory materials better than ordinary derivation models. For illustration, when a material is purely elastic, an integer derivation of order zero is described while an integer derivation of order one is described when it's pure viscous. Straight away, we can imagine describing a viscous-elastic material by a derivation between zero and one. So out of mathematical curiosity and to get closer to the reality of the behavior of human cells, we find ourselves obliged to use models based on fractional derivatives. Moreover, it is confirmed that the fractional-order models are more accurate than the integer-order models.

In this work, based on the same hypotheses of Burgess model (1997) [3], we give a new mathematical formulation under fractional calculus to establish a new partial nonlinear fractional differential equation in order to therefore eliminate or decrease the development of glioma. Our new mathematical model is in line with the clinical data presented by Stupp et al. (2005) [13].

In the literature, there are several types of fractional derivatives. We list the most frequently used, for example, Riemann-Liouville derivative, Caputo derivative, Caputo-Hadamard derivative [9, 14–16], Atangana-Baleanu fractional derivative [17], and Caputo-Fabrizio fractional derivative [18]. Also, there have been some very interesting work published recently using other types of fractional derivatives and their applications [19–23].

In this work, we are interested in the fractional calculus of Caputo. This theory has been presented by Michele Caputo in 1967 [24]. This kind of calculus arose in a natural way to represent the phenomena observed in laboratory and field experiments, where experimental data were used to examine the mathematical theory.

In this work, we used the splitting method. This method allows to solve a mixed problem Parabolic/Hyperbolic by decoupling the parabolic and hyperbolic operators, (for more details, see [25]). *Why we use the splitting method?* A nonlinear mixed problem generated by two completely different operators, (Parabolic/Hyperbolic), can cause difficulties in the numerical simulations. During discretization, the splitting method makes it possible to treat each operator Parabolic and Hyperbolic by an adequate numerical scheme. This method preserves the numerical properties (stability, consistency, $\cdots$) of every scheme utilized for every operator.

This work is outlined in this way. We define and present the time-fractional nonlinear Burgess equation related to the spread of the glioblastoma in the second section. In Sect. 3 we present the discretization of our model. In the last section, we present the numerical solution and the interpretations.

Definition 1 ([26]) The Caputo fractional derivative of order $\alpha > 0, a, \alpha, t \in \mathbb{R}$ can be defined as follows:

$$ {}_a^c D_t^\alpha h(t) = \begin{cases} \frac{1}{\Gamma(m-\alpha)} \int_a^t (t-\tau)^{m-\alpha-1} \frac{d^m}{d\tau^m} h(\tau) d\tau, & m-1 < \alpha < m, \quad m \in \mathbb{N}, \\ \frac{d^m}{d\tau^m} h(t), \quad \alpha = m, \quad m \in \mathbb{N}. \end{cases} $$

When $m = 1 : 0 < \alpha < 1$, then the Caputo fractional derivative of order α of the function h is written as follows:

$$ {}_a^c D_t^\alpha h(t) = \frac{1}{\Gamma(1-\alpha)} \int_a^t (t-\tau)^{-\alpha} \frac{d}{d\tau} h(\tau) d\tau, $$

where the Gamma function $\Gamma(\cdot)$ is given by

$$ \Gamma(z) = \int_0^\infty t^{z-1} e^{-t}\, dt. $$

One interesting reason for that definition of fractional derivatives is that it has characteristics of non-fractional derivatives so they make null the derivative of a constant (see [9]). In our work, we will use the definition when $m = 1$.

2 The Time-Fractional Nonlinear Burgess Equation

In the history, several authors have sought to find relevant mathematical models for modeling the spread of virulent brain tumors, (diseased cells or infected cells), such as glioblastoma. We quote for example [3, 27, 28]. They studied the tumors growth by using spatiotemporal models with a killing term. This model is given as follows:

$$ \frac{\partial \vartheta(z,t)}{\partial t} = \nu(z,t)\Big(\frac{\partial^2 \vartheta(z,t)}{\partial z^2} + \frac{2}{z}\frac{\partial \vartheta(z,t)}{\partial z}\Big) + [q(z,t) - d(z,t)]\vartheta(z,t), \quad (1) $$

$$ \vartheta(z_0, t_0) = \varrho_0, \quad (2) $$

for all $z \in \Omega \subset \mathbb{R}$ and all $t \geq t_0$, where the state $\vartheta(z,t)$ is the condensation of infected cells at the spatiotemporal point (z,t). The coefficient ν is the rate of distribution of diseased cells (diffusion coefficient), and the terms q and d represent, respectively, the growth rate of diseased cells and the cells killing rate. Here, the initial condition $\vartheta(z_0, t_0) = \varrho_0$ is the density of diseased cells at the initial time t_0.

We are now in a position to introduce the time-fractional nonlinear Burgess equation. Indeed, we replace in Eq. (1) the classical time derivative $\frac{\partial\vartheta(z,t)}{\partial t}$ by the time-fractional Caputo derivative ${}_{t_0}^{c}D_t^{\alpha}\vartheta(z,t)$. So, our fractional model that will be examined later is

$$
{}_{t_0}^{c}D_t^{\alpha}\vartheta(z,t) = \nu(z,t)\Big(\frac{\partial^2\vartheta(z,t)}{\partial z^2} + \frac{2}{z}\frac{\partial\vartheta(z,t)}{\partial z}\Big) + [q(z,t) - d(z,t)]\vartheta(z,t), \quad (3)
$$

$$
\vartheta(z_0, t_0) = \varrho_0. \quad (4)
$$

Following [29], we propose the change of variables:

$$
\begin{aligned}
\tau &= 2\nu t, \\
u(z,\tau) &= z\vartheta(z,t), \\
\psi(z,\tau) &= \frac{q(z,t) - d(z,t)}{2\nu},
\end{aligned} \quad (5)
$$

and assuming that ν is constant. In the following, we will rewrite Eq. (3) under the new state $u(z,\tau)$. So, we start with

$$
{}_{0}^{c}D_t^{\alpha}\vartheta(z,t) = \frac{1}{\Gamma(1-\alpha)}\int_0^t (t-s)^{-\alpha}\frac{\partial\vartheta(z,s)}{\partial s}\,ds. \quad (6)
$$

Since, $u(z,\tau) = z\vartheta(z,t)$, so we obtain

$$
\frac{\partial\vartheta(z,s)}{\partial s} = \frac{1}{z}\frac{\partial u(z,\tau)}{\partial s} = \frac{1}{z}\frac{\partial u(z,\tau)}{\partial\tau}\frac{\partial\tau}{\partial s} = \frac{2\nu}{z}\frac{\partial u(z,\tau)}{\partial\tau}. \quad (7)
$$

By substituting (7) in (6) and considering that $\tau = 2\nu s$, we get

$$
\begin{aligned}
{}_{0}^{c}D_t^{\alpha}\vartheta(z,t) &= \frac{1}{z}\Big(\frac{1}{2\nu}\Big)^{-\alpha}\frac{1}{\Gamma(1-\alpha)}\int_0^{\zeta}(\zeta-\tau)^{-\alpha}\frac{\partial u(z,\tau)}{\partial\tau}\,d\tau, \\
{}_{0}^{c}D_t^{\alpha}\vartheta(z,t) &= \frac{1}{z}\Big(\frac{1}{2\nu}\Big)^{-\alpha}D_t^{\alpha}u(z,\tau),
\end{aligned} \quad (8)
$$

where $\zeta = 2\nu t$. Also, we can deduce that

$$
\frac{\partial\vartheta(z,t)}{\partial z} = -\frac{1}{z^2}u(z,\tau) + \frac{1}{z}\frac{\partial u(z,\tau)}{\partial z}, \quad (9)
$$

$$
\frac{\partial^2\vartheta(z,t)}{\partial z^2} = \frac{1}{z}\frac{\partial^2 u(z,\tau)}{\partial z^2} - \frac{2}{z^2}\frac{\partial u(z,\tau)}{\partial z} + \frac{2}{z^3}u(z,\tau), \quad (10)
$$

$$
(q(z,t) - d(z,t))\vartheta(z,t) = \frac{2\nu}{z}\psi(z,\tau)u(z,\tau). \quad (11)
$$

By replacing Eqs. (8)–(11) in Eq. (3) and taking into consideration Eq. (5), after that expression has been simplified, Eq. (3) can be rewritten in this way:

$$ {}_{0}^{c}D_{\tau}^{\alpha}u(z,\tau) = (2\nu)^{1-\alpha}\Big(\frac{1}{2}\frac{\partial^2 u(z,\tau)}{\partial z^2} + \psi(z,\tau)u(z,\tau)\Big). \tag{12} $$

A medical study is conducted in [13], highlighting the advantages of combining treatments for brain cancer, such as chemotherapy and radiation. Some results are indeed determined through the use of temozolomide with radiation treatment. The results show an increase in the patient's life following the decrease in the size of the tumor. Mathematically, this occurs as negative tumor growth; on the other hand, the term $q(z,t) - d(z,t)$ is negative. We, therefore, introduce the nonlinear term modeling the degradation of the glioma:

$$ \psi(z,\tau) = \frac{1}{u^2}(e^{-u} + 0.5e^{-5u}), \tag{13} $$

where $u = u(z,\tau)$. Note that the tumor declines according to the (z,t) position. In Fig. 2 we graph the curve of the evolution of the tumor, modeled by $\psi(z,\tau)$, at each point (z,τ).

In the next, we are interested in the nonlinear fractional Eq. (12). To find the numeric solution of Eq. (12), we use a combination of the splitting method and the implicit finite difference schemes.

3 Splitting Method and Discretization

Let $\Omega =]a,b[$. Our goal in the following is to study the numerical simulation of the nonlinear fractional differential equation defined by

$$ {}_{0}^{c}D_{t}^{\alpha}u = c\frac{\partial^2 u}{\partial x^2} + h(u), \qquad \text{in} \quad \Omega \times (0,T), \tag{14} $$

$$ u(z,0) = u_0(z), \qquad \text{in} \quad \Omega, \tag{15} $$

with h is a nonlinear function of the state u, u_0 is the initial condition, and c is a strictly positive constant, (diffusion coefficient).

To solve numerically the system (14)–(15), we will use a splitting method that will be described below.

Remark 1 The idea of the splitting method is as follows: We divide the mixed problem (14)–(15), (parabolic/hyperbolic), into two problems:

(i)– Parabolic equation:

$$ {}_0^C D_t^\alpha u = c\frac{\partial^2 u}{\partial r^2}, \qquad \text{in} \quad \Omega \times (0, T). \tag{16} $$

(ii)– Hyperbolic equation:

$$ {}_0^C D_t^\alpha u = h(u), \qquad \text{in} \quad \Omega \times (0, T). \tag{17} $$

Then, we solve simultaneously (16) and (17) as described in the following paragraph.

3.1 Splitting Method

We split $\bar{\Omega}$ by x_i into N parts

$$ z_i = a + ih, \quad h = \frac{b-a}{M}, \quad i = 0, \cdots, M. $$

Let $\delta = \frac{T}{N}$ be the step of time and $t^n = n\delta, \forall n = 0, \cdots, N$. Then the following notes are provided:

$$ u(z_i, t^n) = u_i^n, \quad u_0(z_i) = u_{0,i}. $$

The problem (14)–(15) is numerically solved with a time splitting method , see [25], it consists of the next two steps.

(1) On the time interval $[t^n, t^{n+\frac{1}{2}}]$, we resolve the parabolic problem

$$ {}_0^C D_t^\alpha u = c\frac{\partial^2 u}{\partial z^2}, \tag{18} $$

$$ u(z, 0) = u_0(z). \tag{19} $$

The solution at time level $t^{n+\frac{1}{2}}$ is noted $\tilde{u}(z, t^{n+\frac{1}{2}})$.

(2) Then, to get the global solution at time level t^{n+1} to the problem (14)–(15), we resolve the problem:

$$ {}_0^C D_t^\alpha u = h(u), \tag{20} $$

with the initial condition:

$$ \tilde{u}(z, t^{n+\frac{1}{2}}), \tag{21} $$

in the interval $[t^{n+\frac{1}{2}}, t^{n+1}]$ using the solution $\tilde{u}(z, t^{n+\frac{1}{2}})$ of the system (18)–(19). The aforesaid solution is considered as the initial condition at time $t^{n+\frac{1}{2}}$ of the problem (20)–(21). This process is repeated simultaneously on every interval $[t^n, t^{n+1}]$, for all $n \geq 0$.

3.2 Discretization

Discretizing the parabolic fractional problem (18)–(19). For this step, we use an implicit discrete scheme that is based on the discretization of a modified Riemann-Liouville derivative associated to ${}_0^cD_t^\alpha u(z,t)$, at each point (z_i, t^n), for more details see [30]. So, we have

$$\frac{\delta^{-\alpha}}{\Gamma(2-\alpha)}\sum_{j=0}^{k}(u_i^{k+1-j}-u_i^{k-j})b_j = c\frac{u_{i+1}^{k+1}-2u_i^{k+1}+u_{i-1}^{k+1}}{h^2}, \tag{22}$$

where $b_j = (j+1)^{1-\alpha} - j^{1-\alpha}$, $k = 0, \cdots, N-1$ and $i = 1, \cdots, M-1$.

Discretization of the fractional hyperbolic problem (20)–(21). For this step, we use the following scheme described in [31]. For each $j = 0, \cdots, M$, we have

$$u_j^{n+1} = \delta^\alpha h(u_j^n) - \sum_{k=1}^{n+1}\beta_k^\alpha u_j^{n-k+1} + (1+\sum_{k=1}^{n+1}\beta_k^\alpha)u_{0,j}, \tag{23}$$

where $n = 0, \cdots, N-1$, β_k^α is determined by

$$\beta_k^\alpha = (-1)^k\frac{\Gamma(1+\alpha)}{\Gamma(1+\alpha-k)\Gamma(1+k)},$$

where $\Gamma(\cdot)$ is the Gamma function.

4 Numerical Results and Interpretations

In the following numerical examples, we use the MATLAB software for both the numerical implementation and the graphical presentations.

Let $\Omega = [0.001, 1]$. We consider the nonlinear model defined by the system

$${}_0^cD_t^\alpha u = \left(\frac{1}{2}(2\nu)^{1-\alpha}\right)\frac{\partial^2 u}{\partial z^2} + h(u), \qquad \text{in} \quad \Omega\times(0,T), \tag{24}$$

$$u(z,0) = u_0(z), \qquad \text{in} \quad \Omega, \tag{25}$$

where the nonlinear term $h(u)$ is taken, after comparing Eqs. (12) and (14), as follows:

$$h(u) = (2\nu)^{1-\alpha}\psi(z,\tau)u(z,\tau),$$

using (13), we get

$$h(u) = \frac{(2\nu)^{1-\alpha}}{u}(e^{-u} + 0.5e^{-5u}).$$

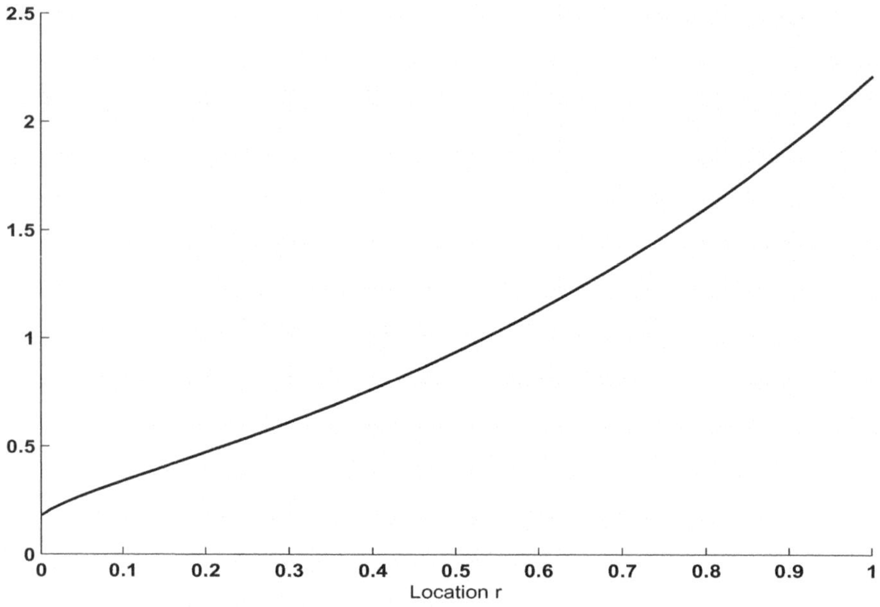

Fig. 1 Evolution of the initial growth-profile $u(z, 0)$

In fact, the nonlinear function $h(u)$ models the decay of the glioma, (h and ψ have the same properties), see Figs. 2 and 3. The initial condition is chosen as follows:

$$u(z, 0) = 0.175 + 0.75z^{0.7}e^{z}. \tag{26}$$

Remark that $u(z, 0)$ is an increasing function. In Fig. 1, we graph the curve of the initial growth-profile $u(z, 0)$.

In all that follows, we consider the following data: $\nu = 0.5$, $\alpha = 0.7$. Consequently, from relation (5), we deduce that $\tau = t$. So, the system (24)–(25) becomes as follows:

$$ {}_{0}^{c}D_{t}^{\alpha} u = \frac{1}{2}\frac{\partial^2 u}{\partial z^2} + \frac{1}{u}(e^{-u} + 0.5e^{-5u}), \tag{27}$$

$$u(z, 0) = 0.175 + 0.75z^{0.7}e^{z}. \tag{28}$$

Note that the initial condition (26) constitutes the initial growth tumor profile when we commence glioma attenuation using radiation and chemotherapy treatments. We can also observe that the initial (26) condition makes the function $h(u)$, models glioma decay, decreases, and this matches the medical phenomenon to study. In Fig. 3, the profile of the concept which models the decline of the glioma is shown. In Fig. 4, we plot the solution of the nonlinear model (27)–(28).

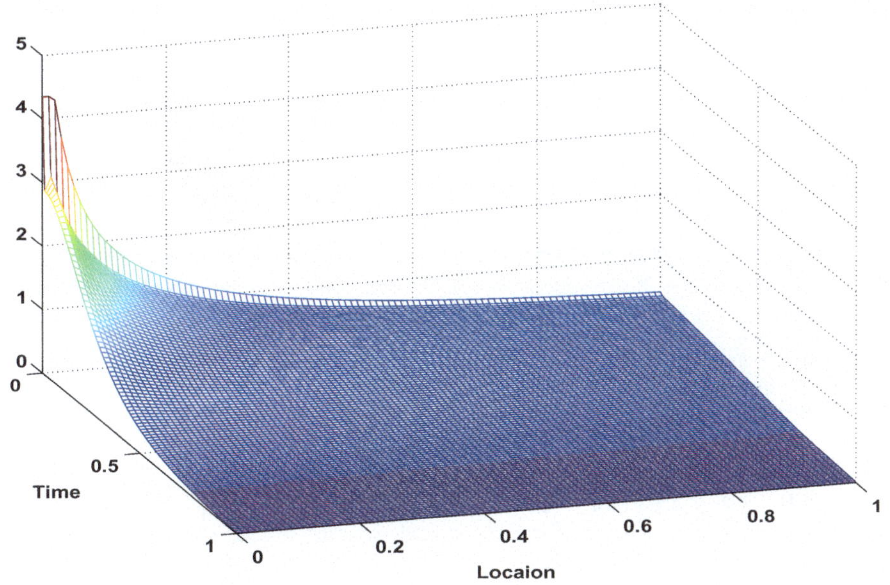

Fig. 2 Evolution of the state $\psi(z, \tau)$ for $(z, \tau) \in \Omega \times [0, 1]$

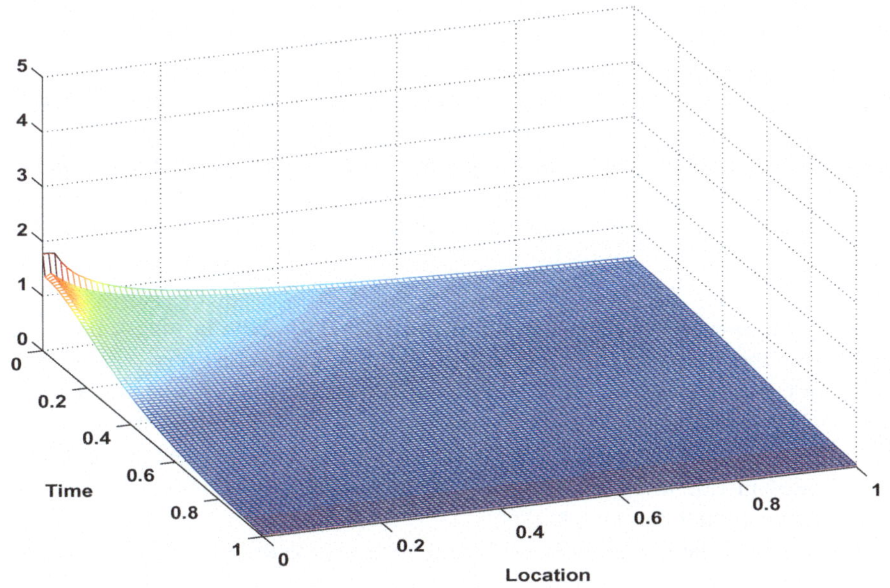

Fig. 3 Evolution of the state $h(u)$ for $(z, \tau) \in \Omega \times [0, 1]$

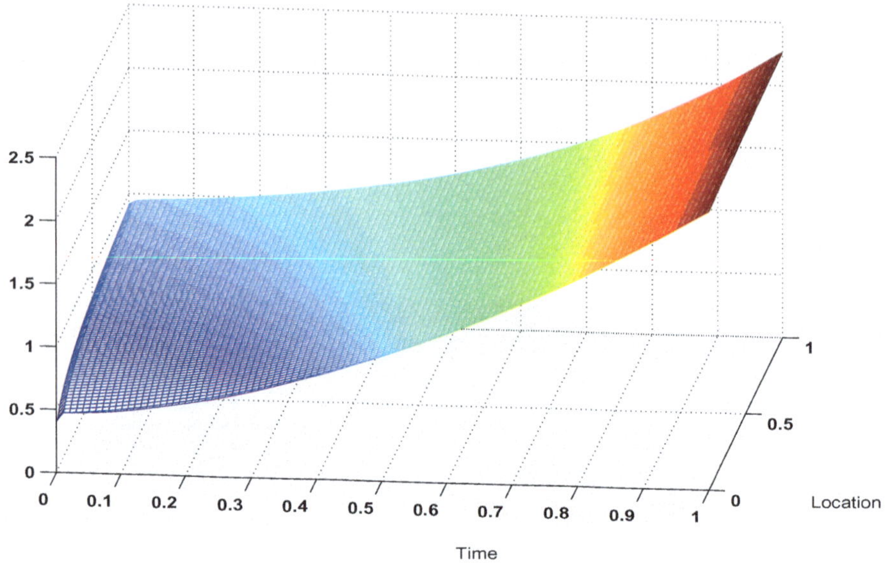

Fig. 4 Evolution of the state $u(z, \tau)$ for $(z, \tau) \in \Omega \times [0, 1]$

Remark 2 We have solved numerically the nonlinear model (27)–(28) and that because of the splitting method introduced in Sect. 3. Contrary to the work done in [4], they have a sufficient initial condition $u(z, 0)$ for the exact system solution of (27)–(28) to be calculated with Adomian decomposition method. We can choose any initial condition by our numerical method, and in practice, this represents the true case. For example, measurement by a medical measuring tool may constitute the initial condition. The Adomian method of decomposition cannot be used in this case.

Using the solution of the nonlinear model (27)–(28) given in Fig. 4, now, the variation of the infected cell density in $\vartheta(z, t)$ can be studied in every $(z, t) \in \Omega \times [0, 1]$. In fact, we graph the tumor growth-profile curve in Fig. 4 when the patient is receiving chemotherapy with themozolomide. On the contrary, in the untreated tumor we have a fast exponential growth-profile [3, 32]. The untreated tumor profile is presented in [3], where the state $\vartheta(z, t)$ is studied with various values of the parameters ν, q, and $d(z, t) = 0$.

The radius of the tumor under medical treatment $(ztreated)$ is compared to the unprocessed radius of a tumor $(zuntreated)$, to show the effectiveness of the medical treatment.

(i) The radius of the untreated tumor is given in [32] by

$$z_{untreated} = 2t\left(\nu q\left|1 - \frac{1}{qt}\ln\left[\frac{\sigma_0}{\varrho_0}(4\pi\nu t)^{\frac{3}{2}}\right]\right|\right)^{0.5}. \tag{29}$$

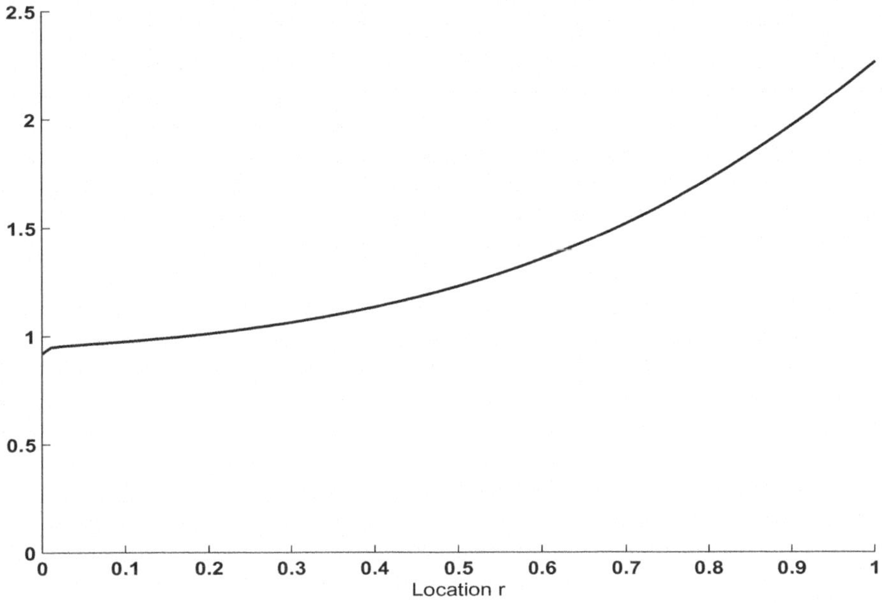

Fig. 5 Evolution of the numerical solution $u_{num}(\tau)$ at time $\tau = 1$

(ii) The radius of the treated tumor: recall that $u(z, \tau) = z\vartheta(z, t)$ on the domain $\Omega \times [0, 1]$. If we assume that the concentration of tumor cells $\vartheta(z, t)$ is constant (is the same) at each point (z, t) in the area $\Omega \times [0, 1]$, the following relation is obtained:

$$\vartheta(z, t) = \vartheta(z_0, t_0) = \varrho_0, \qquad \text{for all} \quad (z, t) \in \Omega \times [0, 1].$$

In this situation, we obtain

$$u(z, \tau) = z\varrho_0. \tag{30}$$

Consequently, we can deduce easily, from (30), that

$$z_{treated} = \frac{1}{\varrho_0} u_{num}(\tau), \tag{31}$$

where $u_{num}(\tau)$ is the numerical solution to the nonlinear model (27)–(28) at the time τ, ($u_{num}(\tau)$ is an approximation to the state $u(z, \tau)$). The numerical solution $u_{num}(\tau)$ is represented in Fig. 5 at time $\tau = 1$.

For the experimental results, we take: $q = 1.5$, $\frac{\sigma_0}{\varrho_0} = 4000$ (see [32]) and $\varrho_0 = 1.348073$. We can now obtain a table 1, in which the values of untreated tumor

Table 1 Comparison between untreated tumor ($z_{untreated}$) and treated tumor ($z_{treated}$)

t(years)	$z_{untreated}$(mm)	$z_{treated}$(mm)
0.1	12.2	12.137
0.2	18.2	11.465
0.3	22.9	11.025
0.4	26.9	10.694
0.5	30.4	10.427
0.6	33.5	10.202
0.7	36.4	10.008
0.8	39	9.837
0.9	41.4	9.684
1.0	43.7	9.546

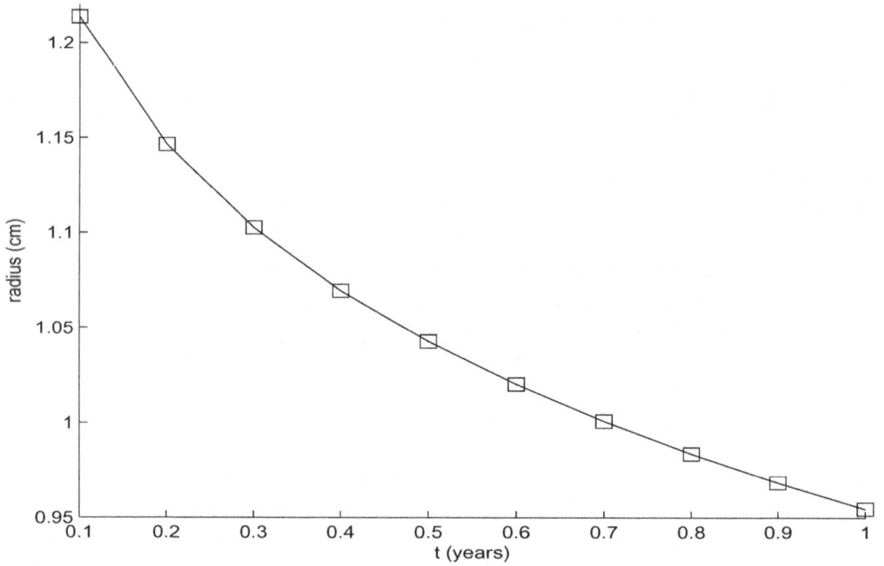

Fig. 6 Evolution of the tumor's radius $z_{treated}$

radius growth $xuntreated$ compared with radius development in treated 1, using the relationship (29) and (31).

In Fig. 6, (or from Table 1), we can observe the decay of the radius under treatment of the tumor by radiotherapy and chemotherapy. On the other hand, in Fig. 7, (or from Table 1), it's very clear that the radius increaseS in time when the tumor is untreated.

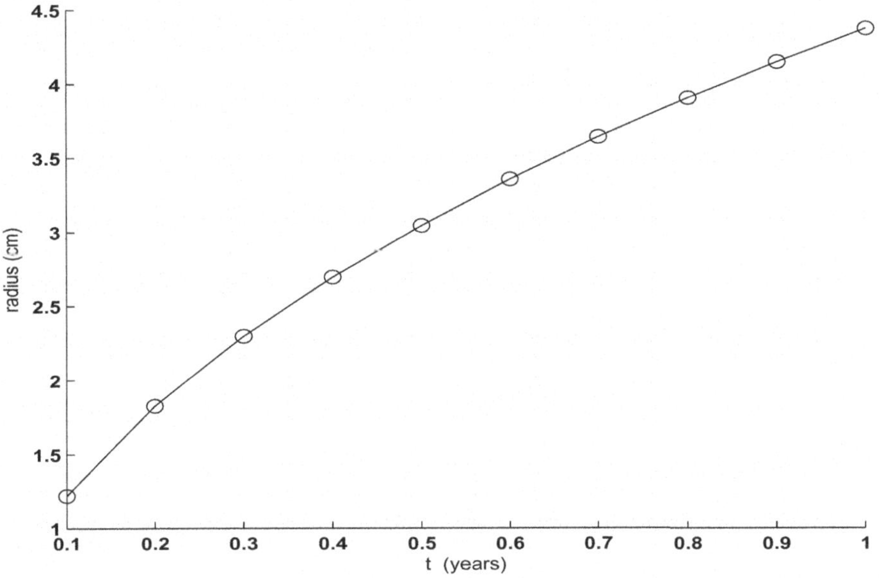

Fig. 7 Evolution of the tumor's radius $z_{untreated}$

5 Conclusion

In this work, we have defined a new fractional model based on the Burgess equation. To solve numerically the obtained model (the nonlinear model (27)–(28)), we have used the splitting method. With this previous mentioned method we are more realistic and more free to choose any initial condition contrary, for instance, to the Adomian decomposition method, where we must choose an adequate initial condition to be able to find the solution.

In fact, this work can be extended to the study of the fractional Burgess model using other types of fractional derivatives such as the Caputo-Hadamard fractional derivative, the Atangana-Baleanu fractional derivative, and the Caputo-Fabrizio fractional derivative.

References

1. P. Tracqui, G.C. Cruywagen, D.E. Woodward, G.T. Bartoo, J.D. Murray, E.C. Jr, Alvord, A mathematical model of glioma growth: the effect of chemotherapy on spatio-temporal growth. Cell Prolif. **28**, 17–31 (1995)
2. D.E. Woodward, J. Cook, P. Tracqui, G.C. Cruywagen, J.D. Murray, E.C.Jr Alvord, A mathematical model of glioma growth: the effect of extent of surgical resection, *Cell Prolif.* **29**(8), 269–288 (1996)

3. P.K. Burgess, P.M. Kulesa, J.D. Murray, E.C. Jr Alvord, The interaction of growth rates and diffusion coefficients in a three-dimensional mathematical model of gliomas. J. Neuropathol. Exp. Neurol. **56**, 704–713 (1997)
4. O. Gonzalez-Gaxiola, R. Bernal-Jaquez, Applying Adomian decomposition method to solve burgess equation with a non-linear source. Int. J. Appl. Computat. Mathe. **3**, 213–224 (2017)
5. M. Benchohra, J.R. Graef, F.Z. Mostefa, Weak solutions for boundary-value problems with nonlinear fractional differential inclusions. Nonlinear Dyn. Syste. Theory. **3**, 227–237 (2011)
6. V. Daftardar-Gejji, S. Bhalekar, Solving multiterm linear and nonlinear diffusion wave equations of fractional order by Adomian decomposition method. Appl. Math. Comput. **202**, 113–120 (2008)
7. D. Baleanu, P.O. Mohammed, M. Vivas-Cortez, Y. Rangel-Oliveros, Some modifications in conformable fractional integral inequalities. Adv. Diff. Equ. **374**, 1–25 (2020)
8. M. Kharrat, Closed-form solution of European option under fractional Heston model. Nonlinear Dyn. Syst. Theory. **18**(2), 191–195 (2018)
9. I. Podlubny, *Fractional Differential Equations Calculus* (Academic, Press, New York, 1999)
10. J.M. Yu, Y.W. Luo, S.B. Zhou, X.R. Lin, Existence and uniqueness for nonlinear multi-variables fractional differential equations. Nonlinear Dyn. Syst. Theory **2**, 213–221 (2011)
11. A. Khalouta, A new analytical series solution with convergence for nonlinear fractional Lienard's equations with caputo fractional derivative. Kyungpook Mathem. J. **62**(3), 583–593 (2022)
12. A. Khalouta, A novel iterative method to solve nonlinear wave-like equations of fractional order with variable coefficients. Revista Colombiana de Matemticas. **56**(1), 13–34 (2022). https://doi.org/10.15446/recolma.v56n1.105612
13. R. Stupp et al., Radiotherapy plus concomitant and adjuvant temozolomide for glioblastoma. N. Engl. J. Med. **352**, 987–996 (2005)
14. A.A. Kilbas, H.M. Srivastava, J.J. Trujillo, *Theory and Applications of Fractional Differential Equations* (Elsevier, Amsterdam, 2006)
15. F. Jarad, T. Abdeljawad, D. Baleanu, Caputo-type modification of the Hadamard fractional derivatives. Adv. Difference Equ. **2012**, 142 (2012). https://doi.org/10.1186/1687-1847-2012-142
16. Y.Y. Gambo, F. Jarad, D. Baleanu, T. Abdeljawad, On Caputo modification of the Hadamard fractional derivatives. Adv. Difference Equ. **2014**, 10 (2014). https://doi.org/10.1186/1687-1847-2014-10
17. A. Atangana, D. Baleanu, New fractional derivatives with nonlocal and non-singular kernel: theory and application to heat transfer model. Therm. Sci. **20**, 757–763 (2016)
18. M. Caputo, M. Fabrizio, A new definition of fractional derivative without singular kernel. Progr. Fract. Differ. Appl. **1**, 73–85 (2015)
19. L. Sadek, A cotangent fractional derivative with the application. Fractal Fract. **7**(6), 444 (2023). https://doi.org/10.3390/fractalfract7060444
20. L. Sadek, T.A. Lazar, On Hilfer cotangent fractional derivative and a particular class of fractional problems. AIMS Mathem. **8**(12), 28334–28352 (2023). https://doi.org/10.3934/math.20231450
21. L. Sadek, A.S. Bataineh, H.T. Alaoui, I. Hashim, The Novel Mittag-LefflerGalerkin method: application to a Riccati differential equation of fractional order. Fractal and Fractional. **7**(4), 302 (2023). https://doi.org/10.3390/fractalfract7040302
22. L. Sadek, A.S. Bataineh, O.R. Isik, H.T. Alaoui, I. Hashim, A numerical approach based on Bernstein collocation method: application to differential Lyapunov and Sylvester matrix equations. Mathematics and Comput. Simul. **212**, 475–488 (2023). https://doi.org/10.1016/j.matcom.2023.05.011
23. L. Sadek, Fractional BDF methods for solving fractional differential matrix equations. Int. J. Appl. Computat. Mathema. **8**(5) (2022). https://doi.org/10.1007/s40819-022-01455-6
24. M. Caputo, Linear models of dissipation whose Q is almost frequency independent II. Fract. Calc. Appl. Anal. **11**, 4–14 (2008). (Reprinted from Geophys. J. R. Astr. Soc. 1967;13:529–539)
25. H. Arfaoui, Stabilization method for the Saint-Venant equations by boundary control. Trans. Inst. Measur. Control **42**(16), 3290–3302 (2020)

26. A.A. Kilbas, H.M. Srivastava, J.J. Trujillo, *Theory and Applications of fractional Differential Equations* (Elsevier Science B.V, Amsterdam, The Netherlands, 2006)
27. J.D. Murray, *Mathematical Biology II: Spatial Models and Biomedical Applications*, 3rd edn. (Springer, New York, 2003)
28. L. Wein, D. Koplow, *Mathematical modeling of brain cancer to identify promising combination treatments* (Preprint, D Sloan School of Management, MIT, 1999)
29. K. Andriopoulos, P.G.L. Leach, A common theme in applied methamatics: an equation connecting applications in economics, medicine and physics. S. Afr. J. Sci. **102**, 66–72 (2006)
30. H. Zhang, F. Liu, I. Turner, Q. Yang, Numerical solution of the time fractional Black-Scholes model governing European options. Comput. Mathem. Appl. **71**(Issue 16), 1772–1783 (2016)
31. E.M.A.M. Mendes, G.H.O. Salgado, L.A. Aguirre, Numerical solution of Caputo fractional differential equations with infinity memory effect at initial condition. Commun. Nonlinear Sci. Numer. Simulat. **69**, 237–247 (2019). https://doi.org/10.1016/j.cnsns.2018.09.022
32. J.D. Murray, Glioblastoma brain tumors: estimating the time from brain tumor initiation and resolution of a patient survival anomaly after treatment protocols. J. Biol. Dyn. **6**(Suppl. 2), 118–127 (2012)

Explicit Form for the Solutions of a System of Difference Equations

Nouressadat Touafek, Mounira Boulouh, Merve Kara, and Yasin Yazlik

Abstract In this chapter, we study a $k+2$-order system of difference equations with powers with respect to the explicit form of the solutions, periodicity and convergence.

Keywords Difference equations · Form of the solutions · Limiting behavior · Periodicity

2010 MSC: 39A10, 39A20, 39A23

1 Introduction

In this chapter, we solve the following system of difference equations of higher order with arbitrary powers given by

$$u_{n+1} = u_{n-k}^{p} \frac{\alpha v_n + \beta v_{n-k-1}^{q}}{\gamma v_n + \delta v_{n-k-1}^{q}}, \quad v_{n+1} = v_{n-k}^{q} \frac{a u_n + b u_{n-k-1}^{p}}{c u_n + d u_{n-k-1}^{p}} \tag{1}$$

N. Touafek (✉) · M. Boulouh
LMAM Laboratory, Faculty of Exact Sciences and Informatics, University of Jijel, 18000 Jijel, Algeria
e-mail: ntouafek@gmail.com

M. Boulouh
e-mail: m.boulouh@hotmail.com

M. Kara
Department of Mathematics, Kamil Ozdag Science Faculty, Karamanoglu Mehmetbey University, 70100 Karaman, Turkey
e-mail: mervekara@kmu.edu.tr

Y. Yazlik
Department of Mathematics, Faculty of Science and Art, Nevsehir Haci Bektas Veli University, 50300 Nevsehir, Turkey
e-mail: yyazlik@nevsehir.edu.tr

M. Kharrat et al. (eds.), *Modeling of Discrete and Continuous Systems*, Forum for Interdisciplinary Mathematics, https://doi.org/10.1007/978-981-97-8715-9_3

where $n \in \mathbb{N}_0, k,\ p,\ q \in \mathbb{N}$, the parameters $a, b, c, d, \alpha, \beta, \gamma, \delta$ and the initial values $u_{-i},\ v_{-i},\ i = -k-1, ..., 0$ are non-zero real numbers. To deal with this system we will distinguish four cases.

First case: $ad = bc$ and $\alpha\delta \neq \beta\gamma$. Second case: $ad \neq bc$ and $\alpha\delta = \beta\gamma$. Third case: $ad = bc$ and $\alpha\delta = \beta\gamma$. Fourth case (general case): $ad \neq bc, \alpha\delta \neq \beta\gamma, a\gamma + c\delta \neq 0$ and $\alpha c + \gamma d \neq 0$.

For solving the general case, we will need the following theorem from [17].

Theorem 1 ([17]) *Consider the following system of difference equations*

$$r_{n+1} = \frac{\alpha_1 t_n + \beta_1}{\gamma_1 t_n + \delta_1}, \quad t_{n+1} = \frac{\alpha_2 r_n + \beta_2}{\gamma_2 r_n + \delta_2}, \quad n \in \mathbb{N}_0, \tag{2}$$

where the parameters $\alpha,\ \beta,\ \gamma,\ \delta$ are real numbers such that $\alpha_2\delta_2 \neq \beta_2\gamma_2$, $\alpha_1\delta_1 \neq \beta_1\gamma_1, \alpha_2\gamma_1 + \gamma_2\delta_1 \neq 0, \alpha_1\gamma_2 + \gamma_1\delta_2 \neq 0$ and the initial values r_0, t_0 are real numbers. Then for all $n \in \mathbb{N}_0$ we have

$$r_{2n} = \frac{r_0(\beta_1\gamma_1 - \alpha_1\delta_1)(\alpha_2\delta_2 - \beta_2\gamma_2)s_{n-1} + ((\alpha_2\alpha_1 + \beta_1\gamma_2)r_0 + \alpha_1\beta_2 + \beta_1\delta_2)s_n}{((\alpha_2\gamma_1 + \gamma_2\delta_1)r_0 - \alpha_2\alpha_1 - \beta_1\gamma_2)s_n + s_{n+1}}$$

$$r_{2n+1} = \frac{(\alpha_1 t_0 + \beta_1)(\beta_1\gamma_1 - \alpha_1\delta_1)(\alpha_2\delta_2 - \beta_2\gamma_2)s_{n-1}}{((\alpha_2\gamma_1 + \gamma_2\delta_1)(\alpha_1 t_0 + \beta_1) - (\alpha_2\alpha_1 + \beta_1\gamma_2)(\gamma_1 t_0 + \delta_1))s_n + (\gamma_1 t_0 + \delta_1)s_{n+1}}$$
$$+ \frac{((\alpha_2\alpha_1 + \beta_1\gamma_2)(\alpha_1 t_0 + \beta_1) + (\alpha_1\beta_2 + \beta_1\delta_2)(\gamma_1 t_0 + \delta_1))s_n}{((\alpha_2\gamma_1 + \gamma_2\delta_1)(\alpha_1 t_0 + \beta_1) - (\alpha_2\alpha_1 + \beta_1\gamma_2)(\gamma_1 t_0 + \delta_1))s_n + (\gamma_1 t_0 + \delta_1)s_{n+1}}$$

$$t_{2n} = \frac{t_0(\beta_1\gamma_1 - \alpha_1\delta_1)(\alpha_2\delta_2 - \beta_2\gamma_2)s_{n-1} + ((\alpha_2\alpha_1 + \beta_2\gamma_1)t_0 + \alpha_2\beta_1 + \beta_2\delta_1)s_n}{((\alpha_1\gamma_2 + \gamma_1\delta_2)t_0 - \alpha_2\alpha_1 - \beta_2\gamma_1)s_n + s_{n+1}}$$

$$t_{2n+1} = \frac{(\alpha_2 r_0 + \beta_2)(\beta_1\gamma_1 - \alpha_1\delta_1)(\alpha_2\delta_2 - \beta_2\gamma_2)s_{n-1}}{((\alpha_1\gamma_2 + \gamma_1\delta_2)(\alpha_2 r_0 + \beta_2) - (\alpha_2\alpha_1 + \beta_2\gamma_1)(\gamma_2 r_0 + \delta_2))s_n + (\gamma_2 r_0 + \delta_2)s_{n+1}}$$
$$+ \frac{((\alpha_2\alpha_1 + \beta_2\gamma_1)(\alpha_2 r_0 + \beta_2) + (\alpha_2\beta_1 + \beta_2\delta_1)(\gamma_2 r_0 + \delta_2))s_n}{((\alpha_1\gamma_2 + \gamma_1\delta_2)(\alpha_2 r_0 + \beta_2) - (\alpha_2\alpha_1 + \beta_2\gamma_1)(\gamma_2 r_0 + \delta_2))s_n + (\gamma_2 r_0 + \delta_2)s_{n+1}}$$

where $(s_n)_{n=0}^{\infty}$ is the sequence defined by

$$s_{n+1} - (\alpha_2\alpha_1 + \beta_2\gamma_1 + \gamma_2\beta_1 + \delta_2\delta_1)s_n + (\alpha_2\delta_2 - \beta_2\gamma_2)(\alpha_1\delta_1 - \beta_1\gamma_1)s_{n-1} = 0,$$

with $s_0 = 0,\ s_1 = 1$.

A lot of contributions are published on difference equations, as example, the following list of works [1–4, 7, 9–11, 13, 14, 16]. Our work is also a contribution to this line of research. Our objective is to solve explicitly System (1) and then use the obtained formulas to deal with the periodicity and the convergence of the solutions. In studying System (1), we are mainly inspired and motivated by the following papers, [5, 6, 8, 12, 15].

From here on, we are interested by well-defined solutions of System (1), i.e., solutions with the following condition:

$$\left(\gamma v_n + \delta v_{n-k-1}^q\right)\left(c u_n + d u_{n-k-1}^p\right) \neq 0,\ n \in \mathbb{N}_0.$$

2 When $ad = bc$ and $\alpha\delta \neq \beta\gamma$

Assume that $ad = bc$ and $\alpha\delta \neq \beta\gamma$. In this case, the second equation of (1) becomes

$$\begin{aligned} v_{n+1} &= \frac{b}{d} v^q_{n-k} \frac{\frac{a}{b} u_n + u^p_{n-k-1}}{\frac{c}{d} u_n + u^p_{n-k-1}} \\ &= \frac{b}{d} v^q_{n-k}, \ n \in \mathbb{N}_0. \end{aligned}$$

Now, using this latest formula, in the first equation of System (1), we get

$$u_{n+2} = u^p_{n-k+1} \frac{\alpha b + \beta d}{\gamma b + \delta d}, \ n \in \mathbb{N}_0.$$

Hence System (1) is reduced to equations

$$\begin{cases} u_{n+1} = A u^p_{n-k}, & n \in \mathbb{N}, \\ v_{n+1} = \frac{b}{d} v^q_{n-k}, & n \in \mathbb{N}_0, \end{cases} \tag{3}$$

where $A = \frac{\alpha b + \beta d}{\gamma b + \delta d}$ and $(\alpha b + \beta d)(\gamma b + \delta d) \neq 0$.

Theorem 2 *The solutions of equation in* (3) *are presented by*

$$\forall n \in \mathbb{N}_0 : u_{(k+1)n+l} = \begin{cases} A^{n+1} u_{l-(k+1)}, & p = 1, \\ A^{\frac{1-p^{n+1}}{1-p}} u^{p^{n+1}}_{l-(k+1)}, & p \neq 1. \end{cases}, \ l = \overline{2, k+2}.$$

$$\forall n \in \mathbb{N}_0 : v_{(k+1)n+l} = \begin{cases} \left(\frac{b}{d}\right)^{n+1} v_{l-(k+1)}, & q = 1, \\ \left(\frac{b}{d}\right)^{\frac{1-q^{n+1}}{1-q}} v^{q^{n+1}}_{l-(k+1)}, & q \neq 1. \end{cases} \quad l = \overline{1, k+1}.$$

Proof From (3), we have

$$\begin{aligned} u_2 &= A u^p_{-k+1} \\ u_3 &= A u^p_{-k+2} \\ &\vdots \\ u_k &= A u^p_{-1} \\ u_{k+1} &= A u^p_0 \\ u_{k+2} &= A u^p_1 \\ u_{k+3} = u_{k+1+2} &= A u^p_2 = A^{1+p} u^{p^2}_{-k+1} \\ &\vdots \\ u_{2k+1} = u_{(k+1)+k} &= A u^p_k = A^{1+p} u^{p^2}_{-1} \end{aligned}$$

$$\begin{aligned}
u_{2k+2} &= u_{k+1+k+1} = Au_{k+1}^{p} = A^{1+p}u_{0}^{p^2}\\
u_{2k+3} &= u_{k+1+k+2} = Au_{k+2}^{p} = A^{1+p}u_{1}^{p^2}\\
u_{2k+4} &= u_{2(k+1)+2} = Au_{k+3}^{p} = A^{1+p+p^2}u_{-k+1}^{p^3}\\
&\vdots\\
u_{3k+2} &= u_{2(k+1)+k} = Au_{2k+1}^{p} = A^{1+p+p^2}u_{-1}^{p^3}\\
u_{3k+3} &= u_{2(k+1)+k+1} = Au_{2k+2}^{p} = A^{1+p+p^2}u_{0}^{p^3}\\
u_{3k+4} &= u_{2(k+1)+k+2} = Au_{2k+3}^{p} = A^{1+p+p^2}u_{1}^{p^3}\\
&\vdots
\end{aligned}$$

By induction, we obtain

$$u_{(k+1)n+l} = A^{1+p+p^2+\ldots+p^n}u_{l-(k+1)}^{p^{n+1}},\ l=\overline{2,k+2},\ n\in\mathbb{N}_0$$

that is

$$\forall n\in\mathbb{N}_0 : u_{(k+1)n+l} = A^{\frac{1-p^{n+1}}{1-p}}u_{l-(k+1)}^{p^{n+1}},\ l=\overline{2,k+2} \tag{4}$$

if $p\neq 1$ and

$$\forall n\in\mathbb{N}_0,\ u_{(k+1)n+l} = A^{n+1}u_{l-(k+1)},\ l=\overline{2,k+2} \tag{5}$$

if $p=1$.

By the same procedure, we get from (3),

$$\begin{aligned}
v_1 &= \frac{b}{d}v_{-k}^{q}\\
v_2 &= \frac{b}{d}v_{-k+1}^{q}\\
&\vdots\\
v_k &= \frac{b}{d}v_{-1}^{q}\\
v_{k+1} &= \frac{b}{d}v_{0}^{q}\\
v_{k+2} = v_{k+1+1} &= \frac{b}{d}v_{1}^{q} = \left(\frac{b}{d}\right)^{1+q}v_{-k}^{q^2}\\
v_{k+3} = v_{k+1+2} &= \frac{b}{d}v_{2}^{q} = \left(\frac{b}{d}\right)^{1+q}v_{-k+1}^{q^2}\\
&\vdots\\
v_{2k+1} = v_{(k+1)+k} &= \frac{b}{d}v_{k}^{q} = \left(\frac{b}{d}\right)^{1+q}v_{-1}^{q^2}
\end{aligned}$$

$$v_{2k+2} = v_{k+1+k+1} = \frac{b}{d} v_{k+1}^{q} = \Big(\frac{b}{d}\Big)^{1+q} v_0^{q^2}$$

$$v_{2k+3} = v_{2(k+1)+1} = \frac{b}{d} v_{k+2}^{q} = \Big(\frac{b}{d}\Big)^{1+q+q^2} v_{-k}^{q^3}$$

$$\vdots$$

$$v_{3k+2} = v_{2(k+1)+k} = \frac{b}{d} v_{2k+1}^{q} = \Big(\frac{b}{d}\Big)^{1+q+q^2} v_{-1}^{q^3}$$

$$v_{3k+3} = v_{2(k+1)+k+1} = \frac{b}{d} v_{2k+2}^{q} = \Big(\frac{b}{d}\Big)^{1+q+q^2} v_0^{q^3}$$

$$\vdots$$

Then, by induction we get

$$v_{(k+1)n+l} = \Big(\frac{b}{d}\Big)^{1+q+q^2+\ldots+q^n} v_{l-(k+1)}^{q^{n+1}}, \ l = \overline{1, k+1}, \ n \in \mathbb{N}_0,$$

which implies

$$v_{(k+1)n+l} = \Big(\frac{b}{d}\Big)^{\frac{1-q^{n+1}}{1-q}} v_{l-(k+1)}^{q^{n+1}}, \ l = \overline{1, k+1}, \ n \in \mathbb{N}_0 \tag{6}$$

if $q \neq 1$ and

$$v_{(k+1)n+l} = \Big(\frac{b}{d}\Big)^{n+1} v_{l-(k+1)}, \ l = \overline{1, k+1}, \ n \in \mathbb{N}_0, \tag{7}$$

if $q = 1$, and this completes the proof.

Assume that $p = q = 1$, then System (3) will be

$$\begin{cases} u_{n+1} = \frac{\alpha b+\beta d}{\gamma b+\delta d} u_{n-k}, & n \in \mathbb{N}, \\ v_{n+1} = \frac{b}{d} v_{n-k}, & n \in \mathbb{N}_0. \end{cases} \tag{8}$$

Here we provide conditions that make the solutions of System (8) periodic.

Theorem 3 *$\alpha + \beta = \gamma + \delta$ and $b = d$ is a necessary and sufficient condition for the solutions of* (8) *to be $k + 1$ periodic.*

Proof Assume that the solutions $(u_n)_{n \geq -k+1}$ and $(v_n)_{n \geq -k}$ of (8) are periodic of period $k + 1$, that is

$$u_n = u_{n-(k+1)}, \ n = 2, 3, \ldots$$

and

$$v_n = v_{n-(k+1)}, \ n = 1, 2, \ldots$$

then,

$$u_2 = u_{-k+1}, \ v_1 = v_{-k}.$$

Using this fact and equation in (8), we get

$$u_2 = u_{-k+1} = \frac{\alpha b + \beta d}{\gamma b + \delta d} u_{-k+1},$$

and

$$v_1 = v_{-k} = \frac{b}{d} v_{-k}.$$

From which we obtain

$$\frac{\alpha b + \beta d}{\gamma b + \delta d} = \frac{b}{d} = 1 \text{ or equivalently } \alpha + \beta = \gamma + \delta \text{ and } b = d.$$

Now, suppose that $\alpha + \beta = \gamma + \delta$ and $b = d$.
Then, from (8), we have

$$u_2 = \frac{\alpha b + \beta d}{\gamma b + \delta d} u_{1-k} = u_{1-k}, \quad v_1 = \frac{b}{d} v_{-k} = v_{-k}.$$

Also, by taking $n = 2$ in (8), we get

$$u_3 = \frac{\alpha b + \beta d}{\gamma b + \delta d} u_{-k+2} = u_{-k+2}, \quad v_2 = \frac{b}{d} v_{-k+1} = v_{-k+1}.$$

Continuing the procedure, we get by induction the result.

The following convergence result follows directly from Theorem 2.

Theorem 4 *For each solution $(u_n)_{n \geq -k+1}$ and $(v_n)_{n \geq -k}$ of System (3), we have*

1. *If $p = q = 1$, then*

$$\lim_{n\to\infty} u_n = \begin{cases} 0, & |A| < 1 \\ \infty, & |A| > 1 \end{cases} \quad \text{and} \quad \lim_{n\to\infty} v_n = \begin{cases} 0, & |b| < |d|, \\ \infty, & |b| > |d|. \end{cases}$$

2. *If $p \neq 1$, $q \neq 1$ then*

$$\lim_{n\to\infty} u_n = \begin{cases} 0, & |A| < 1, \ \max\limits_{l=2,k+2} |u_{l-(k+1)}| < 1, \\ \infty, & |A| > 1, \ \min\limits_{l=2,k+2} |u_{l-(k+1)}| > 1, \end{cases}$$

and

$$\lim_{n\to\infty} v_n = \begin{cases} 0, & |b| < |d|, \quad \max\limits_{l=\overline{1,k+1}} |v_{l-(k+1)}| < 1, \\ \infty, & |b| > |d|, \quad \min\limits_{l=\overline{1,k+1}} |v_{l-(k+1)}| > 1. \end{cases}$$

3 When $ad = bc$ and $\alpha\delta = \beta\gamma$

Assume that $ad = bc$ and $\alpha\delta = \beta\gamma$, then System (1) will be reduced to the following system

$$u_{n+1} = \frac{\beta}{\delta} u_{n-k}^{p}, \quad v_{n+1} = \frac{b}{d} v_{n-k}^{q}, \quad n \in \mathbb{N}_0. \tag{9}$$

For the solutions of this system, we have the following results, which are direct consequences of the two previous sections.

Theorem 5 *For each solution* $(u_n, v_n)_{n\geq -k}$ *of System* (9), *we have for* $n \in \mathbb{N}_0$

$$u_{(k+1)n+l} = \begin{cases} \left(\frac{\beta}{\delta}\right)^{n+1} u_{l-(k+1)}, & p = 1, \\ \left(\frac{\beta}{\delta}\right)^{\frac{1-p^n}{1-p}} u_{l-(k+1)}^{p^{n+1}}, & p \neq 1, \end{cases} \qquad l = \overline{1, k+1}.$$

$$v_{(k+1)n+l} = \begin{cases} \left(\frac{b}{d}\right)^{n+1} v_{l-(k+1)}, & q = 1, \\ \left(\frac{b}{d}\right)^{\frac{1-q^n}{1-q}} v_{l-(k+1)}^{q^{n+1}}, & q \neq 1, \end{cases} \qquad l = \overline{1, k+1}.$$

Now, if $p = q = 1$, System (9) becomes

$$u_{n+1} = \frac{\beta}{\delta} u_{n-k}, \quad v_{n+1} = \frac{b}{d} v_{n-k}, \quad n \in \mathbb{N}_0. \tag{10}$$

Theorem 6 $b = d,\ \beta = \delta \Leftrightarrow$ *every solution of System* (10) *is* $k + 1$ *periodic.*

Theorem 7 *For each solution* $(u_n, y_n)_{n\geq -k}$ *of System* (9). *Then,*

1. *If* $p = q = 1$, *we have*

$$\lim_{n\to\infty} u_n = \begin{cases} 0, & |\beta| < |\delta|, \\ \infty, & |\beta| > |\delta|, \end{cases} \quad \text{and} \quad \lim_{n\to\infty} v_n = \begin{cases} 0, & |b| < |d|, \\ \infty, & |b| > |d|. \end{cases}$$

2. *If* $p \neq 1$ *and* $q \neq 1$, *we have*

$$\lim_{n\to\infty} u_n = \begin{cases} 0, & |\beta| < |\delta| \text{ and } \max\limits_{l=\overline{1,k+1}} |u_{l-(k+1)}| < 1, \\ \infty, & |\beta| > |\delta| \text{ and } \min\limits_{l=\overline{1,k+1}} |u_{l-(k+1)}| > 1, \end{cases}$$

and

$$\lim_{n\to\infty} v_n = \begin{cases} 0, & |b| < |d| \text{ and } \max_{l=\overline{1,k+1}} |v_{l-(k+1)}| < 1, \\ \pm\infty, & |b| > |d| \text{ and } \min_{l=\overline{1,k+1}} |v_{l-(k+1)}| > 1. \end{cases}$$

4 The General Case

In this part, we solve System (1) in the general case.

Theorem 8 *For each solution $(u_n, v_n)_{n\geq -k-1}$ of System* (1)*, we have for $n \in \mathbb{N}_0$*

(i) If k is odd: $k = 2k' + 1,\ k' \in \mathbb{N}_0$, *then*

$$\forall l = \overline{0, k'},\ j \in \{0,1\} : \begin{cases} u_{2[(k'+1)n+l]+j} = u_{2(l-k'-1)+j}^{p^{n+1}} \prod_{i=0}^{n} A_{2[(k'+1)i+l]+j}^{p^{n-i}}, \\ v_{2[(k'+1)n+l]+j} = v_{2(l-k'-1)+j}^{q^{n+1}} \prod_{i=0}^{n} B_{2[(k'+1)i+l]+j}^{q^{n-i}}, \end{cases}.$$

(ii) If k is even: $k = 2k',\ k' \in \mathbb{N}$, *then*

$$u_{2[(2k'+1)n+k'+l+1]} = u_{2(l-k')}^{p^{2(n+1)}} \prod_{i=0}^{n} A_{2[(2k'+1)i+l]+1}^{p^{2(n-i)+1}} A_{2[(2k'+1)i+k'+l+1]}^{p^{2(n-i)}}$$
$$l = \overline{0, k'-1},$$

$$u_{2[(2k'+1)n+l]+1} = u_{2(l-k')}^{p^{2n+1}} \Big(\prod_{i=0}^{n} A_{2[(2k'+1)i+l]+1}^{p^{2(n-i)}}\Big)\Big(\prod_{i=0}^{n-1} A_{2[(2k'+1)i+k'+l+1]}^{p^{2(n-i)-1}}\Big)$$
$$l = \overline{0, k'-1},$$

$$u_{2[(2k'+1)n+k'+l]+1} = u_{2(l-k')-1}^{p^{2(n+1)}} \prod_{i=0}^{n} A_{2[(2k'+1)i+l]}^{p^{2(n-i)}} A_{2((2k'+1)i+k'+l)+1}^{p^{2(n-i)}}$$
$$l = \overline{0, k'},$$

$$u_{2[(2k'+1)n+l]} = u_{2(l-k')-1}^{p^{2n+1}} \Big(\prod_{i=0}^{n} A_{2[(2k'+1)i+l]}^{p^{2(n-i)+1}}\Big)\Big(\prod_{i=0}^{n-1} A_{2[(2k'+1)i+k'+l]+1}^{p^{2(n-i)-1}}\Big)$$
$$l = \overline{0, k'},$$

$$v_{2[(2k'+1)n+k'+l+1]} = v_{2(l-k')}^{q^{2(n+1)}} \prod_{i=0}^{n} B_{2((2k'+1)i+l)+1}^{q^{2(n-i)+1}} B_{2[(2k'+1)i+k'+l+1]}^{q^{2(n-i)}}$$
$$l = \overline{0, k'-1},$$

$$v_{2[(2k'+1)n+l]+1} = v_{2(l-k')}^{q^{2n+1}}\Big(\prod_{i=0}^{n} B_{2[(2k'+1)i+l]+1}^{q^{2(n-i)}}\Big)\Big(\prod_{i=0}^{n-1} B_{2[(2k'+1)i+k'+l+1]}^{p^{2(n-i)-1}}\Big)$$
$$l = \overline{0, k'-1},$$
$$v_{2[(2k'+1)n+k'+l]+1} = v_{2(l-k')-1}^{q^{2(n+1)}} \prod_{i=0}^{n} B_{2[(2k'+1)i+l]}^{q^{2(n-i)}} B_{2[(2k'+1)i+k'+l]+1}^{q^{2(n-i)}}$$
$$l = \overline{0, k'},$$
$$v_{2[(2k'+1)n+l]} = v_{2(l-k')-1}^{q^{2n+1}}\Big(\prod_{i=0}^{n} B_{2[(2k'+1)i+l]}^{q^{2(n-i)+1}}\Big)\Big(\prod_{i=0}^{n-1} B_{2[(2k'+1)i+k'+l]+1}^{q^{2(n-i)-1}}\Big)$$
$$l = \overline{0, k'},$$

where $(A_n)_{n=0}^{\infty}$ *and* $(B_n)_{n=0}^{\infty}$ *are given by*

$$A_{2n} = \frac{A_0(\beta\gamma - \alpha\delta)(ad - bc)s_{n-1} + ((a\alpha + \beta c)A_0 + \alpha b + \beta d)s_n}{((a\gamma + c\delta)A_0 - a\alpha - \beta c)s_n + s_{n+1}}$$
$$A_{2n+1} = \frac{(\alpha B_0 + \beta)(\beta\gamma - \alpha\delta)(ad - bc)s_{n-1}}{((a\gamma + c\delta)(\alpha B_0 + \beta) - (a\alpha + \beta c)(\gamma B_0 + \delta))s_n + (\gamma B_0 + \delta)s_{n+1}}$$
$$+ \frac{((a\alpha + \beta c)(\alpha B_0 + \beta) + (\alpha b + \beta d)(\gamma B_0 + \delta))s_n}{((a\gamma + c\delta)(\alpha B_0 + \beta) - (a\alpha + \beta c)(\gamma B_0 + \delta))s_n + (\gamma B_0 + \delta)s_{n+1}}$$
$$B_{2n} = \frac{B_0(\beta\gamma - \alpha\delta)(ad - bc)s_{n-1} + ((a\alpha + b\gamma)C_0 + a\beta + b\delta)s_n}{((\alpha c + \gamma d)B_0 - a\alpha - b\gamma)s_n + s_{n+1}}$$
$$B_{2n+1} = \frac{(aA_0 + b)(\beta\gamma - \alpha\delta)(ad - bc)s_{n-1}}{((\alpha c + \gamma d)(aA_0 + b) - (a\alpha + b\gamma)(cA_0 + d))s_n + (cA_0 + d)s_{n+1}}$$
$$+ \frac{((a\alpha + b\gamma)(aA_0 + b) + (a\beta + b\delta)(cA_0 + d))s_n}{((\alpha c + \gamma d)(aA_0 + b) - (a\alpha + b\gamma)(cA_0 + d))s_n + (cA_0 + d)s_{n+1}},$$

with

$$A_0 = \frac{u_0}{u_{-k-1}^p}, \; B_0 = \frac{v_0}{v_{-k-1}^q}$$

and the sequence $(s_n)_{n=0}^{\infty}$ *is as in Theorem* (1).

Proof Writing System (1) as follows:

$$\frac{u_{n+1}}{u_{n-k}^p} = \frac{\alpha\frac{v_n}{v_{n-k-1}^q} + \beta}{\gamma\frac{v_n}{v_{n-k-1}^q} + \delta}, \quad \frac{v_{n+1}}{v_{n-k}^q} = \frac{a\frac{u_n}{u_{n-k-1}^p} + b}{c\frac{u_n}{u_{n-k-1}^p} + d},$$

and putting

$$A_n := \frac{u_n}{u_{n-k-1}^p}, \quad B_n := \frac{v_n}{v_{n-k-1}^q}, \quad n \in \mathbb{N}_0, \tag{11}$$

we get the following system:

$$A_{n+1} = \frac{\alpha B_n + \beta}{\gamma B_n + \delta}, \quad B_{n+1} = \frac{a A_n + b}{c A_n + d}. \tag{12}$$

Clearly System (12) is in the form of System (2).
By Theorem (1), we get

$$\begin{aligned}
A_{2n} &= \frac{A_0(\beta\gamma - \alpha\delta)(ad - bc)s_{n-1} + ((a\alpha + \beta c)A_0 + \alpha b + \beta d)s_n}{((a\gamma + c\delta)A_0 - a\alpha - \beta c)s_n + s_{n+1}} \\
A_{2n+1} &= \frac{(\alpha B_0 + \beta)(\beta\gamma - \alpha\delta)(ad - bc)s_{n-1}}{((a\gamma + c\delta)(\alpha B_0 + \beta) - (a\alpha + \beta c)(\gamma B_0 + \delta))s_n + (\gamma B_0 + \delta)s_{n+1}} \\
&+ \frac{(((a\alpha + \beta c)(\alpha B_0 + \beta) + (\alpha b + \beta d)(\gamma B_0 + \delta))s_n}{((a\gamma + c\delta)(\alpha B_0 + \beta) - (a\alpha + \beta c)(\gamma B_0 + \delta))s_n + (\gamma B_0 + \delta)s_{n+1}} \\
B_{2n} &= \frac{B_0(\beta\gamma - \alpha\delta)(ad - bc)s_{n-1} + ((a\alpha + b\gamma)C_0 + a\beta + b\delta)s_n}{((\alpha c + \gamma d)B_0 - a\alpha - b\gamma)s_n + s_{n+1}} \\
B_{2n+1} &= \frac{(aA_0 + b)(\beta\gamma - \alpha\delta)(ad - bc)s_{n-1}}{((\alpha c + \gamma d)(aA_0 + b) - (a\alpha + b\gamma)(cA_0 + d))s_n + (cA_0 + d)s_{n+1}} \\
&+ \frac{((a\alpha + b\gamma)(aA_0 + b) + (a\beta + b\delta)(cA_0 + d))s_n}{((\alpha c + \gamma d)(aA_0 + b) - (a\alpha + b\gamma)(cA_0 + d))s_n + (cA_0 + d)s_{n+1}}
\end{aligned}$$

with $A_0 = \frac{u_0}{u^p_{-k-1}}$ and $B_0 = \frac{v_0}{v^q_{-k-1}}$.

Now, using the relations (11), we get

$$u_0 = A_0 u^p_{-(k+1)}$$

$$u_1 = A_1 u^p_{-k}$$

$$\vdots$$

$$u_k = A_k u^p_{-1}$$

$$u_{k+1} = A_{k+1} A_0^p u^{p^2}_{-(k+1)}$$

$$u_{k+2} = u_{(k+1)+1} = A_{k+2} u_1^p = A_{(k+1)+1} A_1^p u^{p^2}_{-k}$$

$$\vdots$$

$$u_{2k} = u_{(k+1)+k-1} = A_{2k} x^p_{k-1} = A_{(k+1)+k-1} A^p_{k-1} u^{p^2}_{-2}$$

$$u_{2k+1} = u_{(k+1)+k} = A_{2k+1} u_k^p = A_{2k+1} A_k^p u^{p^2}_{-1}$$

$$u_{2k+2} = u_{2(k+1)} = A_{2(k+1)} u_{k+1}^{p} = A_{2(k+1)} A_{k+1}^{p} A_0^{p^2} u_{-(k+1)}^{p^3}$$

$$\vdots$$

$$u_{3k+2} = u_{2(k+1)+k} = A_{2(k+1)+k} u_{2k+1}^{p} = A_{2(k+1)+k} A_{2k+1}^{p} A_k^{p^2} u_{-1}^{p^3}$$

$$u_{3k+3} = u_{3(k+1)} = A_{3(k+1)} u_{2(k+1)}^{p} = A_{3(k+1)} A_{2(k+1)}^{p} A_{k+1}^{p^2} A_0^{p^3} u_{-(k+1)}^{p^4}$$

$$\vdots$$

$$u_{4k+3} = u_{3(k+1)+k} = A_{3(k+1)+k} u_{3k+2}^{p} = A_{3(k+1)+k} A_{2(k+1)+k}^{p} A_{2k+1}^{p^2} A_k^{p^3} u_{-1}^{p^4}$$

$$\vdots$$

and

$$\begin{aligned}
v_0 &= B_0 v_{-(k+1)}^{q} \\
v_1 &= B_1 v_{-k}^{q} \\
&\vdots \\
v_k &= B_k v_{-1}^{q} \\
v_{k+1} &= B_{k+1} B_0^{q} v_{-(k+1)}^{q^2} \\
v_{k+2} = v_{(k+1)+1} = B_{k+2} v_1^{q} &= B_{(k+1)+1} B_1^{q} v_{-k}^{q^2} \\
&\vdots \\
v_{2k} = v_{(k+1)+k-1} = B_{2k} v_{k-1}^{q} &= B_{(k+1)+k-1} B_{k-1}^{q} v_{-2}^{q^2} \\
v_{2k+1} = v_{(k+1)+k} = B_{2k+1} v_k^{q} &= B_{2k+1} B_k^{q} v_{-1}^{q^2} \\
v_{2k+2} = v_{2(k+1)} = B_{2(k+1)} v_{k+1}^{q} &= B_{2(k+1)} B_{k+1}^{q} B_0^{q^2} v_{-(k+1)}^{q^3} \\
&\vdots \\
v_{3k+2} = v_{2(k+1)+k} = B_{2(k+1)+k} v_{2k+1}^{q} &= B_{2(k+1)+k} B_{2k+1}^{q} B_k^{q^2} v_{-1}^{q^3} \\
v_{3k+3} = v_{3(k+1)} = B_{3(k+1)} v_{2(k+1)}^{q} &= B_{3(k+1)} B_{2(k+1)}^{q} B_{k+1}^{q^2} B_0^{q^3} v_{-(k+1)}^{q^4} \\
&\vdots \\
v_{4k+3} = v_{3(k+1)+k} = B_{3(k+1)+k} v_{3k+2}^{q} &= B_{3(k+1)+k} B_{2(k+1)+k}^{q} B_{2k+1}^{q^2} B_k^{q^3} v_{-1}^{q^4} \\
&\vdots
\end{aligned}$$

Hence, by induction we get for all $n \in \mathbb{N}_0$,

$$\begin{cases} u_{(k+1)n+l} &= u_{l-k-1}^{p^{n+1}} \prod_{i=0}^{n} A_{(k+1)i+l}^{p^{n-i}}, \\ v_{(k+1)n+l} &= v_{l-k-1}^{q^{n+1}} \prod_{i=0}^{n} B_{(k+1)i+l}^{q^{n-i}}, \end{cases} \quad l = \overline{0, k}. \tag{13}$$

Here, we distinguish two cases:

(i) k odd: $k = 2k' + 1,\ k' \in \mathbb{N}_0$. Then for all $n \in \mathbb{N}_0$, we have

$$\forall l = \overline{0, k'} \text{ and } j = 0, 1 : \begin{cases} u_{2[(k'+1)n+l]+j} = u_{2(l-k'-1)+j}^{p^{n+1}} \prod_{i=0}^{n} A_{2[(k'+1)i+l]+j}^{p^{n-i}}, \\ v_{2[(k'+1)n+l]+j} = v_{2(l-k'-1)+j}^{q^{n+1}} \prod_{i=0}^{n} B_{2[(k'+1)i+l]+j}^{q^{n-i}}. \end{cases}$$

(ii) k even: $k = 2k',\ k' \in \mathbb{N}$. Here we have two sub-cases:
If n is odd, we find:

$$u_{(2k'+1)(2n+1)+l} = u_{l-2k'-1}^{p^{2(n+1)}} \prod_{i=0}^{n} A_{2(2k'+1)i+l}^{p^{2(n-i)+1}} A_{(2k'+1)(2i+1)+l}^{p^{2(n-i)}}, \tag{14}$$

and

$$v_{(2k'+1)(2n+1)+l} = v_{l-2k'-1}^{q^{2(n+1)}} \prod_{i=0}^{n} B_{2(2k'+1)i+l}^{q^{2(n-i)+1}} B_{(2k'+1)(2i+1)+l}^{q^{2(n-i)}}, \tag{15}$$

where $l = \overline{0, 2k'},\ n \in \mathbb{N}_0$. If n is even, we find:

$$u_{(2k'+1)(2n)+l} = u_{l-2k'-1}^{p^{2n+1}} \Big(\prod_{i=0}^{n} A_{2(2k'+1)i+l}^{p^{2(n-i)}}\Big)\Big(\prod_{i=0}^{n-1} A_{(2k'+1)(2i+1)+l}^{p^{2(n-i)-1}}\Big), \tag{16}$$

and

$$v_{(2k'+1)(2n)+l} = v_{l-2k'-1}^{q^{2n+1}} \Big(\prod_{i=0}^{n} B_{2(2k'+1)i+l}^{q^{2(n-i)}}\Big)\Big(\prod_{i=0}^{n-1} B_{(2k'+1)(2i+1)+l}^{q^{2(n-i)-1}}\Big), \tag{17}$$

where $l = \overline{0, 2k'},\ n \in \mathbb{N}_0$.
Now, with respect to the parity of the parameter l we have, from (14), (15), (16), and (17), the two cases:

a. If l is odd: we get for all $n \in \mathbb{N}_0$ and $l = \overline{0, k' - 1}$,

$$u_{2[(2k'+1)n+k'+l+1]} = u_{2(l-k')}^{p^{2(n+1)}} \prod_{i=0}^{n} A_{2[(2k'+1)i+l]+1}^{p^{2(n-i)+1}} A_{2[(2k'+1)i+k'+l+1]}^{p^{2(n-i)}},$$

$$u_{2[(2k'+1)n+l]+1} = u_{2(l-k')}^{p^{2n+1}} \Big(\prod_{i=0}^{n} A_{2[(2k'+1)i+l]+1}^{p^{2(n-i)}}\Big)\Big(\prod_{i=0}^{n-1} A_{2[(2k'+1)i+k'+l+1]}^{p^{2(n-i)-1}}\Big),$$

$$v_{2[(2k'+1)n+k'+l+1]} = v_{2(l-k')}^{q^{2(n+1)}} \prod_{i=0}^{n} B_{2((2k'+1)i+l)+1}^{q^{2(n-i)+1}} B_{2[(2k'+1)i+k'+l+1]}^{q^{2(n-i)}},$$

$$v_{2[(2k'+1)n+l]+1} = v_{2(l-k')}^{q^{2n+1}} \Big(\prod_{i=0}^{n} B_{2[(2k'+1)i+l]+1}^{q^{2(n-i)}}\Big)\Big(\prod_{i=0}^{n-1} B_{2[(2k'+1)i+k'+l+1]}^{p^{2(n-i)-1}}\Big).$$

b. If l is even: we get for all $n \in \mathbb{N}_0$ and $l = \overline{0, k'}$,

$$u_{2[(2k'+1)n+k'+l]+1} = u_{2(l-k')-1}^{p^{2(n+1)}} \prod_{i=0}^{n} A_{2[(2k'+1)i+l]}^{p^{2(n-i)+1}} A_{2((2k'+1)i+k'+l)+1}^{p^{2(n-i)}},$$

$$u_{2[(2k'+1)n+l]} = u_{2(l-k')-1}^{p^{2n+1}} \Big(\prod_{i=0}^{n} A_{2[(2k'+1)i+l]}^{p^{2(n-i)}}\Big)\Big(\prod_{i=0}^{n-1} A_{2[(2k'+1)i+k'+l]+1}^{p^{2(n-i)-1}}\Big),$$

$$v_{2[(2k'+1)n+k'+l]+1} = v_{2(l-k')-1}^{q^{2(n+1)}} \prod_{i=0}^{n} B_{2[(2k'+1)i+l]}^{q^{2(n-i)+1}} B_{2[(2k'+1)i+k'+l]+1}^{q^{2(n-i)}},$$

$$v_{2[(2k'+1)n+l]} = v_{2(l-k')-1}^{q^{2n+1}} \Big(\prod_{i=0}^{n} B_{2[(2k'+1)i+l]}^{q^{2(n-i)}}\Big)\Big(\prod_{i=0}^{n-1} B_{2[(2k'+1)i+k'+l]+1}^{q^{2(n-i)-1}}\Big).$$

5 The Case $p = q = 1$

This part is devoted to System (1) when $p = q = 1$, in other words the system

$$u_{n+1} = u_{n-k}\frac{\alpha v_n + \beta v_{n-k-1}}{\gamma v_n + \delta v_{n-k-1}}, \quad v_{n+1} = v_{n-k}\frac{a u_n + b u_{n-k-1}}{c u_n + d u_{n-k-1}}, \quad n \in \mathbb{N}_0, \tag{18}$$

here we assume that u_{-i}, v_{-i} , $i = 0, \ldots, k+1$ and $a, b, c, d, \alpha, \beta, \gamma, \delta$ are positive.

The following result shows the limiting behavior of positive solutions of System (18).

Theorem 9 *For each positive solution* $(u_n, v_n)_{n\geq -k-1}$ *of System* (18), *we have*

$$\lim_{n\to\infty} u_n = \begin{cases} 0, & \max\left(\frac{\alpha}{\gamma}, \frac{\beta}{\delta}\right) < 1, \\ +\infty, & \min\left(\frac{\alpha}{\gamma}, \frac{\beta}{\delta}\right) > 1, \end{cases} \quad \textit{and} \quad \lim_{n\to\infty} v_n = \begin{cases} 0, & \max\left(\frac{a}{c}, \frac{b}{d}\right) < 1, \\ +\infty, & \min\left(\frac{a}{c}, \frac{b}{d}\right) > 1. \end{cases}$$

Proof Let $(u_n, v_n)_{n\geq -k-1}$ be a positive solution of System (18). We have for all $n \in \mathbb{N}_0$:

$$u_{n+1} - \frac{\alpha}{\gamma}u_{n-k} = \frac{(\beta\gamma - \alpha\delta)v_{n-k-1}}{\gamma(\gamma v_n + \delta v_{n-k-1})}u_{n-k},$$

$$u_{n+1} - \frac{\beta}{\delta}u_{n-k} = \frac{(\alpha\delta - \beta\gamma)v_n}{\delta(\gamma v_n + \delta v_{n-k-1})}u_{n-k},$$

$$v_{n+1} - \frac{a}{c}v_{n-k} = \frac{(bc - ad)u_{n-k-1}}{c(cu_n + du_{n-k-1})}v_{n-k}$$

and

$$v_{n+1} - \frac{b}{d}v_{n-k} = \frac{(ad - bc)u_n}{d(cu_n + du_{n-k-1})}v_{n-k}.$$

So, it follows that

$$\min\left(\frac{\alpha}{\gamma}, \frac{\beta}{\delta}\right)u_{n-k} < u_{n+1} < \max\left(\frac{\alpha}{\gamma}, \frac{\beta}{\delta}\right)u_{n-k}, \ n \in \mathbb{N}_0, \tag{19}$$

$$\min\left(\frac{a}{c}, \frac{b}{d}\right)v_{n-k} < v_{n+1} < \max\left(\frac{a}{c}, \frac{b}{d}\right)v_{n-k}, \ n \in \mathbb{N}_0. \tag{20}$$

Let

$$\theta := \max\left(\frac{\alpha}{\gamma}, \frac{\beta}{\delta}\right), \ \sigma := \min\left(\frac{\alpha}{\gamma}, \frac{\beta}{\delta}\right), \ \rho := \max\left(\frac{a}{c}, \frac{b}{d}\right), \ \tau := \min\left(\frac{a}{c}, \frac{b}{d}\right).$$

Now, from the inequalities (19) and (20), we get for all $m = \overline{1, n}, \ n \in \mathbb{N}$,

$$u_{-k}\left(\sigma\right)^m < u_{(k+1)m-k} < u_{-k}\left(\theta\right)^m,$$

$$u_{-k+1}\left(\sigma\right)^m < u_{(k+1)m-(k-1)} < u_{-k+1}\left(\theta\right)^m,$$

$$\vdots$$

$$u_{-1}\left(\sigma\right)^m < u_{(k+1)m-1} < u_{-1}\left(\theta\right)^m,$$

$$u_0\left(\sigma\right)^m < u_{(k+1)m} < u_0\left(\theta\right)^m,$$

$$v_{-k}(\tau)^m < v_{(k+1)m-k} < v_{-k}(\rho)^m,$$

$$v_{-k+1}(\tau)^m < v_{(k+1)m-(k-1)} < v_{-k+1}(\rho)^m,$$

$$\vdots$$

$$v_{-1}(\tau)^m < v_{(k+1)m-1} < v_{-1}(\rho)^m,$$

$$v_0(\tau)^m < v_{(k+1)m} < v_0(\rho)^m.$$

Then, for $m = \overline{1, n}$, $n \in \mathbb{N}$ we obtain

$$\min_{i=\overline{0,k}}(u_{-i})\min_m(\sigma)^m < u_n < \max_{i=\overline{0,k}}(u_{-i})\max_m(\theta)^m,$$

$$\min_{i=\overline{0,k}}(v_{-i})\min_m(\tau)^m < v_n < \max_{i=\overline{0,k}}(v_{-i})\max_m(\rho)^m.$$

Or equivalently,

$$\min_{i=\overline{0,k}}(u_{-i})\min_m\left(\min\left(\frac{\alpha}{\gamma},\frac{\beta}{\delta}\right)\right)^m < u_n < \max_{i=\overline{0,k}}(u_{-i})\max_m\left(\max\left(\frac{\alpha}{\gamma},\frac{\beta}{\delta}\right)\right)^m,$$

$$\min_{i=\overline{0,k}}(v_{-i})\min_m\left(\min\left(\frac{a}{c},\frac{b}{d}\right)\right)^m < v_n < \max_{i=\overline{0,k}}(v_{-i})\max_m\left(\max\left(\frac{a}{b},\frac{c}{d}\right)\right)^m.$$

The desired result follows by taking limits in these inequalities.

In the following theorem, we give conditions that make the solutions of System (18) periodic of period $k+1$.

Theorem 10 *Let $(u_n, v_n)_{n\geq-(k+1)}$ be a solution of System (18). Then, a necessary and sufficient condition for this solution to be periodic of period $k+1$ is*

$$(u_0, v_0) = (u_{-(k+1)}, v_{-(k+1)}),\ a+b = c+d \text{ and } \alpha+\beta = \gamma+\delta.$$

Proof Let $(u_n, v_n)_{n\geq-(k+1)}$ be a $k+1$ periodic solution of System (18). Then we have

$$(u_0, v_0) = (u_{-(k+1)}, v_{-(k+1)}),\ u_1 = u_{-k}\frac{\alpha v_0 + \beta v_{-(k+1)}}{\gamma v_0 + \delta v_{-(k+1)}},$$

and

$$v_1 = v_{-k}\frac{au_0 + bu_{-(k+1)}}{cu_0 + du_{-(k+1)}}.$$

Hence, we find

$$\frac{\alpha + \beta}{\gamma + \delta} = 1 \text{ and } \frac{a + b}{c + d} = 1.$$

Now, assume that

$$(u_0, v_0) = (u_{-(k+1)}, v_{-(k+1)}), \ a + b = c + d \text{ and } \alpha + \beta = \gamma + \delta,$$

so,

$$u_1 = u_{-k}\frac{\alpha v_0 + \beta v_{-(k+1)}}{\gamma v_0 + \delta v_{-(k+1)}} \text{ and } v_1 = v_{-k}\frac{au_0 + bu_{-(k+1)}}{cu_0 + du_{-(k+1)}},$$

then, we get $(u_1, v_1) = (u_{-k}, v_{-k})$.
By following the procedure, we obtain

$$u_2 = u_{1-k}\frac{\alpha v_1 + \beta v_{-k}}{\gamma v_1 + \delta v_{-k}} \text{ and } v_2 = v_{1-k}\frac{au_1 + bu_{-k}}{cu_1 + du_{-k}},$$

then, $u_2 = u_{-k+1}$ and $v_2 = v_{-k+1}$.
By induction we get that the solution $(u_n, v_n)_{n \geq -(k+1)}$ is periodic of period $k + 1$.

Competing Interests

The authors have no conflicts of interest to declare that are relevant to the content of this chapter.

References

1. Y. Akrour, M. Kara, N. Touafek, Y. Yazlik, Solutions formulas for some general systems of difference equations. Miskolc Math. Notes **22**(2), 529–555 (2021)
2. E.M. Elabbasy, E.M. Elsayed, Dynamics of a rational difference equation. Chin. Ann. Math. Ser. B **30B**(2), 187–198 (2009)
3. M. Gümüş, The global asymptotic stability of a system of difference equations. J. Difference Equ. Appl. **24**(6), 976–991 (2018)
4. M. Gümüş, R. Abo-Zeid, An explicit formula and forbidden set for a higher order difference equation. J. Appl. Math. Comput. **63**, 133–142 (2020)
5. N. Haddad, N. Touafek, J.F.T. Rabago, Solution form of a higher-order system of difference equations and dynamical behavior of its special case. Math. Meth. Appl. Sci. **40**, 3599–3607 (2017)
6. M. Kara, Y. Yazlik, D.T. Tollu, Solvability of a system of higher order nonlinear difference equations. Hacet. J. Math. Stat. **49**(5), 1566–1593 (2020)
7. M. Kara, Y. Yazlik, Solvability of a nonlinear three-dimensional system of difference equations with constant coefficients. Math. Slovaca **71**(5), 1133–1148 (2021)
8. M. Kara, Y. Yazlik, Solvable three-dimensional system of higher-order nonlinear difference equations. Filomat **36**(10), 3449–3469 (2022)

9. M. Kara, Y. Yazlik, On the solutions of three-dimensional system of difference equations via recursive relations of order two and applications. J. Appl. Anal. Comput. **12**(2), 736–753 (2022)
10. V.L. Kocic, G. Ladas, *Global Behavior of Nonlinear Difference Equations of Higher Order with Applications* (Kluwer Academic Publishers, Dordrecht, 1993)
11. M.R.S. Kulenović, G. Ladas, *Dynamics of Second Order Rational Difference Equations.* (Chapman & Hall/CRC, 2001)
12. L.C. McGrath, C. Teixeira, Existence and behavior of solutions of the rational equation $x_{n+1} = \frac{ax_{n-1}+bx_n}{cx_{n-1}+dx_n}x_n$. Rocky Mountain J. Math. **36**, 649–674 (2006)
13. M. Pituk, More on Poincaré's and Perron's theorems for difference equations. J. Difference Equ. Appl. **8**(3), 201–216 (2002)
14. H. Sedaghat, *Nonlinear Differnce Equations Theory with Applications to Social Science Models.* (Springer Science & Business Media, 2003)
15. D.T. Tollu, Y. Yazlik, N. Taskara, On a solvable nonlinear difference equation of higher order. Turkish J. Math. **42**, 1765–1778 (2018)
16. N. Touafek, E.M. Elsayed, On a third order rational system of difference equations, An Stiint Univ Al I Cuza Iasi. Ser. Noua, Mat. **61**(2), 367–380 (2015)
17. S. Stević, Representation of solutions of bilinear difference equations in terms of generalized Fibonacci sequences. Electron. J. Qual. Theory Differ. Equ. **67**, 1–15 (2014)

Stochastic Integrodifferential Equations with Deviating Argument Driven by Poisson Jumps: Optimal Controls

Essozimna Kpizim, Louk-Man Issaka, Ramkumar Kasinathan, Krichen Meoz, and Mamadou Abdoul Diop

Abstract The goal of this work is to investigate the optimal control for neutral stochastic integrodifferential equations with deviating arguments caused by Poisson jumps in Hilbert space. The fixed-point theorem and stochastic analysis methods define and provide the prerequisites for mild solution outcomes. Additionally, Balder's theorem is used to demonstrate the possibility of optimal control over the proposed issue. Finally, the stochastic partial integrodifferential equations are applied to the theoretical findings.

1 Introduction

The theory of stochastic differential equations can help to represent dynamic situations when it's impossible to predict precisely how the system under study will behave. Financial mathematics, hydrology, medicine, chronic diseases, neurons, and

E. Kpizim (✉)
KPIZIM, Numerical Analysis and Computing Laboratory, Department of Mathematics, Gaston Berger University of Saint-Louis, UFR Applied Sciences and Technologies, B.P 234, Saint-Louis, Senegal
e-mail: kpizimessozimna@gmail.com

L.-M. Issaka
ISSAKA, Institut de Mathématiques de Toulouse, Université Paul Sabatier, 118 route de Narbonne, 31062 Toulouse Cedex 9, France

R. Kasinathan
KASINATHAN, Department of Mathematics, PSG College of Arts and Science, Coimbatore 641 014, India

K. Meoz
MEOZ, Faculty of computer Science and Information Technology, AI-Baha University, AI-Baha 65528, Saudi Arabia
e-mail: moez.krichen@redcad.org

M. A. Diop
DIOP, Numerical Analysis and Computing Laboratory, Department of Mathematics, Gaston Berger University of Saint-Louis, UFR Applied Sciences and Technologies, B.P234 Saint-Louis, Senegal
e-mail: mamadou-abdoul.diop@ugb.edu.sn

M. Kharrat et al. (eds.), *Modeling of Discrete and Continuous Systems*, Forum for Interdisciplinary Mathematics, https://doi.org/10.1007/978-981-97-8715-9_4

telecommunications networks are just a few of the areas where stochastic differential equations can be used to simulate noises that arise in various systems [1, 2]. On the other hand, the control theory also manages to impact these differential or integrodifferential equations by using regulators called cost functions. Hence the optimal control problem. The optimal control problems, a new problem in science and engineering, are frequently used in biomedicine to model cancer chemotherapy and, more recently, in epidemiological models and medicine (see [3–5] and the references therein). There has lately been a lot of interest in researching optimal controls for various nonlinear SDEs and SIEs in infinite dimensional spaces. Yan and Lu [6] have shown the existence of optimal controls for a fractional stochastic partial integrodifferential equation with infinite delay. Researchers Balasubramanian and Tamilalagan [7] have examined the solvability and optimal controls for impulsive fractional stochastic integrodifferential equations in Hilbert space. Recently, Yan Z. and Yan X. [8] have worked about the best ways to control a class of impulsive partial stochastic differential equations with weighted pseudo nearly periodic coefficients. See [9–11] and reference the coefficients in Hilbert spaces for more information on SDEs with Poisson jumps. In this work, we will use the integrodifferential equations developed by Grimmer [12] to model the natural phenomena. These equations use resolvent operators that facilitate theoretical resolution of our work. The integrodifferential equations have various applications, for example, in physics and medicine, to solve the problems of cancer and chronic diseases. In these fields of application, new branches have emerged, and these new topics are delay differential equations. The delay domain allows to take into account the system's historical (it takes into account not only time but also unknown quantities) so that the model is analyzed more accurately from reality. To solve this problem of infinite delay in studying functional differential equations, the researchers Hale and Kato [13] have developed an abstract phase space to consider the system's past. The goal is to bring the system from a given initial state to a specific final state which respects some criteria. The delay is assumed to be either fixed or defined as an integral; the latter situation is known as distributed delay (for more information, see [14–16]). These notions have generalized delay systems called deviated arguments that often appear in research related to medicine, mathematical physics, engineering, economics, and mechanics. These equation systems with side effects, inherited characteristics, or derived arguments have indeed been found in a number of mathematical control models. The fact of using various fixed-point theorems and non-compacticity metrics has permited to word on these domains which have been taken as a model of integer and fractional order differential systems proposed by Rabbani and al. [17]. Furthermore, the applications of ESD that characterize many global issues make it a highly critical field of study. In finance and insurance, Chen and Li have studied events such as corporate failures, policyholder accidents, and stock markets [18]. Cont and Tankov have showed the importance of Poisson jumps in the study of systems of equations to describe how asset and commodity prices evolve. In addition to studying the existence and controllability of nonlocal fractional stochastic differential equations with infinite delay and Poisson jumps, other researchers have also worked on optimal control in epidemiological applications [5, 19, 20]. $\kappa(\vartheta(\gamma), \gamma) \in \mathcal{C}(\mathbb{V}, \mathsf{I})$ seems to be one whose range $\kappa(\mathbb{V}, \mathsf{I})$

is disjoint from I, $\mathsf{A} : D(\mathsf{A}) \subset \mathbb{V} \longrightarrow \mathbb{V}$ is the infinitesimal generator of strongly continuous semigroup $S(\gamma)$, $\gamma \geq 0$ in $\mathbb{V}$. The nonlinear functions $\mathsf{h} : \mathsf{I} \times \mathcal{B} \longrightarrow \mathbb{V}$, $\mathsf{F} : \mathsf{I} \times \mathcal{B} \times \mathbb{V} \longrightarrow \mathbb{V}$, $\sigma : \mathsf{I} \times \mathcal{B} \longrightarrow \mathcal{L}_2^0$, and $\mathsf{g} : \mathsf{I} \times \mathcal{B} \times \mathsf{Z} \longrightarrow \mathbb{V}$ are appropriate mappings and are specified in the next section. $\mathcal{C} = \mathcal{C}([-r,\ 0]; \mathbb{V})$ is the Banach space of all continuous functions $\varphi = \varphi(\gamma) : -\infty < \gamma \leq 0$ is a $\mathcal{F}_0$-measurable. The time history values $\vartheta_\gamma : (-\infty, 0] \longrightarrow \mathbb{V}$ are established by the equations $\vartheta_\gamma(\theta) = \vartheta(\gamma + \theta)$ and $\theta \leq 0$. In $\tilde{\mathcal{N}}(d\gamma, d\eta) = \mathcal{N}(d\gamma, d\eta) - \lambda(d\eta)d\gamma$ the Poisson measure $\tilde{\mathcal{N}}(d\gamma, d\eta)$ denotes the Poisson counting measure. In this paper, we consider the following class

$$
\begin{cases}
d\gamma\big[\vartheta(\gamma) + \mathsf{h}(\gamma, \vartheta_\gamma)\big] = \Big[\mathsf{A}\big[\vartheta(\gamma) + \mathsf{h}(\gamma, \vartheta_\gamma)\big] + \int_0^\gamma \mathsf{B}(\gamma - \varsigma)\big[\vartheta(\varsigma) + \mathsf{h}(\varsigma, \vartheta_\varsigma)\big]d\varsigma \\
\qquad + \mathsf{F}\big(\gamma, \vartheta_\gamma, \vartheta\big(\kappa(\vartheta(\gamma), \gamma)\big)\big) + \mathsf{Q}(\gamma)\mathsf{u}(\gamma)\Big]d\gamma + \sigma(\gamma, \vartheta_\gamma)d\mathsf{W}(\gamma) \\
\qquad + \int_{\mathsf{Z}} \mathsf{g}(\gamma, \vartheta_\gamma, \eta)\tilde{\mathcal{N}}(d\gamma, d\eta), \quad \gamma \in [0, \mathsf{b}] = \mathsf{I} \\
\vartheta(\gamma) = \vartheta_0 = \varphi \in \mathcal{B}, \quad \gamma \in (-\infty, 0],
\end{cases}
\tag{1}
$$

where the deviating argument $\kappa(\vartheta(\gamma), \gamma) \in \mathcal{C}(\mathbb{V}, \mathsf{I})$ seems to be one whose range $\kappa(\mathbb{V}, \mathsf{I})$ is disjoint from I, $\mathsf{A} : D(\mathsf{A}) \subset \mathbb{V} \longrightarrow \mathbb{V}$ is the infinitesimal generator of strongly continuous semigroup $S(\gamma)$, $\gamma \geq 0$ in $\mathbb{V}$. The nonlinear functions $\mathsf{h} : \mathsf{I} \times \mathcal{B} \longrightarrow \mathbb{V}$, $\mathsf{F} : \mathsf{I} \times \mathcal{B} \times \mathbb{V} \longrightarrow \mathbb{V}$, $\sigma : \mathsf{I} \times \mathcal{B} \longrightarrow \mathcal{L}_2^0$, and $\mathsf{g} : \mathsf{I} \times \mathcal{B} \times \mathsf{Z} \longrightarrow \mathbb{V}$ are appropriate mappings and are specified in the next section. $\mathcal{C} = \mathcal{C}([-r,\ 0]; \mathbb{V})$ is the Banach space of all continuous functions $\varphi = \varphi(\gamma) : -\infty < \gamma \leq 0$ is a $\mathcal{F}_0$-measurable. The time history values $\vartheta_\gamma : (-\infty, 0] \longrightarrow \mathbb{V}$ are established by the equations $\vartheta_\gamma(\theta) = \vartheta(\gamma + \theta)$ and $\theta \leq 0$. In $\tilde{\mathcal{N}}(d\gamma, d\eta) = \mathcal{N}(d\gamma, d\eta) - \lambda(d\eta)d\gamma$ the Poisson measure $\tilde{\mathcal{N}}(d\gamma, d\eta)$ denotes the Poisson counting measure. As a result, we now have a new system called the generalized delay differential equations with a deviating argument concerning the state. In addition, a mild solution and the best possible control are examined here using the integrodifferential first method, according to [20].

The following attributes apply to this article. Some fundamental definitions and early findings are provided in Sect. 2 . By employing some properties of resolvent operators, the fixed-point theorem, integrodifferential calculus, and stochastic analysis, Sect. 3 provides the necessary conditions for the existence of mild solutions to the proposed system. Additionally, Sect. 4 contributes to optimum control by employing the Bolza issue that is obtained by applying the Balder theorem. The resulting theoretical conclusions applied to the stochastic integrodifferential equations are contributed by Sect. 5.

2 Preliminaries

We presented some basic concepts, definitions, and lemmas in this section, that are required to examine the results.

Let $\mathcal{C}((-\infty, \mathsf{b}], \mathcal{L}^2(\Omega, \mathbb{V}))$ be the Banach space of all continuous mappings satisfying the condition $\| \vartheta \|^2 < \infty$ from $(-\infty, \mathsf{b}]$ to $\mathcal{L}(\Omega, \mathbb{V})$. Let $\mathcal{C}$ be the closed subspace of all continuous processes ϑ that belong to the space $\mathcal{C}((-\infty, \mathsf{b}], \mathcal{L}^2(\mathcal{F}, \mathbb{V}))$ and are composed of $\mathcal{F}_\gamma$-adapted measurable processes $\{\vartheta(\gamma) : \gamma \in \mathsf{I}\}$ such that ϑ is continuous. We always assume that $\mathcal{C}$ has the norm $\mathbb{E} \| \vartheta \|_{\mathcal{C}}^2 = \left[\sup_{\gamma \in \mathsf{I}} \| \vartheta \|^2 (\gamma)\right]$ in this paper. The space $\mathcal{C}$ provided with the norm topology described above is unmistakably a Banach space.

Let $\mathbb{Y}$ and $\mathbb{X}$ be two Banach spaces such that $\|\vartheta\|_{\mathbb{X}} = \|\mathsf{A}\vartheta\| + \|\vartheta\|$ for $\vartheta \in \mathbb{X}$. A and $\mathsf{B}(t)$ are closed linear operators on $\mathbb{Y}$. Let $\mathcal{C}(\mathbb{R}^+, \mathbb{X})$, $\mathcal{L}(\mathbb{X}, \mathbb{Y})$ stand for the space of all continuous functions from $\mathbb{R}^+$ into $\mathbb{X}$, the set of all bounded linear operators from $\mathbb{X}$ into $\mathbb{Y}$.

Let us consider the following system for further purposes:

$$\begin{cases} \vartheta'(t) = \mathsf{A}\vartheta(t) + \displaystyle\int_0^t \mathsf{B}(t-s)\vartheta(s)ds \quad \text{for } t \geq 0 \\ \vartheta(0) = \vartheta_0 \in \mathbb{X}. \end{cases} \tag{2}$$

Definition 1 ([12]) A resolvent operator for Eq. (2) is a bounded linear operator valued function $\Psi(t) \in \mathcal{L}(\mathbb{Y})$ for $t \geq 0$, having the following properties :

(i) $\Psi(0) = I$ (the identity map of $\mathbb{Y}$) and $\| \Psi(t) \| \leq N e^{\beta t}$ for some constants $N > 0$ and $\beta \in \mathbb{R}$.
(ii) For each $\vartheta \in \mathbb{Y}$, $\Psi(t)\vartheta$ is strongly continuous for $t \geq 0$.
(iii) For $\vartheta \in \mathbb{Y}$, $\Psi(\cdot)\vartheta \in \mathcal{C}^1(\mathbb{R}^+; \mathbb{Y}) \cap \mathcal{C}(\mathbb{R}^+; \mathbb{X})$ and

$$\begin{aligned} \Psi'(t)\vartheta &= \mathsf{A}\Psi(t)\vartheta + \int_0^t \mathsf{B}(t-s)\Psi(s)\vartheta ds \\ &= \Psi(t)\mathsf{A}\vartheta + \int_0^t \Psi(t-s)\mathsf{B}(s)\vartheta ds, \quad \text{for } t \in [0, \mathsf{b}]. \end{aligned}$$

Next, we assume A and $(\mathsf{B}(t))_{t\geq 0}$ satisfy the following conditions:

($\mathbf{R_1}$) The operator A is the infinitesimal generator of a strongly continuous semigroup $(\mathsf{S}(t))_{t\geq 0}$ on $\mathbb{Y}$.
($\mathbf{R_2}$) For all $t \geq 0$, the operator $\mathsf{B}(t)$ is closed and linear from $\mathsf{D}(\mathsf{A})$ to $\mathbb{Y}$ and $\mathsf{B}(t) \in \mathcal{L}(\mathcal{B}, \mathbb{Y})$. For any $\vartheta \in \mathbb{Y}$, the map $t \mapsto \mathsf{B}(t)\vartheta$ is bounded, differentiable and the derivative $t \mapsto \mathsf{B}'(t)\vartheta$ is bounded and uniformly continuous for $t \geq 0$.

Theorem 1 [12] *Assume that* ($\mathbf{R_1}$)-($\mathbf{R_2}$) *hold. Then, there exists a unique resolvent operator of the Cauchy problem* (2).

Now, we present some results on the existence of solutions for the following integrodifferential equation:

$$\begin{cases} \xi'(t) = \mathsf{A}\xi(t) + \int_0^t \mathsf{B}(t-s)\xi(s)ds + \mathsf{f}(t) \text{ for } t \geq 0 \\ \xi(0) = \xi_0 \in \mathbb{X}, \end{cases} \tag{3}$$

where $\mathsf{f} : \mathbb{R}^+ \to \mathbb{X}$ is a continuous function.

Definition 2 A continuous function $\xi : [0, \infty[\to \mathbb{Y}$ is said to be a strict solution for Eq. (3) if

1. $\xi \in \mathcal{C}^1(\mathbb{R}^+, \mathbb{Y}) \cap \mathcal{C}(\mathbb{R}^+, \mathbb{X})$,
2. ξ satisfies Eq. (3) for $t \geq 0$.

Theorem 2 [12] *Suppose that hypotheses* $(\mathbf{R_1})$ *and* $(\mathbf{R_2})$ *hold. If* ξ *is a strict solution of* (3), *then the following variation of constants formula holds.*

$$\xi(t) = \Psi(t)\xi_0 + \int_0^t \Psi(t-s)\mathsf{f}(s)\mathsf{ds}, \text{ for } t \geq 0. \tag{4}$$

Theorem 3 [21] *Let the assumptions* $(\mathbf{R_1})$ *and* $(\mathbf{R_2})$ *be satisfied. Let the* C_0-*semigroup* $(\mathsf{S}(t))_{t\geq 0}$ *generated by* A *be compact for* $t > 0$. *Then the corresponding resolvent operator* $(\Psi(t))_{t\geq 0}$ *of equation* (2) *is also compact for* $t > 0$.

Theorem 4 *Assume* $(\mathbf{R_1})$-$(\mathbf{R_2})$ *hold. Then, resolvent operator* $\Psi(t)$ *of Cauchy problem (3) is continuous for* $t \geq 0$ *on the operator norm, namely, for all* $t_0 > 0$, *it holds that*

$$\lim_{h \longrightarrow 0} \| \Psi(t_0 + h) - \Psi(t_0) \| = 0$$

Lemma 1 *Assume* $(\mathbf{R_1})$ *be satisfied. Then exist* $L = L(\mathsf{b})$ *such that*

$$\| \Psi(\varrho + \varepsilon) - \Psi(\varepsilon)\Psi(\varrho) \|_{\mathcal{L}(\mathbb{X})} \leq L(\varepsilon), \quad 0 < \varepsilon < \varrho < \mathsf{b}$$

For more details on resolvent operators, we refer [12, 22, 23].
For background on the theory of Wiener process denote W, see [22]

Lemma 2 [24] *Let* $G : \mathsf{I} \times \Omega \longrightarrow \mathcal{L}_2^0$ *be a strongly measurable mapping such that* $\int_0^{\mathsf{b}} \mathbb{E} \| G(\gamma) \|_{\mathcal{L}_2^0}^2 d\gamma < \infty$. *Then*

$$\mathbb{E} \left\| \int_0^{\mathsf{b}} G(\varsigma) d\mathsf{W}(\varsigma) \right\|^2 \leq \mathsf{L_W} \int_0^{\mathsf{b}} \mathbb{E} \| G(\varsigma) \|_{\mathcal{L}_2^0}^2 d\varsigma \quad \forall \gamma \in \mathsf{I},$$

where $\mathsf{L_W}$ *is a constant.*

Now, using Hale and Kato [13] invention of an axiomatic formulation of the phase space $\mathcal{B}$, we proceed. For $(-\infty, 0]$-measurable functions from $(-\infty, 0]$ equipped with the seminorm $\| \cdot \|_{\mathcal{B}}$, the axioms of phase space $\mathcal{B}$ are developed. We assume that $\mathcal{B}$ should meet the criteria below.

($\mathbf{A}_1$) The following is true for all $\gamma \in [0, a)$ if $\vartheta : (-\infty, a) \longrightarrow \mathbb{V}$, $a > 0$ is continuous on $[0, a)$ and $\vartheta_0 \in \mathcal{B}$.

(i) ϑ_γ is in $\mathcal{B}$,
(ii) $\| \vartheta(\gamma) \| \leq \tilde{\mathsf{H}} \| \vartheta_\gamma \|_{\mathcal{B}}$,
(iii) $\| \vartheta(\gamma) \|_{\mathcal{B}} \leq \mathsf{K}(\gamma - \kappa) \sup\{\| \vartheta(\varsigma) \| : 0 \leq \varsigma \leq \gamma\} + \mathsf{M}(\gamma - \kappa) \| \vartheta_0 \|_{\mathcal{B}}$,
where $\tilde{\mathsf{H}} > 0$ and $\mathsf{K}, \mathsf{M} : [0, +\infty) \longrightarrow [0, +\infty)$ are constants, K is continuous and M is locally bound. Furthermore, $\vartheta(\cdot)$ has no effect on K, M, and $\tilde{\mathsf{H}}$.

($\mathbf{A}_2$) The space $\mathcal{B}$ is complete.

Lemma 3 [13, 25] *Let $\vartheta : (-\infty, \mathsf{b}] \longrightarrow \mathbb{V}$ be a measurable process with $\mathcal{F}$-adapting such that the $\mathcal{F}_0$-adapted process $\vartheta_0 = \varphi \in \mathcal{B}$*

$$\mathbb{E} \| \vartheta_\gamma \|_{\mathcal{B}} \leq \mathsf{K}_\mathsf{b} \sup_{0 \leq \varsigma \leq \mathsf{b}} \| \vartheta(\varsigma) \| + \mathsf{M}_\mathsf{b} \mathbb{E} \| \varphi \|_{\mathcal{B}},$$

where $\mathsf{M}_\mathsf{b} = \sup_{\gamma \in \mathsf{I}} \{\mathsf{M}(\gamma)\}$, $\mathsf{K}_\mathsf{b} = \sup_{\gamma \in \mathsf{I}} \{\mathsf{K}(\gamma)\}$.

Lemma 4 *(Bochner's theorem).*
If $\| \mathsf{V} \|$ is Lebesgue integrable, then a measurable function $\mathsf{V} : \mathsf{I} \longrightarrow \mathbb{V}$ is Bochner integrable.

Theorem 5 *(Krasnoselskii-Schaefer-type fixed-point theorem [26, 27]). Consider two operators, A and B, so that*

(i) A is a contraction,
(ii) The operator B is fully continuous.

Then, either

(1) $\mathsf{A}\vartheta + \mathsf{B}\vartheta = \vartheta$ has a solution or
(2) $\mathsf{V} = \{\vartheta \in \mathbb{V} : \vartheta = \alpha \mathsf{A}\left(\dfrac{\vartheta}{\alpha}\right) + \alpha \mathsf{B}\vartheta \quad for\ some\ 0 < \alpha < 1\}$ is unbounded.

Throughout in the following sections we denote $\mathsf{M}_\Psi = \sup_{\gamma \in \mathsf{I}} \| \Psi(\gamma) \|$.

3 Main Results

In this part, we establish the existence of the system's mild solution (1). In the whole of this work, it is assumed that $(\mathbf{A}_1) - (\mathbf{A}_2)$ in [22] are true.

Definition 3 A Cádlág stochastic process An $\mathcal{F}_\gamma$-adapted stochastic processes ϑ : $(-\infty, \mathsf{b}] \longrightarrow \mathbb{V}$ is said to be a mild solution of Eq. (1), if $\vartheta(\gamma) = \vartheta_0 = \varphi \in \mathcal{B}$, $\gamma \in (-\infty, 0]$, $\vartheta\mid_{\mathsf{I}} \in \mathcal{L}^2(\mathsf{I}, \mathbb{V})$ and

(i) $\vartheta(\gamma)$ is adapted and measurable on $(\mathcal{F}_\gamma)_{\gamma \geq 0}$.
(ii) For each $\gamma \in [0, \mathsf{b}]$, $\vartheta(\gamma)$ satisfies the following integral equation

$$\begin{aligned}\vartheta(\gamma) &= \Psi(\gamma)\big[\varphi(0) + \mathsf{h}(0, \varphi(0))\big] - \mathsf{h}(\gamma, \vartheta_\gamma) + \int_0^\gamma \Psi(\gamma - \varsigma)\sigma(\varsigma, \vartheta_\varsigma)d\mathsf{W}(\varsigma) \\ &+ \int_0^\gamma \Psi(\gamma - s)\mathsf{Q}(\varsigma)\mathsf{u}(\varsigma)d\varsigma + \int_0^\gamma \int_{\mathsf{Z}} \Psi(\gamma - \varsigma)\mathsf{g}(\varsigma, \vartheta_\varsigma, \eta)\tilde{\mathcal{N}}(d\varsigma, d\eta) \\ &+ \int_0^\gamma \Psi(\gamma - \varsigma)\mathsf{F}\big(s, \vartheta_\varsigma, \vartheta\big(\kappa(\vartheta(\varsigma), \varsigma)\big)\big)d\varsigma.\end{aligned}$$

To bolster the major findings, the ensuing hypotheses are required.

(H₁) $S(\gamma)$ is compact, for $\gamma > 0$.
(H₂) The nonlinear function $\mathsf{F} : \mathsf{I} \times \mathcal{B} \times \mathbb{V} \longrightarrow \mathbb{V}$ satisfies the Lipschitz condition. For every $\vartheta_1, \vartheta_2 \in \mathcal{B}$ and $\vartheta_1', \vartheta_2' \in \mathbb{V}$,

$$\mathbb{E} \parallel \mathsf{F}(\gamma, \vartheta_1, \vartheta_1') - \mathsf{F}(\gamma, \vartheta_2, \vartheta_2') \parallel^2 \leq \mathsf{M}_\mathsf{F}\big[\mathbb{E} \parallel \vartheta_1 - \vartheta_2 \parallel_\mathcal{B}^2 + \mathbb{E} \parallel \vartheta_2' - \vartheta_2' \parallel^2\big].$$

(H₃) Let $\kappa : \mathbb{V} \times \mathbb{R}_+ \longrightarrow \mathbb{R}_+$ satisfy the Lipschitz condition with the constant $\mathsf{M}_\kappa > 0$ such that $\mathbb{E} \mid \kappa(\vartheta_1, \varsigma) - \kappa(\vartheta_2, \varsigma) \mid_{\mathbb{R}_+}^2 \leq \mathsf{M}_\kappa\big[\mathbb{E} \parallel \vartheta_1 - \vartheta_2 \parallel_\mathcal{B}^2\big]$, and $\kappa(\cdot, 0) = 0$.
(H₄) The nonlinear function $\mathsf{h} : \mathsf{I} \times \mathcal{B} \longrightarrow \mathbb{V}$ is continuous. Moreover,

(1) $\mathsf{h}(\cdot, \vartheta)$ is strongly measurable for every $\gamma \in \mathsf{I}$.
(2) For every ϑ_1, $\vartheta_2 \in \mathcal{B}$, $\mathsf{h}(\gamma, \cdot)$ satisfies the following conditions with constants $\mathsf{M}_\mathsf{h}, \mathsf{L}_\mathsf{h} > 0$

$$\mathbb{E} \parallel \mathsf{h}(\gamma, \vartheta_1) - \mathsf{h}(\gamma, \vartheta_2) \parallel^2 \leq \mathsf{M}_\mathsf{h}\mathbb{E} \parallel \vartheta_1 - \vartheta_2 \parallel_\mathcal{B}^2 .$$

$$\mathbb{E} \parallel \mathsf{h}(\gamma, \vartheta) \parallel^2 \leq \mathsf{L}_\mathsf{h}\big(1 + \mathbb{E} \parallel \vartheta \parallel^2\big).$$

(H₅) The nonlinear function $\sigma : \mathsf{I} \times \mathcal{B} \longrightarrow \mathcal{L}_2^0$ satisfies the Lipschitz condition. For every $\vartheta_1, \vartheta_2 \in \mathcal{B}$,

$$\mathbb{E} \parallel \sigma(\gamma, \vartheta_1) - \sigma(\gamma, \vartheta_2) \parallel^2 \leq \mathsf{M}_\sigma\mathbb{E} \parallel \vartheta_1 - \vartheta_2 \parallel_\mathcal{B}^2 .$$

($\mathbf{H}_6$) The function $\mathsf{g} : \mathsf{I} \times \mathcal{B} \times \mathsf{Z} \longrightarrow \mathbb{V}$ satisfies the Lipschitz with respect to the second variable for a.e. $\gamma \in \mathsf{I}$ with the Lipschitz constant $\mathsf{M_g} > 0$. For any $\vartheta_1, \vartheta_2 \in \mathbb{V}$ and $\gamma \in \mathsf{I}$,

$$\int_{\mathsf{Z}} \mathbb{E} \parallel \mathsf{g}(\gamma, \vartheta_1(\gamma), \eta) - \mathsf{g}(\gamma, \vartheta_2(\gamma), \eta) \parallel^2 \lambda(d\eta) \leq \mathsf{M_g}\mathbb{E} \parallel \vartheta_1(\gamma) - \vartheta_2(\gamma) \parallel^2_{\mathcal{B}} .$$

Theorem 6 *Assume that* ($\boldsymbol{H}_1$) $-$ ($\boldsymbol{H}_6$) *hold, then there exists a mild solution to system (1) on* I *provided that*

$$\mathcal{Q} = 3\mathsf{M_h}\mathsf{K_b} + 3\mathsf{M_F}\mathsf{b}^2\mathsf{M}^2_{\Psi}(2\mathsf{M}_{\kappa}\tilde{\sigma} + \mathsf{K_b}) + 3\mathsf{b}^2\mathsf{M_g}\mathsf{M}^2_{\Psi}\mathsf{K_b} < 1.$$

Proof Let $\mathcal{B}_{\mathsf{b}}$ be the set defined by

$$\mathcal{B}_{\mathsf{b}} = \left\{\vartheta : (-\infty, \mathsf{b}] \longrightarrow \mathbb{V} \quad such\ that\ \vartheta \mid_{(-\infty,0]} \in \mathcal{B}, \vartheta \mid_{\mathsf{I}} \in \mathcal{C}\right\}$$

Define the operator $\Upsilon : \mathcal{B}_{\mathsf{b}} \longrightarrow \mathcal{B}_{\mathsf{b}}$ by

$$(\Upsilon\vartheta)(\gamma) = \begin{cases} 0, \qquad for\ \ \gamma \in (-\infty, 0], \\ \Psi(\gamma)\big[\varphi(0) + \mathsf{h}(0, \varphi(0))\big] + \displaystyle\int_0^{\gamma} \Psi(\gamma - \varsigma)\sigma(\varsigma, \vartheta_{\varsigma})d\mathsf{W}(\varsigma) \\ + \displaystyle\int_0^{\gamma} \Psi(\gamma - \varsigma)\mathsf{Q}(\varsigma)\mathsf{u}(\varsigma)d\varsigma + \int_0^{\gamma}\int_{\mathsf{Z}} \Psi(\gamma - \varsigma)\mathsf{g}(\varsigma, \vartheta_{\varsigma}, \eta)\tilde{\mathcal{N}}(d\varsigma, d\eta) \\ \quad -\mathsf{h}(\gamma, \vartheta_{\gamma}) + \displaystyle\int_0^{\gamma} \Psi(\gamma - \varsigma)\mathsf{F}\big(\varsigma, \vartheta_{\varsigma}, \vartheta\big(\kappa(\vartheta(\varsigma), \varsigma)\big)\big)d\varsigma, \quad for\ \ \gamma \in \mathsf{I}. \end{cases} \tag{5}$$

Using ($\mathbf{H}_4$), we get

$$\mathbb{E}\left\|\int_0^{\gamma} \Psi(\gamma - \varsigma)\mathsf{Q}(\varsigma)\mathsf{u}(\varsigma)d\varsigma\right\|^2 \leq \mathsf{M}^2_{\Psi}\mathsf{b} \parallel \mathsf{Q} \parallel^2 \parallel \mathsf{u} \parallel^2_{\mathcal{L}_{\mathcal{F}}} .$$

Then, from the Bochner theorem, it follows that $\Psi(\gamma - \varsigma)\mathsf{Q}(\varsigma)\mathsf{u}(\varsigma)$ is integrable on I.

We have that the nonlinear functions $\mathsf{F}, \mathsf{h}, \sigma$, and g are continuous on I, and the set $\Upsilon(\vartheta)$ is well defined on I. Next, we must show that Υ has a fixed point. For $\varphi \in \mathcal{B}$,

$$\tilde{\varphi} = \begin{cases} \varphi(\gamma), \quad \gamma \in (-\infty, 0], \\ \Psi(\gamma)\varphi(0), \quad \gamma \in \mathsf{I}. \end{cases} \tag{6}$$

Then, $\tilde{\varphi} \in \mathcal{B}_{\mathsf{b}}$. Let $\vartheta(\gamma) = \tilde{\varphi}(\gamma) + \varrho(\gamma), -\infty < \gamma \leq \mathsf{b}$. It is clear that $\vartheta(\gamma)$ satisfies Definition 3 if and only if $\varrho(\gamma)$ satisfies that $\varrho(0) = 0$ and

$$\begin{aligned} \varrho(\gamma) = {} & \Psi(\gamma)\mathsf{h}(0, \varphi(0)) - \mathsf{h}(\gamma, \tilde{\varphi}_{\gamma} + \varrho_{\gamma}) \\ & + \int_0^{\gamma} \Psi(\gamma - \varsigma)\mathsf{F}\big(\varsigma, \varphi_{\varsigma} + \varrho_{\varsigma}, (\tilde{\varphi}_{\varsigma} + \varrho_{\varsigma})\big(\kappa(\tilde{\varphi}(\varsigma) + \varrho(\varsigma), \varsigma)\big)\big)d\varsigma \end{aligned}$$

$$+\int_0^\gamma \Psi(\gamma-\varsigma)\mathsf{Q}(\varsigma)\mathsf{u}(\varsigma)d\varsigma+\int_0^\gamma \Psi(\gamma-\varsigma)\sigma(\varsigma,\tilde{\varphi}_\varsigma+\varrho_\varsigma)d\mathsf{W}(\varsigma)$$
$$+\int_0^\gamma\int_{\mathsf{Z}} \Psi(\gamma-\varsigma)\mathsf{g}(\varsigma,\tilde{\varphi}_\varsigma+\varrho_\varsigma,\eta)\tilde{\mathcal{N}}(d\varsigma,d\eta).$$

Let $\mathcal{B}_{\mathsf{b}}^0=\{\varrho\in\mathcal{B}_{\mathsf{b}},\varrho(0)=0\in\mathcal{B}\}$.
For every $\varrho\in\mathcal{B}_{\mathsf{b}}^0$,

$$\|\varrho\|_{\mathsf{b}}=\|\varrho_0\|_{\mathcal{B}}+\sup_{0\le\gamma\le\mathsf{b}}\mathbb{E}\|\varrho(\gamma)\|=\sup_{0\le\gamma\le\mathsf{b}}\mathbb{E}\|\varrho(\gamma)\|.$$

Thus, $(\mathcal{B}_{\mathsf{b}}^0,\ \|\cdot\|_{\mathsf{b}})$ is a Banach space. The set $\Lambda_\sigma=\{\upsilon\in\mathcal{B}_{\mathsf{b}}^0,\ \|\upsilon\|^2\le\sigma\}$ for some $\sigma\ge0$. Then, obviously, $\Lambda_\sigma\subset\mathcal{B}_{\mathsf{b}}^0$ is uniformly bounded. Moreover, for $\varrho\in\Lambda_\sigma$, we have

$$\begin{aligned}
\|\vartheta(\gamma)\|_{\mathcal{B}}^2 &= \|\tilde{\varphi}_\gamma+\varrho_\gamma\|_{\mathcal{B}}^2\\
&\le 2\big(\|\tilde{\varphi}_\gamma\|_{\mathcal{B}}^2+\|\varrho_\gamma\|_{\mathcal{B}}^2\big),\\
&\le 2\Big(\mathsf{K}_{\mathsf{b}}^2\sup_{\varsigma\in\mathsf{I}}\mathbb{E}\|\varrho(\varsigma)\|^2+\mathsf{M}_{\mathsf{b}}^2\|\varrho_0\|_{\mathcal{B}}^2\\
&\qquad+\mathsf{K}_{\mathsf{b}}^2\sup_{\varsigma\in\mathsf{I}}\mathbb{E}\|\tilde{\varphi}(\gamma)\|^2+\mathsf{M}_{\mathsf{b}}^2\|\tilde{\varphi}_0\|_{\mathcal{B}}^2\Big),\\
&\le 2\mathsf{K}_{\mathsf{b}}^2\big(\sigma+\tilde{\mathsf{H}}^2\|\Psi(\gamma)\varphi(0)\|^2\big)+2\mathsf{M}_{\mathsf{b}}^2\sup_{\varsigma\in\mathsf{I}}\mathbb{E}\|\tilde{\varphi}(\gamma)\|^2,\\
&\le 2\mathsf{K}_{\mathsf{b}}^2\big(\sigma+\tilde{\mathsf{H}}^2\mathsf{M}_\Psi^2\mathcal{Q}\|\varphi\|_{\mathcal{B}}^2\big)+2\mathsf{M}_{\mathsf{b}}^2\|\tilde{\varphi}\|_{\mathcal{B}}^2,\\
&:=\tilde{\sigma}.
\end{aligned}$$

Now, we define the operator $\tilde{\Upsilon}:\mathcal{B}_{\mathsf{b}}^0\longrightarrow\mathcal{B}_{\mathsf{b}}^0$ by

$$(\tilde{\Upsilon}\varrho)(\gamma)=\begin{cases}0, & \gamma\in(-\infty,0],\\
\Psi(\gamma)\mathsf{h}(0,\varphi(0))-\mathsf{h}(\gamma,\tilde{\varphi}_\gamma+\varrho_\gamma)\\
\quad+\int_0^\gamma\Psi(\gamma-\varsigma)\mathsf{F}\big(\varsigma,\tilde{\varphi}_\varsigma+\varrho_\varsigma,(\tilde{\varphi}_\varsigma+\varrho_\varsigma)(\kappa(\tilde{\varphi}(\varsigma)+\varrho(\varsigma),\varsigma))\big)d\varsigma\\
+\int_0^\gamma\Psi(\gamma-\varsigma)\mathsf{Q}(\varsigma)\mathsf{u}(\varsigma)d\varsigma+\int_0^\gamma\Psi(\gamma-\varsigma)\sigma(\varsigma,\tilde{\varphi}_\varsigma+\varrho_\varsigma)d\mathsf{W}(\varsigma)\\
\quad+\int_0^\gamma\int_{\mathsf{Z}}\Psi(\gamma-\varsigma)\mathsf{g}(\varsigma,\tilde{\varphi}_\varsigma+\varrho_\varsigma,\eta)\tilde{\mathcal{N}}(d\varsigma,d\eta)\quad for\ \gamma\in\mathsf{I}.\end{cases}\tag{7}$$

For any $\sigma>0$, note that $\tilde{\Upsilon}$ is well defined on Λ_σ. The operator Υ has a fixed point if and only if $\tilde{\Upsilon}$ also has a fixed point. We start by breaking down $\tilde{\Upsilon}$ into $\tilde{\Upsilon}=\tilde{\Upsilon}_1+\tilde{\Upsilon}_2$. Define

$$\begin{aligned}(\tilde{\Upsilon}_1\varrho)(\gamma) &= \Psi(\gamma)\mathsf{h}(0,\varphi(0)) - \mathsf{h}(\gamma,\tilde{\varphi}_\gamma+\varrho_\gamma)\\ &\quad+\int_0^\gamma \Psi(\gamma-\varsigma)\mathsf{F}\big(\varsigma,\varphi_\varsigma+\varrho_\varsigma,(\tilde{\varphi}_\varsigma+\varrho_\varsigma)\big(\kappa(\tilde{\varphi}(\varsigma)+\varrho(\varsigma),\varsigma)\big)\big)d\varsigma\\ &\quad+\int_0^\gamma\int_{\mathsf{Z}}\Psi(\gamma-\varsigma)\mathsf{g}(\varsigma,\tilde{\varphi}_\varsigma+\varrho_\varsigma,\eta)\tilde{\mathcal{N}}(d\varsigma,d\eta),\end{aligned}$$

and

$$(\tilde{\Upsilon}_2\varrho)(\gamma) = \int_0^\gamma \Psi(\gamma-\varsigma)\mathsf{Q}(\varsigma)\mathsf{u}(\varsigma)d\varsigma + \int_0^\gamma \Psi(\gamma-\varsigma)\sigma(\varsigma,\tilde{\varphi}_\varsigma+\varrho_\varsigma)d\mathsf{W}(\varsigma).$$

We will now demonstrate that the operators $\tilde{\Upsilon}_1$ and $\tilde{\Upsilon}_2$ satisfy each and every requirement of Theorem 5.

Step 1. $\tilde{\Upsilon}_1$ is a contraction on mapping. For $\varrho_1, \varrho_2 \in \Lambda_\sigma$ out $\gamma \in \mathsf{I}$, we get

$$\begin{aligned}\mathbb{E}\left\|(\tilde{\Upsilon}_1\varrho_1)(\gamma)-(\tilde{\Upsilon}_1\varrho_2)(\gamma)\right\|^2 &\le 3\mathbb{E}\left\|\mathsf{h}(\gamma,\tilde{\varphi}_\gamma+\varrho_{1\gamma})-\mathsf{h}(\gamma,\tilde{\varphi}_\gamma+\varrho_{2\gamma})\right\|^2\\ &+3\mathbb{E}\Big\|\int_0^\gamma \Psi(\gamma-\varsigma)\Big[\mathsf{F}\big(\varsigma,\varphi_\varsigma+\varrho_{1\varsigma},(\tilde{\varphi}+\varrho_1)\big(\kappa(\tilde{\varphi}(\varsigma)+\varrho_1(\varsigma),\varsigma)\big)\big)\\ &\qquad -\mathsf{F}\big(\varsigma,\varphi_\varsigma+\varrho_{2\varsigma},(\tilde{\varphi}+\varrho_2)\big(\kappa(\tilde{\varphi}(\varsigma)+\varrho_2(\varsigma),\varsigma)\big)\big)\Big]d\varsigma\Big\|^2\\ &+3\mathbb{E}\Big\|\int_0^\gamma\int_{\mathsf{Z}}\Psi(\gamma-\varsigma)\Big[\mathsf{g}(\varsigma,\tilde{\varphi}_\varsigma+\varrho_{1\varsigma},\eta)-\mathsf{g}(\varsigma,\tilde{\varphi}_\varsigma+\varrho_{2\varsigma},\eta)\Big]\tilde{\mathcal{N}}(d\varsigma,d\eta)\Big\|^2\\ &:= 3\sum_{k=1}^{3} I_k,\end{aligned}$$

where

$$\begin{aligned}I_1 &= \mathbb{E}\left\|\mathsf{h}(\gamma,\tilde{\varphi}_\gamma+\varrho_{1\gamma})-\mathsf{h}(\gamma,\tilde{\varphi}_\gamma+\varrho_{2\gamma})\right\|^2 \le \mathsf{M_h} \parallel \varrho_{1\gamma}-\varrho_{2\gamma}\parallel_{\mathcal{B}}^2\\ &\le \mathsf{M_h}\mathsf{K_b}\sup_{\varsigma\in(0,\gamma)}\mathbb{E}\parallel \varrho_1(\varsigma)-\varrho_2(\varsigma)\parallel_{\mathcal{B}}^2.\end{aligned}$$

After some computations, we get

$$\begin{aligned}I_2 &= \mathbb{E}\Big\|\int_0^\gamma \Psi(\gamma-\varsigma)\Big[\mathsf{F}\big(\varsigma,\varphi_\varsigma+\varrho_{1\varsigma},(\tilde{\varphi}+\varrho_1)\big(\kappa(\tilde{\varphi}(\varsigma)+\varrho_1(\varsigma),\varsigma)\big)\big)\\ &\qquad -\mathsf{F}\big(\varsigma,\varphi_\varsigma+\varrho_{2\varsigma},(\tilde{\varphi}+\varrho_2)\big(\kappa(\tilde{\varphi}(\varsigma)+\varrho_2(\varsigma),\varsigma)\big)\big)\Big]d\varsigma\Big\|^2\\ &\le \mathsf{b}\int_0^\gamma \parallel \Psi(\gamma-\varsigma)\parallel^2 \mathbb{E}\Big\|\Big[\mathsf{F}\big(\varsigma,\varphi_\varsigma+\varrho_{1\varsigma},(\tilde{\varphi}+\varrho_1)\big(\kappa(\tilde{\varphi}(s)+\varrho_1(\varsigma),\varsigma)\big)\big)\end{aligned}$$

$$-\mathsf{F}\big(\varsigma,\varphi_\varsigma+\varrho_{2\varsigma},(\tilde{\varphi}+\varrho_2)\big(\kappa(\tilde{\varphi}(\varsigma)+\varrho_2(\varsigma),\varsigma)\big)\big)\Big]\Big\|^2 d\varsigma$$

$$\begin{aligned}
&\leq \mathsf{b}\mathsf{M}_\Psi^2\int_0^\gamma \mathbb{E}\Big[\parallel \mathsf{F}\big(\varsigma,\varphi_\varsigma+\varrho_{1\varsigma},(\tilde{\varphi}+\varrho_1)\big(\kappa(\tilde{\varphi}(\varsigma)+\varrho_1(\varsigma),\varsigma)\big)\big)\\
&-\mathsf{F}\big(\varsigma,\varphi_\varsigma+\varrho_{1\varsigma},(\tilde{\varphi}+\varrho_1)\big(\kappa(\tilde{\varphi}(\varsigma)+\varrho_2(\varsigma),\varsigma)\big)\big)\\
&-\mathsf{F}\big(\varsigma,\varphi_\varsigma+\varrho_{2\varsigma},(\tilde{\varphi}+\varrho_2)\big(\kappa(\tilde{\varphi}(\varsigma)+\varrho_{2\varsigma}(\varsigma),\varsigma)\big)\big)\\
&+\mathsf{F}\big(\varsigma,\varphi_\varsigma+\varrho_{1\varsigma},(\tilde{\varphi}+\varrho_1)\big(\kappa(\tilde{\varphi}(\varsigma)+\varrho_2(\varsigma),\varsigma)\big)\big)\parallel^2\Big]
\end{aligned}$$

$$\begin{aligned}
&\leq \mathsf{b}\mathsf{M}_\Psi^2\mathsf{M}_\mathsf{F}\int_0^\gamma\Big(\mathbb{E}\parallel\tilde{\varphi}+\varrho_1\parallel^2\parallel\kappa(\tilde{\varphi}(\varsigma)+\varrho_1(\varsigma),\varsigma)-\kappa(\tilde{\varphi}(\varsigma)+\varrho_2(\varsigma),\varsigma)\parallel^2\\
&+\mathbb{E}\parallel\varrho_{1\varsigma}-\varrho_{2\varsigma}\parallel^2+\mathbb{E}\parallel\varrho_1-\varrho_2\parallel^2\parallel\kappa(\tilde{\varphi}(\varsigma)+\varrho_2(\varsigma),\varsigma)\parallel^2\Big)d\varsigma\\
&\leq \mathsf{b}\mathsf{M}_\Psi^2\mathsf{M}_\mathsf{F}\int_0^\gamma\Big(\mathsf{M}_\kappa\mathbb{E}\parallel\tilde{\varphi}+\varrho_1\parallel^2\parallel\varrho_1-\varrho_2\parallel_\mathcal{B}^2+\mathbb{E}\parallel\varrho_{1\varsigma}-\varrho_{2\varsigma}\parallel^2\\
&\qquad+\mathbb{E}\parallel\varrho_1-\varrho_2\parallel^2\parallel\tilde{\varphi}+\varrho_2\parallel^2\Big)d\varsigma\\
&\leq \mathsf{b}\mathsf{M}_\Psi^2\mathsf{M}_\mathsf{F}\int_0^\gamma\Big(\tilde{\sigma}\mathsf{M}_\kappa\mathbb{E}\parallel\varrho_1-\varrho_2\parallel_\mathcal{B}^2+\mathbb{E}\parallel\varrho_{1\varsigma}-\varrho_{2\varsigma}\parallel^2\\
&\qquad+\tilde{\sigma}\mathsf{M}_\kappa\mathbb{E}\parallel\varrho_1-\varrho_2\parallel^2\Big)d\varsigma\\
&\leq \mathsf{M}_\mathsf{F}\mathsf{b}\mathsf{M}_\Psi^2\int_0^\gamma\Big[\mathbb{E}\parallel\varrho_{1\varsigma}-\varrho_{2\varsigma}\parallel_\mathcal{B}^2+2\tilde{\sigma}\mathsf{M}_\kappa\mathbb{E}\parallel\varrho_1-\varrho_2\parallel_\mathcal{B}^2\Big]d\varsigma\\
&\leq \mathsf{M}_\mathsf{F}\mathsf{b}^2\mathsf{M}_\Psi^2\Big[\mathsf{K}_\mathsf{b}\sup_{\varsigma\in(0,\gamma)}\mathbb{E}\parallel\varrho_1(\varsigma)-\varrho_2(\varsigma)\parallel_\mathcal{B}^2+2\mathsf{M}_\kappa\tilde{\sigma}\mathbb{E}\parallel\varrho_1-\varrho_2\parallel_\mathcal{B}^2\Big]\\
&\leq \mathsf{M}_\mathsf{F}\mathsf{b}^2\mathsf{M}_\Psi^2\Big[2\mathsf{M}_\kappa\tilde{\sigma}\mathbb{E}\parallel\varrho_1-\varrho_2\parallel_\mathcal{B}^2+\mathsf{K}_\mathsf{b}\sup_{\varsigma\in(0,\gamma)}\mathbb{E}\parallel\varrho_1(\varsigma)-\varrho_2(\varsigma)\parallel_\mathcal{B}^2\Big].
\end{aligned}$$

Also, note that

$$\begin{aligned}
I_3&=\mathbb{E}\Big\|\int_0^\gamma\int_\mathsf{Z}\Psi(\gamma-\varsigma)\Big[\mathsf{g}(\varsigma,\tilde{\varphi}_\varsigma+\varrho_{1\varsigma},\eta)-\mathsf{g}(\varsigma,\tilde{\varphi}_\varsigma+\varrho_{2\varsigma},\eta)\Big]\tilde{\mathcal{N}}(d\varsigma,d\eta)\Big\|^2\\
&\leq \mathsf{b}\int_0^\gamma\int_\mathsf{Z}\parallel\Psi(\gamma-\varsigma)\parallel^2\mathbb{E}\parallel\mathsf{g}(\varsigma,\tilde{\varphi}_\varsigma+\varrho_{1\varsigma},\eta)\\
&\qquad\qquad-\mathsf{g}(\varsigma,\tilde{\varphi}_\varsigma+\varrho_{2\varsigma},\eta)\parallel^2\lambda(d\eta)d\varsigma\\
&\leq \mathsf{b}\mathsf{M}_\mathsf{g}\mathsf{M}_\Psi^2\int_0^\gamma\mathbb{E}\parallel\varrho_{1\varsigma}-\varrho_{2\varsigma}\parallel_\mathcal{B}^2 d\varsigma\\
&\leq \mathsf{b}^2\mathsf{M}_\mathsf{g}\mathsf{M}_\Psi^2\mathsf{K}_\mathsf{b}\sup_{\varsigma\in(0,\gamma)}\mathbb{E}\parallel\varrho_1(\varsigma)-\varrho_2(\varsigma)\parallel_\mathcal{B}^2.
\end{aligned}$$

In summary, we have

$$\mathbb{E}\left\|(\tilde{\Upsilon}_1\varrho_1)(\gamma)-(\tilde{\Upsilon}_1\varrho_2)(\gamma)\right\|^2 \leq 3\mathsf{M_h}\mathsf{K_b}\sup_{\varsigma\in(0,\gamma)}\mathbb{E}\parallel\varrho_1(\varsigma)-\varrho_2(\varsigma)\parallel_{\mathcal{B}}^2$$
$$+3\mathsf{M_F}\mathsf{b}^2\mathsf{M}_\Psi^2\left[2\mathsf{M}_\kappa\tilde{\sigma}\mathbb{E}\parallel\varrho_1-\varrho_2\parallel_{\mathcal{B}}^2+\mathsf{K_b}\sup_{\varsigma\in(0,\gamma)}\mathbb{E}\parallel\varrho_1(\varsigma)-\varrho_2(\varsigma)\parallel_{\mathcal{B}}^2\right]$$
$$+3\mathsf{b}^2\mathsf{M_g}\mathsf{M}_\Psi^2\mathsf{K_b}\sup_{\varsigma\in(0,\gamma)}\mathbb{E}\parallel\varrho_1(\varsigma)-\varrho_2(\varsigma)\parallel_{\mathcal{B}}^2$$

$$\leq 3\mathsf{M_h}\mathsf{K_b}\sup_{\varsigma\in(0,\gamma)}\mathbb{E}\parallel\varrho_1(\varsigma)-\varrho_2(\varsigma)\parallel_{\mathcal{B}}^2+\mathsf{K_b}\sup_{\varsigma\in(0,\gamma)}\mathbb{E}\parallel\varrho_{1\varsigma}-\varrho_{2\varsigma}\parallel_{\mathcal{B}}^2$$
$$+3\mathsf{M_F}\mathsf{b}^2\mathsf{M}_\Psi^2\left[2\mathsf{M}_\kappa\tilde{\sigma}\sup_{\varsigma\in(0,\gamma)}\mathbb{E}\parallel\varrho_1(\varsigma)-\varrho_2(\varsigma)\parallel_{\mathcal{B}}^2\right]$$
$$+3\mathsf{b}^2\mathsf{M_g}\mathsf{M}_\Psi^2\mathsf{K_b}\sup_{\varsigma\in(0,\gamma)}\mathbb{E}\parallel\varrho_1(\varsigma)-\varrho_2(\varsigma)\parallel_{\mathcal{B}}^2$$
$$\leq 3\mathsf{M_h}\mathsf{K_b}\sup_{\varsigma\in(0,\gamma)}\mathbb{E}\parallel\varrho_1(\varsigma)-\varrho_2(\varsigma)\parallel_{\mathcal{B}}^2$$
$$+3\mathsf{M_F}\mathsf{b}^2\mathsf{M}_\Psi^2\left(2\mathsf{M}_\kappa\tilde{\sigma}+\mathsf{K_b}\right)\sup_{\varsigma\in(0,\gamma)}\mathbb{E}\parallel\varrho_1(\varsigma)-\varrho_2(\varsigma)\parallel_{\mathcal{B}}^2$$
$$+3\mathsf{b}^2\mathsf{M_g}\mathsf{M}_\Psi^2\mathsf{K_b}\sup_{\varsigma\in(0,\gamma)}\mathbb{E}\parallel\varrho_1(\varsigma)-\varrho_2(\varsigma)\parallel_{\mathcal{B}}^2$$
$$\leq\Big[3\mathsf{M_h}\mathsf{K_b}+3\mathsf{M_F}\mathsf{b}^2\mathsf{M}_\Psi^2\left(2\mathsf{M}_\kappa\tilde{\sigma}+\mathsf{K_b}\right)$$
$$+3\mathsf{b}^2\mathsf{M_g}\mathsf{M}_\Psi^2\mathsf{K_b}\Big]\sup_{\varsigma\in(0,\gamma)}\mathbb{E}\parallel\varrho_1(\varsigma)-\varrho_2(\varsigma)\parallel_{\mathcal{B}}^2$$
$$\leq\left[3\mathsf{M_h}\mathsf{K_b}+3\mathsf{M_F}\mathsf{b}^2\mathsf{M}_\Psi^2\left(2\mathsf{M}_\kappa\tilde{\sigma}+\mathsf{K_b}\right)+3\mathsf{b}^2\mathsf{M_g}\mathsf{M}_\Psi^2\mathsf{K_b}\right]\mathbb{E}\parallel\varrho_1-\varrho_2\parallel_{\mathcal{B}}^2$$
$$:=Q\mathbb{E}\parallel\varrho_1-\varrho_2\parallel_{\mathcal{B}}^2,$$

where

$$Q=3\mathsf{M_h}\mathsf{K_b}+3\mathsf{M_F}\mathsf{b}^2\mathsf{M}_\Psi^2\left(2\mathsf{M}_\kappa\tilde{\sigma}+\mathsf{K_b}\right)+3\mathsf{b}^2\mathsf{M_g}\mathsf{M}_\Psi^2\mathsf{K_b}.$$

As a result of the fact that $Q<1$, $\tilde{\Upsilon}_1$ is a contraction mapping.
Step 2: Bounded sets are mapped into bounded sets in Λ_σ using the $\tilde{\Upsilon}_2$ operator. To prove the existence of a positive constant Q_1, it is sufficient to show that, for each $\varrho\in\Lambda_\sigma$, we have $\mathbb{E}\parallel\left(\tilde{\Upsilon}_2\varrho\right)(\gamma)\parallel^2\leq Q_1$. Indeed, for $\gamma\in\mathsf{I}$

$$\mathbb{E}\parallel\left(\tilde{\Upsilon}_2\varrho\right)(\gamma)\parallel^2=2\mathbb{E}\left\|\int_0^\gamma\Psi(\gamma-\varsigma)\mathsf{Qu}(\varsigma)d\varsigma\right.$$

$$+\int_0^{\gamma} \Psi(\gamma-\varsigma)\sigma(\varsigma,\tilde{\varphi}(\varsigma)+\varrho(\varsigma))d\mathsf{W}(\varsigma)\Bigg\|^2$$

$$\leq 2\mathsf{b}\mathbb{E}\left\|\int_0^{\gamma} \Psi(\gamma-\varsigma)\mathsf{Q}\mathsf{u}(\varsigma)\right\|^2 d\varsigma$$

$$+2\mathbb{E}\left\|\int_0^{\gamma} \Psi(\gamma-\varsigma)\sigma(\varsigma,\tilde{\varphi}(\varsigma)+\varrho(\varsigma))d\mathsf{W}(\varsigma)\right\|^2$$

$$\leq 2\mathsf{b}^2\mathsf{M}_{\Psi}^2 \parallel \mathsf{Q} \parallel^2 \parallel \mathsf{u} \parallel_{\mathcal{L}_{\mathcal{F}}}^2 +2\mathsf{b}\mathsf{L}_{\mathsf{W}}\mathsf{L}_{\sigma}\mathsf{M}_{\Psi}^2\mathbb{E}\,\|\sigma(\varsigma,\tilde{\varphi}(\varsigma)+\varrho(\varsigma))\|^2$$

$$\leq 2\mathsf{b}^2\mathsf{M}_{\Psi}^2 \parallel \mathsf{Q} \parallel^2 \parallel \mathsf{u} \parallel_{\mathcal{L}_{\mathcal{F}}}^2 +2\mathsf{b}\mathsf{L}_{\mathsf{W}}\mathsf{L}_{\sigma}\mathsf{M}_{\Psi}^2\tilde{\sigma}$$

$$\leq 2\mathsf{M}_{\Psi}^2\mathsf{b}\big[\mathsf{b}\|\mathsf{Q}\|^2 \parallel \mathsf{u} \parallel_{\mathcal{L}_{\mathcal{F}}^2}^2 +\mathsf{L}_{\mathsf{W}}\mathsf{L}_{\sigma}\tilde{\sigma}\big] := \mathcal{Q}_1.$$

Step 3: The $\tilde{\Upsilon}_2$ operator is completely continuous.
First, we demonstrate how $\tilde{\Upsilon}_2$ maps into an equicontinuous family for Λ_{σ}. Let $\varrho \in \Lambda_{\sigma}$ with $\varepsilon > 0$. Suppose if $0 < \varepsilon < \gamma < \gamma + \tau < \mathsf{b}$ where $\tau > 0$.

$$\mathbb{E}\left\|(\tilde{\Upsilon}_2\varrho)(\gamma+\tau)-(\tilde{\Upsilon}_2\varrho)(\gamma)\right\|^2$$

$$\leq \mathbb{E}\Bigg\|\int_0^{\gamma+\tau} \Psi(\gamma+\tau-\varsigma)\sigma(\varsigma,\tilde{\varphi}(\varsigma)+\varrho(\varsigma))d\mathsf{W}(\varsigma)$$

$$-\int_0^{\gamma} \Psi(\gamma-\varsigma)\sigma(\varsigma,\tilde{\varphi}(\varsigma)+\varrho(\varsigma))d\mathsf{W}(\varsigma)\Bigg\|^2$$

$$\leq \mathbb{E}\Bigg\|\int_0^{\gamma} \Psi(\gamma+\tau-\varsigma)\sigma(\varsigma,\tilde{\varphi}(\varsigma)+\varrho(\varsigma))d\mathsf{W}(\varsigma)$$

$$+\int_{\gamma}^{\gamma+\tau} \Psi(\gamma+\tau-\varsigma)\sigma(\varsigma,\tilde{\varphi}(\varsigma)+\varrho(\varsigma))d\mathsf{W}(\varsigma)$$

$$-\int_0^{\gamma} \Psi(\gamma-\varsigma)\sigma(\varsigma,\tilde{\varphi}(\varsigma)+\varrho(\varsigma))d\mathsf{W}(\varsigma)\Bigg\|^2$$

$$\leq \mathbb{E}\Bigg\|\int_0^{\gamma} [\Psi(\gamma+\tau-\varsigma)-\Psi(\gamma-\varsigma)]\sigma(\varsigma,\tilde{\varphi}(\varsigma)+\varrho(\varsigma))d\mathsf{W}(\varsigma)$$

$$+\int_{\gamma}^{\gamma+\tau} \Psi(\gamma+\tau-\varsigma)\sigma(\varsigma,\tilde{\varphi}(\varsigma)+\varrho(\varsigma))d\mathsf{W}(\varsigma)\Bigg\|^2$$

$$\leq 3\mathsf{L}_{\mathsf{W}}\int_0^{\gamma} \parallel \Psi(\gamma+\tau-\varsigma)-\Psi(\gamma-\varsigma) \parallel^2 \mathbb{E} \parallel \sigma(\varsigma,\tilde{\varphi}(\varsigma)+\varrho(\varsigma)) \parallel_{\mathcal{L}_{\mathcal{F}}^2}^2 d\varsigma$$

$$+3\mathsf{L_W}\int_{\gamma}^{\gamma+\tau}\|\Psi(\gamma+\tau-\varsigma)\|^2\,\mathbb{E}\,\|\sigma(\varsigma,\tilde{\varphi}(\varsigma)+\varrho(\varsigma))\|^2_{\mathcal{L}^2_{\mathcal{F}}}\,d\varsigma$$
$$\leq 3\mathsf{L_W}\sup_{\varsigma\in[0,\gamma]}\|\Psi(\gamma+\tau-\varsigma)-\Psi(\gamma-\varsigma)\|^2\int_0^{\gamma}\mathbb{E}\,\|\sigma(\varsigma,\tilde{\varphi}(\varsigma)+\varrho(\varsigma))\|^2_{\mathcal{L}^2_{\mathcal{F}}}\,d\varsigma$$
$$+3\mathsf{L_W}\mathsf{M}_\Psi^2\int_{\gamma}^{\gamma+\tau}\mathbb{E}\,\|\sigma(\varsigma,\tilde{\varphi}(\varsigma)+\varrho(\varsigma))\|^2_{\mathcal{L}^2_{\mathcal{F}}}\,d\varsigma$$
$$< 3\mathsf{L}_\sigma\mathsf{L_W}\sup_{\varsigma\in[0,\gamma]}\|\Psi(\gamma+\tau-\varsigma)-\Psi(\gamma-\varsigma)\|^2\int_0^{\gamma}\tilde{\sigma}d\varsigma$$
$$+3\mathsf{L}_\sigma\mathsf{L_W}\mathsf{M}_\Psi^2\int_{\gamma}^{\gamma+\tau}\tilde{\sigma}d\varsigma$$
$$\leq 3\mathsf{L}_\sigma\mathsf{L_W}\sup_{\varsigma\in[0,\gamma]}\|\Psi(\gamma+\tau-\varsigma)-\Psi(\gamma-\varsigma)\|^2\,\tilde{\sigma}\gamma+3\mathsf{L}_\sigma\mathsf{L_W}\mathsf{M}_\Psi^2\tau\tilde{\sigma}.$$

The right-hand side of the aforementioned inequality tends to zero as $\tau \longrightarrow 0$, according to Lebesgue's dominated convergence theorem. With this in mind, we draw the conclusion that Υ_2 is continuous from the right in $(0, \mathsf{b}]$. Similar reasoning demonstrates that it is continuous from the left in the interval $(0, \mathsf{b}]$. Les ensembles bornés sont donc transformés en une famille de fonctions équicontinues par $\tilde{\Upsilon}_2$.
Step 4: $\tilde{\Upsilon}_2 \in \Lambda_\sigma$ is relatively compact.
Let $\gamma \in \mathsf{I}$ be fixed and $\varepsilon \in \mathbb{R}$, such that $0 < \varepsilon < \gamma < \mathsf{b}$. For $\varrho \in \mathcal{B}$, we define the operators

$$(\tilde{\Upsilon}_2^{\varepsilon*}\varrho)(\gamma)=\Psi(\varepsilon)\Big[\int_0^{\gamma-\varepsilon}\Psi(\gamma-\varepsilon-\varsigma)\mathsf{Qu}(\varsigma)d\varsigma$$
$$+\int_0^{\gamma-\varepsilon}\Psi(\gamma-\varepsilon-\varsigma)\sigma(\varsigma,\tilde{\varphi}(\varsigma)+\varrho(\varsigma))d\mathsf{W}(\varsigma)\Big]$$

and

$$(\tilde{\Upsilon}_2^{\varepsilon}\varrho)(\gamma)=\int_0^{\gamma-\varepsilon}\Psi(\gamma-\varsigma)\mathsf{Qu}(\varsigma)d\varsigma+\int_0^{\gamma-\varepsilon}\Psi(\gamma-\varsigma)\sigma(\varsigma,\tilde{\varphi}(\varsigma)+\varrho(\varsigma))d\mathsf{W}(\varsigma).$$

$$\mathbb{E}\,\|(\tilde{\Upsilon}_2^{\varepsilon*}\varrho)(\gamma)-(\tilde{\Upsilon}_2^{\varepsilon}\varrho)(\gamma)\|^2$$
$$=\mathbb{E}\Big\|\Psi(\varepsilon)\int_0^{\gamma-\varepsilon}\Psi(\gamma-\varepsilon-\varsigma)\mathsf{Qu}(\varsigma)d\varsigma$$
$$+\Psi(\varepsilon)\int_0^{\gamma-\varepsilon}\Psi(\gamma-\varepsilon-\varsigma)\sigma(\varsigma,\tilde{\varphi}(\varsigma)+\varrho(\varsigma))d\mathsf{W}(\varsigma)$$
$$-\int_0^{\gamma-\varepsilon}\Psi(\gamma-\varsigma)\mathsf{Qu}(\varsigma)d\varsigma-\int_0^{\gamma-\varepsilon}\Psi(\gamma-\varsigma)\sigma(\varsigma,\tilde{\varphi}(\varsigma)+\varrho(\varsigma))d\mathsf{W}(\varsigma)\Big\|^2$$

$$\leq \int_0^{\gamma-\varepsilon} \| \Psi(\varepsilon)\Psi(\gamma-\varepsilon-\varsigma) - \Psi(\gamma-\varsigma) \|^2 \mathbb{E} \| \mathsf{Qu}(\varsigma) \|^2 d\varsigma$$

$$+ \int_0^{\gamma-\varepsilon} \| \Psi(\varepsilon)\Psi(\gamma-\varepsilon-\varsigma) - \Psi(\gamma-\varsigma) \|^2 \mathbb{E} \| \sigma(\varsigma, \tilde{\varphi}(\varsigma) + \varrho(\varsigma)) \|^2 d\mathsf{W}(\varsigma)$$

$$\leq \big(L(\varepsilon)\big)^2 \int_0^{\gamma-\varepsilon} \mathbb{E} \| \mathsf{Qu}(\varsigma) \|^2 d\varsigma + \big(L(\varepsilon)\big)^2 \mathsf{L_W} \int_0^{\gamma-\varepsilon} \mathbb{E} \| \sigma(\varsigma, \tilde{\varphi}(\varsigma) + \varrho(\varsigma)) \|^2_{\mathcal{L}_2^0} d\varsigma$$

$$\leq \big(L(\varepsilon)\big)^2 \| \mathsf{Q} \|^2 \| \mathsf{u} \|^2_{\mathcal{L}_{\mathcal{F}}} + \big(L(\varepsilon)\big)^2 \mathsf{L_W} \int_0^{\gamma-\varepsilon} \mathbb{E} \| \sigma(\varsigma, \tilde{\varphi}(\varsigma) + \varrho(\varsigma)) \|^2_{\mathcal{L}_2^0} d\varsigma$$

$$\leq \big(L(\varepsilon)\big)^2 \Big[\| \mathsf{Q} \|^2 \| \mathsf{u} \|^2 + \mathsf{L_W} \int_0^{\gamma-\varepsilon} \mathbb{E} \| \sigma(\varsigma, \tilde{\varphi}(\varsigma) + \varrho(\varsigma)) \|^2_{\mathcal{L}_2^0} d\varsigma \Big]$$

$$\leq \big(L(\varepsilon)\big)^2 \Big[\| \mathsf{Q} \|^2 \| \mathsf{u} \|^2 + \mathsf{L_\sigma} \mathsf{L_W} \int_0^{\gamma-\varepsilon} \tilde{\sigma} d\varsigma \Big] \underset{\varepsilon \longrightarrow 0}{\longrightarrow} 0.$$

Since $\Psi(\gamma)$ is compact, the set $\Xi_\varepsilon(\gamma) = \{\big(\tilde{\Upsilon}^\varepsilon \varrho\big)(\gamma) : \varrho \in \Lambda_\sigma\}$ is precompact set in $\mathbb{V}$ for every ε with $0 < \varepsilon < \gamma$. Furthermore, for every $\varrho \in \Lambda_\sigma$, we have

$$\mathbb{E} \| \big(\tilde{\Upsilon}_2 \varrho\big)(\gamma) - \big(\tilde{\Upsilon}_2^\varepsilon \varrho\big)(\gamma) \|^2$$

$$= \mathbb{E} \Bigg\| \int_0^\gamma \Psi(\gamma-\varsigma)\mathsf{Qu}(\varsigma)d\varsigma + \int_0^\gamma \Psi(\gamma-\varsigma)\sigma(\varsigma, \tilde{\varphi}(\varsigma) + \varrho(\varsigma))d\mathsf{W}(\varsigma)$$

$$- \int_0^{\gamma-\varepsilon} \Psi(\gamma-\varsigma)\mathsf{Qu}(\varsigma)d\varsigma - \int_0^{\gamma-\varepsilon} \Psi(\gamma-\varsigma)\sigma(\varsigma, \tilde{\varphi}(\varsigma) + \varrho(\varsigma))d\mathsf{W}(\varsigma) \Bigg\|^2$$

$$\leq \mathbb{E} \Bigg\| \int_{\gamma-\varepsilon}^\gamma \Psi(\gamma-\varsigma)\mathsf{Qu}(\varsigma)d\varsigma + \int_{\gamma-\varepsilon}^\gamma \Psi(\gamma-\varsigma)\sigma(\varsigma, \tilde{\varphi}(\varsigma) + \varrho(\varsigma))d\mathsf{W}(\varsigma) \Bigg\|^2$$

$$\leq \int_{\gamma-\varepsilon}^\gamma \mathbb{E} \| \Psi(\gamma-\varsigma)\mathsf{Qu}(\varsigma) \|^2 d\varsigma$$

$$+ \int_{\gamma-\varepsilon}^\gamma \mathbb{E} \| \Psi(\gamma-\varsigma)\sigma(\varsigma, \tilde{\varphi}(\varsigma) + \varrho(\varsigma)) \|^2 d\mathsf{W}(\varsigma)$$

$$\leq \int_{\gamma-\varepsilon}^\gamma \| \Psi(\gamma-\varsigma) \|^2 \mathbb{E} \| \mathsf{Qu}(\varsigma) \|^2 d\varsigma$$

$$+ \int_{\gamma-\varepsilon}^\gamma \| \Psi(\gamma-\varsigma) \|^2 \mathbb{E} \| \sigma(\varsigma, \tilde{\varphi}(\varsigma) + \varrho(\varsigma)) \|^2 d\mathsf{W}(\varsigma)$$

$$\leq \mathsf{M}_\Psi^2 \| \mathsf{Q} \|^2 \| \mathsf{u} \|^2_{\mathcal{L}_{\mathcal{F}}^0} \varepsilon + \mathsf{M}_\Psi^2 \mathsf{L_W} \int_{\gamma-\varepsilon}^\gamma \mathbb{E} \| \sigma(\varsigma, \tilde{\varphi}(\varsigma) + \varrho(\varsigma)) \|^2 d\varsigma$$

$$\leq \mathsf{M}_\Psi^2 \| \mathsf{Q} \|^2 \| \mathsf{u} \|^2_{\mathcal{L}_{\mathcal{F}}^0} \varepsilon + \mathsf{M}_\Psi^2 \mathsf{L_W} \mathsf{L_\sigma} \int_{\gamma-\varepsilon}^\gamma \| \tilde{\varphi}(\varsigma) + \varrho(\varsigma) \|^2 d\varsigma$$

$$\leq \mathsf{M}_\Psi^2 \| \mathsf{Q} \|^2 \| \mathsf{u} \|^2_{\mathcal{L}_{\mathcal{F}}^0} \varepsilon + \mathsf{M}_\Psi^2 \mathsf{L_W} \mathsf{L_\sigma} \int_{\gamma-\varepsilon}^\gamma \tilde{\sigma} d\varsigma$$

$$\leq \mathsf{M}_{\Psi}^2 \parallel \mathsf{Q} \parallel^2 \parallel \mathsf{u} \parallel_{\mathcal{L}_{\mathcal{F}}^0}^2 \varepsilon + \mathsf{M}_{\Psi}^2 \mathsf{L}_{\mathsf{W}} \mathsf{L}_{\sigma} \tilde{\sigma} \varepsilon$$

$$\leq \left(\mathsf{M}_{\Psi}^2 \parallel \mathsf{Q} \parallel^2 \parallel \mathsf{u} \parallel_{\mathcal{L}_{\mathcal{F}}^0}^2 + \mathsf{M}_{\Psi}^2 \mathsf{L}_{\mathsf{W}} \mathsf{L}_{\sigma} \tilde{\sigma} \right) \varepsilon \underset{\varepsilon \longrightarrow 0}{\longrightarrow} 0.$$

As ε tends to zero, the right-hand side of the inequality above tends to zero as well. These are arbitrary close to the sets and precompact

$$\Xi(\gamma) - \{(\tilde{\Upsilon}_2 \varrho)(\gamma) : \varrho \in \Lambda_\sigma\}.$$

Therefore, the set $\{(\tilde{\Upsilon}_2 \varrho)(\gamma) : \varrho \in \Lambda_\sigma\}$ in $\mathbb{V}$ is precompact.
Step 5: $\tilde{\Upsilon}_2 : \mathcal{B}_{\mathsf{b}}^0 \longrightarrow \mathcal{B}_{\mathsf{b}}^0$ is continuous.
Let $\{\varrho_n\}_{n=1}^{\infty}$ be the sequence in $\mathcal{B}_{\mathsf{b}}^0$ with $\varrho_n \longrightarrow \varrho \in \mathcal{B}_{\mathsf{b}}^0$. Then, there is a number $\sigma > 0$ such that $\parallel \varrho_n(\gamma) \parallel < \sigma$, $\forall n$ and $\gamma \in \mathsf{I}$ a.e., so $\varrho_n \in \Lambda_\sigma$ and $\varrho \in \Lambda_\sigma$

$$\sigma(\varsigma, \tilde{\varphi}_\gamma + \varrho_{n_\gamma}) \longrightarrow \sigma(\varsigma, \tilde{\varphi}_\gamma + \varrho_\gamma) \quad as \quad n \longrightarrow \infty,$$

for $\gamma \in \mathsf{I}$. Since

$$\mathbb{E} \parallel \sigma(\varsigma, \tilde{\varphi}_\gamma + \varrho_{n_\gamma}) - \sigma(\varsigma, \tilde{\varphi}_\gamma + \varrho_\gamma) \parallel^2 \leq \mathsf{M}_\sigma \parallel \varrho_{n_\gamma} - \varrho_\gamma \parallel_{\mathcal{B}}^2$$

by dominated convergence theorem,

$$\begin{aligned}
\mathbb{E} \parallel (\tilde{\Upsilon}_2 \varrho_n)(\gamma) - (\tilde{\Upsilon}_2 \varrho)(\gamma) \parallel^2 &\leq \mathbb{E} \left\| \int_0^\gamma \Psi(\gamma - \varsigma) \sigma(\gamma, \tilde{\varphi}_\gamma + \varrho_{n_\gamma}) d\mathsf{W}(\varsigma) \right. \\
&\quad \left. - \int_0^\gamma \Psi(\gamma - \varsigma) \sigma(\gamma, \tilde{\varphi}_\gamma + \varrho_\gamma) d\mathsf{W}(\varsigma) \right\|^2 \\
&\leq \mathbb{E} \left\| \int_0^\gamma \Psi(\gamma - \varsigma) \Big[\sigma(\gamma, \tilde{\varphi}_\gamma + \varrho_{n_\gamma}) - \sigma(\gamma, \tilde{\varphi}_\gamma + \varrho_\gamma) \Big] d\mathsf{W}(\varsigma) \right\|^2 \\
&\leq \mathsf{L}_{\mathsf{W}} \mathsf{M}_{\Psi}^2 \int_0^\gamma \mathbb{E} \parallel \sigma(\gamma, \tilde{\varphi}_\gamma + \varrho_{n_\gamma}) - \sigma(\gamma, \tilde{\varphi}_\gamma + \varrho_\gamma) \parallel^2 d\varsigma \\
&\leq \mathsf{M}_\sigma \mathsf{L}_{\mathsf{W}} \mathsf{M}_{\Psi}^2 \int_0^\gamma \mathbb{E} \parallel \varrho_{n_\gamma} - \varrho_\gamma \parallel^2 d\varsigma \underset{n \longrightarrow \infty}{\longrightarrow} 0.
\end{aligned}$$

Thus, $\tilde{\Upsilon}_2$ is continuous.

This illustrates the complete continuous of $\tilde{\Upsilon}_2$.

Step 6: $\mathsf{V} = \{\varrho \in \mathcal{B}_{\mathsf{b}}^0 : \vartheta = \alpha \tilde{\Upsilon}_1\left(\frac{\varrho}{\alpha}\right) + \alpha \tilde{\Upsilon}_2 \varrho\}$ is bounded set on I for some $\alpha \in (0, 1)$.
Let $\varrho \in \mathsf{V}$

$$\begin{aligned}
\mathbb{E} \parallel \varrho(\gamma) \parallel^2 \leq 6 \Big[& \mathbb{E} \parallel \Psi(\gamma)\mathsf{h}(0, \varphi(0)) \parallel^2 + \mathbb{E} \parallel \mathsf{h}(\gamma, \tilde{\varphi}(\gamma) + \varrho(\gamma)) \parallel^2 \\
& + \mathbb{E} \left\| \int_0^\gamma \Psi(\gamma - \varsigma)\mathsf{F}\big(\varsigma, \varphi(\varsigma) + \varrho(\varsigma), (\tilde{\varphi}(\varsigma) + \varrho(\varsigma))(\kappa(\tilde{\varphi}(\varsigma) + \varrho(\varsigma), \varsigma))\big) \right\|^2 d\varsigma \\
& + \mathbb{E} \left\| \int_0^\gamma \int_{\mathsf{Z}} \Psi(\gamma - \varsigma)\mathsf{g}(\varsigma, \tilde{\varphi}(\varsigma) + \varrho(\varsigma), \eta)\tilde{\mathcal{N}}(d\varsigma, d\eta) \right\|^2 \\
& + \mathbb{E} \left\| \int_0^\gamma \Psi(\gamma - \varsigma)\mathsf{Q}(\varsigma)\mathsf{u}(\varsigma) \right\|^2 d\varsigma + \mathbb{E} \left\| \int_0^\gamma \Psi(\gamma - \varsigma)\sigma(\varsigma, \tilde{\varphi}(\varsigma) + \varrho(\varsigma))d\mathsf{W}(\varsigma) \right\|^2 \Big] \\
\leq 6 \Big[& \mathbb{E} \parallel \Psi(\gamma)\mathsf{h}(0, \varphi(0)) \parallel^2 + \mathbb{E} \parallel \mathsf{h}(\gamma, \tilde{\varphi}(\gamma) + \varrho(\gamma)) \parallel^2 \\
& + \mathsf{b} \int_0^\gamma \parallel \Psi(\gamma - \varsigma) \parallel^2 \mathbb{E} \parallel \mathsf{F}\big(\varsigma, \varphi(\varsigma) + \varrho(\varsigma), (\tilde{\varphi}(\varsigma) \\
& \qquad + \varrho(\varsigma))(\kappa(\tilde{\varphi}(\varsigma) + \varrho(\varsigma), \varsigma))\big) \parallel^2 d\varsigma \\
& + \mathsf{b} \int_0^\gamma \parallel \Psi(\gamma - \varsigma) \parallel^2 \mathbb{E} \parallel \mathsf{Qu}(\varsigma) \parallel^2 d\varsigma \\
& + \int_0^\gamma \parallel \Psi(\gamma - \varsigma) \parallel^2 \mathbb{E} \parallel \sigma(\varsigma, \tilde{\varphi}(\varsigma) + \varrho(\varsigma)) \parallel^2 d\mathsf{W}(\varsigma) \\
& + \mathsf{b} \int_0^\gamma \int_{\mathsf{Z}} \parallel \Psi(\gamma - \varsigma) \parallel^2 \mathbb{E} \parallel \mathsf{g}(\varsigma, \tilde{\varphi}(\varsigma) + \varrho(\varsigma), \eta) \parallel \tilde{\mathcal{N}}(d\varsigma, d\eta) \Big]
\end{aligned}$$

$$\begin{aligned}
\leq 6 \Big[& \mathsf{M}_\Psi^2 \mathsf{L}_\mathsf{h}(1 + \parallel \varphi(0) \parallel^2) + \mathsf{M}_\Psi^2 \mathsf{L}_\mathsf{h}(1 + \mathbb{E} \parallel \tilde{\varphi}(\gamma) + \varrho(\gamma) \parallel^2) \\
& + \mathsf{b}\mathsf{M}_\Psi^2 \int_0^\gamma \mathbb{E} \parallel \mathsf{F}\big(\varsigma, \varphi(\varsigma) + \varrho(\varsigma), (\tilde{\varphi}(\varsigma) + \varrho(\varsigma))(\kappa(\tilde{\varphi}(\varsigma) + \varrho(\varsigma), \varsigma))\big) \parallel^2 d\varsigma \\
& + \mathsf{b}\mathsf{M}_\Psi^2 \int_0^\gamma \mathbb{E} \parallel \mathsf{Qu}(\varsigma) \parallel^2 d\varsigma + \mathsf{M}_\Psi^2 \int_0^\gamma \mathbb{E} \parallel \sigma(\varsigma, \tilde{\varphi}(\varsigma) + \varrho(\varsigma)) \parallel^2 d\mathsf{W}(\varsigma) \\
& + \mathsf{b}\mathsf{M}_\Psi^2 \int_0^\gamma \int_{\mathsf{Z}} \mathbb{E} \parallel \mathsf{g}(\varsigma, \tilde{\varphi}(\varsigma) + \varrho(\varsigma), \eta) \parallel \tilde{\mathcal{N}}(d\varsigma, d\eta) \Big] \\
\leq 6 \Big[& \mathsf{M}_\Psi^2 \mathsf{M}_\mathsf{h}(1 + \parallel \varphi(0) \parallel^2) + \mathsf{M}_\Psi^2 \mathsf{M}_\mathsf{h}(1 + \tilde{\sigma}) - \mathsf{F}\big(\varsigma, 0, (\tilde{\varphi} + \varrho)(\kappa(\tilde{\varphi} + \varrho)(0), 0)\big) \parallel^2 \\
& + \mathsf{b}\mathsf{M}_\Psi^2 \int_0^\gamma \Big(\mathbb{E} \parallel \mathsf{F}\big(\varsigma, \tilde{\varphi}_\varsigma + \varrho_\varsigma, (\tilde{\varphi} + \varrho)(\kappa(\tilde{\varphi}(\varsigma) + \varrho(\varsigma), \varsigma))\big) \\
& \quad + \parallel \mathsf{F}\big(\varsigma, 0, (\tilde{\varphi} + \varrho)(\kappa(\tilde{\varphi} + \varrho)(0), 0)\big) \parallel^2 \Big) d\varsigma \\
& + \mathsf{b}\mathsf{M}_\Psi^2 \parallel \mathsf{Q} \parallel^2 \int_0^\gamma \parallel \mathsf{u}(\varsigma) \parallel^2 d\varsigma + \mathsf{b}\mathsf{M}_\Psi^2 \mathsf{M}_\mathsf{g} \int_0^\gamma \mathbb{E} \parallel \tilde{\varphi}(\varsigma) + \varrho(\varsigma) \parallel^2 d\varsigma
\end{aligned}$$

$$
\begin{aligned}
&\quad +\mathsf{M}_\Psi^2 \mathsf{L_w L_\sigma} \int_0^\gamma \mathbb{E} \parallel \tilde{\varphi}(\varsigma)+\varrho(\varsigma) \parallel^2 d\varsigma \Big] \\
&\leq 6\Big[\mathsf{M}_\Psi^2 \mathsf{L_h}(1+ \parallel \varphi(0) \parallel^2) + \mathsf{M}_\Psi^2 \mathsf{L_h}(1+\tilde{\sigma}) \\
&\quad +\mathsf{bM}_\Psi^2 \mathsf{M_F} \int_0^\gamma \Big(\mathbb{E} \parallel \tilde{\varphi}+\varrho \parallel^2 + \mathsf{M}_\kappa \mathbb{E} \parallel \tilde{\varphi}+\varrho \parallel^2 \mathbb{E} \parallel \tilde{\varphi}(\varsigma)+\varrho(\varsigma) \parallel^2 + \mathsf{F}_0\Big) d\varsigma \\
&\quad +\mathsf{bM}_\Psi^2 \parallel \mathsf{Q} \parallel^2 \parallel u \parallel_{\mathcal{L}_{\mathcal{F}}}^2 + \mathsf{bM}_\Psi^2 \mathsf{M_g} \int_0^\gamma \mathbb{E} \parallel \tilde{\varphi}(\varsigma)+\varrho(\varsigma) \parallel^2 d\varsigma \\
&\quad +\mathsf{M}_\Psi^2 \mathsf{L_w L_\sigma} \int_0^\gamma \mathbb{E} \parallel \tilde{\varphi}(\varsigma)+\varrho(\varsigma) \parallel^2 d\varsigma \Big] \\
&\leq 6\Big[\mathsf{M}_\Psi^2 \mathsf{L_h}(1+ \parallel \varphi(0) \parallel^2) + \mathsf{M}_\Psi^2 \mathsf{L_h}(1+\tilde{\sigma}) \\
&\quad +\mathsf{bM}_\Psi^2 \mathsf{M_F}(1+\mathsf{M}_\kappa \tilde{\sigma}) \int_0^\gamma \mathbb{E} \parallel \tilde{\varphi}(\varsigma)+\varrho(\varsigma) \parallel^2 d\varsigma + \mathsf{bM}_\Psi^2 \mathsf{M_F} \int_0^\gamma \mathsf{F}_0 d\varsigma \\
&\quad +\mathsf{bM}_\Psi^2 \parallel \mathsf{Q} \parallel^2 \parallel \mathsf{u} \parallel_{\mathcal{L}_{\mathcal{F}}}^2 + \mathsf{bM}_\Psi^2 \mathsf{M_g} \int_0^\gamma \mathbb{E} \parallel \tilde{\varphi}(\varsigma)+\varrho(\varsigma) \parallel^2 d\varsigma \\
&\quad +\mathsf{M}_\Psi^2 \mathsf{L_w L_\sigma} \int_0^\gamma \mathbb{E} \parallel \tilde{\varphi}(\varsigma)+\varrho(\varsigma) \parallel^2 d\varsigma \Big] \\
&\leq 6\Big[\mathsf{M}_\Psi^2 \mathsf{L_h}(1+ \parallel \varphi(0) \parallel^2) + \mathsf{M}_\Psi^2 \mathsf{L_h}(1+\tilde{\sigma}) + \mathsf{b}^2 \mathsf{M}_\Psi^2 \mathsf{M_F F_0} + \mathsf{bM}_\Psi^2 \parallel \mathsf{Q} \parallel^2 \parallel \mathsf{u} \parallel_{\mathcal{L}_{\mathcal{F}}}^2 \\
&\quad + \Big[\mathsf{bM}_\Psi^2 \mathsf{M_F}(1+\mathsf{M}_\kappa \tilde{\sigma}) + \mathsf{bM}_\Psi^2 \mathsf{M_g} + \mathsf{M}_\Psi^2 \mathsf{L_w L_\sigma}\Big] \int_0^\gamma \mathbb{E} \parallel \tilde{\varphi}(\varsigma)+\varrho(\varsigma) \parallel^2 d\varsigma \Big] \\
&\leq 6\Big[\mathsf{M}_\Psi^2 \mathsf{L_h}(1+ \parallel \varphi(0) \parallel^2) + \mathsf{M}_\Psi^2 \mathsf{L_h}(1+\tilde{\sigma}) + \mathsf{b}^2 \mathsf{M}_\Psi^2 \mathsf{M_F F_0} + \mathsf{bM}_\Psi^2 \parallel \mathsf{Q} \parallel^2 \parallel \mathsf{u} \parallel_{\mathcal{L}_{\mathcal{F}}}^2 \\
&\quad + \Big[\mathsf{bM}_\Psi^2 \mathsf{M_F}(1+\mathsf{M}_\kappa \tilde{\sigma}) + \mathsf{M}_\Psi^2 \mathsf{M_g} + \mathsf{M}_\Psi^2 \mathsf{L_w L_\sigma}\Big] \mathsf{b}\tilde{\sigma} \Big] \\
&:= \tilde{C}.
\end{aligned}
$$

V is therefore bounded on I. As a result, according to the Krasnoselskii-Schaefer fixed-point theorem, (1) on I has a mild solution.

4 Optimal Control Results

This section tests if the system (1) has an optimal control using Balder's theorem [28]. Let the mild solution of Eq. (1), which corresponds to the control u, be ϑ_{u}. Take into account that Z is a separable reflexive Banach space, with u acting as the control and acquiring the values. The notation $\mathcal{L}^\infty(\mathsf{I}, \mathcal{L}(Z, \mathbb{V}))$ refers to the set of operator-valued functions that are measurable in the strong operator topology and uniformly bounded in I. Let $\mathcal{L}_{\mathcal{F}}^2(\mathsf{I}, Z) \in Z$ is a closed subspace consisting of all measurable and

$\mathcal{F}_\gamma$-adapted, Z-valued stochastic processes satisfying the condition $\int_0^{\mathsf{b}} \| \mathsf{u}(\gamma) \|_Z^2 \, d\gamma < \infty$ and having a norm

$$\| \mathsf{u} \|_{\mathcal{L}^2_{\mathcal{F}}} = \left[\int_0^{\mathsf{b}} \| \mathsf{u}(\gamma) \|_Z^2 \, d\gamma \right]^{1/2}.$$

Let $\mathcal{U}$ closed bounded and convex subset of Z. Introduce the admissible set $\mathcal{U}_{ad} = \{\mathsf{u}(\cdot) \in \mathcal{L}^2_{\mathcal{F}}(\mathsf{I}, Z) : \mathsf{u}(\gamma) \in \mathcal{U}, \gamma \in \mathsf{I}\}$. Assumed are the following.

(H$_7$) $\mathsf{Q} \in \mathcal{L}^\infty(\mathsf{I}, \mathcal{L}(Z, \mathbb{V}))$.
Think about the Bolza problem $(\tilde{P})$ [19, 29, 30], which is to identify the optimal pair $(\vartheta^0, \mathsf{u}^0) \in \mathcal{B} \times \mathcal{U}_{ad}$ such that $\mathcal{J}(\vartheta^0, \mathsf{u}^0) \le \mathcal{J}(\vartheta \mathsf{u}, \mathsf{u})$, $\forall \mathsf{u} \in \mathcal{U}_{ad}$ where the cost functional

$$\mathcal{J}(\vartheta^{\mathsf{u}}, \mathsf{u}) = \mathbb{E} \int_0^{\mathsf{b}} l(\gamma, \vartheta^{\mathsf{u}}_\gamma, \vartheta^{\mathsf{u}}(\gamma), \mathsf{u}(\gamma)) d\gamma + \mathbb{E}\Theta(\vartheta^{\mathsf{u}}(\mathsf{b})). \tag{8}$$

(H$_8$) The functional $\iota : \mathsf{I} \times \mathcal{B} \times \mathbb{V} \times Z \longrightarrow \mathbb{R} \cup \{\infty\}$ is Borel measurable.

(H$_9$) $\iota(\gamma, \cdot, \cdot, \cdot)$ is sequentially lower semicontinuous on $\mathcal{B} \times \mathbb{V} \times Z$, for almost all $\gamma \in \mathsf{I}$.

(H$_{10}$) $\iota(\gamma, \vartheta, z, \cdot)$ is a convex on Z, $\forall \vartheta \in \mathcal{B}$, $z \in \mathbb{V}$, and almost all $\gamma \in \mathsf{I}$.

(H$_{11}$) There exists $c_1, c_2 \ge 0$, $c_3 > 0$ and $\bar{w}$ is nonnegative, $\bar{w} \in \mathcal{L}^1([0, \mathsf{b}], \mathbb{R})$ such that

$$\iota(\gamma, \vartheta, z, \mathsf{u}) \ge \bar{w} + c_1 \| \vartheta \|_{\mathcal{B}} + c_2 \| \vartheta \|_{\mathbb{V}} + c_3 \| \vartheta \|_Z^2$$

(H$_{12}$) $\Theta : \mathbb{V} \longrightarrow \mathbb{R}$ is a continuous, nonnegative function.

Theorem 7 *If the criteria in Theorem 6 are satisfied, Hypotheses* **(*H*$_7$)** $-$ **(*H*$_{12}$)** *hold, and* B *is a highly continuous operator, the Bolza problem* $\tilde{P}$ *has at least one optimal pair on* $\mathcal{B} \times \mathcal{U}_{ad}$.

Proof So, $\inf\{\mathcal{J}(\vartheta_{\mathsf{u}}, \mathsf{u})/\mathsf{u} \in \mathcal{U}_{ad}\} = +\infty$, there is nothing to prove. By assumptions **(H$_7$)** $-$ **(H$_{12}$)**, we have

$$\begin{aligned}\mathcal{J}(\vartheta_{\mathsf{u}}, \mathsf{u}) \ge& \int_0^{\mathsf{b}} \mathbb{E}\bar{w}(\gamma) d\gamma + c_1 \int_0^{\mathsf{b}} \mathbb{E} \| \vartheta_{\gamma \mathsf{u}} \|_{\mathcal{B}} \, d\gamma + c_2 \int_0^{\mathsf{b}} \mathbb{E} \| \vartheta_{\mathsf{u}}(\gamma) \|_{\mathbb{V}} \, d\gamma \quad (9)\\ &+ c_3 \int_0^{\mathsf{b}} \mathbb{E} \| \mathsf{u}(\gamma) \|_Z^2 \, d\gamma + \mathbb{E}\Theta(\vartheta_{\mathsf{u}}(\mathsf{b}))\\ \ge& -\tilde{\gamma},\end{aligned}$$

where $\tilde{\gamma} \ge 0$ is a constant [19, 31].
Hence, $\gamma > -\tilde{\gamma} > -\infty$. Using the infimum definition, a minimizing sequence of feasible pairs can be found $\{(\vartheta^m, \mathsf{u}^m)\} \subset \tilde{\mathcal{U}}_{ad}$, where

$$\tilde{\mathcal{U}}_{ad} = \{(\vartheta, \mathsf{u}) : \vartheta \textit{ a mild solution of } (1) \textit{ corresponding to } \mathsf{u} \in \mathcal{U}_{ad}\}$$

such that $\mathcal{J}(\vartheta^m, \mathsf{u}^m) \underset{m \longrightarrow +\infty}{\longrightarrow} \gamma$. Because $\{\mathsf{u}^m\} \subseteq \mathcal{U}_{ad}$, $m = 1, 2, \cdots, \{\mathsf{u}^m\}$ is a bounded subset of the separable reflexive Banach space $\mathcal{L}^2_{\mathcal{F}}(\mathsf{I}, Z)$, there exists a subsequence, relabeled as $\{\mathsf{u}^m\}$ and $\mathsf{u}^0 \in \mathcal{L}^2_{\mathcal{F}}(\mathsf{I}, Z)$ such that $\mathsf{u}^m \longrightarrow \mathsf{u}^0$ weakly in $\mathcal{L}^2_{\mathcal{F}}(\mathsf{I}, Z)$. Since $\mathcal{U}^0_{ad}$ is closed and compact, by Marzur's lemma, we can conclude that $\mathsf{u}^0 \in \mathcal{U}_{ad}$. The mild solution of (1) corresponding to u^m be $\{\vartheta^m\}$ and the mild solution of (1) corresponding to u^0 be ϑ^0.

$$\vartheta^m(\gamma) = \begin{cases} 0, \quad \text{for } \gamma \in (-\infty, 0] \\ \Psi(\gamma)\big[\varphi(0) + \mathsf{h}(0, \varphi(0))\big] - \mathsf{h}(\gamma, \vartheta^m_\gamma) + \int_0^\gamma \Psi(\gamma - \varsigma)\mathsf{Q}(\varsigma)\mathsf{u}^m(\varsigma)d\varsigma \\ + \int_0^\gamma \int_Z \Psi(\gamma - \varsigma)\mathsf{g}(\varsigma, \vartheta^m_\varsigma, \eta)\tilde{\mathcal{N}}(d\varsigma, d\eta) + \int_0^\gamma \Psi(\gamma - \varsigma)\sigma(\varsigma, \vartheta^m_\varsigma)d\mathsf{W}(\varsigma) \\ + \int_0^\gamma \Psi(\gamma - \varsigma)F\big(\varsigma, \vartheta^m_\varsigma, \vartheta^m(\kappa(\vartheta^m_\varsigma, \varsigma))\big)d\varsigma \quad for \ \gamma \in \mathsf{I}. \end{cases} \tag{10}$$

A similar mild solution of (1) exists for u^0, denoted by ϑ^0, that is,

$$\vartheta^0(\gamma) = \begin{cases} 0, \quad \text{for } \gamma \in (-\infty, 0] \\ \Psi(\gamma)\big[\varphi(0) + \mathsf{h}(0, \varphi(0))\big] - \mathsf{h}(\gamma, \vartheta^0_\gamma) + \int_0^\gamma \Psi(\gamma - \varsigma)\mathsf{Q}(\varsigma)\mathsf{u}^0(\varsigma)d\varsigma \\ + \int_0^\gamma \int_Z \Psi(\gamma - \varsigma)\mathsf{g}(\varsigma, \vartheta^0_\varsigma, \eta)\tilde{\mathcal{N}}(d\varsigma, d\eta) + \int_0^\gamma \Psi(\gamma - \varsigma)\sigma(\varsigma, \vartheta^0_\varsigma)d\mathsf{W}(\varsigma) \\ + \int_0^\gamma \Psi(\gamma - s)\mathsf{F}\big(\varsigma, \vartheta^0_\varsigma, \vartheta^0(\kappa(\vartheta^0_\varsigma, \varsigma))\big)d\varsigma \quad for \ \gamma \in \mathsf{I}. \end{cases} \tag{11}$$

We can confirm the existence of a positive number β such that $\mathbb{E} \parallel \vartheta^m \parallel^2, \mathbb{E} \parallel \vartheta^0 \parallel^2 \leq \beta$. This is done by applying the properties of boundedness of $\{\mathsf{u}^m\}$ and $\{\mathsf{u}^0\}$. For every $\gamma \in \mathsf{I}$, we have

$$\begin{aligned} \mathbb{E} \parallel \vartheta^m(\gamma) - \vartheta^0(\gamma) \parallel^2 \leq 5 \Big\{ & \mathbb{E} \parallel \mathsf{h}(\gamma, \vartheta^m_\gamma) - \mathsf{h}(\gamma, \vartheta^0_\gamma) \parallel^2 \\ & + \mathsf{b} \int_0^\gamma \parallel \Psi(\gamma - \varsigma) \parallel^2 \parallel \mathsf{Q}(\mathsf{u}^m(\varsigma) - \mathsf{u}^0(\varsigma)) \parallel^2 d\varsigma \\ & + \mathsf{b} \int_0^\gamma \parallel \Psi(\gamma - \varsigma) \parallel^2 \mathbb{E} \parallel \mathsf{F}\big(\varsigma, \vartheta^m_\varsigma, \vartheta^m(\kappa(\vartheta^m(\varsigma), \varsigma))\big) \\ & \qquad - \mathsf{F}\big(\varsigma, \vartheta^0_\varsigma, \vartheta^0(\kappa(\vartheta^0(\varsigma), \varsigma))\big) \parallel^2 d\varsigma \\ & + \mathsf{b} \int_0^\gamma \int_Z \parallel \Psi(\gamma - \varsigma) \parallel^2 \mathbb{E} \parallel \mathsf{g}(\varsigma, \vartheta^m_\varsigma, \eta) - \mathsf{g}(\varsigma, \vartheta^0_\varsigma, \eta) \parallel^2 \tilde{\mathcal{N}}(d\varsigma, d\eta) \\ & + \int_0^\gamma \parallel \Psi(\gamma - \varsigma) \parallel^2 \mathbb{E} \parallel \sigma(\varsigma, \vartheta^m_\varsigma) - \sigma(\varsigma, \vartheta^0_\varsigma) \parallel^2 d\mathsf{W}(\varsigma) \Big\} \\ \leq 5 \Big\{ & \mathbb{E} \parallel \mathsf{h}(\gamma, \vartheta^m_\gamma) - \mathsf{h}(\gamma, \vartheta^0_\gamma) \parallel^2 + \mathsf{b}\mathsf{M}^2_\Psi \int_0^\gamma \int_Z \mathbb{E} \parallel \mathsf{g}(\varsigma, \vartheta^m_\varsigma, \eta) \\ & \qquad - \mathsf{g}(\varsigma, \vartheta^0_\varsigma, \eta) \parallel^2 \tilde{\mathcal{N}}(d\varsigma, d\eta) \\ & + \mathsf{b}\mathsf{M}^2_\Psi \int_0^\gamma \mathbb{E} \parallel \mathsf{F}\big(\varsigma, \vartheta^m_\varsigma, \vartheta^m(\kappa(\vartheta^m(\varsigma), \varsigma))\big) - \mathsf{F}\big(\varsigma, \vartheta^0_\varsigma, \vartheta^0(\kappa(\vartheta^0(\varsigma), \varsigma))\big) \parallel^2 d\varsigma \end{aligned}$$

$$
\begin{aligned}
&+\mathsf{bM}_{\Psi}^2 \int_0^{\gamma} \parallel \mathsf{Q}(\mathsf{u}^m(\varsigma) - \mathsf{u}^0(\varsigma)) \parallel^2 d\varsigma \\
&+\mathsf{M}_{\Psi}^2 \mathsf{L_W} \int_0^{\gamma} \mathbb{E} \parallel \sigma(\varsigma, \vartheta_{\varsigma}^m) - \sigma(\varsigma, \vartheta_{\varsigma}^0) \parallel^2 d\varsigma \Big\} \\
\leq\ & 5 \Big\{ \mathsf{M_h}\mathbb{E} \parallel \vartheta_{\gamma}^m - \vartheta_{\gamma}^0 \parallel^2 + \mathsf{bM}_{\Psi}^2 \int_0^{\gamma} \mathbb{E} \parallel \mathsf{F}\big(\varsigma, \vartheta_{\varsigma}^m, \vartheta^m\big(\kappa(\vartheta^m(\varsigma), \varsigma)\big)\big) \\
&-\mathsf{F}\big(\varsigma, \vartheta_{\varsigma}^m, \vartheta^m\big(\kappa(\vartheta^0(\varsigma), \varsigma)\big)\big) - \mathsf{F}\big(\varsigma, \vartheta_{\varsigma}^0, \vartheta^0\big(\kappa(\vartheta^0(\varsigma), \varsigma)\big)\big) \\
&+\mathsf{F}\big(\varsigma, \vartheta_{\varsigma}^m, \vartheta^m\big(\kappa(\vartheta^0(\varsigma), \varsigma)\big)\big) \parallel^2 d\varsigma + \mathsf{bM}_{\Psi}^2 \int_0^{\gamma} \parallel \mathsf{Q}(\mathsf{u}^m(\varsigma) - \mathsf{u}^0(\varsigma)) \parallel^2 d\varsigma \\
&+\mathsf{bM}_{\Psi}^2 \mathsf{M_g} \int_0^{\gamma} \mathbb{E} \parallel \vartheta_{\varsigma}^m - \vartheta_{\varsigma}^0 \parallel^2 d\varsigma + \mathsf{M}_{\Psi}^2 \mathsf{L_h} \mathsf{M}_{\sigma} \int_0^{\gamma} \mathbb{E} \parallel \vartheta_{\varsigma}^m - \vartheta_{\varsigma}^0 \parallel^2 d\varsigma \Big\} \\
\leq\ & 5 \Big\{ \mathsf{M_h}\mathbb{E} \parallel \vartheta_{\gamma}^m - \vartheta_{\gamma}^0 \parallel^2 + \mathsf{M}_{\Psi}^2 \Big[\mathsf{L_W} M_{\sigma} + \mathsf{bM_g} \Big] \int_0^{\gamma} \mathbb{E} \parallel \vartheta_{\varsigma}^m - \vartheta_{\varsigma}^0 \parallel^2 d\varsigma \\
&+\mathsf{bM}_{\Psi}^2 \int_0^{\gamma} \mathbb{E} \parallel F\big(\varsigma, \vartheta_{\varsigma}^m, \vartheta^m\big(\kappa(\vartheta^m(\varsigma), \varsigma)\big)\big) - \mathsf{F}\big(\varsigma, \vartheta^m(\varsigma), \vartheta^m\big(\kappa(\vartheta^0(\varsigma), \varsigma)\big)\big) \\
&-\mathsf{F}\big(\varsigma, \vartheta_{\varsigma}^0, \vartheta^0\big(\kappa(\vartheta^0(\varsigma), \varsigma)\big)\big) + \mathsf{F}\big(\varsigma, \vartheta_{\varsigma}^m, \vartheta^m\big(\kappa(\vartheta^0(\varsigma), \varsigma)\big)\big) \parallel^2 d\varsigma \\
&+\mathsf{bM}_{\Psi}^2 \int_0^{\gamma} \parallel \mathsf{Q}(\mathsf{u}^m(\varsigma) - \mathsf{u}^0(\varsigma)) \parallel^2 d\varsigma \Big\} \\
\leq\ & 5 \Big\{ \mathsf{M_h}\mathbb{E} \parallel \vartheta_{\gamma}^m - \vartheta_{\gamma}^0 \parallel^2 + \mathsf{M}_{\Psi}^2 \Big[\mathsf{L_W M}_{\sigma} + \mathsf{bM_g} \Big] \int_0^{\gamma} \mathbb{E} \parallel \vartheta_{\varsigma}^m - \vartheta_{\varsigma}^0 \parallel^2 d\varsigma \\
&+\mathsf{bM}_{\Psi}^2 \Big[\mathsf{M_F} \int_0^{\gamma} \mathbb{E} \parallel \vartheta^m\big(\kappa(\vartheta^m(\varsigma), \varsigma)\big) - \vartheta^m\big(\kappa(\vartheta^0(\varsigma), \varsigma)\big) \parallel^2 \\
&-\mathbb{E} \parallel \vartheta^m\big(\kappa(\vartheta^0(\varsigma), \varsigma)\big) - \vartheta^0\big(\kappa(\vartheta^0(\varsigma), \varsigma)\big) \parallel^2 \Big] d\varsigma \\
&\qquad +\mathsf{bM}_{\Psi}^2 \int_0^{\gamma} \parallel \mathsf{Q}(\mathsf{u}^m(\varsigma) - \mathsf{u}^0(\varsigma)) \parallel^2 d\varsigma \Big\}
\end{aligned}
$$

$$
\begin{aligned}
\leq\ & 5 \Big\{ \mathsf{M_h}\mathbb{E} \parallel \vartheta_{\gamma}^m - \vartheta_{\gamma}^0 \parallel^2 + \mathsf{M}_{\Psi}^2 \Big[\mathsf{L_W M}_{\sigma} + \mathsf{bM_g} \Big] \int_0^{\gamma} \mathbb{E} \parallel \vartheta_{\varsigma}^m - \vartheta_{\varsigma}^0 \parallel^2 d\varsigma \\
&+\mathsf{bM}_{\Psi}^2 \Big[\mathsf{M_F} \int_0^{\gamma} \mathsf{M}_{\kappa} \mathsf{K_b} \sup_{\gamma \in \mathsf{I}} \mathbb{E} \parallel \vartheta^m \parallel^2 \mathbb{E} \parallel \vartheta^m(\varsigma) - \vartheta^0(\varsigma) \parallel^2 + \mathbb{E} \parallel \vartheta_{\varsigma}^m - \vartheta_{\varsigma}^0 \parallel^2 \\
&+\mathbb{E} \parallel \vartheta^m(\varsigma) - \vartheta^0(\varsigma) \parallel^2 \Big] d\varsigma + \mathsf{bM}_{\Psi}^2 \int_0^{\gamma} \parallel \mathsf{Q}(\mathsf{u}^m(\varsigma) - \mathsf{u}^0(\varsigma)) \parallel^2 d\varsigma \Big\} \\
\leq\ & 5 \Big\{ \mathsf{M_h}\mathbb{E} \parallel \vartheta_{\gamma}^m - \vartheta_{\gamma}^0 \parallel^2 + \mathsf{M}_{\Psi}^2 \Big[\mathsf{L_W M}_{\sigma} + \mathsf{bM_g} \Big] \int_0^{\gamma} \mathbb{E} \parallel \vartheta_{\varsigma}^m - \vartheta_{\varsigma}^0 \parallel^2 d\varsigma \\
&+\mathsf{bM}_{\Psi}^2 \Big[\mathsf{M_F} \int_0^{\gamma} \mathsf{M}_{\kappa} r \mathbb{E} \parallel \vartheta^m(\varsigma) - \vartheta^0(\varsigma) \parallel^2 + \mathsf{K_b} \sup_{\gamma \in \mathsf{I}} \mathbb{E} \parallel \vartheta^m(\varsigma) - \vartheta^0(\varsigma) \parallel^2 \\
&+\mathbb{E} \parallel \vartheta^m(\varsigma) - \vartheta^0(\varsigma) \parallel^2 \Big] d\varsigma + \mathsf{bM}_{\Psi}^2 \int_0^{\gamma} \parallel \mathsf{Q}(\mathsf{u}^m(\varsigma) - \mathsf{u}^0(\varsigma)) \parallel^2 d\varsigma \Big\}
\end{aligned}
$$

$$\leq 5\Bigg\{\mathsf{M_h}\mathbb{E}\parallel \vartheta^m_\gamma - \vartheta^0_\gamma \parallel^2 + \mathsf{M}^2_\Psi\Big[\mathsf{L_w}\mathsf{M}_\sigma + \mathsf{b}\mathsf{M_g}\Big]\int_0^\gamma \mathbb{E}\parallel \vartheta^m_\varsigma - \vartheta^0_\varsigma \parallel^2 d\varsigma$$

$$+\mathsf{b}\mathsf{M}^2_\Psi\mathsf{M_F}\big(r\mathsf{M}_\kappa + \mathsf{K_b} + 1\big)\int_0^\gamma \mathbb{E}\parallel \vartheta^m(\varsigma) - \vartheta^0(\varsigma) \parallel^2 d\varsigma$$

$$+\mathsf{b}\mathsf{M}^2_\Psi\int_0^\gamma \parallel \mathsf{Q}(\mathsf{u}^m(\varsigma) - \mathsf{u}^0(\varsigma)) \parallel^2 d\varsigma\Bigg\}$$

$$\leq 5\Bigg\{\mathsf{K_b}\mathsf{M_h}\sup_{\gamma\in\mathsf{I}}\mathbb{E}\parallel \vartheta^m(\gamma) - \vartheta^0(\gamma) \parallel^2 + \mathsf{M}^2_\Psi\mathsf{K_b}\big(\mathsf{L_w}\mathsf{M}_\sigma$$

$$+\mathsf{b}\mathsf{M_g}\big)\int_0^\gamma \mathbb{E}\parallel \vartheta^m(\varsigma) - \vartheta^0(\varsigma) \parallel^2 d\varsigma$$

$$+\mathsf{b}\mathsf{M}^2_\Psi\mathsf{M_F}\big(\sigma\mathsf{M}_\kappa + \mathsf{K_b} + 1\big)\int_0^\gamma \mathbb{E}\parallel \vartheta^m(\varsigma) - \vartheta^0(\varsigma) \parallel^2 d\varsigma$$

$$+\mathsf{b}\mathsf{M}^2_\Psi\int_0^\gamma \parallel \mathsf{Q}(\mathsf{u}^m(\varsigma) - \mathsf{u}^0(\varsigma)) \parallel^2 d\varsigma\Bigg\}$$

$$\leq 5\mathsf{K_b}\mathsf{M_h}\sup_{\gamma\in\mathsf{I}}\mathbb{E}\parallel \vartheta^m(\gamma) - \vartheta^0(\gamma) \parallel^2 + 6\mathsf{b}\mathsf{M}^2_\Psi\int_0^\gamma \parallel \mathsf{Q}(\mathsf{u}^m(\varsigma) - \mathsf{u}^0(\varsigma)) \parallel^2 d\varsigma$$

$$+5\Big[\mathsf{M}^2_\Psi\mathsf{K_b}\big(\mathsf{L_w}\mathsf{M}_\sigma + \mathsf{b}\mathsf{M_g}\big) + \mathsf{b}\mathsf{M}^2_\Psi\mathsf{M_F}\big(\sigma\mathsf{M}_\kappa$$

$$+\mathsf{K_b} + 1\big)\Big]\int_0^\gamma \mathbb{E}\parallel \vartheta^m(\varsigma) - \vartheta^0(\varsigma) \parallel^2 d\varsigma$$

$$\leq \frac{1}{1-\Sigma_1}\Bigg[\Sigma_2\int_0^\gamma \mathbb{E}\parallel \vartheta^m(\varsigma) - \vartheta^0(\varsigma) \parallel^2 d\varsigma$$

$$+5\mathsf{b}\mathsf{M}^2_\Psi\int_0^\gamma \parallel \mathsf{Q}(\mathsf{u}^m(\varsigma) - \mathsf{u}^0(\varsigma)) \parallel^2 d\varsigma\Bigg]$$

where $\Sigma_1 = 5\mathsf{K_b}\mathsf{M_h}\sup_{\gamma\in\mathsf{I}}\mathbb{E}\parallel \vartheta^m(\gamma) - \vartheta^0(\gamma) \parallel^2$, $\sigma = \mathsf{K}^2_\mathsf{b}\sup_{\gamma\in\mathsf{I}}\mathbb{E}\parallel \vartheta^m(\varsigma) \parallel^2$,
$\Sigma_2 = 5\Big[\mathsf{M}^2_\Psi\mathsf{K_b}\big(\mathsf{L_w}\mathsf{M}_\sigma + \mathsf{b}\mathsf{M_g}\big) + \mathsf{b}\mathsf{M}^2_\Psi\mathsf{M_F}\big(\sigma\mathsf{M}_\kappa + \mathsf{K_b} + 1\big)\Big]$, and $1 - \Sigma_1 \neq 0$.
By the virtue of Gronwall's inequality, there exists $\Sigma_3 > 0$ such that

$$\mathbb{E}\parallel \vartheta^m(\varsigma) - \vartheta^0(\varsigma) \parallel^2 \leq \Sigma_3 \parallel \mathsf{Q}\mathsf{u}^m - \mathsf{Q}\mathsf{u}^0 \parallel^2_{\mathcal{L}^2_\mathcal{F}}.$$

Since Q is strongly continuous, we have $\parallel \mathsf{Q}\mathsf{u}^m - \mathsf{Q}\mathsf{u}^0 \parallel^2_{\mathcal{L}^2_\mathcal{F}} \underset{m\longrightarrow\infty}{\longrightarrow} 0$. Therefore,

$$\mathbb{E}\parallel \vartheta^m(\varsigma) - \vartheta^0(\varsigma) \parallel^2 \underset{m\longrightarrow\infty}{\longrightarrow} 0.$$

It is significant to take not of the fact that the expression ting that $(\mathbf{H}_7) - (\mathbf{H}_{12})$ suggests that the Balder theorem's preconditions are met. This leads us to the conclusion that $(\vartheta_\gamma, \vartheta, \mathsf{u}) \longrightarrow \int_0^\gamma \iota(\gamma, \vartheta_\gamma, \vartheta(\gamma), \mathsf{u}(\gamma))d\gamma$ is sequentially lower semicontinuous on $\mathcal{L}^2_{\mathcal{F}}(\mathsf{I}, Z) \subset \mathcal{L}^1_{\mathcal{F}}(\mathsf{I}, Z)$. Due to the weakly lower semicontinuous nature of $\mathcal{J}$ on $\mathcal{L}^2_{\mathcal{F}}(\mathsf{I}, Z)$ and by hypothesis that $\mathcal{J} > -\infty$, $\mathcal{J}$ reaches its infimum at $\mathsf{u}^0 \in \mathcal{U}_{ad}$, that is,

$$\begin{aligned} \mathcal{T} &= \lim_{m \longrightarrow \infty} \mathbb{E} \int_0^{\mathsf{b}} \iota(\gamma, \vartheta^m_\gamma, \vartheta^m(\gamma), \mathsf{u}^m(\gamma))d\gamma + \lim_{m \longrightarrow \infty} \mathbb{E}\Theta(\vartheta^m(\mathsf{b})) \\ &\geq \mathbb{E} \int_0^{\mathsf{b}} \iota(\gamma, \vartheta^0_\gamma, \vartheta^0(\gamma), \mathsf{u}^0(\gamma))d\gamma + \mathbb{E}\Theta(\vartheta^0(\mathsf{b})) = \mathcal{J}(\vartheta^0, \mathsf{u}^0) \geq \mathcal{T}. \end{aligned}$$

This demonstrates that the optimal control pair is $(\vartheta^0, \mathsf{u}^0)$ since $\mathcal{J}$ reaches its minimum at u^0.

5 Application

Applications of the discovered theoretical results are provided in this section. We consider the following stochastic system:

$$\begin{cases} d\big(\vartheta(\gamma, y) + 2e^{3\gamma}\vartheta(\gamma - \varphi, y)\big) = \dfrac{\partial^2}{\partial y^2}\big[\vartheta(\gamma, y) + 2e^{3\gamma}\vartheta(\gamma - \varphi, y)\big]d\gamma \\ +\Big[\displaystyle\int_0^\gamma b(\gamma - \varsigma)\dfrac{\partial^2}{\partial y^2}\big[\vartheta(\gamma, y) + 2e^{3\varsigma}\vartheta(\varsigma - \varphi, y)\big]d\varsigma \\ +\displaystyle\int_0^1 q(\varsigma, y)\mathsf{u}(\varsigma, y)d\varsigma + \dfrac{\vartheta(\gamma, \sin\gamma \mid \vartheta(\gamma, y) \mid) + \gamma^2\vartheta(\gamma, y)}{5\pi} \\ +\dfrac{e^{\gamma} \mid \vartheta(\gamma - \varphi, y) \mid}{8}\dfrac{\partial \mathsf{B}(\gamma)}{\partial\gamma} \\ +\displaystyle\int_{\mathsf{Z}} \dfrac{e^{\gamma}\cos \mid y\eta \mid\mid \vartheta(\gamma - \varphi, y) \mid}{9(1 + \mid \vartheta(\gamma - \varphi, y) \mid)}\tilde{\mathcal{N}}(d\gamma, d\eta)\Big]d\gamma, \quad \gamma \in [0, 1], \quad y \in [0, \pi], \\ \vartheta(\gamma, 0) = \vartheta(\gamma, y) = 0, \quad \gamma \in [0, 1] \\ \vartheta(\gamma, y) = \varphi(\gamma, y), \quad \gamma \in (-\infty, 0], \quad y \in [0, \pi]. \end{cases} \tag{12}$$

where $\gamma : \mathbb{R}^+ \longrightarrow \mathbb{R}$, $q : [0, 1] \times [0, \pi] \longrightarrow \mathbb{R}$ are continuous functions. Let $\mathbb{V} = \mathbb{Y} = \mathcal{L}^2([0, 1], [0, \pi])$ with the norm $\| \cdot \|$. Define the operator $\mathsf{A} : D(\mathsf{A}) \subseteq \mathbb{V} \longrightarrow \mathbb{V}$ by $\mathsf{A}\vartheta = \dfrac{\partial^2\vartheta}{\partial y^2}$ where $D(\mathsf{A}) = \{\vartheta \in \mathbb{V} : \vartheta, \vartheta'$ are absolutely continuous $\vartheta' \in \mathbb{V}, \vartheta(0) = \vartheta(\pi) = 0\}$.

Let $\mathsf{B} : D(\mathsf{A}) : \mathbb{V} \longrightarrow \mathbb{V}$ be the operator defined by $\mathsf{B}(\gamma)\vartheta = b(\gamma)\mathsf{A}\vartheta$ for $\gamma \geq 0$ and $\vartheta \in D(\mathsf{A})$.

Now, we present C_{g} with the norm

$$\| \vartheta \|_{C_{\mathsf{g}}}= \int_{-\infty}^{0} \mathsf{g}(\varsigma) \sup_{\varsigma \leq \theta \leq 0} \left(\mathbb{E} \| \varphi(\theta) \|^2 \right)^{\frac{1}{2}} d\varsigma$$

where $\mathsf{g}(\varsigma) = e^{2\varsigma}$, for $\varsigma < 0$, $\nu = \int_{-\infty}^{0} \mathsf{g}(\varsigma) d\varsigma = \frac{1}{2}$. For any $\gamma \in [0, 1]$, let $\vartheta(\gamma)(y) - \vartheta(\gamma, y)$, and

$$\begin{aligned}
\mathsf{F}\big(\varsigma, \vartheta\big(\kappa(\vartheta(\varsigma), \varsigma)\big)\big) &= \frac{\vartheta(\gamma, \sin\gamma \mid \vartheta(\gamma, y) \mid) + \gamma^2 \vartheta(\gamma, y)}{5\pi}, \\
\sigma((\gamma, \vartheta(\gamma, y))) &= \frac{e^{\gamma} \mid \vartheta(\gamma - \varphi, y) \mid}{8}, \\
\mathsf{g}(\gamma, \vartheta(\gamma, y), \eta) &= \frac{e^{\gamma} \cos \mid y\eta \mid\mid \vartheta(\gamma - \varphi, y) \mid}{9\big(1 + \mid \vartheta(\gamma - \varphi, y) \mid\big)}, \\
\mathsf{h}(\gamma, \vartheta(\gamma, y)) &= 2e^{3\gamma} \vartheta(\gamma - \varphi, y).
\end{aligned}$$

Then, problem (12) can be rewritten into the abstract form of (1) Moreover, if γ is bounded and $\mathcal{C}^1$ function such that γ' is bounded and uniformly continuous, then $(\mathbf{A}_2)$ in [22] are satisfied, Eq. (3) in [22] has a resolvent operator $(\Psi(\gamma))_{\gamma \geq 0}$ on $\mathbb{V}$. Furthermore, the nonlinear functions F, g and h satisfy the hypotheses $(\mathbf{H}_2) - (\mathbf{H}_8)$.

Thus, the conditions $(\mathbf{A}_1) - (\mathbf{A}_2)$ in [22] and $(\mathbf{H}_4) - (\mathbf{H}_6)$ are satisfied. Moreover, we assume that $\mathsf{M}_{\mathsf{F}} = \frac{1}{25}$, $\mathsf{M}_{\kappa} = 1$, $\mathsf{M}_{\mathsf{g}} = \frac{1}{81}$ and $\mathsf{M}_{\sigma} = \frac{1}{64}$.

Hence, substituting the values obtained above, we get by Theorem 6, the solution of (12) exists if $Q = 3\mathsf{M}_{\mathsf{h}}\mathsf{K}_{\mathsf{b}} + 3\mathsf{M}_{\mathsf{F}}\mathsf{b}^2\mathsf{M}_{\Psi}^2\big(2\mathsf{M}_{\kappa}\tilde{\sigma} + \mathsf{K}_{\mathsf{b}}\big) + 3\mathsf{b}^2\mathsf{M}_{\mathsf{g}}\mathsf{M}_{\Psi}^2\mathsf{K}_{\mathsf{b}} << 1$. Now, consider the cost function as

$$\mathcal{J}(\vartheta^{\mathsf{u}}, \mathsf{u})(y) = \int_0^{\pi} \int_0^1 \mathbb{E} \| \vartheta^{\mathsf{u}}(\gamma, y) \|^2 d\gamma dy + \int_0^{\pi} \int_0^1 \| \mathsf{u}(\gamma, y) \|^2 d\gamma dy.$$

Here $\mathsf{Qu}(\gamma, y) = \int_0^1 q(\varsigma, y)\mathsf{u}(s, y) d\varsigma$. Now

$$\begin{aligned}
\int_0^{\pi} \int_0^1 \| \mathsf{Qu}(\gamma, y) \|^2 &\leq \int_0^{\pi} \int_0^1 \left\| \int_0^1 q(\varsigma, y)\mathsf{u}(\varsigma, y) d\varsigma \right\|^2 d\gamma dy \\
&= M_q^2 \int_0^{\pi} \int_0^1 \int_0^1 \| \mathsf{u}(\gamma, y) \|^2 d\varsigma d\gamma dy \\
&\leq M_q^2 mes(\pi) \times \| \mathsf{u}(\gamma, y) \|^2_{\mathcal{L}^2([0,1],[0,\pi])},
\end{aligned}$$

where $M_q^2 = \max\{\| q(\gamma, y) \|^2 : \gamma \in [0, 1]\}$. In light of this, it may be said that Q is a bounded linear operator in $\mathcal{L}^2([0, 1], [0, \pi])$. In addition, the assumptions on Theorem 7 are satisfied, and then, there exists at least one optimal pair $(\vartheta^{\mathsf{u}}(\gamma, y), \mathsf{u}(\gamma, y))$.

6 Conclusion

This study investigates the optimal control issue in Hilbert space for a class of neutral stochastic integodifferential systems with deviating arguments driven by Poisson jumps. Neutral stochastic integodifferential equations are the ones in question. The Grimmer resolvent operator theory, the fixed-point theorem, and stochastic analytic methods have all been used to construct and illustrate the requirements for a valid solution. Balder's theorem, which has been shown to be accurate, has also been used to show that there is optimal control of the given problem. In the future, one of our goals is to research the approximation of controllability for quasi-linear differential systems that involve varying arguments in a Hilbert space.

References

1. M.A. Ouahra, B. Boufoussi, E.H. Lakhel, Existence and stability for stochastic impulsive neutral partial differential equations driven by Rosenblatt process with delay and Poisson jumps. Commun. Stochastic Anal. **11**(1), 7 (2017)
2. T. Caraballo, C. Ogouyandjou, F.K. Allognissode, M.A. Diop, Existence and exponential stability for neutral stochastic integro-differential equations with impulses driven by a Rosenblatt process. Discrete Continuous Dyn. Syst. B **25**(2), 507 (2020)
3. O.P. Agrawal, General formulation for the numerical solution of optimal problems. Int. J. Control. **50**(2), 627–638 (1989)
4. M.R. Hestenes, *Calculus of Variations and Optimal Control Theory* (Wiley, New York, 1966)
5. A. Ivan, J.J. Nieto, C.J. Silva, D.F. Torres, Ebola model and optimal control with vaccination constraints. J. Ind. Manag. Optim. **14**(2), 427–446 (2018)
6. Z. Yan, F. Lu, Existence of an optimal control for fractional stochastic partial neutral integro-differential equations with infinite delay. J. Nonlinear Sci. Appl. **8**(5), 557–577 (2015)
7. P. Balasubramaniam, P. Tamilalagan, The solvability and optimal controls for impulsive fractional stochastic integro-differential equations via resolvent operators. J. Optim. Theory Appl. **174**(1), 139–155 (2017)
8. Z. Yan, X. Yan, Optimal controls for impulsive partial stochastic differential equations with weighted pseudo almost periodic coefficients. Int. J. Control **94**(1), 111–133 (2021)
9. M.A. Ouahra, B. Boufoussi, E. Lakhel, Existence and stability for stochastic impulsive neutral partial differential equations driven by Rosenblatt process with delay and Poisson jumps. Commun. Stoch. Anal. **11**(1), 99–117 (2017)
10. A. Anguraj, K. Ravikumar, Existence and stability results for impulsive stochastic functional integrodifferential equations with Poisson jumps. J. Appl. Nonlinear Dyn. **8**(3), 407–417 (2019)
11. J. Luo, T. Taniguchi, The existence and uniqueness for non-Lipschitz stochastic neutral delay evolution equations driven by Poisson jumps. Stoch. Dyn. **9**(1), 135–152 (2009)
12. R. Grimmer, Resolvent operators for integral equations in a Banach space. Trans. Amer. Mathem. Soc. **273**, 333–349 (1982)

13. J.K. Hale, J. Kato, Phase space for retarded equations with infinite delay. Funkcialaj Ekvacioj-Serio Internacia. **21**(1), 11–41 (1978)
14. H. Smith, *An introduction to Delay Differential Equations with Applications to the Life Sciences* (Springer, New York, 2011)
15. Y. Hino, S. Murakami, T. Naito, *Functional Differential Equations with Unbounded Delay* (Springer, Berlin, 1991)
16. J.H.J. Kato, Phase space for retarded equations with infinite delay. Funkcial. Ekvac. 21, 11–41 (1978)
17. M. Rabbani, A. Das, B. Hazarika, R. Arab, Measure of noncompactness of a new space of tempered sequences and its application on fractional differential equations. Chaos, Solitons & Fractals **140**, 110221 (2020)
18. P. Jorion, On jump processes in the foreign exchange and stock markets. Rev. Financial Stud. **1**, 427–445 (1988)
19. C. Rajivganthi, P. Muthukumar, Almost automorphic solutions for fractional stochastic differential equations and its optimal control. Optim. Control Appl. Meth. **37**, 663–681 (2016)
20. P. Muthukumar, K. Thiagu, Existence of solutions and approximate controllability of fractional nonlocal stochastic differential equations of order $1<q \leq 2$ with infinite delay and Poisson jumps. Differ Equ. Dyn. Syst. **26**(1–3), 15–36 (2018)
21. W. Desch, R. Grimmer, W. Schappacher, Some considerations for linear integrodifferential equations. J. Mathem. Analy. Appl. **104**(1), 219–234 (1984)
22. M. Dieye, M.A. Diop, K. Ezzinbi, On exponential stability of mild solutions for some stochastic partial integrodifferential equations. Statist. Probability Lett. **123**, 61–76 (2017)
23. G. Chen, R. Grimmer, Semigroups and integral equations. J. Integral Equ. **2**, 133–154 (1980)
24. G. Da Prato, J. Zabezyk, *Stochastic Equations in Infinite Dimensions, Encyclopedia of Mathematics and Its Applications* (Cambridge University Press, Cambridge, UK, 1992)
25. Y. Ren, R. Sakthivel, *Existence, uniqueness and stability of mild solutions for second-order neutral stochastic evolution equations with infinite delay and Poisson jumps.* J Math Phys. **53**, 1–14 2012. https://doi.org/10.1063/1.4739406
26. A. Boudaoui, T. Caraballo, A. Ouahab, Stochastic differential equations with non-instantaneous impulses driven by a fractional Brownian motion. Discrete Contin. Dyn. Syst. **22**(7), 2521–2541 (2017)
27. T.A. Burton, C. Kirk, A fixed point theorem of Krasnoselski-Schaefer type. Math. Nachr. **189**, 23–31 (1998)
28. E. Balder, Necessary and sufficient conditions for L1-strong-weak lower semicontinuity of integral functionals. Nonlinear Anal. Theory Methods Appl. **11**, 1399–1404 (1987)
29. Z. Yan, J. Xiumei, *Optimal controls for fractional stochastic functional differential equations of order* $\alpha \in (1, 2]$. Bull Malays Math. Sci. Soc. **41**(3), 1581–1606 (2016). https://doi.org/10.1007/s40840-016-0415-2
30. Z. Yan, L. Fangxia, The optimal control for a new class of impulsive stochastic evolution integro-differential equations with infinite delay. Int. J. Control. **89**(8), 1592–1612 (2016)
31. P. Tamilalagan, P. Balasubramaniam, The solvability and optimal controls for fractional stochastic differential equations driven by Poisson jumps via resolvent operators. Appl. Math. Optim. **77**(3), 443–462 (2018)

A Review on Fractional Calculus in Modeling of Cancer

Kaushik Dehingia and Bhagya Jyoti Nath

Abstract This study carefully reviews a few important works that use fractional-order differential equations to model tumor–immune interactions and cancer treatment. It has been found that mathematical models with fractional-order derivatives describe tumor–immune dynamics better than integer-order mathematical models in terms of how realistic the scenarios are and how the solutions converge. Also, our review shows that more mathematical study will be required for cancer disease that considers fractional calculus to understand the complex relationship between the tumor and the immune system.

1 Introduction

Cancer is called the disease of the century, and it can be described by abnormal cells' uncontrolled growth, which affects the DNA of cells and makes them inactive. Cancer originates in any part of a human organ through a tumor. Generally, two types of tumors are found in the human body: benign tumors are neutral and cannot spread, while malignant tumors are primarily responsible for creating cancer sites in the body. The evolution of cancer in the body's immune system is very nonlinear and complex. Therefore, mathematical modeling is a powerful tool for the research community to understand tumor–immune complex dynamics.

Successful cancer treatment depends on various factors, viz., the tumor's severity, treatment options based on cancer types and sites, and the strength of the patient's immune system. However, a few treatment strategies, such as surgery, radiation, chemotherapy, hormone therapy, oncolytic virus therapy, immunotherapy, and a combination of two or more treatments, have been developed so far [1–3]. Surgical treatment is successful for solid cancer only. Radiotherapy and chemotherapy work

K. Dehingia
Department of Mathematics, Sonari College, Sonari 785690, Assam, India

B. Jyoti Nath (✉)
Department of Mathematics, Rabindranath Tagore University, Hojai 782435, Assam, India
e-mail: bhagyajyotinath13@gmail.com

M. Kharrat et al. (eds.), *Modeling of Discrete and Continuous Systems*, Forum for Interdisciplinary Mathematics, https://doi.org/10.1007/978-981-97-8715-9_5

best if the cancer is detected at the earliest stage. Virus therapy, immunotherapy, and combining two or more treatments are still in practice.

Many mathematicians and applied researchers have been interested in using fractional calculus to develop mathematical models of real-world phenomena. Fractional calculus has been widely used in many research fields, such as fundamental sciences [4, 5], engineering [6, 7], biology [8–11], finance [12], economics [13], and social sciences [14]. In particular, fractional calculus is mainly used where one may be interested in describing a phenomenon using the memory effect. For example, mathematical models can describe many biological processes, such as how cells interact, grow, divide, etc. It is best to use fractional-order differential equations to describe these processes, as the fractional order gives a more accurate picture than integer-order differential equations due to their non-local nature. Fractional-order derivatives can also define the fractals found in many biological systems.

One of the biological phenomena is tumor–immune interaction and the intervention of drugs to eliminate cancer. We already mentioned that the tumor–immune interaction phenomenon is mostly heterogeneous and nonlinear. In the nineteenth and twentieth centuries, integer-order derivatives were the only way to explain how this interaction happened. However, nowadays, researchers also use the fractional-order model to observe the dynamics of tumor cells concerning the immune system and treatment [15, 16]. The research shows that the fractional-order model can give a more realistic scenario than the integer-order model. In this study, we will review a few significant works to discuss how fractional calculus has been used to describe the tumor–immune association and the role of this association in cancer treatment and eradication processes.

2 Fractional Calculus

Fractional calculus is one of the growing branches of mathematics. In recent decades, it has become an area of intensive research due to its ability to describe any natural phenomenon better as compared to integer-order differential equations. In this section, we will discuss a brief history of fractional calculus and some types of fractional derivatives.

2.1 Early Era of Fractional Calculus

We can define the derivative of integer order as $\frac{d^n f}{dx^n}$, where n is an integer. However, can we define it as the derivative $\frac{d^n f}{dx^n}$ for a non-integer n. At this point, the idea of fractional calculus originated from generalizing the meaning of integer-order derivatives to non-integer-order derivatives.

Leibniz first used the notation $\frac{d^n f}{dx^n}$ to define the nth- order integral derivative for a linear function f [17]. In 1695, L'Hospital wrote a letter to Leibniz raising a question about $\frac{d^n f}{dx^n}$: "*What if* n *is* $\frac{1}{2}$?" On September 30, 1695, Leibniz [18] responded in a letter: "*You can see by that, sir, that one can express by an infinite series a quantity such that* $d^{1/2}\bar{xy}$ *or* $d^{1:2}\bar{xy}$. *Although infinite series and geometry are distant relations, infinite series admits only the use of exponents that are positive and negative integers and does not yet know the use of fractional exponents.*" But in the same letter, he also wrote prophetically: "*This is an apparent paradox, from which one-day useful consequences will be drawn.*" Thus, for the first time, Leibniz and L'Hospital initiated the concept of fractional calculus on that date, which can be considered the birthday of this new type of calculus. Leibniz also mentioned "general order" derivatives in his correspondence with Johann Bernoulli [19]. In another correspondence with John Wallis [20] about the infinite product of $\frac{\pi}{2}$, Leibniz used the notion of $d^{1/2}x$ in order to denote the derivative of order $\frac{1}{2}$.

Following this first inquisition between L'Hospital and Leibniz, different mathematicians in Europe studied fractional calculus. Euler was one of those who thought about the derivative with fraction number [21]. In 1730, Euler wrote: "*For if* n *is a positive integer, then* d^n *can be found by continued differentiation. However, such an assumption is not evident if* n *is a fraction. However, with the help of interpolation, which I have already explained in this dissertation, one may be able to expedite the matter.*" After that, J. L. Lagrange [22] discovered the law of exponents for differential operators of integer order in 1772. Mathematicians later found it interesting as they tried to make the rule more general after fractional calculus was developed. In 1812, P. S. Laplace [23] used an integral to define a fractional derivative. In 1819, S. F. Lacroix [24] answered correctly to the problem raised by Leibniz and L'Hospital for the first time. In his book, Lacroix developed the following formula for the nth derivative of the function $y = x^k$, k being a positive integer:

$$\frac{d^n}{dx^n}x^k = \frac{k!}{(k-n)!}x^{k-n}, k \geq n. \tag{1}$$

For fractional derivative, he replaced the factorial sign in Eq. (1) by the Gamma function Γ to obtain the following formula:

$$D_x^{\gamma} x^{\delta} = \frac{\Gamma(\delta+1)}{\Gamma(\delta-\gamma+1)}x^{\delta-\gamma}, \tag{2}$$

where γ and δ are two fractional numbers. Hence, Lacroix computed the following expression:

$$D_x^{\frac{1}{2}} x = \frac{\Gamma(2)}{\Gamma(\frac{1}{2})} = 2\sqrt{\frac{x}{\pi}}. \tag{3}$$

After Lacroix, the arbitrary order derivative was mentioned by Fourier [25]. He obtained the expression

$$\frac{d^{\upsilon}}{dx^{\upsilon}}f(x)=\frac{1}{2\pi}\int_{-\infty}^{\infty}f(\gamma)d\gamma\int_{-\infty}^{\infty}p^{\upsilon}cos[p(x-\gamma)+\frac{1}{2}\upsilon\pi]dp, \tag{4}$$

and stated that the number υ will be regarded as any quantity whatsoever, positive or negative.

In 1823, fractional calculus was first used by Niels Henrik Abel [26–28] to solve the tautochrone problem, the problem of finding the shape of a curve so that the time of an object's descent is independent of the object's starting point under uniform gravity, by solving an integral equation that arises in this formulation. For many decades, different mathematicians used different definitions to address the idea of non-integer-order derivatives or integrals. Influenced by the works of Abel, Liouville introduced one version of the definition for the non-integer derivative, which developed into the popularly known Riemann–Liouville definition [29–31]. It is observed that this definition gives the same result as obtained by Lacroix for $D_x^{\frac{1}{2}}x$, and the other definitions of the non-integer derivative can be considered a variation of the Riemann–Liouville definition. We will discuss definitions for non-integer or fractional derivatives in the next part.

2.2 Types of Fractional Derivatives

The starting point of Liouville was the following result related to the derivative of integer order m:

$$D_x^m e^{ax}=a^m e^{ax}. \tag{5}$$

Liouville generalized the above result (5) for non-integer-order derivative as follows:

$$D_x^{\alpha} e^{ax}=a^{\alpha} e^{ax}. \tag{6}$$

Then, he expanded the function $f(x)$ by means of Fourier series

$$f(x)=\sum_{n=0}^{\infty}c_n e^{a_n x}, \tag{7}$$

and used this to obtain the following formula for obtaining the arbitrary order derivative of any function $f(x)$:

$$D_x^{\alpha}f(x)=\sum_{n=0}^{\infty}c_n a_n^{\alpha}e^{a_n x}. \tag{8}$$

This can be regarded as Liouville's first formula for fractional derivatives [29, 30]. However, it has some disadvantages, like a restriction on α to make the series convergent. In order to overcome this, Liouville worked on another definition, which is well-known as Liouville's second definition. It was applied to explicit functions of the form x^{-k}, $k > 0$, where he considered the following definite integral related to the gamma function:

$$I = \int_0^{\infty} u^{k-1} e^{-xu} du, k > 0, x > 0, \tag{9}$$

and derived the formula

$$D^u x^{-k} = \frac{(-1)^u \Gamma(k+u)}{\Gamma(k)} x^{-k-u}, k > 0, \tag{10}$$

which is known as Liouville's second definition of a fractional derivative.

Controversy centered on two definitions of fractional calculus between 1835 and 1850. Some mathematicians favored Lacroix's generalization, while others preferred Liouville's definition. However, G. F. Bernhard Riemann made a significant contribution to the field of fractional calculus. Riemann developed the theory of fractional integration during his student days, but unfortunately, these works were published in 1892 after his death [32]. He derived a definition involving a definite integral that was applicable to the generalization of Taylor's series.

$$D^{-u} f(x) = \frac{1}{\Gamma(u)} \int_c^x (x-t)^{u-1} f(t) dt + \Psi(x). \tag{11}$$

The complementary function $\Psi(x)$ was added due to the ambiguity of the lower limit c. This expression was, in fact, the basis for the present-day, widely used Riemann–Liouville definition of fractional calculus. However, in today's definition, this complementary function has yet to be found. In 1869, N. Ya. Sonin [33] used Cauchy's integral formula as a starting point to reach the present day's definition. In 1872, A. V. Letnikov [34] followed the work of Sonin and extended his idea by utilizing a closed contour. Instead of a closed circuit, Laurent used a contour given as an open circuit to give today's definition of the Riemann–Liouville fractional integral as follows:

$$D^{-u} f(x) = \frac{1}{\Gamma(u)} \int_c^x (x-t)^{u-1} f(t) dt, R(u) > 0. \tag{12}$$

It can be observed that this formula (12) is similar to the Riemann integral (10) without the complementary function, and if we choose $c = 0$, then we obtain

Liouville's first definition. Therefore, the formula (12) is termed the Riemann–Liouville integral formula for fractional integral. From calculations, we can obtain the following formula for fractional differentiation:

$$D_x^u f(x) = \frac{1}{\Gamma(m-u)} \frac{d^m}{dx^m} \int_0^x (x-t)^{m-u-1} f(t)dt. \tag{13}$$

For the above Riemann–Liouville definition, Guy Jumarie proposed a simplified version as given below

$$D_x^u f(x) = \frac{1}{\Gamma(m-u)} \frac{d^m}{dx^m} \int_0^x (x-t)^{m-u-1} \{f(t) - f(0)\} dt. \tag{14}$$

Caputo's definition of the fractional derivative is also widely used. The Riemann–Liouville definition of the fractional derivative has one major drawback, i.e., the need for a rather "strange" set of initial conditions for the fractional differential equations with this kind of differential operator. The initial values of certain fractional integrals and derivatives must be known to find the solution to these fractional differential equations. Caputo rewrote the definition of the Riemann–Liouville fractional derivative to fix this problem. Caputo defined a fractional derivative for a function f with $m-1$ absolutely continuous integer-order derivative as follows:

$$D_x^u f(x) = \frac{1}{\Gamma(m-u)} \int_0^x (x-t)^{m-u-1} \frac{d^m f(t)}{dx^m} dt, \; m-1 < u \leq m. \tag{15}$$

Similarly, for Weyl, we have the following definition:

$$D_x^u f(x) = \frac{1}{\Gamma(m-u)} \frac{d^m}{dx^m} \int_0^\infty (x-t)^{m-u-1} f(t)dt. \tag{16}$$

Besides the above definitions, in recent times, some other definitions of fractional-order derivatives such as Caputo–Fabrizio, Liouville–Caputo, and Atangana–Baleanu–Caputo fractional and fractal operators have been proposed [35–37].

3 Tumor–Immune Interaction, Cancer Treatment, and Integer-Order Mathematical Models

In recent decades, the application of mathematical models has been prevalent in cancer research, facilitating a deeper understanding of various aspects such as tumor growth dynamics, immune system responses, and therapeutic strategies [38–42]. This section provides a review of several mathematical models of ordinary derivatives that are commonly employed to investigate the dynamical behavior of tumor–immune systems and assess the impact of treatment on these dynamics.

- **Kuznetsov et al.** [43]
  The cytotoxic T lymphocyte is a crucial constituent of the immune system that possesses the ability to eradicate tumor cells. In [43], a mathematical framework has been proposed to study the responses of cytotoxic T lymphocytes to the growth of an immunogenic tumor. They observed both local and global bifurcations after fitting the model with realistic parameters. A critical threshold value has been identified, beyond which the proliferation of tumor cells becomes irrepressible.
- **Kirschner and Panetta** [44]
  Kirschner and Panetta [44] have performed a mathematical investigation to understand the association among tumor cells, immune-effector cells, and IL-2 cytokine in the presence of ACI therapy and IL-2 therapy. The results of the analysis of the model indicate that ACI therapy may be a better treatment option, either as a monotherapy or in conjunction with IL-2 therapy.
- **Bunimovich-Mendrazitsky et al.** [45]
  Bacillus Calmette-Guérin (BCG) is used as an immunotherapy to help the immune system fight cancer cells in people with bladder cancer. Bunimovich-Mendrazitsky et al. [45] proposed a model that provides a description of the interaction between tumors and the immune system during BCG therapy. According to their findings, the model exhibited a bi-stable characteristic when BCG immunotherapy was applied. It has also been reported that a small amount of BCG is not enough to get rid of the tumor, while a large amount of BCG can cause the immune system to become too active, which can have dangerous side effects for the patient.
- **Castiglione and Piccoli** [46]
  In addition to $CD4^+$ T cells and $CD8^+$ T cells, dendritic cells are recognized as highly effective antigen-presenting cells within the immune system, contributing significantly to the eradication of tumors. Castiglione and Piccoli [46] explore the role of dendritic cells in tumor eradication by using vaccine therapy that controls the injection rate of dendritic cells. The study findings indicate that administering a high dose of the initial vaccination resulted in a reduction in the tumor population. Subsequent vaccinations, characterized by lower doses, were administered in a relatively uniform manner throughout the remaining duration of the treatment.
- **Itik and Banks** [47]
  Itik and Banks [47] proposed a theoretical framework for the progression of cancer cells considering tumor cells, healthy host cells, and effector immune cells. The model exhibited complex and chaotic dynamics characterized by connections reminiscent of the Shilnikov phenomenon.
- **Ku-Carrillo et al.** [48]
  The presence of obesity exerts a substantial influence on the proliferation and development of tumors. Ku-Carrillo et al. [48] developed a mathematical model to understand the effect of obesity on the tumor–immune system. They discussed the detrimental impacts of obesity on cancer and the consequences of consuming low- and high-calorie diets.

In addition, many other integer-order mathematical models have been proposed and investigated in [49–55] to elucidate the dynamics of tumor–immune system interactions and the effects of cancer treatment on tumor growth.

4 Fractional-Order Mathematical Models Applied in Cancer Research

In recent years, interest in fractional-order mathematical modeling in cancer management has increased due to its memory-capturing capability. Here, we will look at a few notable papers on fractional-order mathematical models of cancer and its treatment.

4.1 Fractional-Order Models Describe Tumor–Immune System

Several fractional-order mathematical models have studied the tumor–immune system interaction, which provides sufficient evidence that fractional models can describe the fractional dynamics of cancer cells [56–60]. Ahmed et al. [61] analyzed the dynamics of a tumor–immune system considering Holling type III immune responses to the tumor and showed that system stability depends on the death rate of immune cells. Complex and intricate dynamics of tumor–immune interactions have been observed through investigating a fractional-order cancer system in [62]. In [63], a fractional-order tumor model was analyzed in order to investigate the impact of effector cells and cytotoxic T lymphocytes (CTLs) on the dynamics of tumor cells. Arshad et al. [64] examined the responses of cytotoxic T lymphocytes to the growing tumor cells by analyzing a mathematical model under the Caputo fractional derivative. A rich chaotic dynamics has been observed in a tumor model [65] by changing the order of the fractional derivative. A comparative modeling study of tumor–immune interactions with Caputo and conformable fractional derivative has been performed, which revealed that the conformable fractional derivative model shows a more chaotic and realistic nature [66]. Ucar et al. [67] observed the effect of fractional order on the dynamics of cancer cells and found that as the order of the fractional operators decreases, the number of tumor cells also decreases. The effect of obesity on tumor development has been investigated in [68], which suggests that a low-calorie diet is associated with a reduced tumor burden compared to a high-calorie diet.

Numerous significant outcomes of tumor–immune interactions are described by using fractional-order derivatives. In [69], the interaction among the immune system, cancer cells, IL-12 cytokine, and anti-PD-L1 inhibitors was studied under the Caputo and Caputo–Fabrizio (CF) fractional operators and found that the model behaves more realistically with the Caputo–Fabrizio (CF) operator than with the Caputo operator. The Caputo fractional derivative was used to observe the dynamical behavior of a tumor–macrophage interaction model in [70]. It was observed that the model shows damp oscillations around the positive equilibrium point, and for different values of fractional derivative, the system exhibits a different state. Byun and Jung [71] analyzed a model using a fractional operator for tumor growth inhibition (TGI) and abnormal drug disposition and found that the order of the fractional derivative affected the stability of the model. The dynamical behavior of a fractional tumor–immune model involving the Atangana–Baleanu fractional derivative was studied in [72] and demonstrated how activation of effector immune cells affected the dynamics of tumor cells. It was reported in [73] that both normal cells and macrophages are very important in stopping the growth of tumors and in the process that makes macrophages. Debbouche et al. [74] analyzed a fractional-order cancer model for commensurate and incommensurate cases to discuss the chaotic dynamics of cancer growth. An HIV-1 and cancer co-infected model via the Caputo fractional operator has been investigated in [75] and observed the chaotic behavior of the model. The effect of Caputo fractional order on cancer dynamics has been investigated in [76] and revealed that a decrease in the order of the fractional derivative leads to an increase in the chaotic behavior of cancer dynamics. In [77], the interactions between tumor, healthy tissue, and effector immune cells were examined by analyzing a mathematical model under the Caputo fractal–fractional operator. They have observed multiple nonlinear phenomena within the model, including limit cycles, pitchforks, period-doubling, quasi-periodic, and inverse period-doubling bifurcations.

4.2 *Fractional-Order Models Applied in Cancer Treatment*

Fractional calculus has been applied in the modeling of administering cancer drugs also [78]. Sohail et al. [79] developed a fractional-order model to elucidate the dynamics of tumor–immune interactions when TIL or LAK cells are present, along with IL-2 therapy. They have revealed how the tumor mass changes as the tumor's antigenicity and fractional-order derivative change. A cancer-obesity model under the Caputo fractional operator has been investigated in [80] to find an optimal treatment schedule for cancer chemo-immune treatment. Moreover, they have obtained the impact of fractional order and the decay rate of chemotherapeutic drugs on the cost function. In [81], a fractional-order BCG treatment model for bladder cancer was examined and demonstrated the substantial impact of BCG treatment in the management of bladder cancer. Baleanu et al. [82] used fractional differential equations (FDEs) to study the dynamics of a tumor–immune surveillance system. They have also developed an optimal control strategy for analyzing the impact of chemotherapy on the model, resulting in a decrease in the growth rate of the untreated tumor cell population. The effect of fractional Caputo derivative on an immunotherapy bladder cancer model has been investigated in [83], where the BCG vaccine is used as a cancer treatment. They observed the memory effect of the fractional operator by changing the values of fractional derivatives, which significantly affected the model's dynamics. In [84], the effect of oncolytic virotherapy has been investigated by analyzing a fractional-order cancer treatment model. Sweilam et al. [85] proposed a fractional-order model with the Atangana–Baleanu–Caputo operator to obtain an optimal cancer treatment strategy combining immunotherapy and anti-angiogenic therapy. Breast cancer progression during immunochemotherapy treatment was modeled using a fractional-order differential neural network [86]. A fractional-order model was proposed in [87] to investigate the efficacy of drugs used for preventing cancer cells, boosting the immune system, and ketogenic diets in combating cancer and showed that the treatment was effective if the tumor density was detected at an early stage. Arfan et al. [88] investigated the stability of a fractional-order model pertaining to tumor–immune–drug interactions and depicted that the system exhibits different stages while changing its parameters and fractional derivative. A Caputo-type fractional-order cancer treatment model under radiotherapy has been analyzed in [89], and the bifurcation analysis has been performed by selecting the value of fractional order as the bifurcating parameter. Owolabi et al. [90] discussed a fractal–fractional partial differential equation model to observe the dynamics of HER2+ breast cancer cells in the presence of optimally administered trastuzumab and paclitaxel therapies.

5 Discussions

Fractional calculus has undergone numerous advancements since its inception at the end of the seventeenth century and up to the present day. Even though fractional calculus was initially developed in a mathematical context, it has recently been used to model and study a wide range of natural and artificial phenomena. Mathematical models have been used to describe various biological and physical phenomena. Differential equations, or systems of differential equations, are well-known mathematical tools for mathematical modeling. Since the beginning of the mathematical modeling era, differential equations with ordinary and partial derivatives have been heavily used. But nowadays, the research community uses delay differential equations and differential equations with fractional-order derivatives.

Along with other diseases, fractional calculus concepts are also used to propose and study mathematical models for controlling the unrestrained growth of tumors that cause cancer, a threat to today's human population. Cancer, a disease caused by the uncontrolled growth of abnormal cells, threatens human civilization. So, researchers in various fields used mathematical models to understand tumor–immune dynamics. In cancer research, ordinary and partial differential equation models are used a lot more often and are preferred. However, with the development of fractional calculus, researchers are now studying fractional differential equation models. Compared to models consisting of ordinary and partial differential equations, fractional differential equations give more accurate results. However, this area of the mathematical model is still developing; hence, different numerical methods for fractional differential equations are also in the development stage. Also, the advancement of this area will bring a new revolution to different areas of science, as fractional calculus applications have been found in many fields besides mathematics.

6 Conclusions

In this study, a systematic review of the use of fractional calculus in cancer modeling has been presented. We have begun with a brief history of the development of this subject and then presented the definitions of some commonly used fractional derivative operators like Riemann–Liouville, Caputo, Fourier fractional operators, etc. Before reviewing fractional-order models for cancer, we presented some well-known mathematical models for cancer modeling and discussed the models. From this study, we can conclude that fractional-order derivative models for cancer more accurately describe how tumors and immune systems work together, and their solutions converge more quickly than other mathematical models.

Acknowledgements Not applicable.

Ethics Declarations

Competing Interests There are no competing interests regarding this chapter.

Ethics Approval There is no ethical issue in this work. All the authors actively participated in this research and approved it for publication.

References

1. P. Das, S. Das, R.K. Upadhyay, P. Das, Optimal treatment strategies for delayed cancer-immune system with multiple therapeutic approach. Chaos Solit. Fractals **136**, 109806 (2020). https://doi.org/10.1016/j.chaos.2020.10980
2. J. Malinzi, K.B. Basita, S. Padidar, H.A. Adeola, Prospect for application of mathematical models in combination cancer treatments. Info. Med. Unlocked **23**, 100534 (2021). https://doi.org/10.1016/j.imu.2021.100534
3. K. Dehingia, S.W. Yao, K. Sadri, A. Das, H.K. Sarmah, A. Zeb, M. Inc, A study on cancer-obesity-treatment model with quadratic optimal control approach for better outcomes. Results Phys. **42**, 105963 (2022). https://doi.org/10.1016/j.rinp.2022.105963
4. D. Baleanu, R.P. Agarwal, Fractional calculus in the sky. Adv. Differ. Equ. 117 2021. https://doi.org/10.1186/s13662-021-03270-7
5. K. Cao, Y.Q. Chen, *Fractional Order Crowd Dynamics: Cyber-human System Modeling and Control, Fractional Calculus in Applied Sciences and Engineering*, vol. 4 (de Gruyter, Berlin, 2018)
6. Z. Korpinar, M. Inc, D. Baleanu, On the fractional model of Fokker-Planck equations with two different operator. AIMS Mathem. **5**(1), 236–248 (2020). https://doi.org/10.3934/math.2020015

7. M. Yavuz, N. Sene, Fundamental calculus of the fractional derivative defined with Rabotnov exponential kernel and application to nonlinear dispersive wave model. J. Ocean Eng. Sci. **6**(2), 196–205 (2021). https://doi.org/10.1016/j.joes.2020.10.004
8. F. A. Rihan, Numerical modeling of fractional-order biological systems. Abstract Appl. Anal., 816803 (2013). https://doi.org/10.1155/2013/816803
9. F. Özköse, R. Habbireeh, M. T. Şenel, A novel fractional order model of SARS-CoV-2 and Cholera disease with real data. J. Comput. Appl. Math., 423 (2023). https://doi.org/10.1016/j.cam.2022.114969
10. F. Özköse, Fractional mathematical modelling of the spread of rotavirus disease. J. Inst. Sci. Techn. **39**(2), 253–270 (2023)
11. M. Yavuz, F. Özköse, M. Susam, M. Kalidass, A new modeling of fractional-order and sensitivity analysis for hepatitis-B disease with real data. Fractal Fract. **7**(2), 165 (2023). https://doi.org/10.3390/fractalfract7020165
12. M. Diouf, N. Sene, Analysis of the financial chaotic model with the fractional derivative operator. Complexity, 9845031 (2020). https://doi.org/10.1155/2020/9845031
13. V. E. Tarasov, Mathematical economics: application of fractional calculus. MDPI, Basel (2020). https://doi.org/10.3390/books978-3-03936-119-9
14. D. Baleanu, A.M. Lopes, Handbook of fractional calculus with applications: applications in engineering, life and social Sciences, 7, *De Gruyter*, Berlin (2019)
15. K.M. Owolabi, A. Shikongo, A. Atangana, Fractal Fractional Derivative Operator Method on MCF-7 Cell Line Dynamics. *Methods of Mathematical Modelling and Computation for Complex Systems. Studies in Systems, Decision and Control*, 373. Springer, Cham (2022). https://doi.org/10.1007/978-3-030-77169-0_13
16. K.M. Owolabi, A. Shikongo, Fractional operator method on a multi-mutation and intrinsic resistance model. Alexandria Eng. J. **59**(4), 1999–2013 (2020). https://doi.org/10.1016/j.aej.2019.12.033
17. G.H. Pertz, C.J. Gerhardt (eds.), *Leibnizens gesammelte Werke, Lebinizens mathematische Schriften, Erste Abtheilung,Band II*, pages 301.302. Dritte Folge Mathematik (Erster Band). A. Asher & Comp., Briefwechsel zwischen Leibniz, Hugens van Zulichem und dem Marquis de l'Hospital, 1849
18. K.S. Miller, B. Ross, *An Introduction to the Fractional Calculus and Fractional Differential Equations* (Wiley, New York, 1993)
19. G.W. Leibniz, Letter from Hanover; Germany; to G.F.A. L'Hopital, September 30; 1695; in Mathematische Schriften, 1849; reprinted 1962, Olms verlag; Hidesheim; Germany; Vol. 2, PP. 301–302 (1965)
20. G.W. Leibniz, Letter from Hanover, Germany, to John Wallis, May 28, 1697, in Mathematishe Schriften; reprinted, Olms Verlag, Hildesheim. Germany **4**, 25 (1962)
21. L. Euler, De progressionibus transcendendibus seu quarum termini generales algebraicae dari nequent. Comm. Acad. Sci. Petropolitanae **5**, 36.57,1738
22. J.L. Lagrange, Sur un nouvelle espece de calcul relatif a la differentiation et a l'integration des quantities variables, *Oeuvres de Lagrange*, Vol 3, Gauthier-Villars, Paris, 441–476 (1849)
23. P.S. Laplace, *Theorie analytique des probabilities* (Courcier, Paris, 1820)
24. S.F. Lacroix, *Traite Du Calcul Differential et du Calcul Integral, 2nd.* Vol.3, Paris Courcier, 409–410 (1819)
25. J. Fourier, *Théorie analytique de la chaleur* (Paris,1822)
26. R. Gorenflo, S. Vessella, *Abel Integral Equations: Analysis and Applications* (Lecture Notes in Mathematics, Springer, Berlin Heidelberg, 1991)
27. N.H. Abel, Auosung einer mechanischen Aufgabe. J. fur reine und angew. Math. **1**, 153–157 (1826)
28. N.H. Abel, Solution de quelques problemes a l'aide d'integrales dÂ´enies. Oeuvres Completes **1**, 16–18 (1881)
29. J. Liouville, Memoire sur quelques questions de geometrie et de mecanique, et sur un noveau genre de calcul pour resoudre ces questions. J. l'Ecole Roy.Polytechn. **13**, 1–69 (1832a)

30. J. Liouville, Memoire sur le calcul des differentielles a indices quelconques. J. l'Ecole Roy. Polytechn. **13**, 71–162 (1832b)
31. J. Liouville, Memoire sur le theoreme des fonctions complementaires. J. fur reine und angew. Math. **11**, 1–19 (1834)
32. B. Riemann, Versuch einer allgemeinen Auffassung der Integration und Differentiation, Gesammelte Mathematische Werke und Wissenschaftlicher Nachlass, Teubner, Leipzig, Dover. New York **1953**, 331–344 (1876)
33. N.Y. Sonin, On differentiation with arbitrary index. Moscow Matem. Sbornik **6**(1), 1–38 (1869)
34. A.V. Letnikov, An explanation of the concepts of the theory of differentiation of arbitrary index (Russian). Moscow Matem. Sbornik **6**, 413–445 (1872)
35. K.S. Miller, B. Ross, *An Introduction to the Fractional Calculus and Fractional Differential Equations* (Wiley, New York, NY, USA, 1993)
36. A.A. Kilbas, H.M. Srivastava, J.J. Trujillo, *Theory and Applications of Fractional Differential Equations*, North-Holland Mathematics Studies, vol. 204. (Elsevier, Amsterdam, The Netherlands, 2006)
37. D. Baleanu, K. Diethelm, E. Scalas, J.J. Trujillo, *Fractional Calculus Models and Numerical Methods* (World Scientific, Singapore, 2009)
38. C. DeLisi, A. Rescigno, Immune surveillance and neoplasia-I: a minimal mathematical model. Bull. Math. Biol. **39**, 201–221 (1977)
39. H.N. Moore, N.K. Li, A mathematical model for chronic myelogenous leukemia (CML) and T cell interaction. J. Theor. Biol. **227**, 513–523 (2004)
40. K. Dehingia, H.K. Sarmah, M.B. Jeelani, A brief review on cancer research and its treatment through mathematical modelling. Ann. Cancer Res. Ther. **29**(1), 34–40 (2021). https://doi.org/10.4993/acrt.29.41
41. P. Das, S. Mukherjee, P. Das, An investigation on Michaelis - Menten kinetics based complex dynamics of tumor - immune interaction, Chaos. Solit. Fractals **128**, 297–305 (2019). https://doi.org/10.1016/j.chaos.2019.08.006
42. P. Das, S. Mukherjee, P. Das, S. Banerjee, Characterizing chaos and multifractality in noise-assisted tumor-immune interplay. Nonlinear Dyn. **101**, 675–685 (2020). https://doi.org/10.1007/s11071-020-05781-6
43. V.A. Kuznetsov, L.A. Makalkin, M.A. Taylor, A.S. Perelson, Nonlinear dynamics of immunogenic tumors: parameter estimation and global bifurcation analysis. Bull. Math. Biol. **56**(2), 295–321 (1994). https://doi.org/10.1016/S0092-8240(05)80260-5
44. D. Kirschner, J.C. Panetta, Modeling immunotherapy of the tumor-immune interaction. J. Math. Biol. **37**, 235–252 (1998). https://doi.org/10.1007/s002850050127
45. S. Bunimovich-Mendrazitsky, E. Shochat, L. Stone, Mathematical model of BCG immunotherapy in superficial bladder cancer. Bull. Math. Biol. **69**(6), 1847–1870 (2007). https://doi.org/10.1007/s11538-007-9195-z
46. F. Castiglione, B. Piccoli, Cancer immunotherapy, mathematical modeling and optimal control. J. Theoret. Biol. **247**, 723–732 (2007)
47. M. Itik, S.P. Banks, Chaos in a three-dimensional cancer model. Int. J. Bifurc. Chaos **20**, 71–79 (2010)
48. R.A. Ku-Carrillo, S.E. Delgadillo, B.M. Chen-Charpentier, A mathematical model for the effect of obesity on cancer growth and on the immune system response. Appl. Math. Model. **40**(7), 4908–4920 (2016)
49. L. Arlotti, A. Gamba, M. Lachowicz, A kinetic model of tumor/immune system cellular interactions. J. Theor. Med. **4**(1), 39–50 (2002)
50. H. Dritschel, S.L. Waters, A. Roller, H.M. Byrne, A mathematical model of cytotoxic and helper T cell interactions in a tumour microenvironment. Lett. Biomath. **5**(sup1), S36–S68 (2018). https://doi.org/10.1080/23737867.2018.1465863
51. M. Robertson-Tessi, A. El-Kareh, A. Goriely, A mathematical model of tumor-immune interactions. J. Theor. Biol. **294**, 56–73 (2012). https://doi.org/10.1016/j.jtbi.2011.10.027
52. L.G. de Pillis, A.E. Radunskaya, A mathematical tumor model with immune resistance and drug therapy: an optimal control approach. Comput. Math. Methods Med. **3**(2), 79–100, 318436 (2001). https://doi.org/10.1080/10273660108833067

53. L.G. de Pillis, A.E. Radunskaya, The dynamics of an optimally controlled tumor model: a case study. Math. Comput. Modell. **37**, 1221–1244 (2003)
54. P. Das, R.K. Upadhyay, P. Das, D. Ghosh, Exploring dynamical complexity in a time-delayed tumor-immune model. Chaos **30**(12) (2020). https://doi.org/10.1063/5.0025510
55. P. Das, P. Das, S. Mukherjee, Stochastic dynamics of Michaelis–Menten kinetics based tumor-immune interactions. Physica A, 541 (2020). https://doi.org/10.1016/j.physa.2019.123603
56. H. Nasrolahpour, Fractional dynamics in bioscience and biomedicine and the physics of cancer (2017), bioRxiv preprint https://doi.org/10.1101/214197
57. F.A. Rihan, A. Hashish, F. Al-Maskari, M. Sheek-Hussein, E. Ahmed, M.B. Riaz, R. Yafia, Dynamics of tumor-immune system with fractional-order. J. Tumor Res. **2**, 109 (2016)
58. B. Ghanbari, On the modeling of the interaction between tumor growth and the immune system using some new fractional and fractional-fractal operators. Adv. Differ. Equ. **2020**, 585 (2020). https://doi.org/10.1186/s13662-020-03040-x
59. S. Ahmad, A. Ullah, T. Abdeljawad, A. Akgül, N. Mlaiki, Analysis of fractal-fractional model of tumor-immune interaction. Results Phys. **25**, 104178 (2021)
60. M. Farman, A. Ahmad, A. Akgül, M.U. Saleem, K.S. Nisar, V. Vijayakumar, Dynamical behavior of tumor-immune system with fractal-fractional operator. AIMS Mathem. **7**(5), 8751–8773 (2022). https://doi.org/10.3934/math.2022489
61. E. Ahmed, A.H. Hashis, F.A. Rihan, On fractional order cancer model. J. Fractional Calculus Appl. **3**(2), 1–6 (2012)
62. I. N'Doye, H. Voos, M. Darouach, Chaos in a fractional-order cancer system, *European Control Conference (ECC), June 24–27, 2014* (Strasbourg, France, 2014)
63. S. Arshad, A. Sohail, S. Javed, Dynamical study of fractional order tumor model. Int. J. Computat. Methods **12**(5), 1550032 (2015). https://doi.org/10.1142/S0219876215500322
64. S. Arshad, D. Baleanu, J. Huang, Y. Tang, M.M.A. Qurashi, Dynamical analysis of fractional order model of immunogenic tumors. Adv. Mech. Eng. **8**(7), 1–13 (2016). https://doi.org/10.1177/1687814016656704
65. J.F. Gömez-Aguilar, M.G. Löpez-Löpez, V.M. Alvarado-Martïnez, D. Baleanu, H. Khan, Chaos in a cancer model via fractional derivatives with exponential decay and Mittag-Leffler law. Entropy **19**, 681 (2017). https://doi.org/10.3390/e19120681
66. E. Balcı, Ã–. Öztürk, S. Kartal, Dynamical behaviour of fractional order tumor model with Caputo and conformable fractional derivative. Chaos, Solitons and Fractals **123**, 43–51 (2019). https://doi.org/10.1016/j.chaos.2019.03.032
67. E. Ucar, N. Özdemir, E. Altun, Fractional order model of immune cells influenced by cancer cells. Math. Model. Nat. Phenom. **14**, 308 (2019). https://doi.org/10.1051/mmnp/2019002
68. S. Arshad, T. A. Yıldız, D. Baleanu, Y. Tang, The role of obesity in fractional order tumor-immune model. U.P.B. Sci. Bull., Series A **82**(2) (2020)
69. E. Ucar, N. Özdemir, A fractional model of cancer-immune system with Caputo and Caputo-Fabrizio derivatives. Eur. Phys. J. Plus **136**, 43 (2021). https://doi.org/10.1140/epjp/s13360-020-00966-9
70. I. Öztürk, F. Özköse, Stability analysis of fractional order mathematical model of tumor-immune system interaction. Chaos, Solitons and Fractals **133**, 109614 (2020). https://doi.org/10.1016/j.chaos.2020.109614
71. J.H. Byun, I.H. Jung, A fractional-order tumor growth inhibition model in PKPD. East Asian Math. J. **36**(1), 081–090 (2020). https://doi.org/10.7858/eamj.2020.008
72. S. Kumar, A. Kumar, B. Samet, J.F. Gömez-Aguilar, M.S. Osman, A chaos study of tumor and effector cells in fractional tumor-immune model for cancer treatment. Chaos, Solitons and Fractals **141**, 110321 (2020). https://doi.org/10.1016/j.chaos.2020.110321
73. F. Özköse, S. Yilmaz, M. Yavuz, I. Öztürk, M.T. Senel, B.S. Bagci, M. Dogan, Ö. Önal, A fractional modeling of tumor-immune system interaction related to lung cancer with real data. Eur. Phys. J. Plus **137**, 40 (2022). https://doi.org/10.1140/epjp/s13360-021-02254-6
74. N. Debbouche, A. Ouannas, G. Grassi, A.B.A. Al-Hussein, F.R. Tahir, K.M. Saad, H. Jahanshahi, A.A. Aly, Chaos in cancer tumor growth model with commensurate and incommensurate fractional-order derivatives. Comput. Mathem. Methods Med., 5227503 (2022). https://doi.org/10.1155/2022/5227503

75. P.A. Naik, K.M. Owolabi, M. Yavuz, J. Zu, Chaotic dynamics of a fractional order HIV-1 model involving AIDS-related cancer cells. Chaos, Solitons and Fractals **140**, 110272 (2020). https://doi.org/10.1016/j.chaos.2020.110272
76. P.A. Naik, J. Zu, M. Naik, Stability analysis of a fractional-order cancer model with chaotic dynamics. Int. J. Biomathematics **14**(6) (2021). https://doi.org/10.1142/S1793524521500467
77. L. Xuan, S. Ahmad, A. Ullah, S. Saifullah, A. Akgül, H. Qu, Bifurcations, stability analysis and complex dynamics of Caputo fractal-fractional cancer model. Chaos, Solitons, Fractals **159**, 112113 (2022). https://doi.org/10.1016/j.chaos.2022.112113
78. S.S. Ahrabi, Fractional model of cancer immunotherapy and its optimal control, *Ph.D. thesis. Sapienza-University of Rome, 2018*
79. A. Sohail, S. Arshad, S. Javed, Numerical analysis of fractional-order tumor model. Int. J. Biomathematics **8**(5), 1550069 (2015). https://doi.org/10.1142/S1793524515500692
80. T.A. Yıldız, S. Arshad, D. Baleanu, Optimal chemotherapy and immunotherapy schedules for a cancer-obesity model with Caputo time fractional derivative. Math. Meth. Appl. Sci., 1–18 (2018). https://doi.org/10.1002/mma.5298
81. I.A. Baba, A fractional-order bladder cancer model with BCG treatment effect. Comput. Appl. Mathem. (2019). https://doi.org/10.1007/s40314-019-0810-z
82. D. Baleanu, A. Jajarmi, S. S. Sajjadi, D. Mozyrska, A new fractional model and optimal control of a tumor-immune surveillance with non-singular derivative operator. Chaos **29**, 083127 (2019). https://doi.org/10.1063/1.5096159
83. M. Farman, A. Akgül, A. Ahmad, S. Imtiaz, Analysis and dynamical behavior of fractional-order cancer model with vaccine strategy. Math. Meth. Appl. Sci., 1–12 (2020). https://doi.org/10.1002/mma.6240
84. A. Abu-Rqayiq, M. Zannon, On dynamics of fractional-order oncolytic virotherapy models. J. Math. Computer Sci. **20**, 79–87 (2020). https://doi.org/10.22436/jmcs.020.02.01
85. N.H. Sweilam, S.M. Al-Mekhlafi, T. Assiri, A. Atangana, Optimal control for cancer treatment mathematical model using Atangana–Baleanu–Caputo fractional derivative. Adv. Differ. Equ., 334 (2020). https://doi.org/10.1186/s13662-020-02793-9
86. Z. Sabir, M. Munawar, M.A. Abdelkawy, M.A.Z. Raja, C. Ünlü, M.B. Jeelani, A.S. Alnahdi, Numerical investigations of the fractional-order mathematical model underlying immune-chemotherapeutic treatment for breast cancer using the neural networks. Fractal and Fractional **6**, 184 (2022). https://doi.org/10.3390/fractalfract6040184
87. A. Yousef, F. Bozkurt, T. Abdeljawad, Mathematical modeling of the immune-chemotherapeutic treatment of breast cancer under some control parameters. Adv. Differ. Equ. **2020**, 696 (2020). https://doi.org/10.1186/s13662-020-03151-5
88. M. Arfan, K. Shah, A. Ullah, M. Shutaywi, P. Kumam, Z. Shah, On fractional order model of tumor dynamics with drug interventions under nonlocal fractional derivative. Results Phys. **21**, 103783 (2021). https://doi.org/10.1016/j.rinp.2020.103783
89. M.F. Farayola, S. Shafie, F.M. Siam, Stability analysis of radiotherapy cancer treatment model with fractional derivative. Sains Malaysiana **49**(11), 2833–2846 (2020). https://doi.org/10.17576/jsm-2020-4911-22
90. K.M. Owolabi, A. Shikongo, Fractal fractional operator method on HER2+ breast cancer dynamics. Int. J. Appl. Comput. Math. **7**, 85 (2021). https://doi.org/10.1007/s40819-021-01030-5

Conformable Fractional Taylor Series Algorithm for Solving Temporal Fractional Reaction–Diffusion–Convection Equation

Ali Khalouta

Abstract In 2022, the author developed a new combination method to solve temporal fractional reaction–diffusion–convection equation in the sense of the Caputo fractional derivative [13]. The motivation of this chapter is to propose a new algorithm to solve the same equation but involving a conformable fractional derivative. This algorithm is called the conformable fractional Taylor series algorithm (CFTSA). Three numerical examples of temporal fractional reaction–diffusion–convection equation were used to validate the proposed algorithm. The advantage of CFTSA is that it can be used directly to solve nonlinear conformable fractional partial differential equations without being linearized, discretized, or perturbation. In addition, this algorithm provides lower computing cost and faster convergence rate. As a result, we confirm that in the future, the current algorithm can be applied to several nonlinear conformable fractional problems.

1 Introduction

The concept of fractional calculus is debuted over 327 years, which was first proposed in 1695 by Leibniz and L'Hopital. Fractional calculus which is the generalization of integer-order calculus, due to the memory property, has many applications in various fields such as engineering, physical sciences, elasticity, electrochemistry, diffusion process, statistics, transport theory, signal processing, and biological model.

In the literature, many mathematicians have proposed several definitions of fractional order derivative and studied their advantages and disadvantages, for solving differential equations involving fractional order derivatives of unknown functions which is also called fractional differential equations. These definitions include, Riemann–Liouville [8], Liouville-Caputo [15], Hadamard [16], Hilfer [10], Caputo–Fabrizio [9], and Atangana-Baleanu [4].

A. Khalouta (✉)
Laboratory of Fundamental and Numerical Mathematics, Department of Mathematics, Faculty of Sciences, Ferhat Abbas Sétif University 1, 19000 Sétif, Algeria
e-mail: nadjibkh@yahoo.fr; ali.khalouta@univ-setif.dz

M. Kharrat et al. (eds.), *Modeling of Discrete and Continuous Systems*, Forum for Interdisciplinary Mathematics, https://doi.org/10.1007/978-981-97-8715-9_6

In 2014, Khalil et al. [12] presented a novel definition of fractional derivative, so-called conformable fractional derivative, which is largely compatible with the classical derivative .Consequently, it has been acknowledged so swiftly and has been an important subject for many mathematical scientists and researchers.

During the past years, many powerful techniques have been established and developed in the literature for solving linear and nonlinear fractional differential equations involving the conformable fractional derivative, including the conformable fractional Laplace transform method (CFLTM) [18], conformable fractional natural transform method (CFNTM) [3], Shehu conformable fractional transform method (SCFTM) [6], conformable fractional power series method (CFPSM) [17], conformable fractional Sumudu decomposition method (CFSDM) [7], conformable fractional reduced differential transform method (CFRDTM) [19], conformable fractional variational iteration method (CFVIM) [21], conformable homotopy analysis method (CHAM) [2], conformable sine-Gordon expansion method (CSGEM) [11].

In this chapter, we propose a new algorithm known as conformable fractional Taylor series algorithm (CFTSA) for determining a solution to temporal fractional reaction–diffusion–convection equation described as

$$\mathfrak{T}_{\varsigma}^{\varrho}\psi = \left(a(\psi)\psi_{\xi}\right)_{\xi} + b(\psi)\psi_{\xi} + c(\psi), \tag{1}$$

under

$$\psi(\xi, 0) = \psi_0(\xi). \tag{2}$$

Here $\mathfrak{T}_{\varsigma}^{\varrho}$ is the conformable temporal fractional operator of order $0 < \varrho \leq 1$ of the function ψ where $\psi = \left\{\psi(\xi, \varsigma), (\xi, \varsigma) \in \mathbb{R} \times \mathbb{R}^{+}\right\}$ is an unknown function and the functions $a(\psi)$, $b(\psi)$ and $c(\psi)$ denote the diffusion term, the convection term and the reaction term respectively. In the case where $\varrho = 1$, this equation becomes the classical reaction–diffusion–convection equation.

The reaction–diffusion–convection equation is one of the important nonlinear partial differential equations that can be applied in many areas of sciences and engineering such as biology modeling, physics, chemistry, astrophysics, hydrology, and medicine.

This chapter is structured as follows. In Sect. 2, the basic definitions and some important theorems based on conformable fractional derivative are presented. In Sect. 3, the methodology of the suggested algorithm to provide the representation of CFTS-solution for Eqs. (1)–(2) is introduced and prove its convergence. In Sect. 4, three numerical examples are given to show that the numerical results confirm our theoretical findings. In Sect. 5, our results are presented and discussed in the form of graphs and tables. Section 6, includes the concluding observations.

2 Basic Definitions

This section reviews some essential concepts of conformable fractional calculus theory that will be used in this work.

Definition 1 [12] The conformable fractional derivative of order ϱ for a function $\psi \in C\left(\mathbb{R}^{+}\right)$ is defined as

$$(\mathfrak{T}^{\varrho}\psi)(\varsigma) = \frac{d^{\varrho}}{d\varsigma^{\varrho}}\psi(\varsigma) = \lim_{\varepsilon\to 0}\frac{\psi^{([\varrho]-1)}(\varsigma+\varepsilon\varsigma^{[\varrho]-\varrho}) - \psi^{([\varrho]-1)}(\varsigma)}{\varepsilon}, \varsigma > 0,$$

where $n-1<\varrho\le n, n\in\mathbb{N}$ and $\left[\varrho\right]$ is the smallest integer greater than or equal to ϱ. Provided $\mathfrak{T}^{\varrho}\psi(0) = \lim_{\varsigma\to 0^{+}}\mathfrak{T}^{\varrho}\psi(\varsigma)$ is $n-$differentiable and $\lim_{\varsigma\to 0^{+}}\mathfrak{T}^{\varrho}\psi(\varsigma)$ exists.

The most important and useful rule is that: If $\psi(\varsigma)$ is an $n-$differentiable function at $\varsigma>0$ and $n-1<\varrho\le n$, $n\in\mathbb{N}$, then

$$(\mathfrak{T}^{\varrho}\psi)(\varsigma) = \varsigma^{([\varrho]-\varrho)}\psi^{[\varrho]}(\varsigma).$$

As a special case, if $0<\varrho\le 1$, then we have

$$(\mathfrak{T}^{\varrho}\psi)(\varsigma) = \frac{d^{\varrho}}{d\varsigma^{\varrho}}\psi(\varsigma) = \lim_{\varepsilon\to 0}\frac{\psi(\varsigma+\varepsilon\varsigma^{1-\varrho}) - \psi(\varsigma)}{\varepsilon}, \varsigma > 0,$$

provided $\mathfrak{T}^{\varrho}\psi(0) = \lim_{\varsigma\to 0^{+}}\mathfrak{T}^{\varrho}\psi(\varsigma)$ is $\varrho-$differentiable and $\lim_{\varsigma\to 0^{+}}\mathfrak{T}^{\varrho}\psi(\varsigma)$ exists.

Theorem 1 [5] *Let* $0<\varrho\le 1$ *and* $\psi,\varphi\in C\left(\mathbb{R}^{+}\right)$ *be* $\varrho-$*differentiable at a point* $\varsigma>0$. *Then*

(1) $(\mathfrak{T}^{\varrho}(\lambda\psi+\delta\varphi))(\varsigma) = \lambda\left(\mathfrak{T}^{\varrho}\psi\right)(\varsigma) + \delta\left(\mathfrak{T}^{\varrho}\varphi\right)(\varsigma)$, *for all* $\lambda,\delta\in\mathbb{R}$.
(2) $\mathfrak{T}^{\varrho}(\gamma)=0$, *for all* $\gamma\in\mathbb{R}$.
(3) $\mathfrak{T}^{\varrho}(\varsigma^{p}) = p\varsigma^{p-\varrho}$, *for all* $p\in\mathbb{R}$.
(4) $(\mathfrak{T}^{\varrho}\psi\varphi)(\varsigma) = \psi(\varsigma)\left(\mathfrak{T}^{\varrho}\varphi\right)(\varsigma) + \varphi(\varsigma)\left(\mathfrak{T}^{\varrho}\psi\right)(\varsigma)$.
(5) $(\mathfrak{T}^{\varrho}\psi/\varphi)(\varsigma) = \left(\varphi(\varsigma)\left(\mathfrak{T}^{\varrho}\psi\right)(\varsigma) - \psi(\varsigma)\left(\mathfrak{T}^{\varrho}\varphi\right)(\varsigma)\right)/\varphi^{2}(\varsigma), \varphi(\varsigma)\neq 0$.
(6) If ψ *is differentiable, then* $\left(\mathfrak{T}^{\varrho}\psi\right)(\varsigma) = \varsigma^{1-\alpha}\frac{d}{d\varsigma}\psi(\varsigma)$.

Theorem 2 [5] *Let* $\psi,\varphi\in C\left(\mathbb{R}^{+}\right)$ *be* $\varrho-$*differentiable functions where* $0<\varrho\le 1$ *and* $\chi(\varsigma)=\psi(\varphi(\varsigma))$. *Then* $\chi(\varsigma)$ *is* $\varrho-$*differentiable and for all* ς *with* $\varsigma\neq 0$ *and* $\varphi(\varsigma)\neq 0$ *we have*

$$(\mathfrak{T}^{\varrho}\chi)(\varsigma) = (\mathfrak{T}^{\varrho}\psi)(\varphi(\varsigma)).(\mathfrak{T}^{\varrho}\varphi)(\varsigma).\varphi(\varsigma)^{\varrho-1}.$$

Definition 2 [1] Let $\partial^{k}\psi/\partial\varsigma^{k}$ and $\partial^{k}\psi/\partial\xi^{k}$ defined on $\mathbb{R}\times\mathbb{R}^{+}$ for $k=1,2,\ldots,n-1$. Then the conformable temporal fractional derivative of order ϱ with $n-1<\varrho\le n$ of $\psi\in C\left(\mathbb{R}\times\mathbb{R}^{+}\right)$ is defined as

$$\left(\mathfrak{T}_{\varsigma}^{\varrho}\psi\right)(\xi,\varsigma)=\frac{\partial^{\varrho}}{\partial\varsigma^{\varrho}}\psi(\xi,\varsigma)=\lim_{\varepsilon\to 0}\frac{\psi_{\varsigma}^{(n-1)}(\xi,\varsigma+\varepsilon\varsigma^{n-\varrho})-\psi_{\varsigma}^{(n-1)}(\xi,\varsigma)}{\varepsilon},\ \varsigma\geq 0.$$

3 Methodology of the CFTSA and Its Convergence

Theorem 3 *According to the CFTSA, the solution of Eqs. (1)–(2) can be presented as an infinite expansion as follows:*

$$\psi(\xi,\varsigma)=\sum_{k=0}^{\infty}\psi_k(\xi)\frac{\varsigma^{k\varrho}}{\varrho^k k!},\ \varsigma\geq 0,0<\varrho\leq 1, \tag{3}$$

where $\psi_k(\xi)$ *denotes the coefficients functions of series (3).*

Proof To prove this result, suppose that the solution of Eqs. (1) and (2) takes the form of the following infinite series

$$\psi(\xi,\varsigma)=\sum_{k=0}^{\infty}\psi_k(\xi)\frac{\varsigma^{k\varrho}}{\varrho^k k!}.$$

Thus, the nth-CFTS approximate solution can be defined as follows:

$$\psi_n(\xi,\varsigma)=\sum_{k=0}^{n}\psi_k(\xi)\frac{\varsigma^{k\varrho}}{\varrho^k k!}=\psi_0(\xi)+\sum_{k=1}^{n}\psi_k(\xi)\frac{\varsigma^{k\varrho}}{\varrho^k k!}. \tag{4}$$

Applying the operator $\mathfrak{T}_{\varsigma}^{\varrho}$ on Eq. (4), and using Theorem 1, we get the following formula

$$\left(\mathfrak{T}_{\varsigma}^{\varrho}\psi_n\right)(\xi,\varsigma)=\sum_{k=0}^{n-1}\psi_{k+1}(\xi)\frac{\varsigma^{k\varrho}}{\varrho^k k!}. \tag{5}$$

Substituting Eqs. (4) and (5) in Eq. (1), we get the following iterative relation

$$\begin{aligned}0=&\sum_{k=0}^{n-1}\psi_{k+1}(\xi)\frac{\varsigma^{k\varrho}}{\varrho^k k!}-\left(a\left(\sum_{k=0}^{n}\psi_k(\xi)\frac{\varsigma^{k\varrho}}{\varrho^k k!}\right)\left(\sum_{k=0}^{n}\psi_k'(\xi)\frac{\varsigma^{k\varrho}}{\varrho^k k!}\right)\right)_{\xi}\\&-b\left(\sum_{k=0}^{n}\psi_k(\xi)\frac{\varsigma^{k\varrho}}{\varrho^k k!}\right)\left(\sum_{k=0}^{n}\psi_k'(\xi)\frac{\varsigma^{k\varrho}}{\varrho^k k!}\right)-c\left(\sum_{k=0}^{n}\psi_k(\xi)\frac{\varsigma^{k\varrho}}{\varrho^k k!}\right).\end{aligned}$$

To find the value of unknown coefficients $\psi_n(\xi), n=1,2,3,\ldots,$ we solve the following equation

$$\mathfrak{T}_{\varsigma}^{(n-1)\varrho}\left\{\mathfrak{L}(\xi,\varsigma,\varrho,n)\right\}\downarrow_{\varsigma=0}=0,$$

where

$$\mathfrak{L}(\xi,\varsigma,\varrho,n)=\sum_{k=0}^{n-1}\psi_{k+1}(\xi)\frac{\varsigma^{k\varrho}}{\varrho^k k!}-\left(a\left(\sum_{k=0}^{n}\psi_k(\xi)\frac{\varsigma^{k\varrho}}{\varrho^k k!}\right)\left(\sum_{k=0}^{n}\psi_k'(\xi)\frac{\varsigma^{k\varrho}}{\varrho^k k!}\right)\right)_\xi$$
$$-b\left(\sum_{k=0}^{n}\psi_k(\xi)\frac{\varsigma^{k\varrho}}{\varrho^k k!}\right)\left(\sum_{k=0}^{n}\psi_k'(\xi)\frac{\varsigma^{k\varrho}}{\varrho^k k!}\right)-c\left(\sum_{k=0}^{n}\psi_k(\xi)\frac{\varsigma^{k\varrho}}{\varrho^k k!}\right).$$

and $\mathfrak{T}_\varsigma^{(n-1)\varrho}$ denotes the sequential conformable temporal fractional derivative of order $n-1$ that is defined by

$$\mathfrak{T}_\varsigma^{(n-1)\varrho}\{\mathfrak{L}(\xi,\varsigma,\varrho,n)\}\downarrow_{t=0}=\underbrace{\mathfrak{T}_\varsigma^{\varrho}.\mathfrak{T}_\varsigma^{\varrho}...\mathfrak{T}_\varsigma^{\varrho}}_{(n-1)-times}\{\mathfrak{L}(\xi,\varsigma,\varrho,n)\}\downarrow_{\varsigma=0}.$$

Now, to find the value of the first unknown coefficient $\psi_1(\xi)$, we substitute $n=1$ into the value of $\mathfrak{L}(\xi,\varsigma,\varrho,n)$, we get

$$\mathfrak{L}(\xi,\varsigma,\varrho,1)=\psi_1(\xi)-\left(a\left(\psi_0(\xi)+\psi_1(\xi)\frac{\varsigma^\varrho}{\varrho}\right)\left(\psi_0'(\xi)+\psi_1'(\xi)\frac{\varsigma^\varrho}{\varrho}\right)\right)_\xi$$
$$-b\left(\psi_0(\xi)+\psi_1(\xi)\frac{\varsigma^\varrho}{\varrho}\right)\left(\psi_0'(\xi)+\psi_1'(\xi)\frac{\varsigma^\varrho}{\varrho}\right)-c\left(\psi_0(\xi)+\psi_1(\xi)\frac{\varsigma^\varrho}{\varrho}\right).$$

Solving the equation $\mathfrak{L}(\xi,0,\varrho,1)=0$, gives

$$\psi_1(\xi)=\left(a\left(\psi_0(\xi)\right)\psi_0'(\xi)\right)_\xi+b\left(\psi_0(\xi)\right)\psi_0'(\xi)+c\left(\psi_0(\xi)\right).$$

To find the value of the second unknown coefficient $\psi_2(\xi)$, we substitute $n=2$ into the value of $\mathfrak{L}(\xi,\varsigma,\varrho,n)$, we get

$$\mathfrak{L}(\xi,\varsigma,\varrho,2)=\psi_1(\xi)+\psi_2(\xi)\frac{\varsigma^\varrho}{\varrho}$$
$$-\left(a\left(\psi_0(\xi)+\psi_1(\xi)\frac{\varsigma^\varrho}{\varrho}+\psi_2(\xi)\frac{\varsigma^{2\varrho}}{\varrho^2 2!}\right)\left(\psi_0'(\xi)+\psi_1'(\xi)\frac{\varsigma^\varrho}{\varrho}+\psi_2'(\xi)\frac{\varsigma^{2\varrho}}{\varrho^2 2!}\right)\right)_\xi$$
$$-b\left(\psi_0(\xi)+\psi_1(\xi)\frac{\varsigma^\varrho}{\varrho}+\psi_2(\xi)\frac{\varsigma^{2\varrho}}{\varrho^2 2!}\right)\left(\psi_0'(\xi)+\psi_1'(\xi)\frac{\varsigma^\varrho}{\varrho}+\psi_2'(\xi)\frac{\varsigma^{2\varrho}}{\varrho^2 2!}\right)$$
$$-c\left(\psi_0(\xi)+\psi_1(\xi)\frac{\varsigma^\varrho}{\varrho}+\psi_2(\xi)\frac{\varsigma^{2\varrho}}{\varrho^2 2!}\right).$$

From Theorems 1 and 2 and solving $\mathfrak{T}_{\varsigma}^{\varrho}\left\{\mathfrak{L}(\xi,\varsigma,\varrho,2)\right\}\downarrow_{\varsigma=0}=0$, we get

$$\begin{aligned}\psi_2(\xi) &= \left(a\left(\psi_0(\xi)\right)\psi_1'(\xi) + a'\left(\psi_0(\xi)\right)\psi_0'(\xi)\psi_1(\xi)\right)_{\xi}\\ &\quad + b\left(\psi_0(\xi)\right)\psi_1'(\xi) + b'\left(\psi_0(\xi)\right)\psi_0'(\xi)\psi_1(\xi) + c'\left(\psi_0(\xi)\right)\psi_1(\xi).\end{aligned}$$

To find the value of the third unknown coefficient $\psi_3(\xi)$, we substitute $n = 3$ into the value of $\mathfrak{L}(\xi,\varsigma,\varrho,n)$, we get

$$\begin{aligned}\mathfrak{L}(\xi,\varsigma,\varrho,3) &= \psi_1(\xi) + \psi_2(\xi)\frac{\varsigma^{\varrho}}{\varrho} + \psi_3(\xi)\frac{\varsigma^{2\varrho}}{\varrho^2 2!}\\ &\quad - \left(a\left(\psi_0(\xi) + \psi_1(\xi)\frac{\varsigma^{\varrho}}{\varrho} + \psi_2(\xi)\frac{\varsigma^{2\varrho}}{\varrho^2 2!} + \psi_3(\xi)\frac{\varsigma^{3\varrho}}{\varrho^3 3!}\right)\right.\\ &\quad \times \left.\left(\psi_0'(\xi) + \psi_1'(\xi)\frac{\varsigma^{\varrho}}{\varrho} + \psi_2'(\xi)\frac{\varsigma^{2\varrho}}{\varrho^2 2!} + \psi_3'(\xi)\frac{\varsigma^{3\varrho}}{\varrho^3 3!}\right)\right)_{\xi}\\ &\quad - b\left(\psi_0(\xi) + \psi_1(\xi)\frac{\varsigma^{\varrho}}{\varrho} + \psi_2(\xi)\frac{\varsigma^{2\varrho}}{\varrho^2 2!} + \psi_3(\xi)\frac{\varsigma^{3\varrho}}{\varrho^3 3!}\right)\\ &\quad \times \left(\psi_0'(\xi) + \psi_1'(\xi)\frac{\varsigma^{\varrho}}{\varrho} + \psi_2'(\xi)\frac{\varsigma^{2\varrho}}{\varrho^2 2!} + \psi_3'(\xi)\frac{\varsigma^{3\varrho}}{\varrho^3 3!}\right)\\ &\quad - c\left(\psi_0(\xi) + \psi_1(\xi)\frac{\varsigma^{\varrho}}{\varrho} + \psi_2(\xi)\frac{\varsigma^{2\varrho}}{\varrho^2 2!} + \psi_3(\xi)\frac{\varsigma^{3\varrho}}{\varrho^3 3!}\right).\end{aligned}$$

From Theorems 1 and 2 and solving $\mathfrak{T}_{\varsigma}^{2\varrho}\left\{\mathfrak{L}(\xi,\varsigma,\varrho,3)\right\}\downarrow_{\varsigma=0}=0$, we get

$$\begin{aligned}\psi_3(\xi) &= (a(\psi_0(\xi))\psi_2'(\xi) + a'\left(\psi_0(\xi)\right)(\psi_0'(\xi)\psi_2(\xi)\\ &\quad + 2\psi_1(\xi)\psi_1'(\xi)) + a''\left(\psi_0(\xi)\right)\psi_0'(\xi)\psi_1^2(\xi))_{\xi}\\ &\quad + b\left(\psi_0(\xi)\right)\psi_2'(\xi) + b'\left(\psi_0(\xi)\right)(\psi_0'(\xi)\psi_2(\xi)\\ &\quad + 2\psi_1(\xi)\psi_1'(\xi)) + b''\left(\psi_0(\xi)\right)\psi_0'(\xi)\psi_1^2(\xi)\\ &\quad + c''\left(\psi_0(\xi)\right)\psi_1^2(\xi) + c'\left(\psi_0(\xi)\right)\psi_2(\xi).\end{aligned}$$

To find the value of fourth unknown coefficient $\psi_4(\xi)$, consider $\mathfrak{L}(\xi,\varsigma,\varrho,4)$ and solving

$$\mathfrak{T}_{\varsigma}^{3\varrho}\left\{\mathfrak{L}(\xi,\varsigma,\varrho,4)\right\}\downarrow_{\varsigma=0}=0.$$

and so on.

In general, to find the value of lth unknown coefficient $\psi_l(\xi)$, we solve

$$\mathfrak{T}_{\varsigma}^{(l-1)\varrho}\left\{\mathfrak{L}(\xi,\varsigma,\varrho,l)\right\}\downarrow_{\varsigma=0}=0.$$

where

$$\mathfrak{L}(\xi,\varsigma,\varrho,l)=\sum_{k=0}^{l-1}\psi_{k+1}(\xi)\frac{\varsigma^{k\varrho}}{\varrho^k k!}-\left(a\left(\sum_{k=0}^{l}\psi_k(\xi)\frac{\varsigma^{k\varrho}}{\varrho^k k!}\right)\left(\sum_{k=0}^{l}\psi_k'(\xi)\frac{\varsigma^{k\varrho}}{\varrho^k k!}\right)\right)_\xi$$
$$-b\left(\sum_{k=0}^{l}\psi_k(\xi)\frac{\varsigma^{k\varrho}}{\varrho^k k!}\right)\left(\sum_{k=0}^{l}\psi_k'(\xi)\frac{\varsigma^{k\varrho}}{\varrho^k k!}\right)-c\left(\sum_{k=0}^{l}\psi_k(\xi)\frac{\varsigma^{k\varrho}}{\varrho^k k!}\right).$$

Finally, the CFTS-solution $\psi(\xi,\varsigma)$ of Eqs. (1)–(2) can be expressed in an infinite expansion as follows:

$$\begin{aligned}\psi(\xi,\varsigma)&=\lim_{n\to\infty}\psi_n(\xi,\varsigma)\\&=\lim_{n\to\infty}\sum_{k=0}^{n}\psi_k(\xi)\frac{\varsigma^{k\varrho}}{\varrho^k k!}\\&=\sum_{k=0}^{\infty}\psi_k(\xi)\frac{\varsigma^{k\varrho}}{\varrho^k k!}.\end{aligned}$$

Theorem 4 *Let $\psi_n(\xi,\varsigma)$ and $\psi(\xi,\varsigma)$ be respectively the CFTS-approximate and exact solutions of Eq. (1) under condition (2). If there exists $\varpi\in(0,1)$ so that*

$$\|\psi_{n+1}(\xi,\varsigma)\|\le\varpi\,\|\psi_n(\xi,\varsigma)\|\,,\forall(\xi,\varsigma)\in\mathbb{R}\times\mathbb{R}^+,\tag{6}$$

with

$$\|\psi_0(\xi)\|<\infty,$$

for all $\xi\in\mathbb{R}$.

Then, $\psi_n(\xi,\varsigma)$ converges to $\psi(\xi,\varsigma)$ as soon as $n\to\infty$.

Proof From inequality (6), we have

$$\begin{aligned}\|\psi_{n+1}(\xi,\varsigma)\|\le&\varpi\,\|\psi_n(\xi,\varsigma)\|\le\varpi^2\,\|\psi_{n-1}(\xi,\varsigma)\|\le\varpi^3\,\|\psi_{n-2}(\xi,\varsigma)\|\\\le&\cdots\le\varpi^{n+1}\,\|\psi_0(\xi)\|\,.\end{aligned}$$

Therefore, we obtain

$$\sum_{m=n+1}^{\infty}\|\psi_m(\xi,\varsigma)\|\le\sum_{m=n+1}^{\infty}\varpi^m\,\|\psi_0(\xi)\|=\|\psi_0(\xi)\|\sum_{m=n+1}^{\infty}\varpi^m.$$

And so, we have

$$\begin{aligned}\|\psi(\xi,\varsigma)-\psi_n(\xi,\varsigma)\| &= \left\|\sum_{m=n+1}^{\infty}\psi_m(\xi,\varsigma)\right\| \\ &\le \sum_{m=n+1}^{\infty}\|\psi_m(\xi,\varsigma)\| \\ &\le \|\psi_0(\xi)\|\sum_{m=n+1}^{\infty}\varpi^m \\ &= \frac{\varpi^{n+1}}{1-\varpi}\|\psi_0(\xi)\| \to 0,\end{aligned}$$

when $n\to\infty$.

4 Numerical Examples

This section examines the validity and accuracy of the CFTSA through three different numerical examples.

Example 1 Consider the following temporal fractional reaction–diffusion–convection equation

$$\mathfrak{T}_{\varsigma}^{\varrho}\psi = \psi_{\xi\xi} + \psi\psi_{\xi} + \psi - \psi^2, 0 < \varrho \le 1, \tag{7}$$

under

$$\psi(\xi,0) = \psi_0(\xi) = 1 + e^{\xi}. \tag{8}$$

Here $\psi = \psi(\xi,\varsigma)$ for all $(\xi,\varsigma)\in\mathbb{R}\times\mathbb{R}^+$.

According to the methodology of the CFTSA presented in Sect. 3, the nth-CFTS approximate solution of Eqs. (7)–(8) is given by

$$\psi_n(\xi,\varsigma) = \psi_0(\xi) + \sum_{k=1}^{n}\psi_k(\xi)\frac{\varsigma^{k\varrho}}{\varrho^k k!}, \tag{9}$$

and the value of unknown coefficients $\psi_k(\xi)$ are

$$\psi_k(\xi) = e^{\xi},$$

for all $k = 1, 2, 3, \ldots$

Hence, Eq. (9) can be written as follows:

$$\begin{aligned}\psi_n(\xi,\varsigma) &= 1+e^{\xi}+e^{\xi}\frac{\varsigma^{\varrho}}{\varrho}+e^{\xi}\frac{\varsigma^{2\varrho}}{\varrho^2 2!}+e^{\xi}\frac{\varsigma^{3\varrho}}{\varrho^3 3!}\\ &\quad +e^{\xi}\frac{\varsigma^{4\varrho}}{\varrho^4 4!}+e^{\xi}\frac{\varsigma^{5\varrho}}{\varrho^5 5!}+\cdots+e^{\xi}\frac{\varsigma^{n\varrho}}{\varrho^n n!}\\ &= 1+e^{\xi}\left(1+\frac{\varsigma^{\varrho}}{\varrho}+\frac{\varsigma^{2\varrho}}{\varrho^2 2!}+\frac{\varsigma^{3\varrho}}{\varrho^3 3!}+\frac{\varsigma^{4\varrho}}{\varrho^4 4!}+\frac{\varsigma^{5\varrho}}{\varrho^5 5!}+\cdots+\frac{\varsigma^{n\varrho}}{\varrho^n n!}\right)\\ &= 1+\sum_{k=0}^{n} e^{\xi}\frac{\varsigma^{k\varrho}}{\varrho^k k!}.\end{aligned}$$

Therefore, the CFTS-solution $\psi(\xi,\varsigma)$ of Eqs. (7)–(8) can be expressed in an infinite expansion as follows:

$$\psi(\xi,\varsigma) = \lim_{n\to\infty}\psi_n(\xi,\varsigma) = 1+\sum_{k=0}^{\infty} e^{\xi}\frac{\varsigma^{k\varrho}}{\varrho^k k!}.$$

In particular at $\varrho = 1$, the CFTS-solution of Eqs. (7)–(8) becomes

$$\psi(\xi,\varsigma) = 1+e^{\xi+\varsigma},$$

which is fully compatible with the result obtained by Aboodh variational iteration method (AVIM) [13].

Example 2 Consider the following temporal fractional reaction–diffusion–convection equation

$$\mathfrak{T}_{\varsigma}^{\varrho}\psi = \left(\psi\psi_{\xi}\right)_{\xi} + 3\psi\psi_{\xi} + 2(\psi-\psi^2), 0<\varrho\le 1, \tag{10}$$

under

$$\psi(\xi,0) = \psi_0(\xi) = 2\sqrt{e^{\xi}-e^{-4\xi}}. \tag{11}$$

Here $\psi = \psi(\xi,\varsigma)$ for all $(\xi,\varsigma)\in\mathbb{R}\times\mathbb{R}^{+}$.

According to the methodology of the CFTSA presented in Sect. 3, the nth-CFTS approximate solution of Eqs. (10)–(11) is given by

$$\psi_n(\xi,\varsigma) = \psi_0(\xi)+\sum_{k=1}^{n}\psi_k(\xi)\frac{\varsigma^{k\varrho}}{\varrho^k k!}, \tag{12}$$

and the value of unknown coefficients $\psi_k(\xi)$ are

$$\psi_k(\xi) = 2^{k+1}\sqrt{e^{\xi}-e^{-4\xi}},$$

for all $k = 1, 2, 3, \ldots$

Hence, Eq. (12) can be written as follows:

$$\begin{aligned}\psi_n(\xi,\varsigma) &= 2\sqrt{e^{\xi}-e^{-4\xi}} + 2^2\sqrt{e^{\xi}-e^{-4\xi}}\frac{\varsigma^{\varrho}}{\varrho} + 2^3\sqrt{e^{\xi}-e^{-4\xi}}\frac{\varsigma^{2\varrho}}{\varrho^2 2!}\\ &+ 2^4\sqrt{e^{\xi}-e^{-4\xi}}\frac{\varsigma^{3\varrho}}{\varrho^3 3!} + 2^5\sqrt{e^{\xi}-e^{-4\xi}}\frac{\varsigma^{4\varrho}}{\varrho^4 4!}\\ &+ 2^6\sqrt{e^{\xi}-e^{-4\xi}}\frac{\varsigma^{5\varrho}}{\varrho^5 5!} + \cdots + 2^{n+1}\sqrt{e^{\xi}-e^{-4\xi}}\frac{\varsigma^{n\varrho}}{\varrho^n n!}2\sqrt{e^{\xi}-e^{-4\xi}}\\ &= \left(1+\frac{2\varsigma^{\varrho}}{\varrho}+\frac{(2\varsigma^{\varrho})^2}{\varrho^2 2!}+\frac{(2\varsigma^{\varrho})^3}{\varrho^3 3!}+\frac{(2\varsigma^{\varrho})^4}{\varrho^4 4!}+\frac{(2\varsigma^{\varrho})^5}{\varrho^5 5!}+\cdots+\frac{(2\varsigma^{\varrho})^n}{\varrho^n n!}\right)\\ &= \sum_{k=0}^{n} 2\sqrt{e^{\xi}-e^{-4\xi}}\frac{(2\varsigma^{\varrho})^k}{\varrho^k k!}.\end{aligned}$$

Therefore, the CFTS-solution $\psi(\xi,\varsigma)$ of Eqs. (10)–(11) can be expressed in an infinite expansion as follows:

$$\psi(\xi,\varsigma) = \lim_{n\to\infty}\psi_n(\xi,\varsigma) = \sum_{k=0}^{\infty} 2\sqrt{e^{\xi}-e^{-4\xi}}\frac{(2t^{\alpha})^k}{\alpha^k k!}.$$

In particular at $\varrho = 1$, the CFTS-solution of Eqs. (10)–(11) becomes

$$\psi(\xi,\varsigma) = 2e^{2\varsigma}\sqrt{e^{\xi}-e^{-4\xi}}, \tag{13}$$

which is fully compatible with the result obtained by Aboodh variational iteration method (AVIM) [13].

Example 3 Consider the following temporal fractional reaction–diffusion–convection equation

$$\mathfrak{T}_{\varsigma}^{\varrho}\psi = \psi_{\xi\xi} - \psi_{\xi} + \psi\psi_{\xi} + \psi - \psi^2, 0 < \varrho \le 1, \tag{14}$$

under

$$\psi(\xi,0) = \psi_0(\xi) = e^{\xi}. \tag{15}$$

Here $\psi = \psi(\xi,\varsigma)$ for all $(\xi,\varsigma) \in \mathbb{R}\times\mathbb{R}^+$.

According to the methodology of the CFTSA presented in Sect. 3, the nth-CFTS approximate solution of Eqs. (14)–(15) is given by

$$\psi_n(\xi,\varsigma) = \psi_0(\xi) + \sum_{k=1}^{n}\psi_k(\xi)\frac{\varsigma^{k\varrho}}{\varrho^k k!}, \tag{16}$$

and the value of unknown coefficients $\psi_k(\xi)$ are

$$\psi_k(\xi) = e^{\xi},$$

for all $k = 1, 2, 3, \ldots$

Hence, Eq. (16) can be written as follows:

$$\begin{aligned}
\psi_n(\xi, \varsigma) &= e^{\xi} + e^{\xi}\frac{\varsigma^{\varrho}}{\varrho} + e^{\xi}\frac{\varsigma^{2\varrho}}{\varrho^2 2!} + e^{\xi}\frac{\varsigma^{3\varrho}}{\varrho^3 3!} + e^{\xi}\frac{\varsigma^{4\varrho}}{\varrho^4 4!} + e^{\xi}\frac{\varsigma^{5\varrho}}{\varrho^5 5!} + \cdots + e^{\xi}\frac{\varsigma^{n\varrho}}{\varrho^n n!} \\
&= e^{\xi}\left(1 + \frac{\varsigma^{\varrho}}{\varrho} + \frac{\varsigma^{2\varrho}}{\varrho^2 2!} + \frac{\varsigma^{3\varrho}}{\varrho^3 3!} + \frac{\varsigma^{4\varrho}}{\varrho^4 4!} + \frac{\varsigma^{5\varrho}}{\varrho^5 5!} + \cdots + \frac{\varsigma^{n\varrho}}{\varrho^n n!}\right) \\
&= \sum_{k=0}^{n} e^{\xi}\frac{\varsigma^{k\varrho}}{\varrho^k k!}.
\end{aligned}$$

Therefore, the CFTS-solution $\psi(\xi, \varsigma)$ of Eqs. (14)–(15) can be expressed in an infinite expansion as follows

$$\psi(\xi, \varsigma) = \lim_{n\to\infty} \psi_n(\xi, \varsigma) = \sum_{k=0}^{\infty} e^{\xi}\frac{\varsigma^{k\varrho}}{\varrho^k k!}.$$

In particular at $\varrho = 1$, the CFTS-solution of Eqs. (14)–(15) becomes

$$\psi(\xi, \varsigma) = e^{\xi+\varsigma}, \tag{17}$$

which is fully compatible with the result obtained by Aboodh variational iteration method (AVIM) [13].

5 Results and Discussion

Our results are presented graphically and in the form of tables. Figures 1, 2, 3, 4, 5 and 6 show the performance of the 5th-CFTS approximate and exact solutions of Eqs. (7), (10) and (14), at various values of fractional order of ϱ. In addition, we give the numerical values of the 5th CFTS-approximate solutions and exact solution for

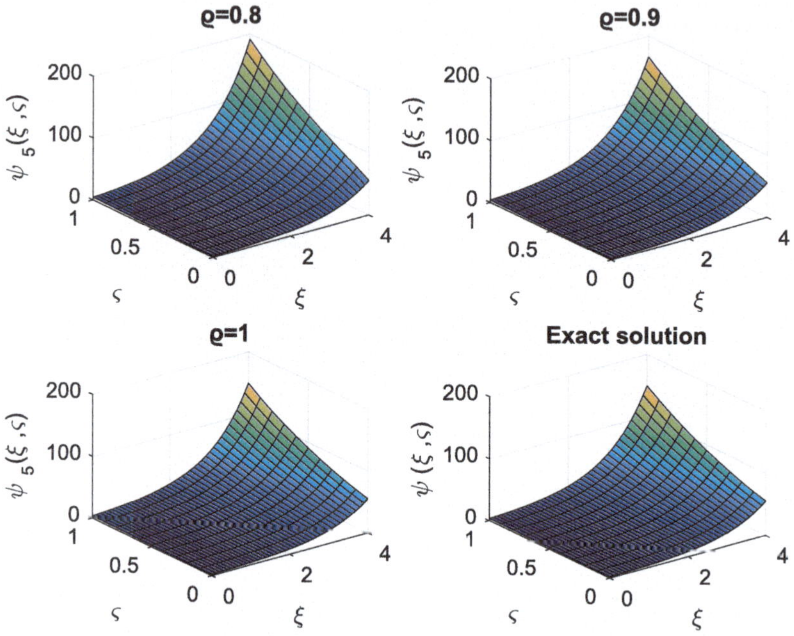

Fig. 1 3D graphs of the 5th CFTS-solution and exact solution for Example 1

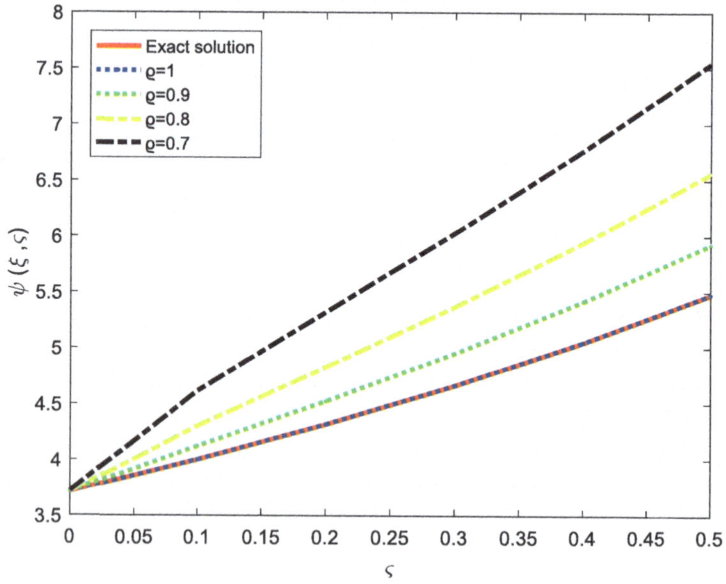

Fig. 2 2D graphs of the 5th CFTS-solution and exact solution for Example 1

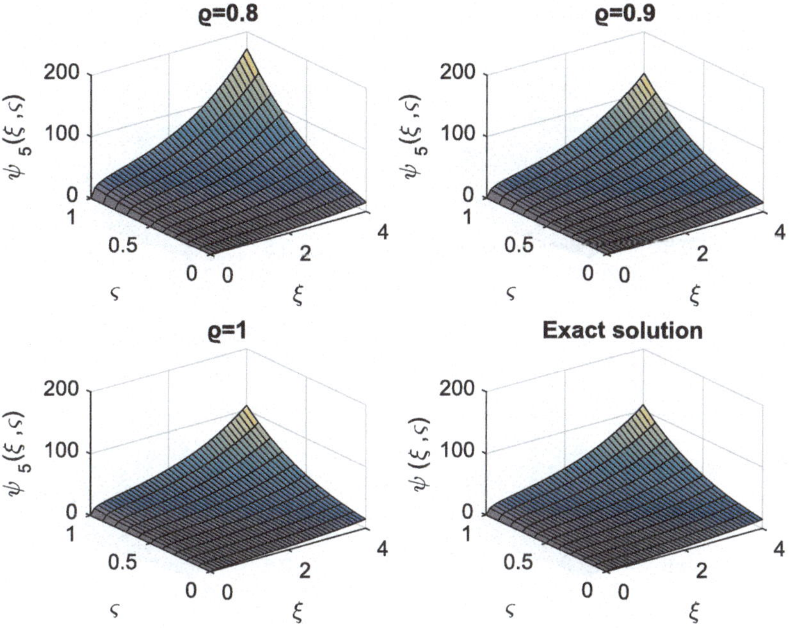

Fig. 3 3D graphs of the 5th CFTS-solution and exact solution for Example 2

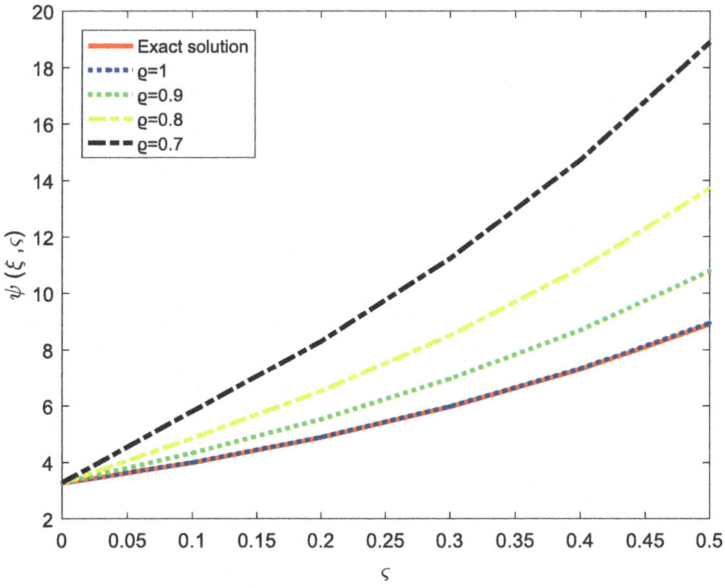

Fig. 4 2D graphs of the 5th CFTS-solution and exact solution for Example 2

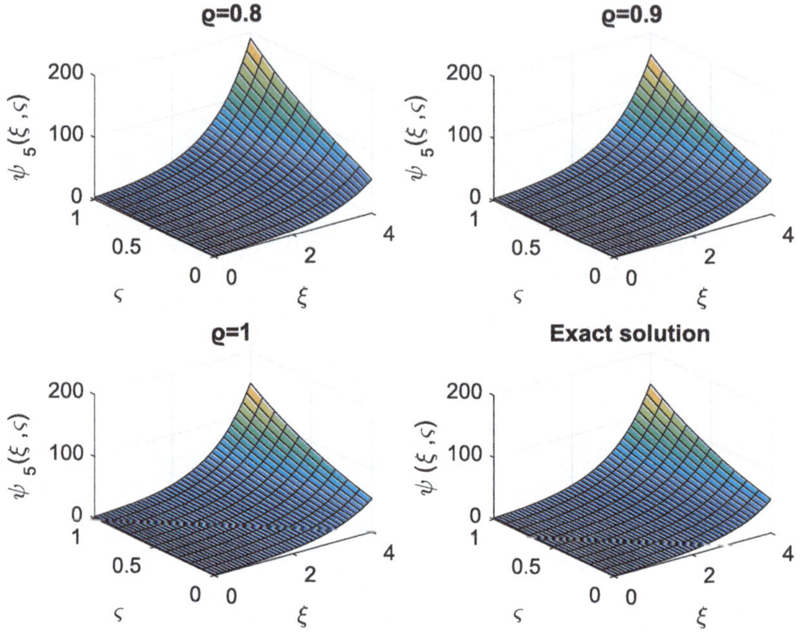

Fig. 5 3D graphs of the 5th CFTS-solution and exact solution for Example 3

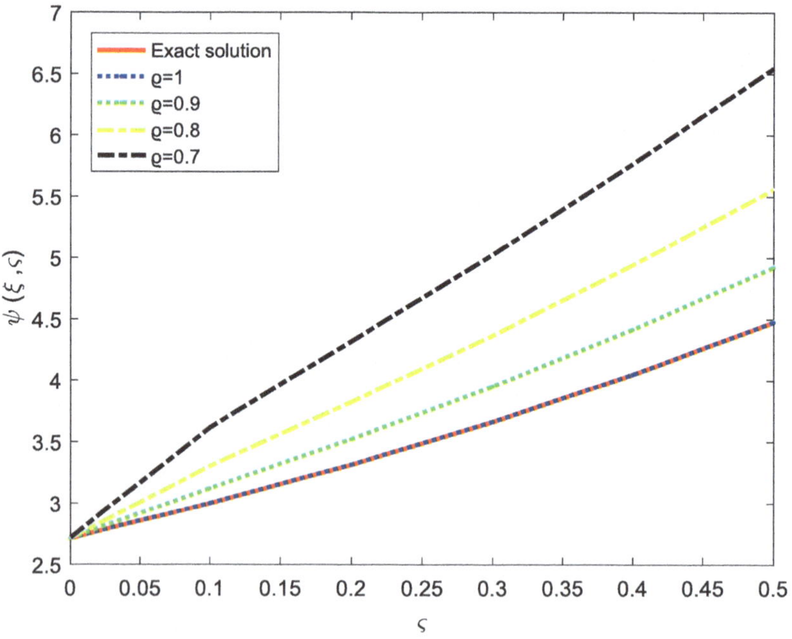

Fig. 6 2D graphs of the 5th CFTS-solution and exact solution for Example 3

Table 1 Numerical values of the 5th CFTS-solution and exact solution for different values of ϱ and ξ for Example 1

ς	ξ	$\varrho = 0.6$	$\varrho = 0.7$	$\varrho = 0.8$	$\varrho = 0.9$	$\varrho = 1$	ψ_{exact}
	0.2	2.3568	2.2929	2.2604	2.2431	2.2337	2.2337
	0.4	2.6572	2.5791	2.5394	2.5183	2.5068	2.5068
0.01	0.6	3.0242	2.9288	2.8802	2.8545	2.8404	2.8404
	0.8	3.4723	3.3558	3.2965	3.2651	3.2479	3.2479
	1	4.0197	3.8774	3.8050	3.7666	3.7456	3.7456

Table 2 Comparison of the proposed method with RPSM and VITM in terms of absolute error for Example 1

ς	ξ	Caputo	Caputo-Fabrizio	Conformable
		$\varrho = 1(RPSM)$	$\varrho = 1(VITM)$	$\varrho = 1(CFTSA)$
	0.2	1.6988×10^{-15}	1.6988×10^{-15}	1.6988×10^{-15}
	0.4	2.0749×10^{-15}	2.0749×10^{-15}	2.0749×10^{-15}
0.01	0.6	2.5343×10^{-15}	2.5343×10^{-15}	2.5343×10^{-15}
	0.8	3.0955×10^{-15}	3.0955×10^{-15}	3.0955×10^{-15}
	1	3.7808×10^{-15}	3.7808×10^{-15}	3.7808×10^{-15}

Table 3 Numerical values of the 5th CFTS-solution and exact solution for different values of ϱ and ξ for Example 2

ς	ξ	$\varrho = 0.6$	$\varrho = 0.7$	$\varrho = 0.8$	$\varrho = 0.9$	$\varrho = 1$	ψ_{exact}
	0.2	2.1159	1.9691	1.8713	1.8204	1.7929	1.7929
	0.4	2.7349	2.5451	2.4187	2.3529	2.3174	2.3174
0.01	0.6	3.1686	2.9487	2.8022	2.7260	2.6848	2.6848
	0.8	3.5593	3.3123	3.1478	3.0622	3.0159	3.0159
	1	3.9568	3.6822	3.4993	3.4041	3.3527	3.3527

different values of ξ and fractional order of ϱ at $\varsigma = 0.01$ in Tables 1, 3 and 6. The absolute error comparison of the CFTSA with RPSM [20] and VITM [14] at $\varrho = 1$ is shown in Tables 2, 4 and 5. It is clearly observed from the figures and tables that the results in instances of fractional values of ϱ converge to the results in case of $\varrho = 1$. Moreover, the CFTS-approximate solutions are consistent with the exact solutions at $\varrho = 1$ and this confirms the efficiency and accuracy of the proposed algorithm.

Table 4 Comparison of the proposed method with RPSM and VITM in terms of absolute error for Example 2

ς	ξ	Caputo	Caputo-Fabrizio	Conformable
		$\varrho = 1\ (RPSM)$	$\varrho = 1\ (VITM)$	$\varrho = 1\ (CFTSA)$
0.01	0.2	1.5666×10^{-13}	1.5666×10^{-13}	1.5666×10^{-13}
	0.4	2.0249×10^{-13}	2.0249×10^{-13}	2.0249×10^{-13}
	0.6	2.3459×10^{-13}	2.3459×10^{-13}	2.3459×10^{-13}
	0.8	2.6353×10^{-13}	2.6353×10^{-13}	2.6353×10^{-13}
	1	2.9295×10^{-13}	2.9295×10^{-13}	2.9295×10^{-13}

Table 5 Comparison of the proposed method with RPSM and VITM in terms of absolute error for Example 3

ς	ξ	Caputo	Caputo-Fabrizio	Conformable
		$\varrho = 1\ (RPSM)$	$\varrho = 1\ (VITM)$	$\varrho = 1\ (CFTSA)$
0.01	0.2	1.6988×10^{-15}	1.6988×10^{-15}	1.6988×10^{-15}
	0.4	2.0749×10^{-15}	2.0749×10^{-15}	2.0749×10^{-15}
	0.6	2.5343×10^{-15}	2.5343×10^{-15}	2.5343×10^{-15}
	0.8	3.0955×10^{-15}	3.0955×10^{-15}	3.0955×10^{-15}
	1	3.7808×10^{-15}	3.7808×10^{-15}	3.7808×10^{-15}

Table 6 Numerical values of the 5th CFTS-solution and exact solution for different values of ϱ and ξ for Example 3

ς	ξ	$\varrho = 0.6$	$\varrho = 0.7$	$\varrho = 0.8$	$\varrho = 0.9$	$\varrho = 1$	ψ_{exact}
0.01	0.2	1.3055	1.2720	1.2527	1.2410	1.2337	1.2337
	0.4	1.5945	1.5537	1.5300	1.5157	1.5068	1.5068
	0.6	1.9475	1.8977	1.8688	1.8513	1.8404	1.8404
	0.8	2.3787	2.3178	2.2826	2.2612	2.2479	2.2479
	1	2.9054	2.8310	2.7879	2.7618	2.7456	2.7456

Remark 1 We define $E_n(t)$ as the absolute error between the exact solution $\psi(\xi, \varsigma)$ and the nth-CFTS approximate solution $\psi_n(\xi, \varsigma)$ as follows:

$$E_n(\xi, \varsigma) = |\psi(\xi, \varsigma) - \psi_n(\xi, \varsigma)|,$$

for all $n = 0, 1, 2, \ldots$

6 Conclusions

In this chapter, a new algorithm called conformable fractional Taylor series algorithm (CFTSA) is introduced to find a solution to temporal fractional reaction–diffusion–convection equation. We took three different numerical examples of temporal fractional reaction–diffusion–convection equations to confirm the validity of the CFTSA. The current analysis shows the effectiveness of the studied algorithm. We can conclude from the obtained results that it is a very reliable, simple, and powerful algorithm for finding approximate and analytical solutions of nonlinear fractional partial differential equations. Finally, we confirm that in the future, this algorithm will be taken into account in order to deal with other complex fractional models that arise in applied sciences.

References

1. T. Abdeljawad, On conformable fractional calculus. Int. J. Comput. Appl. Math. **279**, 57–66 (2015)
2. O. Acan, O. Firat, Y. Keskin, Conformable variational iteration method, conformable fractional reduced differential transform method and conformable homotopy analysis method for non-linear fractional partial differential equations. Waves Random Complex Media. **8**, 1–19 (2018)
3. Z. Al-Zhour, N. Al-Mutairi, F. Alrawajeh, R. Alkhasawneh, New theoretical results and applications on conformable fractional Natural transform. Ain Shams Eng. J. **12**, 927–933 (2021)
4. A. Atangana, D. Baleanu, New fractional derivatives with non-local and non-singular kernel: theory and application to heat transfer model. Thermal Sci. **20**(2), 763–769 (2016)
5. A. Atangana, D. Baleanu, A. Alsaedi, New properties of conformable derivative. Open Math. **13**, 889–898 (2015)
6. M.E. Benattia, K. Belghaba, Shehu conformable fractional transform, theories and applications. J. Sci. Eng. **12**(1), 024–032 (2021)
7. M.E. Benattia, K. Belghaba, Solution of Newel-Whitehead-Segel equation using conformable fractional Sumudu decomposition method. J. Sci. Arts. **21**(2), 479–486 (2021)
8. P.L. Butzer, U. Westphal, An introduction to fractional calculus. Appl. Fract. Calc. Phys. 1–85 (2000)
9. M. Caputo, M. Fabrizio, A new definition of fractional derivative without singular kernel. Progr. Fract. Differ. Appl. **1**(2), 73–85 (2015)
10. R. Hilfer, *Applications of Fractional Calculus in Physics* (World Scientific, Singapore, 2000)
11. N. Kadkhoda, H. Jafari, An analytical approach to obtain exact solutions of some space-time conformable fractional differential equations. Adv. Differ. Equ. **2019**(428) (2019). https://doi.org/10.1186/s13662-019-2349-0
12. R. Khalil, M. Al Horani, A. Yousef, M. Sababheh, A new definition of fractional derivative. Int. J. Comput. Appl. Math. **264**, 65–70 (2014)
13. A. Khalouta, Exact solution of nonlinear time-fractional reaction-diffusion-convection equation via a new coupling method. Int. J. Nonlinear Anal. Appl. **13**(2), 333–344 (2022)
14. A. Khalouta, On the solutions of nonlinear Caputo-Fabrizio fractional partial differential equations arising in applied mathematics. J. Prime Res. Math. **18**(2), 42–54 (2022)
15. A.A. Kilbas, H.M. Srivastava, J.J. Trujillo, *Theory and Application of Fractional Differential Equations* (Elsevier, Amsterdam, 2006)
16. L. Ma, Ch. Li, On Hadamard fractional calculus. Fractals **25**(3), 1750033 (2017)

17. M.I. Liaqat, A. Khan, A. Akgül, Adaptation on power series method with conformable operator for solving fractional order systems of nonlinear partial differential equations. Chaos Solitons Fractals. **157**, 111984 (2022)
18. M. Molaei, F.D. Saei, M. Javidi, Y. Mahmoudi, New analytical methods for solving a class of conformable fractional differential equations by fractional Laplace transform. Comput. Methods Differ. Equ. **10**(2), 396–407 (2022)
19. S.S. Omorodion, Conformable fractional reduced differential transform method for solving linear and nonlinear time-fractional Swift-Hohenberg (S-H) equation. Int. J. Sci. Res. Math. Stat. Sci. **8**(6), 20–29 (2021)
20. A. Shidfar, A. Babaei, A. Molabahrami, M. Alinejadmofradi, Approximate analytical solutions of nonlinear reaction-diffusion-convection problem. Math. Comput. Model. **53**, 261–268 (2011)
21. A. Zulfiqar, J. Ahmad, Comparative study of two techniques on some nonlinear problems based using conformable derivative. Nonlinear Eng. **9**, 470–482 (2020)

Modified Some n-Point Fractional Formulae with Richardson Extrapolation

Iqbal M. Batiha, Iqbal Jebril, and Shameseddin Alshorm

Abstract This chapter intends to introduce some new n-point fractional formulas for finding an approximation for Caputo fractional differentiator. These formulas are regarded as generalizations of some n-point classical formulas for finding good approximations for the first and second classical derivatives. This would enable us to study some functions behaviors in accordance with various fractional-order values. For the purpose of achieving a high degree of precision for the acquired approaches, we employ the so-called Richardson extrapolation approach, furthermore, to demonstrate the dependability and validity of our approximations, we provide several numerical simulations via tables and graphs.

1 Introduction

Isaac Newton is indeed known for his pioneering work on differential equations during the fourth quarter of the seventeenth century. However, it is worth noting that many of the techniques used in difference equations had already been developed by earlier mathematicians. Thomas Harriot, an English mathematician, astronomer, and navigator, lived from 1561 to 1621. He produced substantial contributions to the evolution of algebra and was known for his work on finite differences. Harriot's work on finite differences laid the foundation for the study of difference equations, which are equations that relate the time or space to the function's values at various discrete nodes. Harriot's approach to solving problems using finite differences was innovative and paved the way for the later development of difference equations by Newton and other mathematicians. Henry Briggs, also an English mathematician, lived from 1561 to 1630, and he is primarily known for his work on logarithms. Briggs collaborated with John Napier, a Scottish mathematician, in the development of logarithms, which are mathematical tools used to simplify complex calculations

I. M. Batiha (✉) · I. Jebril · S. Alshorm
Department of Mathematics, Al Zaytoonah University of Jordan, Amman 11733, Jordan
e-mail: i.batiha@zuj.edu.jo

I. M. Batiha
Nonlinear Dynamics Research Center (NDRC), Ajman University, Ajman, UAE

M. Kharrat et al. (eds.), *Modeling of Discrete and Continuous Systems*, Forum for Interdisciplinary Mathematics, https://doi.org/10.1007/978-981-97-8715-9_7

involving multiplication and division. Briggs extended Napier's work by introducing what is known as "Briggsian logarithms," which are based on a logarithmic scale with base 10. Briggs also developed methods for approximating logarithmic and trigonometric functions, which were later used in solving several difference and differential problems, including those involving non-integer order derivatives, see the references [1–4] for more information.

The concept of fractional dynamical systems, also known as fractional-order systems, refers to systems that are described by differential equations with non-integer order derivatives [5–8]. While Newton's work focused on difference equations, the study of fractional-order systems originated later and might be dated back to several mathematicians and physicists works in the eighteenth and nineteenth centuries, including Augustin-Louis Cauchy, Joseph Liouville, and Victor Grünwald. However, it was not until the twentieth century that the study of fractional calculus and fractional-order systems gained significant attention and became an important area of research in applied mathematics, physics, engineering, and other fields see [9–12]. Caputo and Riemann-Liouville are two common types of fractional operators used in fractional calculus, a discipline of mathematics which extends the concept of integration and differentiation to the fractional orders. Caputo differentiator is a type of fractional derivative that is commonly used in many applications [13–15]. It is defined as the convolution of the ordinary derivative through a fractional-order kernel and is particularly suitable for initial value problems [16, 17]. The Caputo derivative is widely used in areas such as fractional differential equations, fractional control systems, and fractional calculus of variations. It has the advantage of being well-defined for functions that are not necessarily differentiable at the initial point, making it particularly useful in modeling real-world systems with memory or history-dependent behaviors [18–21]. The Riemann-Liouville integrator and differentiator are another type of fractional-order operators that are widely used in fractional calculus. The Riemann-Liouville integrator is defined as the convolution of a function with a fractional-order kernel and is a general form of the ordinary integral to its fractional-order case [22–25]. The Riemann-Liouville differentiator is determined as the convolution of a function with a fractional-order kernel and is a general form of the ordinary derivative to its fractional-order case. The Riemann-Liouville operators are particularly suitable for boundary value problems and have applications in areas such as fractional differential equations, fractional control systems, and time-series analysis see [26–29].

2 Preliminaries

This part aims to remember essential notions associated with fractional calculus comprising Caputo differentiator, Riemann-Liouville integrator, and additional related theories.

Definition 1 The Riemann-Liouville integrator of a function $f(t)$ of order $\mu > 0$ is determined by

$$J^{\mu} f(t) = \frac{1}{\Gamma(\mu)} \int_0^t f(\tau)(t-\tau)^{\mu-1}\, d\tau, \quad t > 0,\ \mu > 0. \tag{1}$$

In the lines below, we remember certain characteristics connected with the Riemann-Liouville integrator:

$$J^0 f(t) = f(t). \tag{2}$$

$$J^{\mu}(t-a)^{\gamma} = \frac{\Gamma(\gamma+1)}{\Gamma(\mu+\gamma+1)}(t-a)^{\mu+\gamma}, \quad \gamma \geq -, a \in \mathbb{R}. \tag{3}$$

$$J^{\mu} J^{\beta} f(t) = J^{\beta} J^{\mu} f(t), \quad \mu, \beta \geq 0. \tag{4}$$

$$J^{\mu} J^{\beta} f(t) = J^{\mu+\beta} f(t), \quad \mu, \beta \geq 0. \tag{5}$$

Definition 2 The Caputo differentiator of the function $f(t)$ of order $\alpha > 0$ is determined by

$$D_*^{\alpha} f(t) = \begin{cases} \frac{1}{\Gamma(n-\alpha)} \int_0^t (t-\tau)^{-\alpha+n-1} \frac{d^m f(\tau)}{d\tau^m}\, d\tau, & m-1 < \alpha < m, \\ \frac{d^m f(t)}{dt^m}, & \alpha = m, \end{cases} \tag{6}$$

where $m = \lceil \alpha \rceil$ and $t > 0$.

In the lines below, we remember certain characteristics connected with the Caputo differentiator:

1. For the two constants μ and ω, D_*^{α} is a linear operator, i.e.,

$$D_*^{\alpha}(\mu f(t) + \omega k(t)) = \mu D_*^{\alpha}(f(t)) + \omega D_*^{\alpha}(k(t)).$$

2. We have

$$D_*^{\alpha}(t-a)^{\rho} = \begin{cases} \frac{\Gamma(\rho+1)}{\Gamma(\rho-\alpha+1)}(t-a)^{\rho-\alpha}, & \rho > \alpha - 1, \\ 0, & otherwise \end{cases},$$

   where $a \in \mathbb{R}$.
3. $D_*^{\alpha} c = 0$, where c is constant.
4. If $m - 1 < \alpha \leq m$ such that $m \in \mathbb{N}$, then we have

$$J^{\alpha} D_*^{\alpha} f(t) = f(t) - \sum_{i=1}^{n} f^i(0^+)\frac{t^i}{i!},$$
$$D_*^{\alpha} J^{\alpha} f(t) = f(t), \ t > 0. \tag{7}$$

Definition 3 The Caputo differentiator might be expressed in relation to the Riemann-Liouville integrator as follows:

$$D_*^{\alpha} f = J^{m-\alpha} D^m f, \tag{8}$$

where $m = \lceil \alpha \rceil$ in which $\alpha \in \mathbb{R}^+$.

Definition 4 [30] Through the two points (x_0, y_0) and (x_1, y_1), the linear Lagrange interpolating polynomial can be determined by

$$P(x) = L_0(x) f(x_0) + L_1(X) f(x_1) = \frac{x - x_1}{x_0 - x_1} f(x_0) + \frac{x - x_0}{x_1 - x_0} f(x_1).$$

where

$$L_0(x) = \frac{x - x_1}{x_0 - x_1} \quad and \quad L_1(x) = \frac{x - x_0}{x_1 - x_0}.$$

Theorem 1 [30] *Consider $f \in C^{n+1}[a, b]$ and $x_0, x_1, \ldots, x_n$ are $(n+1)$ distinct numbers in some interval $I = [a, b]$. The function $f(x)$ might be describe by*

$$f(x) = \sum_{k=0}^{n} f(x_k) L_k(x) + \frac{(x - x_0) \cdots (x - x_n)}{(n+1)!} f^{(n+1)}(\xi(x)), \tag{9}$$

for an unknown $\xi \in (a, b)$.

3 Fractional Central Formulas

This section presents some modified n-point fractional central formulas for approximating Caputo fractional derivative operator. This would enable one to study the behavior of a function in relation to various numerical or fractional values of α.

3.1 Modified 3-Point Fractional Formula

This part presents a modified version of the classical 3-point central formula called the modified 3-point fractional central formula for finding an approximation for Caputo differentiator.

Theorem 2 *Consider $f \in C^3[a, b]$ and x_0, x_1, x_2 are distinct points in the interval $[a, b]$ in which $a = x_0 < x_1 = x_0 + h < x_2 = x_0 + 2h = b$ with $h > 0$. The modified 3-point fractional formula for finding an approximation for Caputo differentiator is defined by*

$$
\begin{aligned}
D_*^{\alpha} f(x) &= \frac{x^{2-\alpha}}{h^2\Gamma(3-\alpha)}\Big(f(x_0) - 2f(x_1) + f(x_2)\Big) \\
&- \frac{x^{1-\alpha}}{2h^2\Gamma(2-\alpha)}\Big(f(x_0)(x_1+x_2) - 2f(x_1)(x_0+x_2) + f(x_2)(x_0+x_1)\Big) \\
&+ \frac{f^{(3)}(\xi)}{6}\left(\frac{6}{\Gamma(4-\alpha)}x^{3-\alpha} - \frac{2(x_0+x_1+x_2)}{\Gamma(3-\alpha)}x^{2-\alpha} + \frac{(x_0x_1+x_0x_2+x_1x_2)}{\Gamma(2-\alpha)}x^{1-\alpha}\right)
\end{aligned}
\tag{10}
$$

where $\xi \in (a, b)$ *and* $x \in [a, b]$.

Proof For $n = 2$ in Eq. (9), we obtain

$$
f(x) = \sum_{k=0}^{2} f(x_k)L_k(x) + \frac{(x-x_0)(x-x_1)(x-x_2)}{3!} f^{(3)}(\xi(x)).
$$

By simplifying the summation, we can have

$$
\begin{aligned}
f(x) &= f(x_0)L_0(x) + f(x_1)L_1(x) + f(x_2)L_2(x) \\
&+ \frac{1}{6} f^{(3)}(\xi(x))(x-x_0)(x-x_1)(x-x_2).
\end{aligned}
$$

By using Definition 4, we can get

$$
\begin{aligned}
f(x) &= f(x_0)\frac{(x-x_1)(x-x_2)}{(x_0-x_1)(x_0-x_2)} + f(x_1)\frac{(x-x_0)(x-x_2)}{(x_1-x_0)(x_1-x_2)} \\
&+ f(x_2)\frac{(x-x_0)(x-x_1)}{(x_2-x_0)(x_2-x_1)} + \frac{1}{6} f^{(3)}(\xi(x))(x-x_0)(x-x_1)(x-x_2).
\end{aligned}
\tag{11}
$$

Pulling out the common factors gives

$$
\begin{aligned}
f(x) &= \frac{f(x_0)}{(x_0-x_1)(x_0-x_2)}(x^2 - (x_1+x_2)x + x_1x_2) \\
&+ \frac{f(x_1)}{(x_1-x_0)(x_1-x_2)}(x^2 - (x_0+x_2)x + x_0x_2) \\
&+ \frac{f(x_2)}{(x_2-x_0)(x_2-x_1)}(x^2 - (x_0+x_1)x + x_0x_1) \\
&+ \frac{f^{(3)}(\xi(x))}{6}(x-x_0)(x-x_1)(x-x_2).
\end{aligned}
$$

By using $x_1 = x_0 + h$ and $x_2 = x_0 + 2h$, one might express $f(x)$ in the form

$$
\begin{aligned}
f(x) = {} & \frac{f(x_0)}{(-h)(-2h)}(x^2 - (x_1 + x_2)x + x_1x_2) \\
& + \frac{f(x_1)}{(h)(-h)}(x^2 - (x_0 + x_2)x + x_0x_2) \\
& + \frac{f(x_2)}{(h)(2h)}(x^2 - (x_0 + x_1)x + x_0x_1) \\
& + \frac{f^{(3)}(\xi(x))}{6}(x^3 - (x_0 + x_1 + x_2)x^2 \\
& + (x_0x_1 + x_0x_2 + x_1x_2)x - x_0x_1x_2).
\end{aligned}
$$

It is feasible to establish the Caputo fractional derivative given in (6) to be as

$$
\begin{aligned}
D_*^{\alpha} f(x) = {} & \frac{f(x_0)}{2\,h^2}\left[\frac{2x^{2-\alpha}}{\Gamma(-\alpha+3)} - \frac{(x_1+x_2)x^{1-\alpha}}{\Gamma(-\alpha+2)}\right] \\
& + \frac{f(x_1)}{h^2}\left[\frac{2x^{2-\alpha}}{\Gamma(-\alpha+3)} - \frac{(x_0+x_2)x^{1-\alpha}}{\Gamma(-\alpha+2)}\right] \\
& + \frac{f(x_2)}{2\,h^2}\left[\frac{2x^{2-\alpha}}{\Gamma(-\alpha+3)} - \frac{(x_0+x_1)x^{1-\alpha}}{\Gamma(-\alpha+2)}\right] \\
& + \frac{f^{(3)}(\xi)}{\sigma}\left[\frac{6x^{3-\alpha}}{\Gamma(-\alpha+4)} - \frac{(x_0+x_1+x_2)2x^{2-\alpha}}{\Gamma(-\alpha+3)} + \frac{(x_0x_1+x_2x_1+x_0x_2)x^{1-\alpha}}{\Gamma(-\alpha+2)}\right].
\end{aligned}
$$

Simplifying the above formula gives

$$
\begin{aligned}
D_*^{\alpha} f(x) = {} & \frac{f(x_0)}{2h^2}\left[\frac{2}{\Gamma(3-\alpha)}x^{2-\alpha} - \frac{(x_1+x_2)}{\Gamma(2-\alpha)}x^{1-\alpha}\right] \\
& - \frac{f(x_1)}{h^2}\left[\frac{2}{\Gamma(3-\alpha)}x^{2-\alpha} - \frac{(x_0+x_2)}{\Gamma(2-\alpha)}x^{1-\alpha}\right] \\
& + \frac{f(x_2)}{2h^2}\left[\frac{2x^{2-\alpha}}{\Gamma(3-\alpha)} - \frac{(x_0+x_1)}{\Gamma(2-\alpha)}x^{1-\alpha}\right] \\
& + \frac{f^{(3)}(\xi)}{6}\left[\frac{6}{\Gamma(4-\alpha)}x^{3-\alpha} - \frac{2(x_0+x_1+x_2)}{\Gamma(3-\alpha)}x^{2-\alpha} + \frac{(x_0x_1+x_0x_2+x_1x_2}{\Gamma(2-\alpha)}x^{1-\alpha}\right].
\end{aligned}
$$

Therefore, the above equation may be written as follows:

$$
\begin{aligned}
D_*^\alpha f(x) &= \frac{(2x^{2-\alpha})}{(2h^2)\Gamma(3-\alpha)} f(x_0) - \frac{(2x^{2-\alpha})}{h^2\Gamma(3-\alpha)} f(x_1) \\
&+ \frac{(2x^{2-\alpha})}{2h^2\Gamma(3-\alpha)} f(x_2) - \frac{((x_1+x_2)x^{1-\alpha})}{2h^2\Gamma(2-\alpha)} f(x_0)) \\
&+ \frac{((x_0+x_2)x^{1-\alpha})}{h^2\Gamma(2-\alpha)} f(x_1) - \frac{((x_0+x_1)x^{1-\alpha})}{2h^2\Gamma(2-\alpha)} f(x_2) \\
&- \frac{f^{(3)}(\xi)}{6}\left[\frac{6x^{3-\alpha}}{\Gamma(4-\alpha)} - \frac{(x_0+x_1+x_2)(2x^{2-\alpha})}{\Gamma(3-\alpha)} + \frac{(x_0x_1+x_2x_1+x_0x_2)x^{1-\alpha}}{\Gamma(2-\alpha)}\right],
\end{aligned}
$$

which immediately gives the desired result. □

Lemma 1 *Based on Theorem 2, the classical central formula can be derived by setting $x = x_1$ and $\alpha = 1$. In other words, we can have*

$$
f\prime(x_1) = \frac{f(x_2) - f(x_0)}{2h} - \frac{h^2}{6} f^{(3)}(\xi),
$$

for $\xi \in (a, b)$.

Remark 1 If one substitutes $x = x_0$ in Theorem 2, we can derive the traditional formulation of the forward formula, which is similar to the approach used in Lemma 1.

Remark 2 If one substitutes $x = x_2$ in Theorem 2, we can derive the traditional formulation of the backward formula, which is similar to the approach used in Lemma 1.

3.2 *Modified 5-Point Fractional Formula*

This portion presents a modified version of the classical 5-point central formula called the modified 5-point fractional central formula for finding an approximation for Caputo differentiator.

Theorem 3 *Consider $f \in C^5[a, b]$ and x_0, x_1, x_2, x_3, x_4 are distinct points in the interval $[a, b]$ in which $a = x_0 < x_1 = x_0 + h < x_2 = x_0 + 2h < x_3 = x_0 + 3h < x_4 = x_0 + 4h = b$ with $h > 0$. The Caputo differentiator of the function $f(x)$ of order α might be approximated by the following formula:*

$$
\begin{aligned}
D_*^{\alpha} f(x) = & \left[\frac{24}{\Gamma(5-\alpha)} \sum_{k=0}^{4} \frac{f(x_k)}{A_k} x^{4-\alpha}\right] - \left[\frac{6}{\Gamma(4-\alpha)} \sum_{k=0}^{4} \frac{a_k f(x_k)}{A_k} x^{3-\alpha}\right] \\
& + \left[\frac{2}{\Gamma(3-\alpha)} \sum_{k=0}^{4} \frac{b_k f(x_k)}{A_k} x^{2-\alpha}\right] - \left[\frac{1}{\Gamma(2-\alpha)} \sum_{k=0}^{4} \frac{c_k f(x_k)}{A_k} x^{1-\alpha}\right] \\
& + \frac{f^{(5)}(\xi)}{5!}\left(\frac{120 x^{5-\alpha}}{\Gamma(6-\alpha)} - \frac{24 a_5 x^{4-\alpha}}{\Gamma(5-\alpha)} + \frac{6 b_5 x^{3-\alpha}}{\Gamma(4-\alpha)} - \frac{2 c_5 x^{2-\alpha}}{\Gamma(3-\alpha)} + \frac{d_5 x^{1-\alpha}}{\Gamma(2-\alpha)}\right),
\end{aligned}
\tag{12}
$$

where $\xi \in (a, b)$ and $x \in [a, b]$.

Proof By utilizing (9) with supposing that $n = 4$, we obtain

$$
f(x) = \sum_{k=0}^{4} f(x_k) L_k(x) + \frac{(x - x_0)(x - x_1)(x - x_2)(x - x_3)(x - x_4)}{5!} f^{(5)}(\xi(x)). \tag{13}
$$

By simplifying the summation, we get

$$
\begin{aligned}
f(x) = & f(x_0)L_0(x) + f(x_1)L_1 + f(x_2)L_2(x) + f(x_3)L_3(x) + f(x_4)L_4(x) \\
& + \frac{(x - x_0)(x - x_1)(x - x_2)(x - x_3)(x - x_4)}{5!} f^{(5)}(\xi(x)).
\end{aligned}
\tag{14}
$$

With the use of (13), we obtain

$$
\begin{aligned}
f(x) = & f(x_0) \frac{(x - x_1)(x - x_2)(x - x_3)(x - x_4)}{(x_0 - x_1)(x_0 - x_2)(x_0 - x_3)(x_0 - x_4)} \\
& + f(x_1) \frac{(x - x_0)(x - x_2)(x - x_3)(x - x_4)}{(x_1 - x_0)(x_1 - x_2)(x_1 - x_3)(x_1 - x_4)} \\
& + f(x_2) \frac{(x - x_0)(x - x_1)(x - x_3)(x - x_4)}{(x_2 - x_0)(x_2 - x_1)(x_2 - x_3)(x_2 - x_4)} \\
& + f(x_3) \frac{(x - x_0)(x - x_1)(x - x_2)(x - x_4)}{(x_3 - x_0)(x_3 - x_1)(x_3 - x_2)(x_3 - x_4)} \\
& + f(x_4) \frac{(x - x_0)(x - x_1)(x - x_2)(x - x_3)}{(x_4 - x_0)(x_4 - x_1)(x_4 - x_2)(x_4 - x_3)} \\
& + \frac{(x - x_0)(x - x_1)(x - x_2)(x - x_3)(x - x_4)}{5!} f^{(5)}(\xi(x)).
\end{aligned}
\tag{15}
$$

Now, for (15), we take the following assumptions:

(1) For x^3, we assume

$$\begin{aligned} a_0 &= (x_1 + x_2 + x_3 + x_4), \\ a_1 &= (x_0 + x_2 + x_3 + x_4), \\ a_2 &= (x_0 + x_1 + x_3 + x_4), \\ a_3 &= (x_0 + x_1 + x_2 + x_4), \\ a_4 &= (x_0 + x_1 + x_2 + x_3). \end{aligned}$$

(2) For x^2, we assume

$$\begin{aligned} b_0 &= (x_1x_2 + x_1x_3 + x_1x_4 + x_2x_3 + x_2x_4 + x_3x_4), \\ b_1 &= (x_0x_2 + x_0x_3 + x_0x_4 + x_2x_3 + x_2x_4 + x_3x_4), \\ b_2 &= (x_0x_1 + x_0x_3 + x_0x_4 + x_1x_3 + x_1x_4 + x_3x_4), \\ b_3 &= (x_0x_1 + x_0x_2 + x_0x_4 + x_1x_2 + x_1x_4 + x_2x_4), \\ b_4 &= (x_0x_1 + x_0x_2 + x_0x_3 + x_1x_2 + x_1x_3 + x_2x_3). \end{aligned}$$

(3) For x, we assume

$$\begin{aligned} c_0 &= (x_1x_2x_3 + x_1x_2x_4 + x_1x_3x_4 + x_2x_3x_4), \\ c_1 &= (x_0x_2x_3 + x_0x_2x_4 + x_0x_3x_4 + x_2x_3x_4), \\ c_2 &= (x_0x_1x_3 + x_0x_1x_4 + x_0x_3x_4 + x_1x_3x_4), \\ c_3 &= (x_0x_1x_2 + x_0x_1x_4 + x_0x_2x_4 + x_1x_2x_4), \\ c_4 &= (x_0x_1x_2 + x_0x_1x_3 + x_0x_2x_3 + x_1x_2x_3). \end{aligned}$$

(4) For the constants, we assume

$$\begin{aligned} d_0 &= (x_1x_2x_3x_4), \\ d_1 &= (x_0x_2x_3x_4), \\ d_2 &= (x_0x_1x_3x_4), \\ d_3 &= (x_0x_1x_2x_4), \\ d_4 &= (x_0x_1x_2x_3). \end{aligned}$$

(5) For error term, we assume

$$\begin{aligned} a_5 =&(x_0 + x_1 + x_2 + x_3 + x_4), \\ b_5 =&(x_0x_1 + x_0x_2 + x_0x_3 + x_0x_4 + x_1x_2 + x_1x_3 + x_1x_4 + x_2x_3 + x_2x_4 + x_3x_4), \\ c_5 =&(x_0x_1x_2 + x_0x_1x_3 + x_0x_1x_4 + x_0x_2x_3 + x_0x_2x_4 + x_0x_3x_4 + x_1x_2x_3 \\ &+ x_1x_2x_4 + x_1x_3x_4 + x_2x_3x_4), \\ d_5 =&(x_0x_1x_2x_3 + x_0x_1x_2x_4 + x_0x_1x_3x_4 + x_0x_2x_3x_4 + x_1x_2x_3x_4), \\ k =&(x_0x_1x_2x_3x_4). \end{aligned}$$

(6) For the denominator, we assume

$$A_0 = {x_0}^4 - (x_1 + x_2 + x_3 + x_4)x_0^3 + (x_1x_2 + x_1x_3 + x_1x_4 + x_2x_3 + x_2x_4 + x_3x_4)x_0^2 - (x_1x_2x_3 + x_1x_2x_4 + x_1x_3x_4 + x_2x_3x_4)x_0 + (x_1x_2x_3x_4),$$

$$A_1 = {x_1}^4 - (x_0 + x_2 + x_3 + x_4)x_1^3 + (x_0x_2 + x_0x_3 + x_0x_4 + x_2x_3 + x_2x_4 + x_3x_4)x_1^2 - (x_0x_2x_3 + x_0x_2x_4 + x_0x_3x_4 + x_2x_3x_4)x_1 + (x_0x_2x_3x_4),$$
$$A_2 = {x_2}^4 - (x_0 + x_1 + x_3 + x_4)x_2^3 + (x_0x_1 + x_0x_3 + x_0x_4 + x_1x_3 + x_1x_4 + x_3x_4)x_2^2 - (x_0x_1x_3 + x_0x_1x_4 + x_0x_3x_4 + x_1x_3x_4)x_2 + (x_0x_1x_3x_4),$$
$$A_3 = {x_3}^4 - (x_0 + x_1 + x_2 + x_4)x_3^3 + (x_0x_1 + x_0x_2 + x_0x_4 + x_1x_2 + x_1x_4 + x_2x_4)x_3^2 - (x_0x_1x_2 + x_0x_1x_4 + x_0x_2x_4 + x_1x_2x_4)x_3 + (x_0x_1x_2x_4),$$
$$A_4 = {x_4}^4 - (x_0 + x_1 + x_2 + x_3)x_4^3 + (x_0x_1 + x_0x_2 + x_0x_3 + x_1x_2 + x_1x_3 + x_2x_3)x_4^2 - (x_0x_1x_2 + x_0x_1x_3 + x_0x_2x_3 + x_1x_2x_3)x_4 + (x_0x_1x_2x_3).$$

This would immediately yield

$$\begin{aligned} f(x) = {} & f(x_0)\left(\frac{x^4 - a_0x^3 + b_0x^2 - c_0x + d_0}{A_0}\right) \\ & + f(x_1)\left(\frac{x^4 - a_1x^3 + b_1x^2 - c_1x + d_1}{A_1}\right) \\ & + f(x_2)\left(\frac{x^4 - a_2x^3 + b_2x^2 - c_2x + d_2}{A_2}\right) \\ & + f(x_3)\left(\frac{x^4 - a_3x^3 + b_3x^2 - c_3x + d_3}{A_3}\right) \\ & + f(x_4)\left(\frac{x^4 - a_4x^3 + b_4x^2 - c_4x + d_4}{A_4}\right) \\ & + f^{(5)}(\xi(x))\left(\frac{x^5 - a_5x^4 + b_5x^3 - c_5x^2 + d_5x - k}{5!}\right). \end{aligned} \tag{16}$$

After simple simplification of (16), we have

$$
\begin{aligned}
f(x) &= \left[\frac{f(x_0)}{A_0} + \frac{f(x_1)}{A_1} + \frac{f(x_2)}{A_2} + \frac{f(x_3)}{A_3} + \frac{f(x_4)}{A_4}\right]x^4 \\
&- \left[\frac{a_0 f(x_0)}{A_0} + \frac{a_1 f(x_1)}{A_1} + \frac{a_2 f(x_2)}{A_2} + \frac{a_3 f(x_3)}{A_3} + \frac{a_4 f(x_4)}{A_4}\right]x^3 \\
&+ \left[\frac{b_0 f(x_0)}{A_0} + \frac{b_1 f(x_1)}{A_1} + \frac{b_2 f(x_2)}{A_2} + \frac{b_3 f(x_3)}{A_3} + \frac{b_4 f(x_4)}{A_4}\right]x^2 \\
&- \left[\frac{c_0 f(x_0)}{A_0} + \frac{c_1 f(x_1)}{A_1} + \frac{c_2 f(x_2)}{A_2} + \frac{c_3 f(x_3)}{A_3} + \frac{c_4 f(x_4)}{A_4}\right]x \\
&+ \left[\frac{d_0 f(x_0)}{A_0} + \frac{d_1 f(x_1)}{A_1} + \frac{d_2 f(x_2)}{A_2} + \frac{d_3 f(x_3)}{A_3} + \frac{d_4 f(x_4)}{A_4}\right] \\
&+ f^{(5)}(\xi(x))\left[\frac{x^5 - a_5 x^4 + b_5 x^3 - c_5 x^2 + d_5 x - k}{5!}\right],
\end{aligned}
\tag{17}
$$

which consequently yields

$$
\begin{aligned}
f(x) &= \sum_{k=0}^{4} \frac{f(x_k)}{A_k} x^4 - \sum_{k=0}^{4} \frac{a_k f(x_k)}{A_k} x^3 + \sum_{k=0}^{4} \frac{b_k f(x_k)}{A_k} x^2 - \sum_{k=0}^{4} \frac{c_k f(x_k)}{A_k} x \\
&+ \sum_{k=0}^{4} \frac{d_k f(x_k)}{A_k} + f^{(5)}(\xi(x))\left[\frac{x^5 - a_5 x^4 + b_5 x^3 - c_5 x^2 + d_5 x - k}{5!}\right].
\end{aligned}
\tag{18}
$$

Now, operating the Caputo derivative on (18) gives

$$
\begin{aligned}
D_*^{\alpha} f(x) &= \sum_{k=0}^{4} \frac{f(x_k)\Gamma(5)}{A_k \Gamma(5-\alpha)} x^{4-\alpha} - \sum_{k=0}^{4} \frac{a_k f(x_k)\Gamma(4)}{A_k \Gamma(4-\alpha)} x^{3-\alpha} \\
&+ \sum_{k=0}^{4} \frac{b_k f(x_k)\Gamma(3)}{A_k \Gamma(3-\alpha)} x^{2-\alpha} - \sum_{k=0}^{4} \frac{c_k f(x_k)\Gamma(2)}{A_k \Gamma(2-\alpha)} x^{1-\alpha} \\
&+ \frac{f^{(5)}(\xi)}{5!} \times \left[\frac{\Gamma(6)}{\Gamma(6-\alpha)} x^{5-\alpha} - \frac{a_5 \Gamma(5)}{\Gamma(5-\alpha)} x^{4-\alpha} \right. \\
&\left. + \frac{b_5 \Gamma(4)}{\Gamma(4-\alpha)} x^{3-\alpha} - \frac{c_5 \Gamma(3)}{\Gamma(3-\alpha)} x^{2-\alpha} + \frac{d_5 \Gamma(2)}{\Gamma(2-\alpha)} x^{1-\alpha}\right].
\end{aligned}
\tag{19}
$$

Hence, after simplifying (19), we can then obtain the desired result. □

Corollary 1 *By taking similar assumptions provided in Theorem 3, one might produce the modified 5-point fractional formula for finding an approximation for Caputo differentiator, which has the form*

$$
\begin{aligned}
D_*^{\alpha} f(x_1) &= \frac{24}{\Gamma(5-\alpha)} \sum_{k=0}^{4} \frac{f(x_k)}{A_k} x_1^{4-\alpha} - \frac{6}{\Gamma(4-\alpha)} \sum_{k=0}^{4} \frac{a_k f(x_k)}{A_k} x_1^{3-\alpha} \\
&+ \frac{2}{\Gamma(3-\alpha)} \sum_{k=0}^{4} \frac{b_k f(x_k)}{A_k} x_1^{2-\alpha} - \frac{1}{\Gamma(2-\alpha)} \sum_{k=0}^{4} \frac{c_k f(x_k)}{A_k} x_1^{1-\alpha} \\
&+ \frac{f^{(5)}(\xi)}{5!} \left[\frac{120}{\Gamma(6-\alpha)} x_1^{5-\alpha} - \frac{24a_5}{\Gamma(5-\alpha)} x_1^{4-\alpha} + \frac{6b_5}{\Gamma(4-\alpha)} x_1^{3-\alpha} \right. \\
&\left. - \frac{2c_5}{\Gamma(3-\alpha)} x_1^{2-\alpha} + \frac{d_5}{\Gamma(2-\alpha)} x_1^{1-\alpha} \right],
\end{aligned} \tag{20}
$$

where $\xi \in (a, b)$.

Corollary 2 *Based on Corollary 1, if* $\alpha = 1$, *then one obtains the conventional 5-point central formula for finding an approximation for* $f'(x_1)$. *In other words, we have*

$$
\begin{aligned}
f'(x_1) &= \frac{1}{12h}[f(x_0 - 2h) - 8f(x_0 - h) \\
&+ 8f(x_0 + h) - f(x_0 + 2h)] + \frac{h^4}{30} f^{(5)}(\xi),
\end{aligned} \tag{21}
$$

where $\xi \in (a, b)$.

3.3 A Fractional Central Formula for $D_*^{2\alpha}$

In this subsection, we intend to provide an approximate expression for the Caputo differentiator of order 2α (*i.e.* $D_*^{2\alpha}$) by applying a modified 5-point fractional central formula.

Corollary 3 *By taking similar assumptions provided in Theorem 3, then the modified 5-point fractional central formula for finding an approximation for the Caputo differentiator of order* 2α *can be approximated by*

$$
\begin{aligned}
D_*^{2\alpha} f(x_1) &= \left[\frac{24}{\Gamma(5-2\alpha)} \sum_{k=0}^{4} \frac{f(x_k)}{A_k} x_1^{4-2\alpha}\right] - \left[\frac{6}{\Gamma(4-2\alpha)} \sum_{k=0}^{4} \frac{a_k f(x_k)}{A_k} x_1^{3-2\alpha}\right] \\
&+ \left[\frac{2}{\Gamma(3-2\alpha)} \sum_{k=0}^{4} \frac{b_k f(x_k)}{A_k} x_1^{2-2\alpha}\right] - \left[\frac{1}{\Gamma(2-2\alpha)} \sum_{k=0}^{4} \frac{c_k f(x_k)}{A_k} x_1^{1-2\alpha}\right] \\
&+ \frac{f^{(5)}(\xi)}{5!}\left[\frac{120}{\Gamma(6-2\alpha)} x_1^{5-2\alpha} - \frac{24 a_5}{\Gamma(5-2\alpha)} x_1^{4-2\alpha} + \frac{6 b_5}{\Gamma(4-2\alpha)} x_1^{3-2\alpha} \right. \\
&\left. - \frac{2 c_5}{\Gamma(3-2\alpha)} x_1^{2-2\alpha} + \frac{d_5}{\Gamma(2-2\alpha)} x_1^{1-2\alpha}\right],
\end{aligned}
\tag{22}
$$

where $\xi \in (a, b)$.

4 Richardson Extrapolation

Richardson's extrapolation is typically used for the purpose of obtaining great precision results. In order to obtain a full overview about the applications or even the history of Richardson's extrapolation, the reader may refer to the paper by D. C. Joyce [31]. Besides, the reader may also refer to the paper in [30] that analyzes and describes some existing higher-order Richardson extrapolation formulae. However, in the following content, we present our scheme for Richardson extrapolation that proposed on the basis of the fractional central formulas defined above [32, 33]. Without losing generality, we assume that $\alpha = 1$ in (10) to obtain

$$D_*^{\alpha} f(x) = \frac{f(x+h) - f(x-h)}{2h} - \frac{h^2}{6} f^{(3)}(\xi), \tag{23}$$

where $\xi \in (a, b)$ and $x \in (a, b)$. In what follows, we shall denote $f'(x)$ as Kv, $\frac{f(x+h)-f(x-h)}{2h}$ as $Yv(h)$ and $e = \frac{-f^{(3)}(\xi)}{6}$. This would immediately imply

$$Kv = (Yv)(h) + eh^2. \tag{24}$$

If h is substituted by $\frac{h}{2}$ in (24), we obtain

$$Kv = (Yv)\left(\frac{h}{2}\right) + e\left(\frac{h}{2}\right)^2.$$

Multiplying the above equality by -4 yields

$$-4Kv = -4(Yv)\left(\frac{h}{2}\right) - eh^2. \tag{25}$$

Table 1 Richardson extrapolation of 4th-order approximation

O(h^2)	O(h^4)	O(h^6)	O(h^8)
$F_1(h)$	–	–	–
$F_1(\frac{h}{2})$	$F_2(h)$	–	–
$F_1(\frac{h}{4})$	$F_2(\frac{h}{2})$	$F_3(h)$	–
$F_1(\frac{h}{8})$	$F_2(\frac{h}{4})$	$F_3(\frac{h}{2})$	$F_4(h)$

By adding (24) to (25), we have

$$-3Kv = (Yv)(h) - (Yv)\left(\frac{h}{2}\right) - 3(Yv)\left(\frac{h}{2}\right).$$

This consequently implies

$$Kv = (Yv)\left(\frac{h}{2}\right) + \frac{(Yv)(\frac{h}{2}) - (Yv)(h)}{3}.$$

Assuming $Kv = F_2(h)$ and $F_1(h) = (Yv)(h)$ leads to the following assertion:

$$F_2(h) = F_1\left(\frac{h}{2}\right) + \frac{F_1(\frac{h}{2}) - F_1(h)}{3}.$$

If we keep moving in a similar manner, we can then generate Table 1 mentioned below. Such a procedure describes the methodology of Richardson extrapolation that could be used to produce a high accuracy approximation for the desired formulae.

For instance, we can have for F_1, the following expressions:

$$F_1(h) = \frac{f(x+h) - f(x-h)}{2h},$$

$$F_1\left(\frac{h}{2}\right) = \frac{h(x+\frac{h}{2}) - f(x-\frac{h}{2})}{h},$$

$$F_1\left(\frac{h}{4}\right) = \frac{f(x+\frac{h}{4}) - f(x-\frac{h}{4})}{\frac{h}{2}},$$

$$F_1\left(\frac{h}{8}\right) = \frac{f(x+\frac{h}{8}) - f(x-\frac{h}{8})}{\frac{h}{4}}.$$

In the same way, for F_2, we can have

$$F_2(h) = F_1\left(\frac{h}{2}\right) + \frac{F_1(\frac{h}{2}) - F_1(h)}{3},$$

$$F_2\left(\frac{h}{2}\right) = F_1\left(\frac{h}{4}\right) + \frac{F_1(\frac{h}{4}) - F_1(\frac{h}{2})}{3},$$

$$F_2\left(\frac{h}{4}\right) = F_1\left(\frac{h}{8}\right) + \frac{F_1(\frac{h}{8}) - F_1(\frac{h}{4})}{3}.$$

Also for F_3, we obtain

$$F_3(h) = F_2\left(\frac{h}{2}\right) + \frac{F_2(\frac{h}{2}) - F_2(h)}{15},$$

$$F_3\left(\frac{h}{2}\right) = F_2\left(\frac{h}{4}\right) + \frac{F_2(\frac{h}{4}) - F_2(\frac{h}{2})}{15}.$$

Finally, for F_4, we have

$$F_4(h) = F_3\left(\frac{h}{2}\right) + \frac{F_3(\frac{h}{2}) - F_3(h)}{63}.$$

Similarly, we can get

$$F_i(h) = F_{i-1}\left(\frac{h}{2}\right) + \frac{F_{i-1}(\frac{h}{2}) - F_{i-1}(h)}{4^{i-1} - 1}, \tag{26}$$

for each $i = 2, 3, \ldots$.

5 Illustrative Examples

This section targets to evaluate the performance of the derived n-point fractional formulae with Richardson extrapolation, where $n = \{3, 5\}$. Tables and figures are presented to demonstrate the reliability and effectiveness of these formulas.

5.1 Case 1: When n = 3

Example 1 Consider the main function here has the form $f(x) = \frac{1}{4}\sqrt{x^3} - 8\sqrt{x}$. Then, it is obvious that the exact value of this derivative is $f'(\frac{1}{2}) = -5.3917$. Accordingly, if one operates the Caputo differentiator to this function, we get

Table 2 Numerical results produced by modified 3-point fractional formula with $\alpha = 1$

h	O(h²)	O(h⁴)	O(h⁶)	O(h⁸)	O(h¹⁰)	O(h¹²)	O(h¹⁴)	O(h¹⁶)
0.0100	−5.3336	0	0	0	0	0	0	0
0.0050	−5.3624	−5.3720	0	0	0	0	0	0
0.0025	−5.3770	−5.3818	−5.3825	0	0	0	0	0
0.0013	−5.3843	−5.3868	−5.3871	−5.3872	0	0	0	0
0.0006	−5.3880	−5.3892	−5.3894	−5.3894	−5.3894	0	0	0
0.0003	−5.3898	−5.3905	−5.3905	−3.3906	−5.3906	−5.3906	0	0
0.0002	−5.3908	−5.3911	−5.3911	−5.3911	−5.3911	−5.3911	−5.3911	0
0.0001	−5.3912	−5.3914	−5.3914	−5.3914	−5.3914	−5.3914	−5.3914	−5.3914

$$D_*^{\alpha} f(x) = \frac{\Gamma(\frac{5}{2})}{4\Gamma(\frac{5}{2}-\alpha)} x^{\frac{3}{2}-\alpha} - \frac{8\Gamma(\frac{3}{2})}{\Gamma(\frac{3}{2}-\alpha)} x^{\frac{1}{2}-\alpha}.$$

Now, by using the order of extrapolation of $N = 8$ and the step size of $h = 0.01$, we produce the approximated results presented in Table 2 that are generated with the use of the modified 3-point fractional central formula for approximating Caputo differentiator $D_*^1 f(\frac{1}{2})$, (i.e., when $\alpha = 1$).

On the other hand, by assuming that the fractional-order value $\alpha = 0.98$ and by evaluating the same function reported above at $x_0 = \frac{1}{2}$, we can have

$$D_*^{0.98} f\left(\frac{1}{2}\right) = \frac{\Gamma(\frac{5}{2})}{4\Gamma(\frac{5}{2}-0.98)} \left(\frac{1}{2}\right)^{\frac{3}{2}-0.98} - \frac{8\Gamma(\frac{3}{2})}{\Gamma(\frac{3}{2}-0.98)} \left(\frac{1}{2}\right)^{\frac{1}{2}-0.98} = -5.5355.$$

Actually, the value -5.5355 depicts the exact value of $D_*^{0.98} f\left(\frac{1}{2}\right)$. Now, with the use of the modified 3-point fractional central formula for approximating Caputo differentiator $D_*^{0.98} f\left(\frac{1}{2}\right)$, (i.e., when $\alpha = 0.98$), we can produce the approximated results presented in Table 3. Clearly, the generated results confirmed the validly of the proposed formula for approximating the derivative in its fractional-order case.

For more clarification, we present in Fig. 1, two numerical comparisons between the exact formula and the proposed modified 3-point fractional formula.

Example 2 Consider the main function has the form $f(x) = 1 - \frac{x^2}{2!} + \frac{x^4}{4!} - \frac{x^6}{6!} + \frac{x^8}{8!} - \frac{x^{10}}{10!}$. Then, it is obvious that the exact value of this derivative is $f'(\frac{\pi}{6}) = -0.5000$. Accordingly, if one operates the Caputo differentiator to this function, we get

$$D_*^{\alpha} f(x) = -\frac{x^{2-\alpha}}{\Gamma(3-\alpha)} + \frac{x^{4-\alpha}}{\Gamma(5-\alpha)} - \frac{x^{6-\alpha}}{\Gamma(7-\alpha)} + \frac{x^{8-\alpha}}{\Gamma(9-\alpha)} - \frac{x^{10-\alpha}}{\Gamma(11-\alpha)}.$$

Table 3 Numerical results produced by modified 3-point fractional formula with $\alpha = 0.98$

h	O(h^2)	O(h^4)	O(h^6)	O(h^8)	O(h^{10})	O(h^{12})	O(h^{14})	O(h^{16})
0.0100	−5.3790	0	0	0	0	0	0	0
0.0050	−5.4069	−5.4162	0	0	0	0	0	0
0.0025	−5.4211	−5.4258	−5.4264	0	0	0	0	0
0.0013	−5.4282	−5.4306	−5.4309	−5.4310	0	0	0	0
0.0006	−5.4318	−5.4330	−5.4331	−5.4332	−5.4332	0	0	0
0.0003	−5.4336	−5.4342	−5.4342	−5.4343	−5.4343	−5.4343	0	0
0.0002	−5.4345	−5.4348	−5.4348	−5.4348	−5.4348	−5.4348	−5.4348	0
0.0001	−5.4349	−5.4351	−5.4351	−5.4351	−5.4351	−5.4351	−5.4351	−5.4351

Table 4 Numerical results produced by modified 3-point fractional formula with $\alpha = 1$

h	O(h^2)	O(h^4)	O(h^6)	O(h^8)	O(h^{10})	O(h^{12})	O(h^{14})	O(h^{16})
0.0100	−0.5086	0	0	0	0	0	0	0
0.0050	−0.5043	−0.5029	0	0	0	0	0	0
0.0025	−0.5022	−0.5014	−0.5013	0	0	0	0	0
0.0013	−0.5011	−0.5007	−0.5007	−0.5007	0	0	0	0
0.0006	−0.5005	−0.5004	−0.5003	−0.5003	−0.5003	0	0	0
0.0003	−0.5003	−0.5002	−0.5002	−0.5002	−0.5002	−0.5002	0	0
0.0002	−0.5001	−0.5001	−0.5001	−0.5001	−0.5001	−0.5001	−0.5001	0
0.0001	−0.5001	−0.5000	−0.5000	−0.5000	−0.5000	−0.5000	−0.5000	−0.5000

Now, by using the order of extrapolation of $N = 8$ and the step size of $h = 0.01$, we produce the approximated results presented in Table 4 that are generated with the use of the modified 3-point fractional central formula for approximating Caputo differentiator $D_*^\alpha f(\frac{\pi}{6})$, (i.e., when $\alpha = 1$).

On the other hand, by assuming that the fractional-order value $\alpha = 0.35$ and by evaluating the same function reported above at $x_0 = \frac{\pi}{6}$, we can have

$$D_*^{0.35} f\left(\frac{\pi}{6}\right) = -\frac{\left(\frac{\pi}{6}\right)^{1.65}}{\Gamma(2.65)} + \frac{\left(\frac{\pi}{6}\right)^{3.65}}{\Gamma(4.65)} - \frac{\left(\frac{\pi}{6}\right)^{5.65}}{\Gamma(6.65)} + \frac{\left(\frac{\pi}{6}\right)^{7.65}}{\Gamma(8.65)} - \frac{\left(\frac{\pi}{6}\right)^{9.65}}{\Gamma(10.65)}.$$

In fact, the value −0.2250 represents the exact value of $D_*^{0.35} f\left(\frac{\pi}{6}\right)$. Now, with the use of the modified 3-point fractional central formula for approximating Caputo differentiator $D_*^{0.35} f\left(\frac{\pi}{6}\right)$, (i.e., when $\alpha = 0.35$), we can produce the approximated results presented in Table 5. Clearly, the generated results confirmed the validly of the proposed formula for approximating the derivative in its fractional-order case.

For more clarification, we present in Fig. 2, two graphical comparisons between the exact formula and the proposed modified 3-point fractional formula.

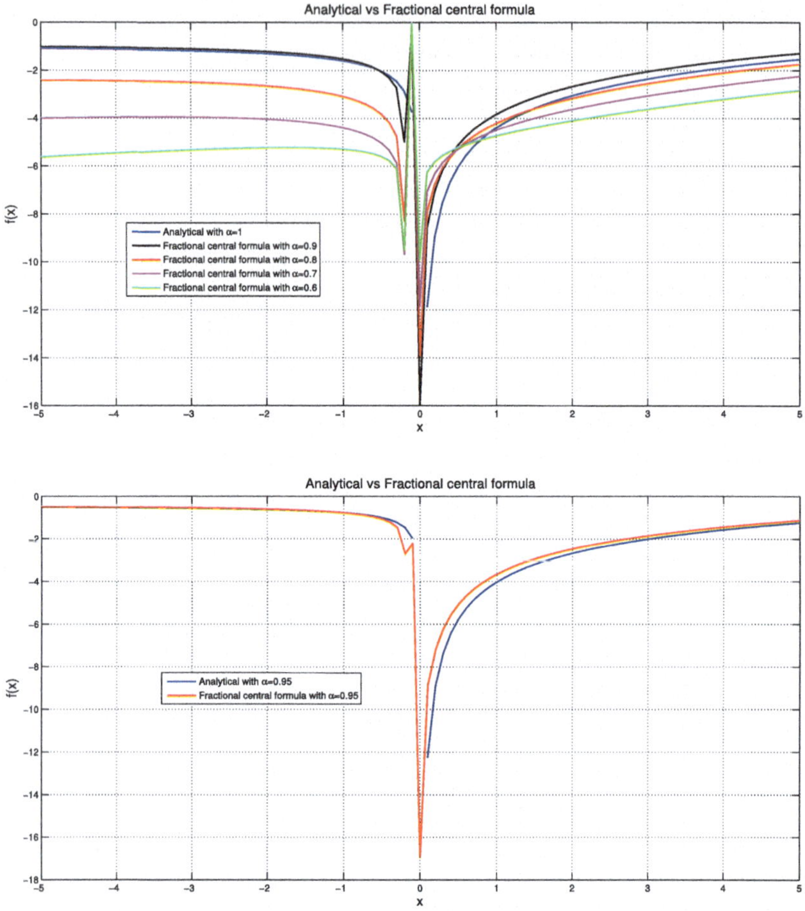

Fig. 1 Comparison results between the exact formula and modified 3-point fractional formula for various values of α

5.2 *Case 2: When n = 5*

Example 3 Consider the main function has the form $f(x) = x^3 + 4x^2 - 2$. Then, it is obvious that the exact value of this derivative is $f'(1) = 11$. Accordingly, if one operates the Caputo differentiator to this function, we get

$$D_*^\alpha f(x) = \frac{6}{\Gamma(4-\alpha)}x^{3-\alpha} + \frac{8}{\Gamma(3-\alpha)}x^{2-\alpha}.$$

Now, by using the order of extrapolation of $N = 7$ and the step size of $h = 0.2$, we produce the approximated results presented in Table 6 that are generated with

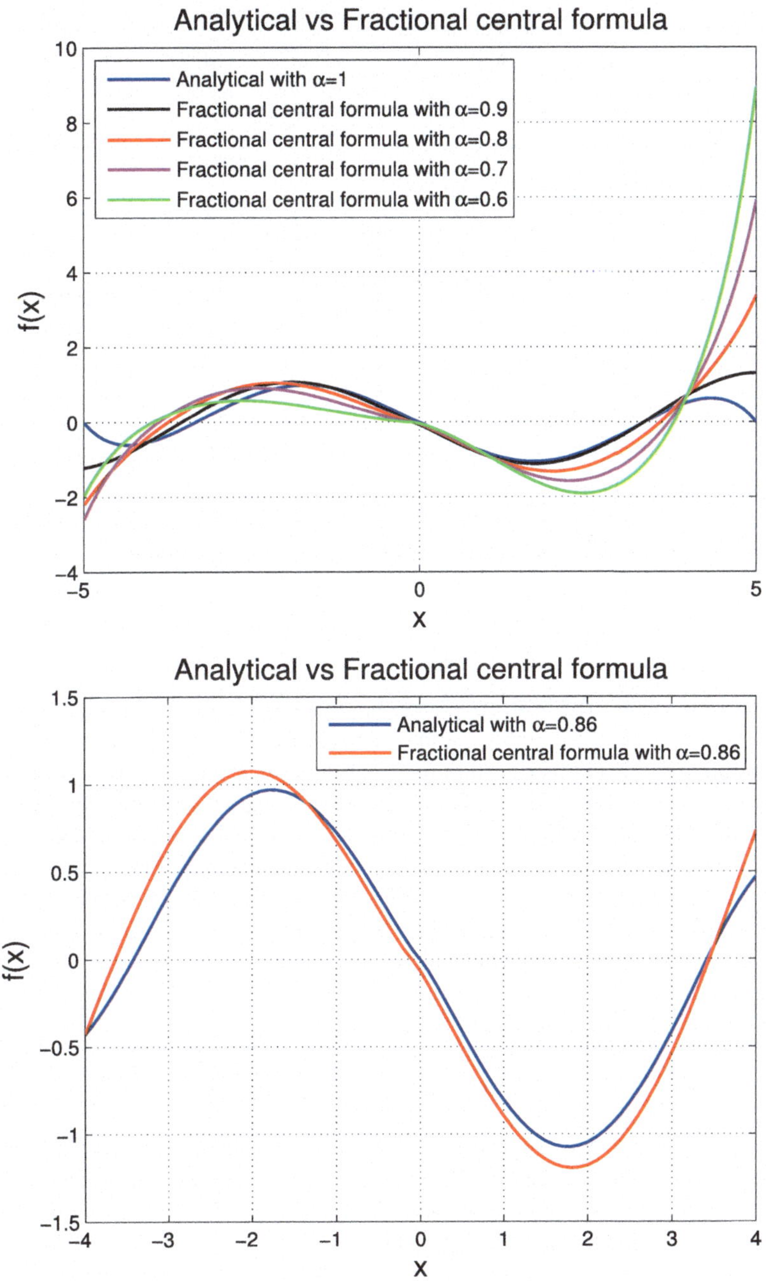

Fig. 2 Comparison results between the exact formula and modified 3-point fractional formula for various values of α

Table 5 Numerical results produced by modified 3-point fractional formula with $\alpha = 0.35$

h	O(h^2)	O(h^4)	O(h^6)	O(h^8)	O(h^{10})	O(h^{12})	O(h^{14})	O(h^{16})
0.0100	−0.2420	0	0	0	0	0	0	0
0.0050	−0.2382	−0.2369	0	0	0	0	0	0
0.0025	−0.2363	−0.2357	−0.2356	0	0	0	0	0
0.0013	−0.2354	−0.2351	−0.2350	−0.2350	0	0	0	0
0.0006	−0.2349	−0.2348	−0.2347	−0.2347	−0.2347	0	0	0
0.0003	−0.2347	−0.2346	−0.2346	−0.2346	−0.2346	−0.2346	0	0
0.0002	−0.2346	−0.2345	−0.2345	−0.2345	−0.2345	−0.2345	−0.2345	0
0.0001	−0.2345	−0.2345	−0.2345	−0.2345	−0.2345	−0.2345	−0.2345	−0.2345

the use of the modified 5-point fractional central formula for approximating Caputo differentiator $D_*^{\alpha} f(1)$, (i.e., when $\alpha = 1$).

On the other hand, by assuming that the fractional-order value $\alpha = 0.90$ and by evaluating the same function reported above at $x_0 = 1$, we can have

$$D_*^{0.90} f(x) = \frac{6}{\Gamma(3.1)}(1)^{2.1} + \frac{8}{\Gamma(2.1)}(1)^{1.1} = 10.3749.$$

In fact, the value 10.3749 represents the exact value of $D_*^{0.90} f(1)$. Now, with the use of the modified 5-point fractional central formula for approximating Caputo differentiator $D_*^{0.90} f\ (1)$, (i.e., when $\alpha = 0.90$), we can produce the approximated results presented in Table 7. Clearly, the generated results confirmed the validly of the proposed formula for approximating the derivative in its fractional-order case.

For more clarification, we present in Fig. 3, two graphical comparisons between the exact formula and the proposed modified 5-point fractional formula.

In the same regard, one can easily find the exact value of the second derivative of the function f at $x_0 = 1$, i.e., $f''(1) = 14$. Accordingly, if one operates the Caputo differentiator once again to this function, we get

$$D_*^{2\alpha} f(x) = \frac{6}{\Gamma(4-2\alpha)} x^{3-2\alpha} + \frac{8}{\Gamma(3-2\alpha)} x^{2-2\alpha}.$$

Now, by using the order of extrapolation of $N = 7$ and the step size of $h = 0.2$, we produce the approximated results presented in Table 8 that are generated with the use of the modified 5-point fractional central formula for approximating Caputo differentiator $D_*^{2\alpha} f(1)$, (i.e., when $\alpha = 1$).

On the other hand, by assuming that the fractional-order value $\alpha = 0.75$ and by evaluating the same function reported above at $x_0 = 1$, we can have

$$D_*^{1.5} f(1) = \frac{6}{\Gamma(2.5)}(1)^{1.5} + \frac{8}{\Gamma(1.5)}(1)^{0.5} = 13.5406.$$

Table 6 Numerical results produced by modified 5-point fractional formula with $\alpha = 1$

h	O(h^2)	O(h^4)	O(h^6)	O(h^8)	O(h^{10})	O(h^{12})	O(h^{14})
0.2000	13.9200	0	0	0	0	0	0
0.1000	12.4300	11.9333	0	0	0	0	0
0.0500	11.7075	11.4667	11.4356	0	0	0	0
0.0250	11.3519	11.2333	11.2178	11.2143	0	0	0
0.0125	11.1755	11.1167	11.1089	11.1071	11.1067	0	0
0.0063	11.0878	11.0585	11.0546	11.0538	11.0536	11.0535	0
0.0031	11.0396	11.0235	11.0212	11.0207	11.0205	11.0205	11.0205

Table 7 Numerical results produced by modified 5-point fractional formula with $\alpha = 0.90$

h	O(h^2)	O(h^4)	O(h^6)	O(h^8)	O(h^{10})	O(h^{12})	O(h^{14})
0.2000	13.3462	0	0	0	0	0	0
0.1000	11.8250	11.8248	0	0	0	0	0
0.0500	11.0912	11.0910	11.0909	0	0	0	0
0.0250	10.7315	10.7315	10.7312	10.7310	0	0	0
0.0125	10.6745	10.6745	10.6745	10.6745	10.6744	0	0
0.0063	10.4536	10.4536	10.4536	10.4536	10.4535	10.4533	0
0.0031	14.1713	14.4055	14.4055	14.4055	14.4055	14.4050	10.4010

Table 8 Numerical results produced by modified 5-point fractional formula for 2α with $\alpha = 1$

h	O(h^2)	O(h^4)	O(h^6)	O(h^8)	O(h^{10})	O(h^{12})	O(h^{14})
0.2000	15.2000	0	0	0	0	0	0
0.1000	14.7000	14.6000	0	0	0	0	0
0.0500	14.5000	14.4000	14.3000	0	0	0	0
0.0250	14.1800	14.1700	14.1600	14.1500	0	0	0
0.0125	14.0992	14.0992	14.0988	14.0986	14.0747	0	0
0.0063	14.0355	14.0356	14.0355	14.0355	14.0353	14.0353	0
0.0031	14.0307	14.0307	14.0307	14.0307	14.0307	14.0307	14.0306

In fact, the value 13.5406 represents the exact value of $D_*^{1.5} f(1)$. Now, with the use of the modified 5-point fractional central formula for approximating Caputo differentiator $D_*^{1.5} f(1)$, (i.e., when $\alpha = 0.75$), we can produce the approximated results presented in Table 9. Clearly, the generated results confirmed the validly of the proposed formula for approximating the derivative in its fractional-order case.

For more clarification, we present in Fig. 4, two graphical comparisons between the exact formula and the proposed modified 5-point fractional formula.

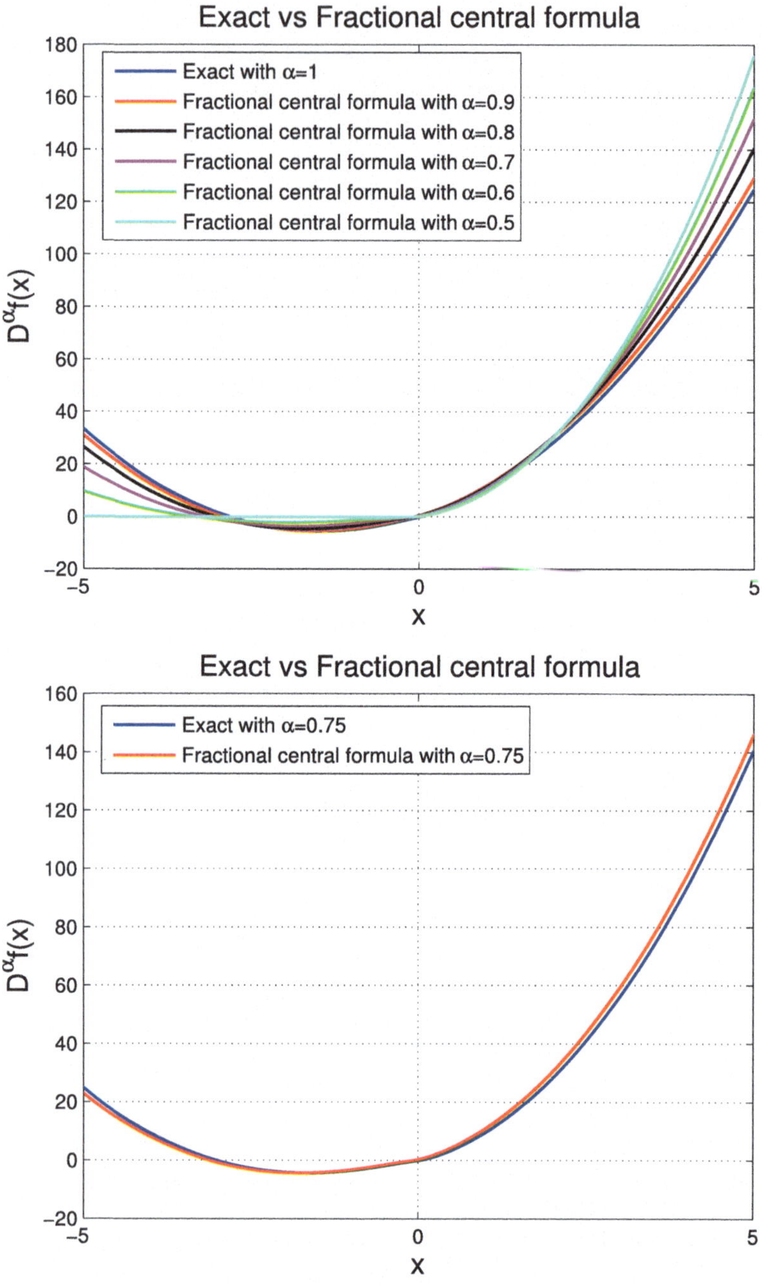

Fig. 3 Comparison results between the exact formula and modified 5-point fractional formula for various values of α

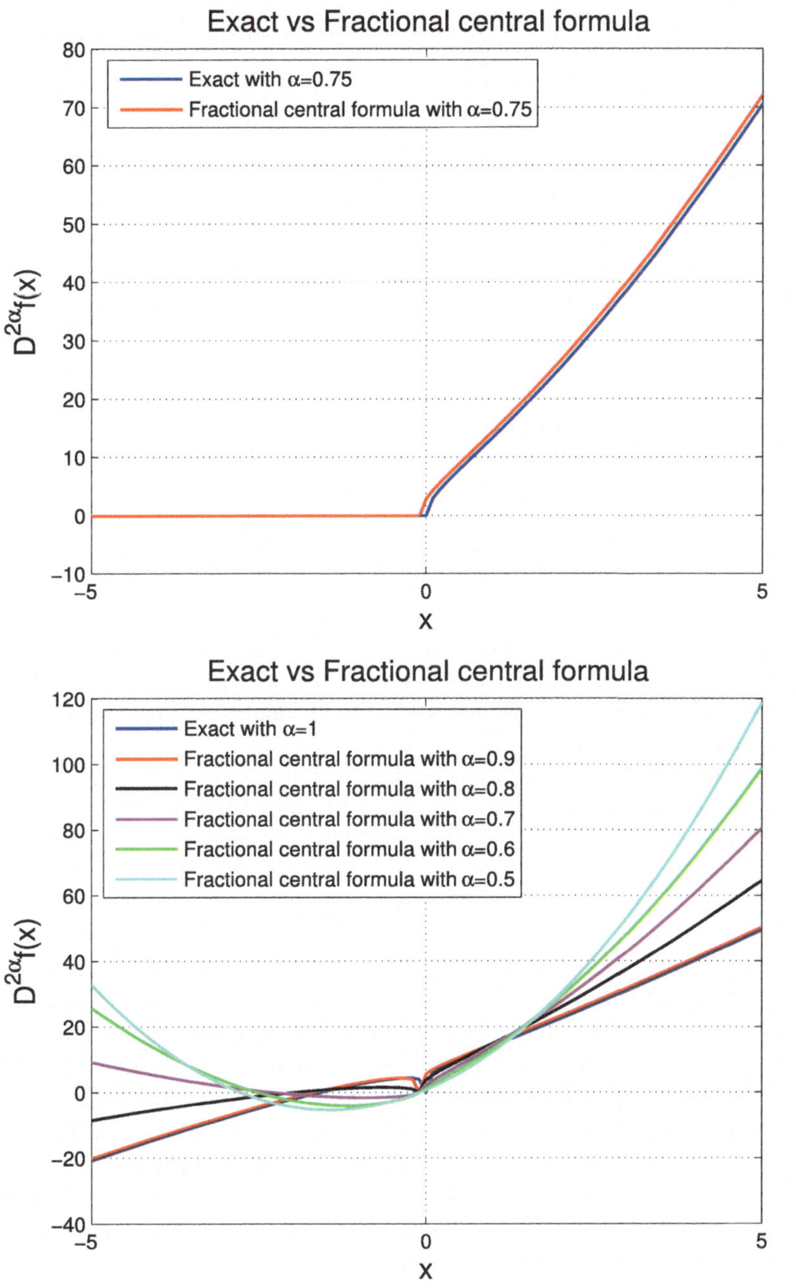

Fig. 4 Comparison results between the exact formula and modified 5-point fractional formula for various values of α

Table 9 Numerical results produced by modified 5-point fractional formula for 2α with $\alpha = 0.75$

h	O(h^2)	O(h^4)	O(h^6)	O(h^8)	O(h^{10})	O(h^{12})	O(h^{14})
0.2000	15.8218	0	0	0	0	0	0
0.1000	14.6750	14.6748	0	0	0	0	0
0.0500	14.1066	14.1065	14.1062	0	0	0	0
0.0250	13.8220	13.8218	13.8218	13.8217	0	0	0
0.0125	13.7756	13.7755	13.7755	13.7755	13.7751	0	0
0.0063	13.6002	13.6002	13.6002	13.6002	14.6002	13.6001	0
0.0031	14.0307	13.5602	13.5601	13.5601	13.5600	13.5600	13.5600

Example 4 Consider the main function has the form $f(x) = \frac{1}{3}x^{\frac{7}{2}} - \frac{1}{2}x^{\frac{5}{2}}$. Then, it is obvious that the exact value of this derivative is $f'(1) = -0.0833$. Accordingly, if one operates the Caputo differentiator to this function, we get

$$D_*^{\alpha} f(x) = \frac{\Gamma(\frac{9}{2})}{3\Gamma(\frac{9}{2}-\alpha)} x^{\frac{7}{2}-\alpha} - \frac{\Gamma(\frac{7}{2})}{2\Gamma(\frac{7}{2}-\alpha)} x^{\frac{5}{2}-\alpha}.$$

Now, by using the order of extrapolation of $N = 7$ and the step size of $h = 0.2$, we produce the approximated results presented in Table 10 that are generated with the use of the modified 5-point fractional central formula for approximating Caputo differentiator $D_*^{\alpha} f(1)$, (i.e., when $\alpha = 1$).

On the other hand, by assuming that the fractional-order value $\alpha = 0.75$ and by evaluating the same function reported above at $x_0 = 1$, we can have

$$D_*^{0.75} f(1) = \frac{\Gamma(\frac{9}{2})}{3\Gamma(3.75)}(1)^{2.75} - \frac{2\Gamma(\frac{7}{2})}{\Gamma(2.75)}(1)^{1.75} = -0.1565.$$

In fact, the value -0.1565 represents the exact value of $D_*^{0.75} f(1)$. Now, with the use of the modified 5-point fractional central formula for approximating Caputo differentiator $D_*^{0.75} f(1)$, (i.e., when $\alpha = 0.75$), we can produce the approximated results presented in Table 11. Clearly, the generated results confirmed the validly of the proposed formula for approximating the derivative in its fractional-order case.

For more clarification, we present in Fig. 5, two graphical comparisons between the exact formula and the proposed modified 5-point fractional formula.

In the same regard, one can easily find the exact value of the second derivative of the function f at $x_0 = 1$, i.e., $f''(1) = 1.0417$. Accordingly, if one operates the Caputo differentiator once again to this function, we get

$$D_*^{2\alpha} f(x) = \frac{\Gamma(\frac{9}{2})}{3\Gamma(\frac{9}{2}-2\alpha)} x^{\frac{7}{2}-2\alpha} - \frac{\Gamma(\frac{7}{2})}{2\Gamma(\frac{7}{2}-2\alpha)} x^{\frac{5}{2}-2\alpha}.$$

Table 10 Numerical results produced by modified 5-point fractional formula with $\alpha = 1$

h	O(h^2)	O(h^4)	O(h^6)	O(h^8)	O(h^{10})	O(h^{12})	O(h^{14})
0.2000	0.1971	0	0	0	0	0	0
0.1000	0.0384	−0.0144	0	0	0	0	0
0.0500	−0.0269	−0.0487	−0.0510	0	0	0	0
0.0250	−0.0562	−0.0660	−0.0671	−0.0674	0	0	0
0.0125	−0.0700	−0.0747	−0.0752	−0.0754	−0.0754	0	0
0.0063	−0.0768	−0.0790	−0.0793	−0.0794	−0.0794	−0.0794	0
0.0031	−0.0798	−0.0809	−0.0810	−0.0810	−0.0810	−0.0810	−0.0810

Table 11 Numerical results produced by modified 5-point fractional formula with $\alpha = 0.75$

h	**O(h^2)**	**O(h^4)**	**O(h^6)**	**O(h^8)**	**O(h^{10})**	**O(h^{12})**	**O(h^{14})**
0.2000	0.0439	0	0	0	0	0	0
0.1000	−0.0690	−0.0688	0	0	0	0	0
0.0500	−0.1131	−0.1130	−0.1126	0	0	0	0
0.0250	−0.1317	−0.1317	−0.1317	−0.1316	0	0	0
0.0125	−0.1562	−0.1562	−0.1562	−0.1562	−0.1561	0	0
0.0063	−0.1562	−0.1562	−0.1562	−0.1562	−0.1562	−0.1562	0
0.0031	−0.1564	−0.1563	−0.1563	−0.1563	−0.1563	−0.1563	−0.1564

Now, by using the order of extrapolation of $N = 7$ and the step size of $h = 0.2$, we produce the approximated results presented in Table 12 that are generated with the use of the modified 5-point fractional central formula for approximating Caputo differentiator $D_*^{2\alpha} f(1)$, (i.e., when $\alpha = 1$).

On the other hand, by assuming that the fractional-order value $\alpha = 0.85$ and by evaluating the same function reported above at $x_0 = 1$, we can have

$$D_*^{1.7} f(1) = \frac{\Gamma(\frac{9}{2})}{3\Gamma(\frac{9}{2} - 1.7)}(1)^{\frac{7}{2}-1.7} - \frac{\Gamma(\frac{7}{2})}{2\Gamma(\frac{7}{2} - 1.7)}(1)^{\frac{5}{2}-1.7} = 0.5286.$$

In fact, the value 0.5286 represents the exact value of $D_*^{1.7} f(1)$. Now, with the use of the modified 5-point fractional central formula for approximating Caputo differentiator $D_*^{1.7} f(1)$, (i.e., when $\alpha = 0.85$), we can produce the approximated results presented in Table 13. Clearly, the generated results confirmed the validly of the proposed formula for approximating the derivative in its fractional-order case.

For more clarification, we present in Fig. 6, two graphical comparisons between the exact formula and the proposed modified 5-point fractional formula.

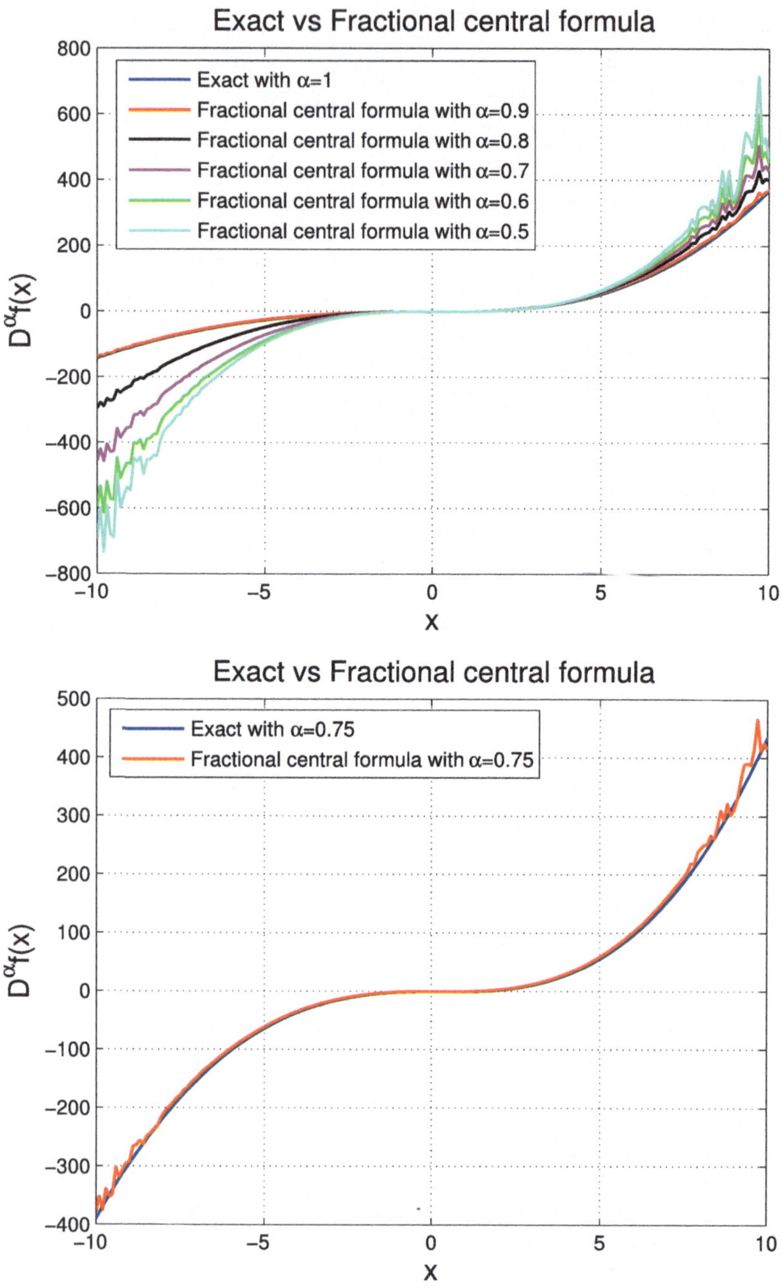

Fig. 5 Comparison results between the exact formula and modified 5-point fractional formula for various values of α

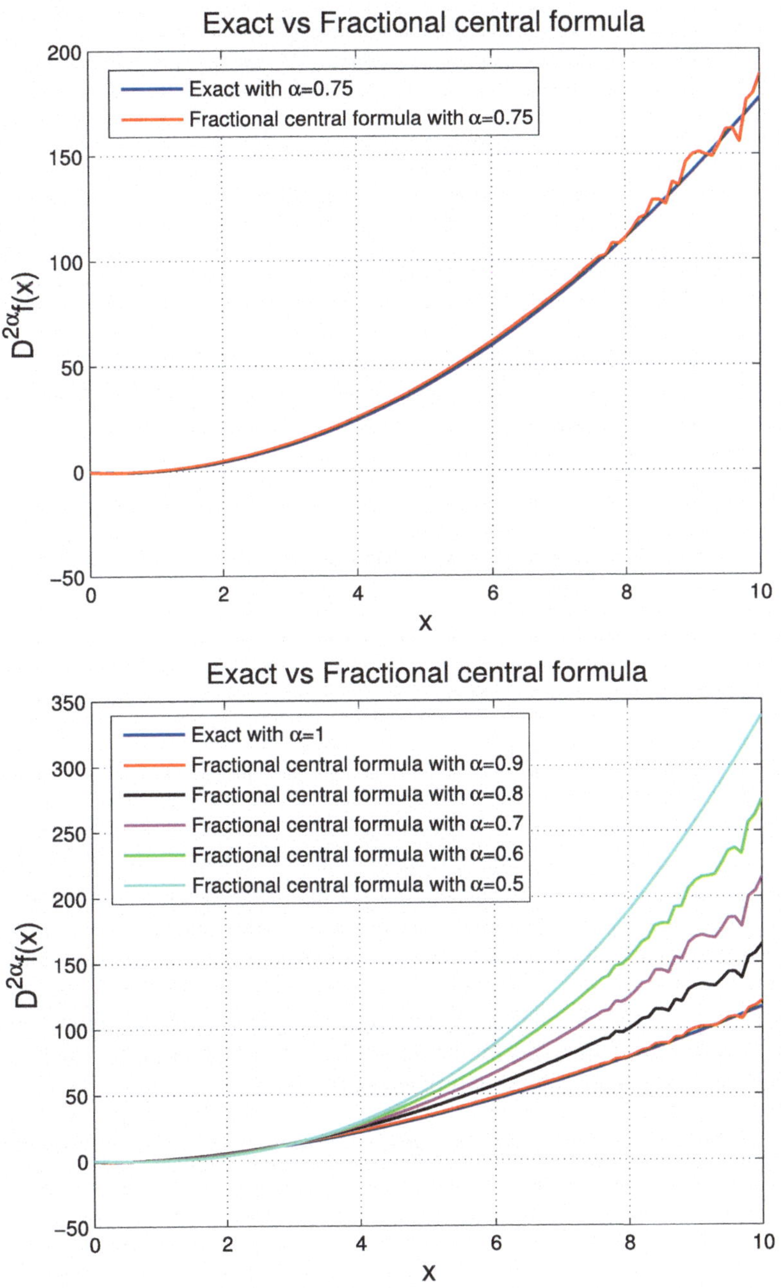

Fig. 6 Comparison results between the exact formula and modified 5-point fractional formula for various values of α

Table 12 Numerical results produced by modified 5-point fractional formula for 2α with $\alpha = 1$

h	O(h^2)	O(h^4)	O(h^6)	O(h^8)	O(h^{10})	O(h^{12})	O(h^{14})
0.2000	1.7808	0	0	0	0	0	0
0.1000	1.3990	1.3985	0	0	0	0	0
0.0500	1.2171	1.2170	1.2168	0	0	0	0
0.0250	1.1286	1.1285	1.1285	1.1284	0	0	0
0.0125	1.0850	1.0850	1.0850	1.0850	1.0849	0	0
0.0063	1.0634	1.0634	1.0634	1.0634	1.0633	1.0633	0
0.0031	1.0497	1.0497	1.0497	1.0497	1.0497	1.0497	1.0497

Table 13 Numerical results produced by modified 5-point fractional formula for 2α with $\alpha = 0.85$

h	O(h^2)	O(h^4)	O(h^6)	O(h^8)	O(h^{10})	O(h^{12})	O(h^{14})
0.2000	1.1301	0	0	0	0	0	0
0.1000	0.8080	0.8075	0	0	0	0	0
0.0500	0.6596	0.6596	0.6595	0	0	0	0
0.0250	0.5891	0.5891	0.5890	0.5888	0	0	0
0.0125	0.5732	0.5732	0.5731	0.5731	0.5731	0	0
0.0063	0.5392	0.5392	0.5391	0.5391	0.5391	0.5390	0
0.0031	0.5291	0.5291	0.5291	0.5291	0.5290	0.5290	0.5290

6 Conclusion

In conclusion, we have presented some fresh n-point fractional formulae to approximate the Caputo fractional derivative operator. These formulae, which are considered extensions of a few n-point classical formulas for estimating the first and second classical derivatives, have allowed us to examine the behavior of particular functions with respect to various fractional-order values. We use the so-called Richardson extrapolation technique to attain high accuracy for the acquired approximations, and we also present various numerical simulations in the form of tables and graphs to show the reliability and validity of our approximations.

Competing Interests The authors have no conflicts of interest to declare that are relevant to the content of this chapter.

References

1. A.V. Aho, J.E. Hopcroft, J.D. Ullman, *The Design and Analysis of Computer Algorithms* (Addison-Wesley, Boston, 1974)
2. A.C. Aitken, On interpolation by iteration of proportional parts, without the use of differences. Proc. Edinburgh Math. Soc. **3**, 56–76 (1932)
3. M.D. Ortigueira, A.G. Batista, On the relation between the fractional Brownian motion and the fractional derivatives. Phys. Lett. A **372**, 958–968 (2008)
4. O. Axelsson, V.A. Barker, *Finite Element Solution of Boundary Value Problems: Theory and Computation* (Academic Press, Orlando, 1984)
5. R.B. Albadarneh, I.M. Batiha, A. Adwai, N. Tahat, A.K. Alomari, Numerical approach of Riemann-Liouville fractional derivative operator. Int. J. Electr. Comput. Eng. **11**, 5367–5378 (2021)
6. A.A. Khennaoui, A. Ouannas, I.M. Batiha, V.T. Pham, Chaos in fractional-order difference systems, in *Emerging Methodologies and Applications in Modelling, Fractional Order Systems*, ed. by A.G. Radwan, F.A. Khanday, L.A. Said (Academic Press, Orlando, 2022), pp. 257–286
7. A. Ouannas, A.A. Khennaoui, I.M. Batiha, V.T. Pham, Stabilization of different dimensional fractional chaotic maps, in *Emerging Methodologies and Applications in Modelling, Fractional Order Systems*, ed. by A.G. Radwan, F.A. Khanday, L.A. Said (Academic Press, Orlando, 2022), pp. 123–155
8. A.A. Khennaoui, A. Ouannas, I.M. Batiha, V.T. Pham, Hidden chaotic attractors in fractional-order discrete-time systems, in *Advanced Studies in Complex Systems, Fractional Order Systems and Applications in Engineering*, ed. by D. Baleanu, V.E. Balas, P. Agarwal (Academic Press, Orlando, 2023), pp. 227–243
9. I.M. Batiha, S. Alshorm, A. Al-Husban, R. Saadeh, G. Gharib, S. Momani, The n-point composite fractional formula for approximating Riemann-Liouville integrator. Symmetry **15**, 938 (2023)
10. W. Alshanti, A. Alshanty, R. Khalil, Atomic solution for both ordinary and fractional abstract Cauchy problem in tensor product form, Gulf. J. Math. **18**(1), 189–196 (2024)
11. J. Oudetallah, Z. Chebana, T.E. Oussaeif, A. Ouannas, I.M. Batiha, Theoretical study of explosion phenomena for a semi-parabolic problem, in *Mathematics and Computation. IACMC 2022. Springer Proceedings in Mathematics & Statistics*, vol. 418, ed. by D. Zeidan, J.C. Cortés, A. Burqan, A. Qazza, J. Merker, G. Gharib (Springer, Singapore, 2023)
12. I.M. Batiha, N. Djenina, A. Ouannas, T.E. Oussaeif, Fractional-order SEIR Covid-19 model: discretization and stability analysis, in *Mathematics and Computation. IACMC 2022. Springer Proceedings in Mathematics & Statistics*, vol. 418, ed. by D. Zeidan, J.C. Cortés, A. Burqan, A. Qazza, J. Merker, G. Gharib (Springer, Singapore, 2023)
13. R.B. Albadarneh, A.M. Adawi, S. Al-Sa'di, I.M. Batiha, S. Momani, A pro rata definition of the fractional-order derivative, in *Mathematics and Computation. IACMC 2022. Springer Proceedings in Mathematics & Statistics*, vol. 418, ed. by D. Zeidan, J.C. Cortés, A. Burqan, A. Qazza, J. Merker, G. Gharib (Springer, Singapore, 2023)
14. I.M. Batiha, A. Bataihah, A.A. Al-Nana, S. Alshorm, I.H. Jebril, A. Zraiqat, A numerical scheme for dealing with fractional initial value problem. Int. J. Innov. Comput. Inform. Control **19**, 763–774 (2023)
15. S. Momani, I.M. Batiha, A. Hioual, A. Ouannas, Fractional neural networks: finite time stability and its application to synchronization, in *International Conference on Fractional Differentiation and Its Applications (ICFDA), Ajman, United Arab Emirates* (2023), pp. 1–5
16. R.C. Karoun, A. Ouannas, M.A. Horani, T. Ziar, I.M. Batiha, Z. Dibi, Chaos in the fractional variable order discrete-time neural networks, in *International Conference on Fractional Differentiation and Its Applications (ICFDA), Ajman, United Arab Emirates* (2023), pp. 1–5
17. S.B. Ahmed, A. Ouannas, M.A. Horani, A.A. Khennaoui, I.M. Batiha, Chaotic attractors in quadratic discrete Tinkerbell system with non-commensurate fractional variable-orders:

complexity, chaos and entropy, in *International Conference on Fractional Differentiation and Its Applications (ICFDA), Ajman, United Arab Emirates* (2023), pp. 1–4
18. I.M. Batiha, N. Djenina, A. Ouannas, T.E. Oussaeif, L.B. Aoua, S. Momani, Control of chaos in incommensurate fractional order discrete system, in *International Conference on Fractional Differentiation and Its Applications (ICFDA), Ajman, United Arab Emirates* (2023), pp. 1–5
19. I.M. Batiha, A.A. Abubaker, I.H. Jebril, S.B. Al-Shaikh, K. Matarneh, New algorithms for dealing with fractional initial value problems. Axioms **12**, 488 (2023)
20. I.M. Batiha, O. Talafha, O.Y. Ababneh, S. Alshorm, S. Momani, Handling a commensurate, incommensurate, and singular fractional-order linear time-invariant system. Axioms **12**, 771 (2023)
21. S. Alshorm, I.M. Batiha, I. Jebril, A. Dababneh, Handling systems of incommensurate fractional order equations using improved fractional Euler method, in *International Conference on Information Technology (ICIT), Amman, Jordan* (2023), pp. 657–660
22. I.M. Batiha, S. Alshorm, A. Zraiqat, M. Alia, I. Jebril, Numerical solution for incommensurate system of fractional order differential equations, in *International Conference on Information Technology (ICIT), Amman, Jordan* (2023), pp. 652–656
23. I.M. Batiha, A.A. Abubaker, I.H. Jebril, S.B. Al-Shaikh, K. Matarneh, M. Almuzini, A mathematical study on a fractional-order SEIR Mpox model: analysis and vaccination influence. Algorithms **16**, 418 (2023)
24. I.M. Batiha, N. Djenina, A. Ouannas, A stabilization of linear incommensurate fractional-order difference systems. AIP Conf. Proc. **2849**, 030013 (2023)
25. R.B. Albadarneh, A. Abbes, A. Ouannas, I.M. Batiha, T.E. Oussaeif, On chaos in the fractional-order discrete-time macroeconomic systems. AIP Conf. Proc. **2849**, 030014 (2023)
26. B. Maayah, A. Moussaoui, S. Bushnaq, O. Abu Arqub, The multistep Laplace optimized decomposition method for solving fractional-order coronavirus disease model (COVID-19) via the Caputo fractional approach. Demonstratio Mathematica **55**, 963–977 (2022)
27. O. Abu Arqub, B. Maayah, Adaptive the Dirichlet model of mobile/immobile advection/dispersion in a time-fractional sense with the reproducing kernel computational approach: Formulations and approximations. Int. J. Modern Phys. B **37**, 2350179 (2022)
28. O. Abu Arqub, H. Alsulami, M. Alhodaly, Numerical Hilbert space solution of fractional Sobolev equation in 1+ 1-dimensional space. Math. Sci. **2022**, 1–12 (2022)
29. M. Arshad, D. Lu, J. Wang, (N+ 1)-dimensional fractional reduced differential transform method for fractional order partial differential equations. Commun. Nonlinear Sci. Numer. Simul. **48**, 509–519 (2017)
30. R.L. Burden, J.D. Faires, *Numerical Analysis* (Brookscole, Boston, 2011)
31. D.C. Joyce, Survey of extrapolation processes in numerical analysis. SIAM Rev. **13**, 435–490 (1971)
32. I.M. Batiha, S. Alshorm, A. Ouannas, S. Momani, O.Y. Ababneh, M. Albdareen, Modified three-point fractional formulas with Richardson extrapolation. Mathematics **10**, 3489 (2022)
33. I.M. Batiha, S. Alshorm, I. Jebril, A. Zraiqat, Z. Momani, S. Momani, Modified 5-point fractional formula with Richardson extrapolation. AIMS Math. **8**, 9520–9534 (2023)

Analysis of Alzheimer's Disease Using Nonlocal Operator

Sonal Jain

Abstract This study outlines the creation of a fractional partial differential equation-based mathematical model for Alzheimer's disease. The dynamics of the spread of Alzheimer's disease within a population can be described mathematically. To get numerical solutions, one uses Newton's polynomial interpolation method. To show the effectiveness of the suggested approach, several numerical simulations are provided.

Keywords Alzheimer's disease · Atangana–Baleanu derivative · Mathematical modeling · Numerical solution

1 Introduction

The degenerative disorder Alzheimer's disease (AD) results in the degeneration and death of brain cells [6]. This results in the most typical cause of dementia, which is a steady deterioration in mental, behavioural, and social abilities that impairs the person's capacity for independent functioning [10]. It is believed that the aberrant protein buildup in and around brain cells is what causes AD. OAmyloid, a protein that creates plaques surrounding brain cells, is one of the involved proteins. Tau protein, which accumulates to form tangles within brain cells, is another protein implicated in the process. The memory and cognitive abilities are truly destroyed by this disease, which is progressive and irreversible and leads to brain disorders. However, it is unclear how this illness spreads. Therefore, it is crucial to create models that represent the dynamics of the disease. Some powerful and successful models offer a partial explanation of the dynamics of AD taking into account neurons, astrocytes, treatments, and numerous other aspects.

In several disciplines, partial and ordinary differential equations are utilised to describe actual events that take place every day. They are frequently employed

S. Jain (✉)
Faculty of Data Science and Analytics, Sir Padampat Singhania University, Udaipur, Rajasthan 313601, India
e-mail: sonaljainmaths@gmail.com

M. Kharrat et al. (eds.), *Modeling of Discrete and Continuous Systems*, Forum for Interdisciplinary Mathematics, https://doi.org/10.1007/978-981-97-8715-9_8

because they represent variations or changes in time and place and can be combined with a wide range of other factors to model a particular issue. In classical mechanics, where Markovian processes with initial starting points and generator operators are used to forecast the future in time and space, some of these differential operators are able to partially reproduce complex circumstances. A number of mathematical models can be used to replicate a specific real-world problem, especially when the modellers are unsure of how the real-world problem will change. In science, technology, and engineering, fractional partial differential equations have been applied to actual physical systems. Recently, in an effort to better match the dynamics of these models with reality, scientists have applied fractional derivatives to scientific and disease models utilising wavelets, framelets, and other methods [4, 6, 10, 11, 17].

Before researchers were able to investigate the corresponding fractional models, numerical methods had to advance. For the purpose of solving fractional ordinary differential equations, particularly nonlinear ones, a number of numerical approaches have been presented [7–9]. To the best of the authors' knowledge, the Adams-Bashforth approach, which was created using Lagrange polynomial interpolation, is the most extensively used scheme [15, 17]. Recent research has revealed that, on average, for a large variety of polynomial functions, Newton's technique outperforms Lagrange polynomials [12, 13]. Atangana and Seda have presented a numerical technique that works with both integer and fractional ordinary differential systems, replacing the Adams-Bashforth scheme's Lagrange polynomial interpolation with Newtonian quadratic interpolation [2]. The authors derived iterative numerical formulas for the standard and fractal versions of the Caputo, Caputo–Fabrizio, and Atangana–Baleanu fractional derivatives. This method was applied to chaotic systems and showed promising results [4, 11], and has been extended to partial differential equations with integer and non-integer orders [1].

We might be able to comprehend the dynamics of AD using a fractional mathematical model of the disease. Since the spread of infectious diseases in people depends on a variety of conditions, its spread cannot be explained by straightforward mathematical equations. In order to account for the impacts of nonlocalities, a relatively new mathematical model is created in this study employing the idea of fractional operators [5].

Atangana and Baleanu established a nonlocal fractional derivative with a nonsingular kernel [2]. The Mittag–Leffler function was introduced to compute fractional derivatives as follows:

Definition 1 For $u(t) \in H^1(x, y)$, $y > x$, $\delta \in [0, 1]$, the Atangana–Baleanu (AB) fractional derivative in the Caputo sense of the function $u(t)$ is given by:

$$ {}_a^{ABC}D_t^{\delta} u(t) = \frac{B(\delta)}{1-\delta} \int_a^t \frac{d}{d\tau} u(\tau) E_\alpha \left[-\delta \frac{(t-\tau)^\delta}{1-\delta} \right] d\tau, \quad (1) $$

where $B(0) = B(1) = 1$ and $H^1(x, y)$ is a Hilbert space. Based on the generalised Mittag-Leffler function, Atangana and Baleanu [2] have proposed differential and integral operators. The intention was to present nonsingular nonlocal kernels for fractional differential and integral operators.

Definition 2 The AB fractional integral of the function $\mathrm{u}(t)$ is defined as

$$ {}^{AB}_{a}I_t^{\alpha}[\mathrm{u}(t)] = \frac{1-\delta}{AB(\delta)}\mathrm{u}(t) + \frac{\delta}{AB(\delta)\Gamma(\delta)}\int_a^t \mathrm{u}(\tau)(t-\tau)^{\delta-1}d\tau. \tag{2}$$

A mathematical model of AD utilising AB fractional differential operators is presented in this chapter. Section 2 presents the mathematical model of AD with non local fractional derivatives. An integro-fractional stochastic model of AD is given in Sect. 3. Section 4 introduces some numerical solutions of AD with fractional derivatives. Finally, Sect. 5 summarizes the conclusions to this study.

2 Mathematical Model of AD Using AB Derivative

We take into consideration the AD [16] mathematical model. The 18 nonlinear partial differential equations that make up the model are as follows:

$$\frac{\partial C^i_\beta}{\partial t} = \left(\lambda^i_\beta(1+R) - d_{C^i_\beta}C^i_\beta\right)N/N_0 \tag{3}$$

$$\begin{aligned}\frac{\partial C^0_\beta}{\partial t} &= C^i_\beta\left|\frac{\partial N}{\partial t}\right| + \frac{\lambda_N N}{N_0} + \frac{\lambda_A A}{C_0}\\ &- \left(d^0_{A_\beta \hat{M}}\left(\hat{N}_1 + \hat{N}_2\theta\right) + d^0_{A_\beta M}\left(N_1 + \hat{N}_2\theta\right)\right)\frac{C^0_\beta}{C^0_\beta + \bar{K}_{C^0_\beta}}\end{aligned} \tag{4}$$

$$\frac{\partial \mathcal{A}'}{\partial t} = \left(\lambda_{\mathcal{A}'0} + \lambda_{\mathcal{A}'}R - d_{\mathcal{A}'}\mathcal{A}'\right)N/N_0 \tag{5}$$

$$\frac{\partial \mathcal{B}_i}{\partial t} = \left(\lambda_{F\mathcal{A}'} + d_{\mathcal{B}_i}\mathcal{B}_i\right)N/N_0 \tag{6}$$

$$\frac{\partial \mathcal{B}_0}{\partial t} = \mathcal{B}_i\left|\frac{\partial N}{\partial t}\right| - f_{\mathcal{B}_0}\mathcal{B}_0 \tag{7}$$

$$\frac{\partial N}{\partial t} = \frac{-d_{NF}\mathcal{B}_i N}{\mathcal{B}_i + K_{\mathcal{B}_i}} - \frac{d_{NT}\mathcal{U}_\alpha N}{(\mathcal{U}_\alpha + K_{\mathcal{U}_\alpha})(K_{\mathcal{K}_{10}} + \gamma\mathcal{K}_{10})} \tag{8}$$

$$\frac{\partial A}{\partial t} = \lambda_{AA_0^\beta} A_0^\beta + \lambda_{A\mathcal{U}_\alpha} \mathcal{U}_\alpha - d_A A \tag{9}$$

$$\begin{aligned}\frac{\partial \mathcal{M}_d}{\partial t} &= \frac{d_{NF}\mathcal{B}_i N}{\mathcal{B}_i + K_{\mathcal{B}_i}} + \frac{d_{NT}\mathcal{U}_\alpha N K_{\mathcal{K}_{10}}}{(\mathcal{U}_\alpha + K_{\mathcal{U}_\alpha})(K_{\mathcal{K}_{10}} + \gamma \mathcal{K}_{10})} \\ &- d_{\mathcal{M}_d M}(\mathcal{N}_1 + \mathcal{N}_2)\frac{\mathcal{M}_d}{\mathcal{M}_d + \bar{K}\mathcal{M}_d} \\ &- d_{\mathcal{M}_d \hat{M}}\left(\hat{\mathcal{N}}_1 + \hat{\mathcal{N}}_2\right)\frac{\mathcal{M}_d}{\mathcal{M}_d + \bar{K}\mathcal{M}_d}\end{aligned} \tag{10}$$

$$\frac{\partial C_0}{\partial t} = \lambda_{A_0} A_0^\beta - d_{C_0} C_0 + D_{C_0} \triangle C_0 \tag{11}$$

$$\frac{\partial H}{\partial t} = \lambda_H N_d - d_H H + D_H \triangle H \tag{12}$$

$$\begin{aligned}\frac{\partial \mathcal{N}_1}{\partial t} &= \left(\frac{M_G^0 \lambda_{MF}\mathcal{B}_0}{\mathcal{B}_0 + K_{\mathcal{B}_0}} + \frac{M_G^0 \lambda_{MA} C_0}{C_0 + K_{C_0}}\right)\frac{\beta\epsilon_1}{\beta\epsilon_1 + \epsilon_2} \\ &- \nabla \cdot (\mathcal{N}_1 \nabla H) + \frac{\mathcal{U}_\beta \mathcal{N}_1 \lambda_{\mathcal{N}_1 \mathcal{U}_\beta}}{\mathcal{U}_\beta + K_{\mathcal{U}_\beta}} - d_{\mathcal{N}_1}\mathcal{N}_1\end{aligned} \tag{13}$$

$$\begin{aligned}\frac{\partial \mathcal{N}_2}{\partial t} &= \left(\frac{M_G^0 \lambda_{MF}\mathcal{B}_0}{\mathcal{B}_0 + K_{\mathcal{B}_0}} + \frac{M_G^0 \lambda_{MA} C_0}{C_0 + K_{C_0}}\right)\frac{\beta\epsilon_2}{\beta\epsilon_1 + \epsilon_2} \\ &- \nabla \cdot (\mathcal{N}_2 \nabla H) + \frac{\mathcal{U}_\beta \mathcal{N}_2 \lambda_{\mathcal{N}_2 \mathcal{U}_\beta}}{\mathcal{U}_\beta + K_{\mathcal{U}_\beta}} - d_{\mathcal{N}_2}\mathcal{N}_2\end{aligned} \tag{14}$$

$$\frac{\partial \hat{\mathcal{N}}_1}{\partial t} = \alpha(P)(M_0 - \hat{M})\frac{\beta\epsilon_1}{\beta\epsilon_1 + \epsilon_2} - \nabla \cdot (\hat{\mathcal{N}}_1 \nabla C_0) + \frac{\mathcal{U}_\beta \hat{\mathcal{N}}_1 \lambda_{\hat{\mathcal{N}}_1 \mathcal{U}_\beta}}{\mathcal{U}_\beta + K_{\mathcal{U}_\beta}} d_{\hat{\mathcal{N}}_1}\hat{\mathcal{N}}_1 \tag{15}$$

$$\frac{\partial \hat{\mathcal{N}}_2}{\partial t} = \alpha(P)(M_0 - \hat{M})\frac{\beta\epsilon_2}{\beta\epsilon_1 + \epsilon_2} - \nabla \cdot (\hat{\mathcal{N}}_2 \nabla C_0) + \frac{\mathcal{U}_\beta \hat{\mathcal{N}}_2 \lambda_{\hat{\mathcal{N}}_2 \mathcal{U}_\beta}}{\mathcal{U}_\beta + K_{\mathcal{U}_\beta}} d_{\hat{\mathcal{N}}_2}\hat{\mathcal{N}}_2 \tag{16}$$

$$\frac{\partial \mathcal{U}_\beta}{\partial t} = \lambda_{\mathcal{U}_\beta M}\mathcal{N}_2 + D_{\mathcal{U}_\beta} \triangle \mathcal{U}_\beta + \lambda_{\mathcal{U}_\beta \hat{M}}\hat{\mathcal{N}}_2 - d_{\mathcal{U}_\beta}\mathcal{U}_\beta \tag{17}$$

$$\frac{\partial \mathcal{K}_{10}}{\partial t} = \lambda_{\mathcal{K}_{10}\mathcal{N}_1}\mathcal{N}_2 + D_{\mathcal{K}_{10}} \triangle \mathcal{U}_\beta + \lambda_{\mathcal{K}_{10}\hat{M}}\hat{\mathcal{N}}_2 - d_{\mathcal{K}_{10}}\mathcal{K}_{10} \tag{18}$$

$$\frac{\partial \mathcal{U}_\alpha}{\partial t} = \lambda_{\mathcal{U}_\alpha \mathcal{N}_1}\mathcal{N}_1 + D_{\mathcal{U}_\alpha} \triangle \mathcal{U}_\alpha + \lambda_{\mathcal{U}_\alpha \hat{\mathcal{N}}_1}\hat{\mathcal{N}}_1 - d_{\mathcal{U}_\alpha}\mathcal{U}_\alpha \tag{19}$$

$$\frac{\partial P}{\partial t} = \lambda_{PA} A + D_P \triangle P_{\lambda P \mathcal{N}_2}\mathcal{N}_2 - d_p P. \tag{20}$$

where

$$\epsilon_1 = \frac{\mathcal{U}_\alpha}{\mathcal{U}_\alpha + K_\alpha}, \quad \epsilon_2 = \frac{\mathcal{K}_{10}}{\mathcal{K}_{10} + K_{\mathcal{K}_{10}}}, \quad \hat{M} = \hat{\mathcal{N}}_1 + \hat{\mathcal{N}}_2,$$

$$\alpha(P) = \frac{\alpha P}{P + K_P}, \quad R(t) = \begin{cases} 0.01 R_0(t), & 0 \le t \le 100, \\ R_0, & t > 100. \end{cases} \tag{21}$$

The AD model is subject to the following periodic boundary conditions:

$$\begin{aligned}
&C_\beta^0(0, z, t) = C_\beta^0(1, z, t); \; C_\beta^0(y, 0, t) = C_\beta^0(y, 1, t); \\
&h(0, z, t) = h(1, z, t); \; h(y, 0, t) = h(y, 1, t); \; \mathcal{U}_\beta(0, z, t) = \mathcal{U}_\beta(1, z, t); \\
&\mathcal{U}_\beta(y, 0, t) = \mathcal{U}_\beta(y, 1, t); \; \mathcal{K}_{10}(0, z, t) = \mathcal{K}_{10}(1, z, t); \\
&\mathcal{K}_{10}(z, 0, t) = \mathcal{K}_{10}(z, 1, t); \; \mathcal{U}_\alpha(0, z, t) = \mathcal{U}_\alpha(1, z, t); \\
&\mathcal{U}_\alpha(y, 0, t) = \mathcal{U}_\alpha(y, 1, t); \; P(0, z, t) = P(1, z, t); \; P(y, 0, t) = P(y, 1, t);
\end{aligned}$$

and the following initial conditions:

$$\begin{aligned}
&C_\beta(y, z, 0) = \frac{1}{10^6}; \; C_\beta^0(y, z, 0) = \frac{1}{10^8}; \; \mathcal{A}'(y, z, 0) = \frac{1.37}{10^{10}}; \; \mathcal{B}_0(y, z, 0) = \frac{3.36}{10^{11}}; \\
&N(y, z, 0) = 0.14; \; A(y, z, 0) = 0.14; \; \mathcal{N}_1(y, z, 0) = 0.02; \\
&\mathcal{N}_2(y, z, 0) = 0.02; \; \hat{\mathcal{N}}_1 = (y, z, 0) = 0; \\
&\hat{\mathcal{N}}_2(y, z, 0) = 0, \; \mathcal{M}_d(y, z, 0) = 0; \; H(y, z, 0) = \frac{1.3}{10^{11}}; \\
&\mathcal{U}_\beta(y, z, 0) = \frac{1}{10^6}; \; \mathcal{U}_\alpha(y, z, 0) = \frac{2}{10^5}; \\
&\mathcal{K}_{10}(y, z, 0) = \frac{1}{10^5}; \; P(y, z, 0) = \frac{5}{10^9}; \; C_0(y, z, 0) = 0.
\end{aligned} \tag{22}$$

Here the given symbol C_β^i is represented as Amyloid-β within neurons, C_β^0 is Extracellular amyloid β peptides outside neurons, $\mathcal{A}'$ is given as Hyperphosphorylated tau protein, $NFT(\mathcal{B}_i)$ shows Neuronfibrillary tangle within neurons, $NFT(\mathcal{B}_0)$ represent Neuronfibrillary tangle outside neurons, N gives the value of Live neurons, A represents Astrocytes, $ROS(R)$ gives Reactive oxygen species, $APP(A_P)$ represents Amyloid precursor protein, $TNf - \alpha(\mathcal{U}_\alpha)$ shows Tumor necrosis factor alpha, $IL - 10(I_10)$ represent Interleukin 10, $MG(M_G)$ represent Microglias, $GSK - 3G$ give the value of Glycogen synthase kinase-type 3, $A\beta0(C_0)$ shows Amyloid β oligomer (soluble), $TGF - \beta(\mathcal{U}_\beta)$ represents Transforming growth factor beta, P represents MCP-1, $\mathcal{N}_1$ represents Proinflammatory microglias, $\hat{\mathcal{N}}_1$ shows Peripheral proinflammatory macrophages, $\mathcal{N}_2$ represents Anti-inflammatory microglias, $\hat{\mathcal{N}}_2$ shows Peripheral anti-inflammatory macrophages, $\mathcal{M}_d$ represents Dead neurons and H represents High mobility group box 1 (HMGB1).

3 AD Model Numerical Solution Using AB Derivative

This section provides numerical solutions to the AD model that were derived using the Newton's polynomial interpolation and the Adams-Bashforth technique. Assume that the stochastic functions can be distinguished with regard to t.

$$dC_i^{'}(t) = C_i^{'}(t)dt. \tag{23}$$

Consider the differential operator with the AB derivative

$$\begin{aligned} C_\beta^i(t) - C_\beta^i(0) &= \frac{1-\chi}{AB(\chi)} f_1(C_\beta^i, z, t) \\ &+ \frac{\chi}{AB(\chi)\Gamma(\chi)} \int_0^t f_1(C_\beta^i, z, \tau)(t-\tau)^{\chi-1} d\tau \\ &+ \frac{1-\chi}{AB(\chi)} G_1(C_\beta^i, t) B_1^{'}(t) \\ &+ \frac{\chi}{AB(\chi)\Gamma(\chi)} \int_0^t G_1(C_\beta^i, \tau) B_1^{'}(C_\beta^i, \tau)(t-\tau)^{\chi-1} d\tau. \end{aligned} \tag{24}$$

$$\begin{aligned} C_\beta^0(t) - C_\beta^0(0) &= \frac{1-\chi}{AB(\chi)} f_2(C_\beta^0, C_\beta^i, t) \\ &+ \frac{\chi}{AB(\chi)\Gamma(\chi)} \int_0^t f_2(C_\beta^0, C_\beta^i, \tau)(t-\tau)^{\chi-1} d\tau \\ &+ \frac{1-\chi}{AB(\chi)} G_2(C_\beta^0, t) B_2^{'}(t) \\ &+ \frac{\chi}{AB(\chi)\Gamma(\chi)} \int_0^t G_2(C_\beta^0, \tau) B_2^{'}(C_\beta^0, \tau)(t-\tau)^{\chi-1} d\tau. \end{aligned} \tag{25}$$

$$\begin{aligned}\mathcal{A}'(t) - \mathcal{A}'(0) &= \frac{1-\chi}{AB(\chi)} f_3(\mathcal{A}', C_\beta^0, t) \\ &+ \frac{\chi}{AB(\chi)\Gamma(\chi)} \int_0^t f_3(\mathcal{A}', C_\beta^0, \tau)(t-\tau)^{\chi-1} d\tau \\ &+ \frac{1-\chi}{AB(\chi)} G_3(\mathcal{A}', C_\beta^0, t) B_3'(t) \\ &+ \frac{\chi}{AB(\chi)\Gamma(\chi)} \int_0^t G_3(\mathcal{A}', \tau) B_3'(\mathcal{A}', \tau)(t-\tau)^{\chi-1} d\tau.\end{aligned} \tag{26}$$

$$\begin{aligned}\mathcal{B}_i(t) - \mathcal{B}_i(0) &= \frac{1-\chi}{AB(\chi)} f_4(\mathcal{B}_i, \mathcal{A}', t) \\ &+ \frac{\chi}{AB(\chi)\Gamma(\chi)} \int_0^t f_4(\mathcal{B}_i, \mathcal{A}', \tau)(t-\tau)^{\chi-1} d\tau \\ &+ \frac{1-\chi}{AB(\chi)} G_4(\mathcal{B}_i, \mathcal{A}', t) B_4'(t) \\ &+ \frac{\chi}{AB(\chi)\Gamma(\chi)} \int_0^t G_4(\mathcal{B}_i, \tau) B_4'(\mathcal{B}_i, \tau)(t-\tau)^{\chi-1} d\tau.\end{aligned} \tag{27}$$

$$\begin{aligned}\mathcal{B}_0(t) - \mathcal{B}_0(0) &= \frac{1-\chi}{AB(\chi)} f_5(\mathcal{B}_0, \mathcal{B}_i, t) \\ &+ \frac{\chi}{AB(\chi)\Gamma(\chi)} \int_0^t f_5(\mathcal{B}_0, \mathcal{B}_i, \tau)(t-\tau)^{\chi-1} d\tau \\ &+ \frac{1-\chi}{AB(\chi)} G_5(\mathcal{B}_0, \mathcal{B}_i, t) B_5'(t) \\ &+ \frac{\chi}{AB(\chi)\Gamma(\chi)} \int_0^t G_5(\mathcal{B}_0, \tau) B_5'(\mathcal{B}_0, \tau)(t-\tau)^{\chi-1} d\tau.\end{aligned} \tag{28}$$

A similar process can be applied to the other constraints.

The functions f_1 and G_1 are approximated using the Newton polynomial, as suggested by Atangana and Seda [1]. This yields

$$
\begin{aligned}
C^{i}_{\beta}(t_{n+1}) &= C^{i}_{\beta}(0) + \frac{1-\chi}{AB(\chi)} f_1\left(C^{i}_{\beta}(t_n), z(t_n), t_n\right) + \frac{\chi(\Delta t)^{\chi}}{AB(\chi)\Gamma(\chi+1)} \\
&\cdot \sum_{j=2}^{n} f_1\left(C^{i}_{\beta j-2}, z_{j-2}, t_{j-2}\right)\Pi + \frac{\chi(\Delta t)^{\chi}}{AB(\chi)\Gamma(\chi+2)} \\
&\cdot \sum_{j=2}^{n} \left(f_1(C^{i}_{\beta j-1}, z_{j-1}, t_{j-1}) - f_1(C^{i}_{\beta j-2}, z_{j-2}, t_{j-2})\right)\Sigma \\
&+ \frac{\chi(\Delta t)^{\chi}}{2AB(\chi)\Gamma(\chi+3)} \cdot \sum_{j=2}^{n} \Big[f_1(C^{i}_{\beta j}, z_j, t_j) - 2f_1(C^{i}_{\beta j-1}, z_{j-1}, t_{j-1}) \\
&+ f_1\left(C^{i}_{\beta j-2}, z_{j-2}, t_{j-2}\right)\Big]\Delta \\
&+ \frac{1-\chi}{AB(\chi)} G_1\left(C^{i}_{\beta}(t_n), z(t_n), t_n\right)\left\{B^{'}_{1}(t_{n+1}) - B^{'}_{1}(t_n)\right\} + \frac{\chi(\Delta t)^{\chi-1}}{AB(\chi)\Gamma(\chi+1)} \\
&\cdot \sum_{j=2}^{n} G_1\left(C^{i}_{\beta j-2}, z_{j-2}, t_{j-2}\right)\left\{B^{'}_{1}(t_{j-1}) - B^{'}_{1}(t_{j-2})\right\}\Pi + \frac{\chi(\Delta t)^{\chi-1}}{AB(\chi)\Gamma(\chi+2)} \\
&\sum_{j=2}^{n} \left(G_1(C^{i}_{\beta j-1}, z_{j-1}, t_{j-1})\left\{B^{'}_{1}(t_j) - B^{'}_{1}(t_{j-1})\right\} - G_1(C^{i}_{\beta j-2}, z_{j-2}, t_{j-2})\right) \\
&\cdot \left\{B^{'}_{1}(t_{j-1}) - B^{'}_{1}(t_{j-2})\right\}\Sigma + \frac{\chi(\Delta t)^{\chi-1}}{2AB(\chi)\Gamma(\chi+3)} \\
&\cdot \sum_{j=2}^{n} \Big[G_1(C^{i}_{\beta j}, z_j, t_j)\left\{B^{'}_{1}(t_{j+1}) - B^{'}_{1}(t_j)\right\} \\
&-2G_1(C^{i}_{\beta j-1}, z_{j-1}, t_{j-1})\left\{B^{'}_{1}(t_j) - B^{'}_{1}(t_{j-1})\right\} \\
&+G_1\left(C^{i}_{\beta j-2}, z_{j-2}, t_{j-2}\right)\left\{B^{'}_{1}(t_{j-1}) - B^{'}_{1}(t_{j-2})\right\}\Big]\Delta,
\end{aligned}
\tag{29}
$$

where

$$
\Sigma = \begin{bmatrix} (n-j+1)^{\chi}(n-j+3+2\chi) \\ -(n-j)^{\chi}(n-j+3+3\chi) \end{bmatrix},
$$

$$
\Delta = \begin{bmatrix} (n-j+1)^{\chi}\begin{bmatrix} 2(n-j)^2 + (3\chi+10)(n-j) \\ +2\chi^2 + 9\chi + 12 \end{bmatrix} \\ -(n-j)^{\chi}\begin{bmatrix} 2(n-j)^2 + (5\chi+10)(n-j) \\ 6\chi^2 + 18\chi + 12 \end{bmatrix} \end{bmatrix},
$$

and

$$
\Pi = [(n-j+1)^{\chi} - (n-j)^{\chi}].
$$

A neuro degenerate condition called AD is characterized by a steady decline in cognitive ability. By using the numerical method to extracellular amyloid-beta peptide restrictions outside of neurons.

Numerous researchers think that pre-clinical research will yield the most promising results for future therapies to reduce or stop AD progression and maintain brain function. In order to determine which people are in these early phases and should receive therapies when they become available, bio-marker tests will be crucial. The following solution is produced when the hyperphosphorylated tau protein's function is applied to the fractional derivative:

$$
\begin{aligned}
C_{\beta}^{0}(t_{n+1}) &= C_{\beta}^{0}(0) + \frac{1-\chi}{AB(\chi)} f_2\left(C_{\beta}^{0}(t_n), C_{\beta}^{i}(t_n), t_n\right) + \frac{\chi(\Delta t)^{\chi}}{AB(\chi)\Gamma(\chi+1)} \\
&\cdot \sum_{j=2}^{n} f_2\left(C_{\beta j-2}^{0}, C_{\beta j-2}^{i}, t_{j-2}\right) \Pi + \frac{\chi(\Delta t)^{\chi}}{AB(\chi)\Gamma(\chi+2)} \\
&\cdot \sum_{j=2}^{n} \left(f_2(C_{\beta j-1}^{0}, C_{\beta j-1}^{i}, t_{j-1}) - f_2(C_{\beta j-2}^{0}, C_{\beta j-2}^{i}, t_{j-2})\right) \Sigma + \frac{\chi(\Delta t)^{\chi}}{2AB(\chi)\Gamma(\chi+3)} \\
&\cdot \sum_{j=2}^{n} \left[f_2(C_{\beta j}^{0}, C_{\beta j}^{i}, t_j) - 2f_2(C_{\beta j-1}^{0}, C_{\beta j-1}^{i}, t_{j-1}) + f_2\left(C_{\beta j-2}^{0}, C_{\beta j-2}^{i}, t_{j-2}\right)\right] \Delta \\
&+ \frac{1-\chi}{AB(\chi)} G_2\left(C_{\beta}^{0}(t_n), C_{\beta}^{i}(t_n), t_n\right) \left\{B_2^{'}(t_{n+1}) - B_2^{'}(t_n)\right\} + \frac{\chi(\Delta t)^{\chi-1}}{AB(\chi)\Gamma(\chi+1)} \\
&\cdot \sum_{j=2}^{n} G_2\left(C_{\beta j-2}^{0}, C_{\beta j-2}^{i}, t_{j-2}\right) \left\{B_2^{'}(t_{j-1}) - B_2^{'}(t_{j-2})\right\} \Pi \\
&+ \frac{\chi(\Delta t)^{\chi-1}}{AB(\chi)\Gamma(\chi+2)} \sum_{j=2}^{n} \Bigg(G_2(C_{\beta j-1}^{0}, C_{\beta j-1}^{i}, t_{j-1}) \left\{B_2^{'}(t_j) - B_2^{'}(t_{j-1})\right\} \\
&- G_2(C_{\beta j-2}^{0}, C_{\beta j-2}^{i}, t_{j-2})\Bigg) \left\{B_2^{'}(t_{j-1}) - B_2^{'}(t_{j-2})\right\} \Sigma \\
&+ \frac{\chi(\Delta t)^{\chi-1}}{2AB(\chi)\Gamma(\chi+3)} \sum_{j=2}^{n} \Bigg[G_2(C_{\beta j}^{0}, C_{\beta j}^{i}, t_j) \left\{B_2^{'}(t_{j+1}) - B_2^{'}(t_j)\right\} \\
&- 2G_2(C_{\beta j-1}^{0}, C_{\beta j-1}^{i}, t_{j-1}) \left\{B_2^{'}(t_j) - B_2^{'}(t_{j-1})\right\} \\
&+ G_2\left(C_{\beta j-2}^{0}, C_{\beta j-2}^{i}, t_{j-2}\right) \left\{B_2^{'}(t_{j-1}) - B_2^{'}(t_{j-2})\right\} \Bigg] \Delta,
\end{aligned} \tag{30}
$$

$$\mathcal{A}'(t_{n+1}) = \mathcal{A}'(0) + \frac{1-\chi}{AB(\chi)} f_3\left(\mathcal{A}'(t_n), C^0_\beta(t_n), t_n\right)$$

$$+ \frac{\chi(\Delta t)^\chi}{AB(\chi)\Gamma(\chi+1)} \sum_{j=2}^{n} f_3\left(\mathcal{A}'_{j-2}, C^0_{\beta j-2}, t_{j-2}\right)$$

$$\Pi + \frac{\chi(\Delta t)^\chi}{AB(\chi)\Gamma(\chi+2)} \sum_{j=2}^{n} \left(f_3(\mathcal{A}'_{j-1}, C^0_{\beta j-1}, t_{j-1}) - f_3(\mathcal{A}'_{j-2}, C^0_{\beta j-2}, t_{j-2})\right)\Sigma$$

$$+ \frac{\chi(\Delta t)^\chi}{2AB(\chi)\Gamma(\chi+3)} \sum_{j=2}^{n} \Big[f_3(\mathcal{A}'_j, C^0_{\beta j}, t_j) - 2f_3(\mathcal{A}'_{j-1}, C^0_{\beta j-1}, t_{j-1})$$

$$+ f_3\left(\mathcal{A}'_{j-2}, C^0_{\beta j-2}, t_{j-2}\right)\Big]\Delta$$

$$+ \frac{1-\chi}{AB(\chi)} G_3\left(\mathcal{A}'(t_n), C^0_\beta(t_n), t_n\right)\left\{B'_3(t_{n+1}) - B'_3(t_n)\right\}$$

$$+ \frac{\chi(\Delta t)^{\chi-1}}{AB(\chi)\Gamma(\chi+1)} \sum_{j=2}^{n} G_3\left(\mathcal{A}'_{j-2}, C^0_{\beta j-2}, t_{j-2}\right)$$

$$\left\{B'_3(t_{j-1}) - B'_3(t_{j-2})\right\}\Pi + \frac{\chi(\Delta t)^{\chi-1}}{AB(\chi)\Gamma(\chi+2)}$$

$$\cdot \sum_{j=2}^{n} \Big(G_3(\mathcal{A}'_{j-1}, C^0_{\beta j-1}, t_{j-1}) \left\{B'_3(t_j) - B'_3(t_{j-1})\right\}$$

$$- G_3(\mathcal{A}'_{j-2}, C^0_{\beta j-2}, t_{j-2})\Big) \left\{B'_3(t_{j-1}) - B'_3(t_{j-2})\right\}\Sigma + \frac{\chi(\Delta t)^{\chi-1}}{2AB(\chi)\Gamma(\chi+3)}$$

$$\sum_{j=2}^{n} \Big[G_3(\mathcal{A}'_j, C^0_{\beta j}, t_j)\left\{B'_3(t_{j+1}) - B'_3(t_j)\right\} - 2G_3(\mathcal{A}'_{j-1}, C^0_{\beta j-1}, t_{j-1})$$

$$\left\{B'_3(t_j) - B'_3(t_{j-1})\right\} + G_3\left(\mathcal{A}'_{j-2}, C^0_{\beta j-2}, t_{j-2}\right)\left\{B'_3(t_{j-1}) - B'_3(t_{j-2})\right\}\Big]\Delta. \tag{31}$$

The NFTs in neurons $\mathcal{B}_i$ are formed from the hyperphosphorylated tau proteins. They are released to the extracellular space when the neurons die. From the constraints on NFTs within neurons, the solutions using fractional derivatives are as follows:

$$
\begin{aligned}
&\mathcal{B}_i(t_{n+1}) = \mathcal{B}_i(0) + \frac{1-\chi}{AB(\chi)} f_4\left(\mathcal{B}_i(t_n), \mathcal{A}'(t_n), t_n\right) \\
&+ \frac{\chi(\Delta t)^{\chi}}{AB(\chi)\Gamma(\chi+1)} \sum_{j=2}^{n} f_4\left(F_{i(j-2)}, \mathcal{A}'_{j-2}, t_{j-2}\right) \Pi \\
&+ \frac{\chi(\Delta t)^{\chi}}{AB(\chi)\Gamma(\chi+2)} \sum_{j=2}^{n} \left(f_4(F_{i(j-1)}, \mathcal{A}'_{j-1}, t_{j-1}) \right. \\
&\left. - f_4(F_{i(j-2)}, \mathcal{A}'_{j-2}, t_{j-2}) \right) \Sigma + \frac{\chi(\Delta t)^{\chi}}{2AB(\chi)\Gamma(\chi+3)} \\
&\cdot \sum_{j=2}^{n} \left[f_4(F_{ij}, \tau'_j, t_j) - 2 f_4(F_{i(j-1)}, \tau'_{j-1}, t_{j-1}) + f_4\left(F_{i(j-2)}, \tau'_{j-2}, t_{j-2}\right)\right] \Delta \\
&+ \frac{1-\chi}{AB(\chi)} G_4\left(\mathcal{B}_i(t_n), \tau'(t_n), t_n\right)\left\{B'_4(t_{n+1}) - B'_4(t_n)\right\} + \frac{\chi(\Delta t)^{\chi-1}}{AB(\chi)\Gamma(\chi+1)} \\
&\cdot \sum_{j=2}^{n} G_4\left(F_{i(j-2)}, \tau'_{j-2}, t_{j-2}\right)\left\{B'_4(t_{j-1}) - B'_4(t_{j-2})\right\} \Pi + \frac{\chi(\Delta t)^{\chi-1}}{AB(\chi)\Gamma(\chi+2)} \\
&\cdot \sum_{j=2}^{n} \left(G_4(F_{i(j-1)}, \tau'_{j-1}, t_{j-1})\left\{B'_4(t_j) - B'_4(t_{j-1})\right\} - G_4(F_{i(j-2)}, \tau'_{j-2}, t_{j-2}) \right) \\
&\left\{B'_4(t_{j-1}) - B'_4(t_{j-2})\right\} \Sigma + \frac{\chi(\Delta t)^{\chi-1}}{2AB(\chi)\Gamma(\chi+3)} \\
&\sum_{j=2}^{n} \left[G_4(F_{ij}, \tau'_j, t_j)\left\{B'_4(t_{j+1}) - B'_4(t_j)\right\} - 2G_4(F_{i(j-1)}, \tau'_{j-1}, t_{j-1}) \right. \\
&\left. \cdot \left\{B'_4(t_j) - B'_4(t_{j-1})\right\} + G_4\left(F_{i(j-2)}, \tau'_{j-2}, t_{j-2}\right)\left\{B'_4(t_{j-1}) - B'_4(t_{j-2})\right\}\right] \Delta.
\end{aligned} \tag{32}
$$

A large number of studies suggest that there is a correlation between synaptotoxic amyloid species, synapse loss, and the cognitive deterioration that clinically characterizes AD patients. These studies provide better insights into the kinetic processes underlying eventual fibril/plaque formation. For NFTs outside neurons, the solutions using fractional derivatives are as follows:

$$
\begin{aligned}
\mathcal{B}_0(t_{n+1}) &= \mathcal{B}_0(0) + \frac{1-\chi}{AB(\chi)} f_5\left(\mathcal{B}_0(t_n), \mathcal{B}_i(t_n), t_n\right) \\
&+ \frac{\chi(\Delta t)^{\chi}}{AB(\chi)\Gamma(\chi+1)} \sum_{j=2}^{n} f_5\left(\mathcal{B}_{0j-2}, F_{i(j-2)}, t_{j-2}\right) \\
\Pi &+ \frac{\chi(\Delta t)^{\chi}}{AB(\chi)\Gamma(\chi+2)} \sum_{j=2}^{n} \left(f_5(\mathcal{B}_{0j-1}, F_{i(j-1)}, t_{j-1}) - f_5(\mathcal{B}_{0j-2}, F_{i(j-2)}, t_{j-2})\right) \Sigma \\
&+ \frac{\chi(\Delta t)^{\chi}}{2AB(\chi)\Gamma(\chi+3)} \sum_{j=2}^{n} \Big[f_5(\mathcal{B}_{0j}, F_{ij}, t_j) - 2f_5(\mathcal{B}_{0j-1}, F_{i(j-1)}, t_{j-1}) \\
&+ f_5\left(\mathcal{B}_{0j-2}, F_{i(j-2)}, t_{j-2}\right) \Big] \Delta + \frac{1-\chi}{AB(\chi)} G_5\left(\mathcal{B}_{0n}, \mathcal{B}_i(t_n), t_n\right) \\
&\left\{B_5'(t_{n+1}) - B_5'(t_n)\right\} + \frac{\chi(\Delta t)^{\chi-1}}{AB(\chi)\Gamma(\chi+1)} \sum_{j=2}^{n} G_5\left(\mathcal{B}_{0j-2}, F_{i(j-2)}, t_{j-2}\right) \\
&\left\{B_5'(t_{j-1}) - B_5'(t_{j-2})\right\} \Pi + \frac{\chi(\Delta t)^{\chi-1}}{AB(\chi)\Gamma(\chi+2)} \\
&\sum_{j=2}^{n} \Big(G_5(\mathcal{B}_{0j-1}, F_{i(j-1)}, t_{j-1}) \left\{B_5'(t_j) - B_5'(t_{j-1})\right\} \\
&- G_5(\mathcal{B}_{0j-2}, F_{i(j-2)}, t_{j-2}) \Big) \left\{B_5'(t_{j-1}) - B_5'(t_{j-2})\right\} \Sigma + \frac{\chi(\Delta t)^{\chi-1}}{2AB(\chi)\Gamma(\chi+3)} \\
&\cdot \sum_{j=2}^{n} \Big[G_5(\mathcal{B}_{0j}, F_{ij}, t_j) \left\{B_5'(t_{j+1}) - B_5'(t_j)\right\} \\
&- 2G_5(\mathcal{B}_{0j-1}, F_{i(j-1)}, t_{j-1}) \left\{B_5'(t_j) - B_5'(t_{j-1})\right\} \\
&+ G_5\left(\mathcal{B}_{0j-2}, F_{i(j-2)}, t_{j-2}\right) \left\{B_5'(t_{j-1}) - B_5'(t_{j-2})\right\} \Big] \Delta.
\end{aligned} \tag{33}
$$

The remaining constraints can be handled in a similar way.

4 Numerical Simulations

The numerical iteration scheme developed in earlier parts is used in this section to describe numerical simulations. Numerous parameters and fractional orders are used in the numerical simulations. The findings are shown in Figs. 1, 2, 3, 4, 5, 6, 7 and 8. The results for various levels of randomness are shown in these graphs, which show how fractional derivatives increase randomness in compared to classical derivatives. Due of the system's complexity and abundance of equations, we only present four solutions.

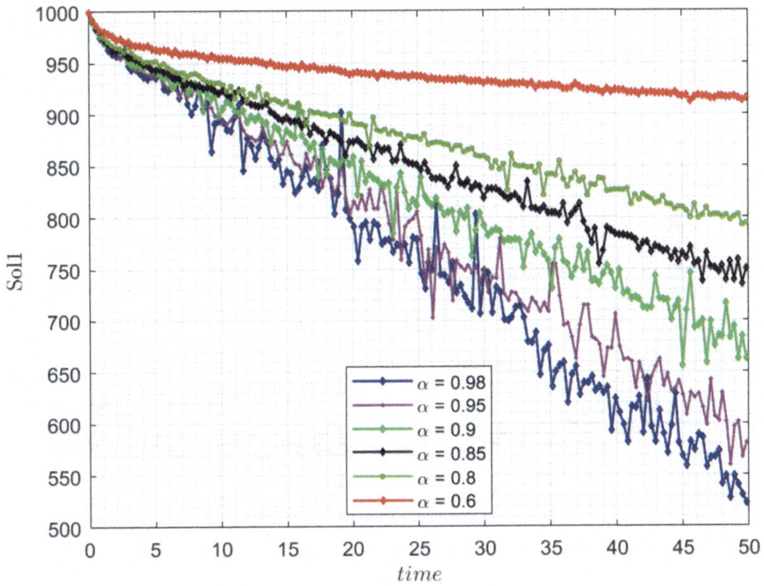

Fig. 1 Amyloid-β inside neurons is simulated numerically using the AB fractional derivative

Figures 1, 2, 3, 4 show solutions under initial conditions of 1000, 100, 1, and 1, respectively. The corresponding randomness densities are 0.001, 0.06, 0.007, and 0.008; for Figs. 5, 6, 7, 8, randomness densities of 0.001, 0.001, 0.002, and 0.001 are used, respectively.

5 Conclusion

This study used fractional differential and integral operators, as well as Brownian motion, to describe the complicated spread of AD. The model was solved using a Newtonian polynomial-based numerical approach. The findings provide us with crucial information that will help us better understand the dynamics of AD. This may provide researchers with some insights into future practical medical trials for a variety of purposes.

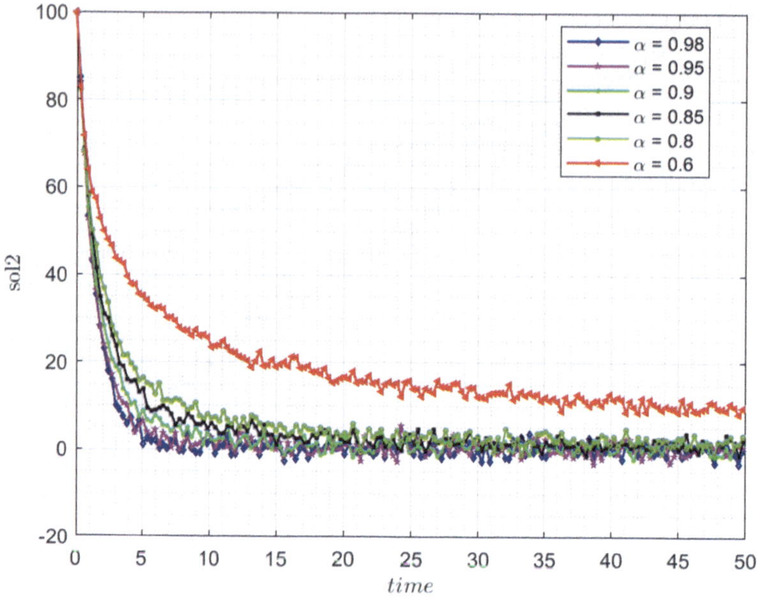

Fig. 2 Numerical computing for amyloid-β within dead neurons with AB derivative

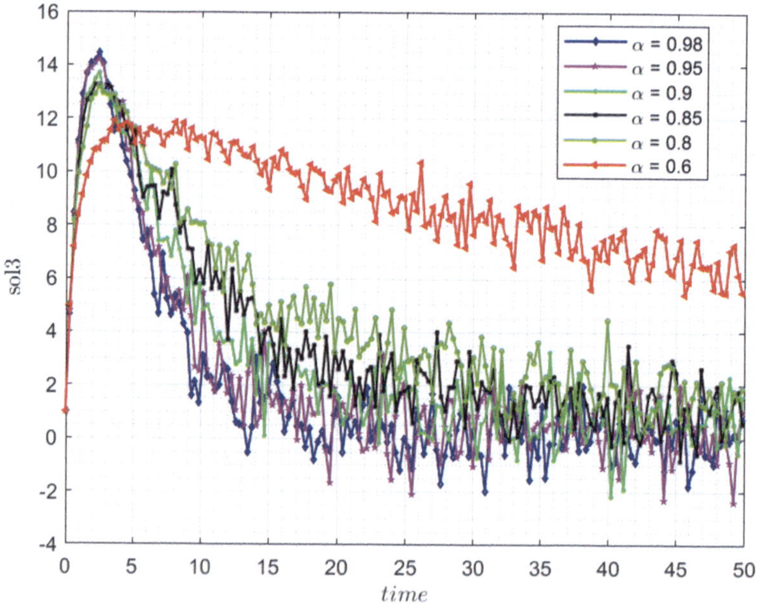

Fig. 3 Numerical computing for extracellular amyloid-β peptides outside neurons with AB derivative

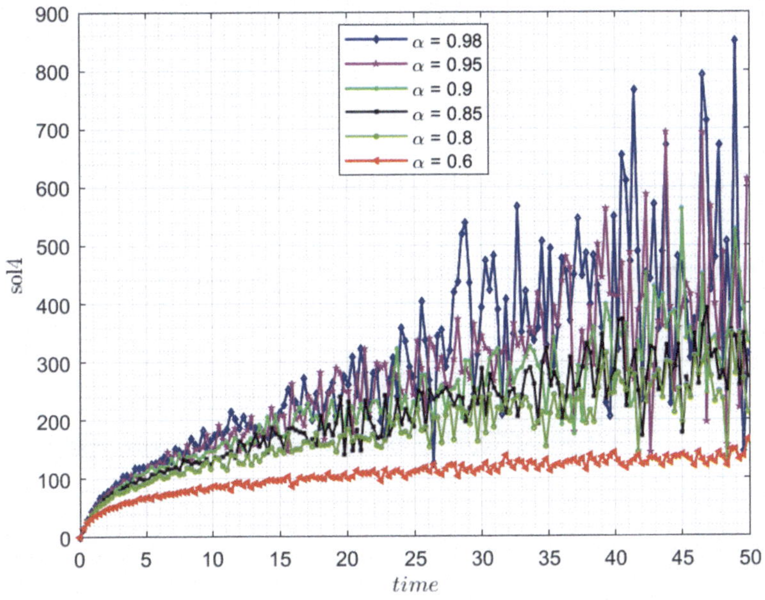

Fig. 4 Numerical computing for transforming growth factor beta with AB derivative

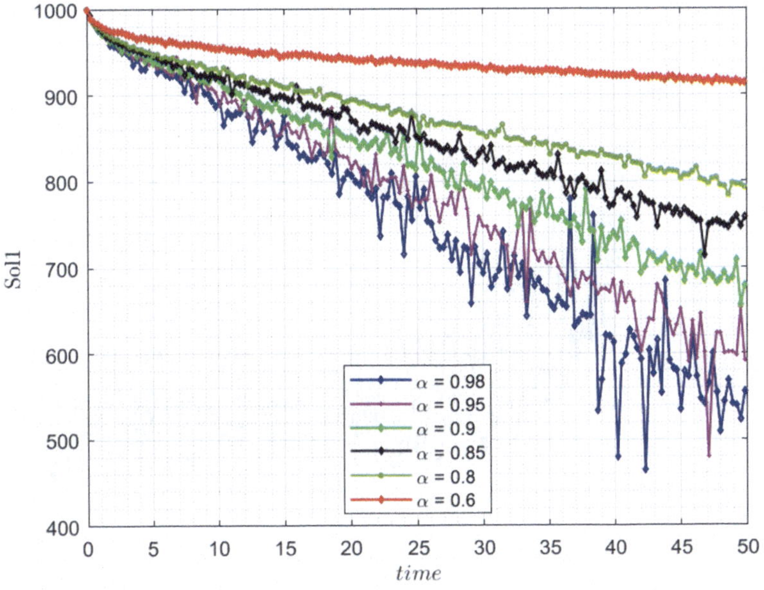

Fig. 5 Numerical computing for amyloid-β with neurons with integer-number derivative

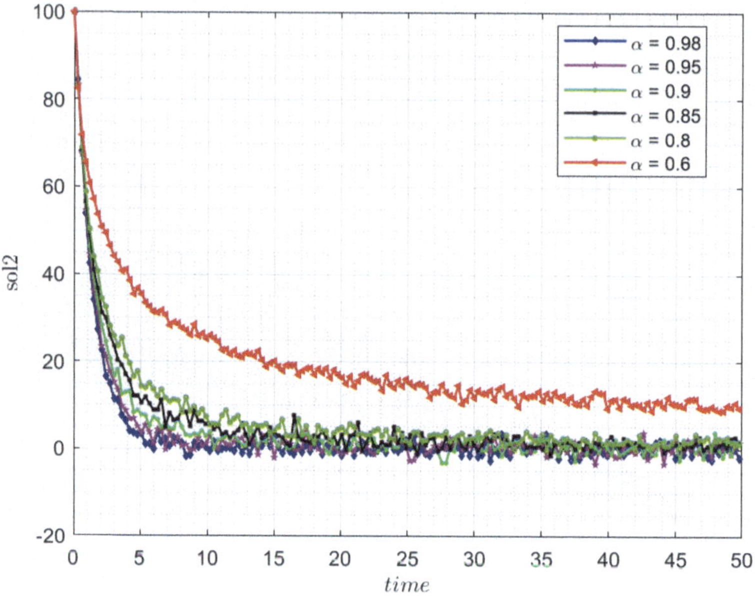

Fig. 6 Numerical computing for amyloid-β with dead neurons with integer-number derivative

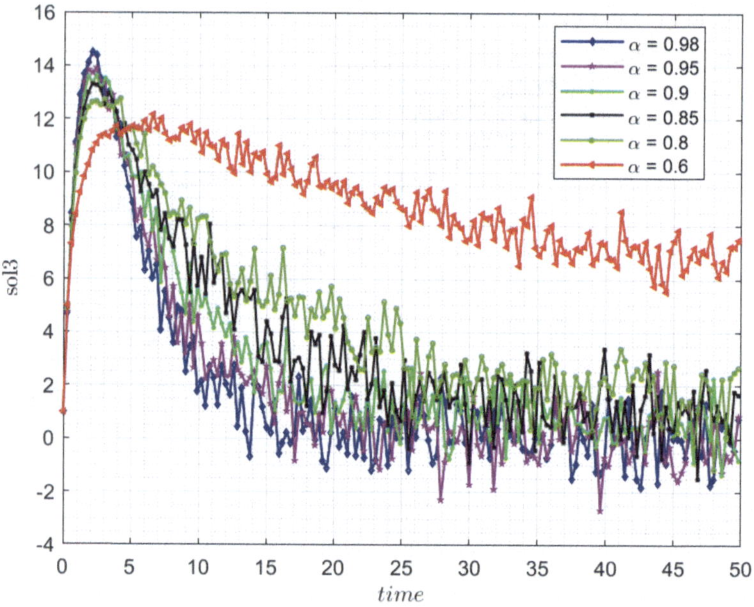

Fig. 7 Numerical computing for extracellular amyloid-β peptides outside neurons with integer-number derivative

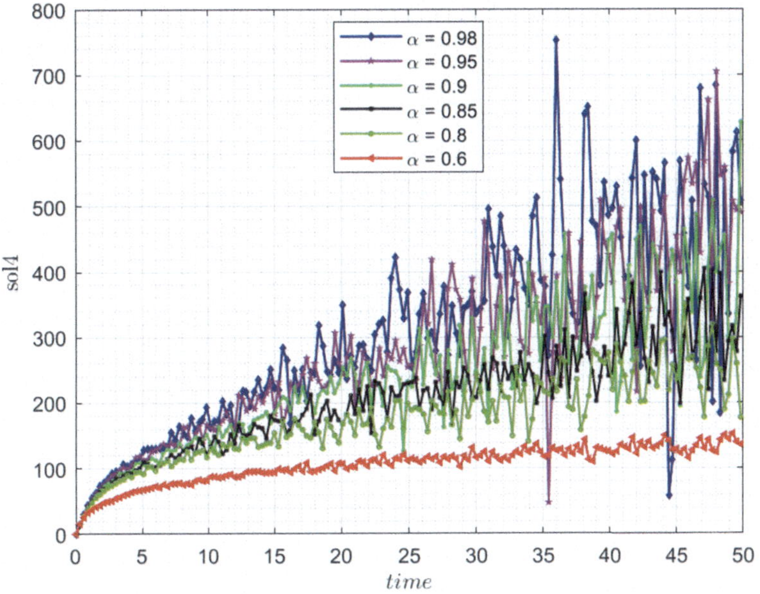

Fig. 8 Numerical computing for transforming growth factor beta for integer-number derivative

References

1. A. Atangana, S. Igret Araz, *New Numerical Scheme with Newton Polynomial: Theory, Methods and Applications* (Academic Press, Elsevier, 2021)
2. A. Atangana, Fractal-fractional differentiation and integration: connecting fractal calculus and fractional calculus to predict complex system. Chaos Soliton Fract. **102**, 397–407 (2017)
3. A. Martin Prince, *World Alzheimer Report 2015: the Global Impact of Dementia: An Analysis of Prevalence, Incidence, Cost and Trends* (Alzheimer's Disease International, 2015)
4. A. Atangana, S. Jain, A new numerical approximation of the fractal ordinary differential equation. Eur. Phys. J. Plus **133**, 37 (2018). https://doi.org/10.1140/epjp/i2018-11895-1
5. Alzheimer's Association, Alzheimer's disease facts and figures. Alzheimer's Dementia **15**(2019), 321–387 (2019)
6. C. Janus, D. Westaway, Transgenic mouse models of Alzheimer's disease. Physiol. Behav. **73**(5), 873–886 (2001)
7. K. Diethelm, A.D. Freed, The FracPECE subroutine for the numerical solution of differential equations of fractional order. Forschung und wissenschaftliches Rechnen **1998**, 57–71 (1999)
8. N.J. Ford, A.C. Simpson, The numerical solution of fractional differential equations: speed versus accuracy. Numer. Algorithms **26**(4), 333–346 (2001)
9. Z. Odibat, S. Momani, An algorithm for the numerical solution of differential equations of fractional order. J. Appl. Math. Inform. **26**(1–2), 15–27 (2008)
10. Q. Zhao, X. Tang, Effects of Huperzine A on an acetylcholinesterase isoform in vitro: comparison with tacrine, donepezil, rivastigmine and physostigmine. Eur. J. Pharmacol. **455**, 101–107 (2002)
11. S. Jain, Numerical analysis for the fractional diffusion and fractional Buckmaster's equation by two step Adams-Bashforth method. Eur. Phys. J. Plus **133**, 19 (2018). https://doi.org/10.1140/epjp/i2018-11854-x

12. R.B. Srivastava, S. Shukla, *Numerical Accuracies of Lagrange's and Newton Polynomial Interpolation: Numerical Accuracies of Interpolation Formulas* (Lambert Academic Publishing, 2012)
13. R.B. Srivastava, P.K. Srivastava, Comparison of Lagrange's and Newton's interpolating polynomials. J. Exp. Sci. **3**(1), 01–04 (2012)
14. R. Brookmeyer, Forecasting the global burden of Alzheimer's disease. Alzheimer's Dementia **3**(3), 186–191 (2007)
15. T. Zhang, J. Jin, T. Jiang, The decoupled Crank-Nicolson/Adams-Bashforth scheme for the Boussinesq equations with nonsmooth initial data. Appl. Math. Comput. **337**, 234–266 (2018)
16. W. Hao, A. Friedman, Mathematical model on Alzheimer's disease. BMC Syst. Biol. **10**, 108 (2016). https://doi.org/10.1186/s12918-016-0348-2
17. Z. Zizhang, S. Jain, Mathematical model of Ebola and Covid-19 with fractional differential operators: non-Markovian process and class for virus pathogen in the environment. Chaos Soliton Fract. **140**, 110175 (2020)

A Stable Relaxation Approach for Valuing European Options Across Diverse Time-Fractional Models

Mohamed Kharrat

Abstract This chapter introduces and elucidates several models for the valuation of European options. We explore two scenarios: first, when the interest rate is stochastic, and second, when volatility follows a stochastic process. The first case involves the utilization of the fractional Vasicek model, the fractional Cox-Ingersoll-Ross model, and, lastly, the fractional Heston model. The dependability, appropriateness of fit, and stability of our methodology are rigorously examined through theoretical studies conducted to validate our models. The outcomes of this study have been disseminated in (Arfaou and Kharrat in Filomat 37(9):2685–2697, 2023 [3], Kharrat in Filomat 35(10), 2021 [20], Kharrat and Arfaoui in Comput Econ 61:1745–1763, 2023 [21] and Kharrat in Methodol Comput Appl Probab 25(50), 2023 [22]).

1 Introduction and Background

Valuing financial derivatives, particularly options, stands out as a prominent challenge in the realm of mathematical finance. European options, in particular, have gained widespread popularity within the global financial markets. In recent decades, a multitude research papers have delved into the intricacies of the evaluation of options, exploring various models and employing diverse methodologies to address this complex problem for instance [5, 6, 15, 17]. The most renowned models include the Black and Scholes model [6] and the Heston model [15]. The former is built on the premise that the underlying asset price follows a log-normal distribution conditioned on the current stock price, where the volatility is constant. In contrast to the model of Black and Scholes, where volatility remains constant, the Heston model [15] assumes stochastic volatility. Given the significance of volatility dynamics in formulating strategies for arbitrage and hedging, a model with constant volatility falls short in capturing the reality of financial markets. However, when the interest

M. Kharrat (✉)
Mathematics Department, College of Science, Jouf University, Sakaka, Saudi Arabia
e-mail: mkharrat@ju.edu.sa

M. Kharrat et al. (eds.), *Modeling of Discrete and Continuous Systems*, Forum for Interdisciplinary Mathematics, https://doi.org/10.1007/978-981-97-8715-9_9

rate is stochastic, the Vasicek Model [29] becomes a necessary choice. Consequently, pricing options under stochastic models becomes more crucial and imperative.

In the forthcoming definition, we outline the Caputo time-fractional derivative.

Definition 1 [23]. The Caputo time fractional derivative of order $\gamma > 0$, where a, γ, and t are real numbers, is defined as follows:

$$
{}^{c}D_{a}^{\gamma} f(t) = \begin{cases} \frac{1}{\Gamma(m-\gamma)} \int\limits_{a}^{t} (t-\tau)^{m-\gamma-1} \frac{d^m}{d\tau^m} f(\tau) d\tau, & m-1 < \gamma < m, \quad m \in \mathbb{N}, \\ \frac{d^m}{d\tau^m} f(t), \quad \gamma = m, \quad m \in \mathbb{N}. & \end{cases} \tag{1}
$$

When $m = 1 : 0 < \gamma < 1$, then the Caputo fractional derivative of order γ of the function f reduces to

$$
{}^{c}D_{a}^{\gamma} f(t) = \frac{1}{\Gamma(1-\gamma)} \int\limits_{a}^{t} (t-\tau)^{-\gamma} \frac{d}{d\tau} f(\tau) d\tau,
$$

with $\Gamma(\cdot)$ is the Gamma function given by

$$
\Gamma(x) = \int\limits_{0}^{\infty} t^{x-1} e^{-t}\, dt.
$$

This definition of a fractional derivative is noteworthy for several reasons, one of which is that it preserves properties of non-fractional derivatives, such as rendering the derivative of a constant null (refer to [26]). In this study, we apply this definition with $m = 1$.

Definition 2 [11] The one-parameter Mittag-Leffler function is defined as:

$$
E_{\zeta}(\sigma) = \sum_{n=0}^{\infty} \frac{\sigma^n}{\Gamma(\zeta n + 1)}, \quad Re(\zeta) > 0, \ \sigma \in \mathbb{C},
$$

where the gamma function noted by $\Gamma(\cdot)$.

Definition 3 [11] The Mittag-Leffler function of two parameters is defined as:

$$
E_{\zeta,\xi}(\sigma) = \sum_{n=0}^{\infty} \frac{\sigma^n}{\Gamma(\zeta n + \xi)}, \quad Re(\zeta) > 0, \ \xi > 0, \ \sigma \in \mathbb{C}.
$$

This chapter is structured in the following manner. In Sect. 2, we present the time fractional Vasicek model and the stability of the solution. In Sect. 3 we study the time fractional Heston model and we analyse the stability of the related solution. In the last section we conclude.

2 Time-Fractional Vasicek Model

Below, we present the Vasicek model [29]. We denote by S_t the asset price driven by the ensuing dynamics:

$$dS_t = r_t S_t dt + \sigma S_t dW_t^S \tag{2}$$

where r_t represents the interest rate, which adheres to the subsequent dynamics:

$$dr_t = (\mu_1 + \mu_2 r_t) + \eta dW_t^r \tag{3}$$

assuming the volatility σ to be constant, where W_t^S and W_t^r are two correlated Brownian motions i.e. $W_t^S = \sqrt{1-\rho^2} B_t^1 + B_t^2$ and $W_t^r = B_t^2$ where B is Standard 2- dimensional Brownian motion and $\rho \in]-1, 1[$ and the parameters $\eta, -\mu_2$ and $\frac{-\mu_1}{\mu_2}$ represent respectively the volatility of the interest rate r_t, the speed of the reversion and the long-term mean level. Consider ϑ as the price of a European option. By employing standard hedging techniques and applying Ito's lemma, we obtain:

$$\begin{aligned} &\frac{\partial \vartheta}{\partial \tau} + \frac{1}{2}\sigma^2 S^2 \frac{\partial^2 \vartheta}{\partial S^2} + \rho\sigma\eta S \frac{\partial^2 \vartheta}{\partial S \partial r} + \frac{1}{2}\eta^2 \frac{\partial^2 \vartheta}{\partial r^2} r S \frac{\partial \vartheta}{\partial S} \\ &+ (\mu_1 + \mu_2 r)\frac{\partial \vartheta}{\partial r} - r\vartheta = 0. \end{aligned} \tag{4}$$

Fractional calculus is utilized across multiple disciplines [1, 4, 9, 10, 26, 28, 31]. For instance, fractional derivative models have demonstrated an improved ability to characterize shape-memory materials compared to full derivative models. When a material is purely elastic, it is described by an integer derivative of order zero, and when it is purely viscous, it is described by an integer derivative of order one. Consequently, driven by mathematical curiosity and a desire to better align with the intricacies of financial markets, there is a compelling need to employ models grounded in fractional derivatives. This trend has gained traction in the field of mathematical finance [20, 30, 31], particularly for addressing challenges related to option pricing. As an illustration, the studies outlined in [18, 32] focus specifically on evaluating European options.

From this standpoint, utilizing the splitting method and within the fractional Vasicek model, we introduce a novel approach for pricing European options. This method enables the resolution of a mixed Hyperbolic/Parabolic problem by separating the parabolic and hyperbolic operators (for additional information, refer to [2]). In numerical simulations, dealing with a nonlinear mixed issue created by two separate operators (Parabolic/Hyperbolic) can be difficult. Through discretization, the splitting method allows the treatment of each Parabolic and Hyperbolic operator with an appropriate numerical scheme, preserving the numerical properties of each scheme used for the respective operators.

Under the time fractional Vasicek model, the price dynamics of the European put option can be stated as follows using the Caputo time fractional order derivative ${}^{c}D_{\tau}^{\gamma}$ with order $0 < \gamma < 1$:

$$\begin{aligned} {}^{c}D_{\tau}^{\gamma}\vartheta + \frac{1}{2}\sigma^2 S^2 \frac{\partial^2 \vartheta}{\partial S^2} + \rho\sigma\eta S \frac{\partial^2 \vartheta}{\partial S \partial r} + \frac{1}{2}\eta^2 \frac{\partial^2 \vartheta}{\partial r^2} \\ + rS\frac{\partial \vartheta}{\partial S} + (\mu_1 + \mu_2 r)\frac{\partial \vartheta}{\partial r} - r\vartheta = 0, \end{aligned} \tag{5}$$

$\forall\ (S, \tau, r) \in \mathcal{Q}_{\infty}^{T} = (0, \infty) \times (0, \infty) \times (0, T)$ and with:

$$\vartheta(S, T, r) = (K - S)^{+}, \tag{6}$$

$$\lim_{S \to 0} \vartheta(S, \tau, r) = K, \qquad \lim_{S \to \infty} \vartheta(S, \tau, r) = 0, \tag{7}$$

$$\lim_{r \to 0} \vartheta(S, \tau, r) = \lim_{r \to \infty} \vartheta(S, \tau, r) = (K - S)^{+}, \tag{8}$$

If $t = T - \tau$ so we get ${}^{c}D_{\tau}^{\gamma}\vartheta = -{}^{c}D_{t}^{\gamma}\vartheta$, with the new variable $(S, t, r) \in \mathcal{Q}_{\infty}^{T}$, the previous problem (5)-(8) is given by:

$$\begin{aligned} {}^{c}D_{t}^{\gamma}\vartheta = \frac{1}{2}\sigma^2 S^2 \frac{\partial^2 \vartheta}{\partial S^2} + \rho\sigma\eta S \frac{\partial^2 \vartheta}{\partial S \partial r} + \frac{1}{2}\eta^2 \frac{\partial^2 \vartheta}{\partial r^2} + \\ rS\frac{\partial \vartheta}{\partial S} + (\mu_1 + \mu_2 r)\frac{\partial \vartheta}{\partial r} - r\vartheta, \end{aligned} \tag{9}$$

with the new boundary conditions:

$$\vartheta(S, 0, r) = (K - S)^{+}, \tag{10}$$

$$\lim_{S \to 0} \vartheta(S, t, r) = K, \qquad \lim_{S \to \infty} \vartheta(S, t, r) = 0, \tag{11}$$

$$\lim_{r \to 0} \vartheta(S, t, r) = \lim_{r \to \infty} \vartheta(S, t, r) = (K - S)^{+}, \tag{12}$$

2.1 Splitting Method

In the upcoming section, we suggest the utilization of a splitting method [2]. This approach maintains the numerical characteristics (consistency, stability, quality, ...) of each scheme employed for individual operators.

The time interval $[0, T]$ is discretized into $(N_t + 1)$ uniformly spaced points by doing the following:

$$\Delta t = \frac{T}{N_t} \quad \text{where} \quad t^k = k\Delta t, \quad \text{for all} \quad k = 0, \dots, N_t.$$

As a result, we have:

$$[0, T] = \bigcup_{k=0}^{N_t-1} [t^k, t^{k+1}].$$

Let consider this approximation:

$$\vartheta(S, r, t^k) \approx \vartheta^k(S, r),$$

$\forall\ k = 0, \dots, N_t$ and $(S, r) \in (0, \infty) \times (0, \infty)$.

The splitting technique entails solving the Eq. (9) for each interval $[t^k, t^{k+1}]\ \forall\ k = 0, \cdots, N_t - 1$, Observe the following steps:

(i) First solving:

$$\begin{aligned} {}^cD_t^\gamma \vartheta &= \rho\sigma\eta S\frac{\partial^2\vartheta}{\partial S\partial r}, \quad \text{on} \quad [t^k, t^{k+\frac{1}{3}}], \\ &\text{with I.C. at } t^k: \quad \vartheta^k. \end{aligned} \tag{13}$$

So, the solution at the time step $t^{k+\frac{1}{3}}$ indicated by $\vartheta^{k+\frac{1}{3}}$.

(ii) After that, we solve:

$$\begin{aligned} {}^cD_t^\gamma \vartheta &= rS\frac{\partial\vartheta}{\partial S} + (\mu_1 + \mu_2 r)\frac{\partial\vartheta}{\partial r} - r\vartheta, \quad \text{on} \quad [t^{k+\frac{1}{3}}, t^{k+\frac{2}{3}}], \\ &\text{with I.C. at } t^{k+\frac{1}{3}}: \quad \vartheta^{k+\frac{1}{3}}, \end{aligned} \tag{14}$$

The solution at time step $t^{k+\frac{2}{3}}$ is indicated by $\vartheta^{k+\frac{2}{3}}$.

(iii) We solve:

$$\begin{aligned} {}^cD_t^\gamma \vartheta &= \frac{1}{2}\sigma^2 S^2\frac{\partial^2\vartheta}{\partial S^2} + \frac{1}{2}\eta^2\frac{\partial^2\vartheta}{\partial r^2}, \quad \text{on} \quad [t^{k+\frac{2}{3}}, t^k], \\ &\text{with I.C. at } t^{k+\frac{2}{3}}: \quad \vartheta^{k+\frac{2}{3}}. \end{aligned} \tag{15}$$

Proceeding iteratively until $T = t^{N_t}$, we concurrently resolve the equations (13), (14), and (15).

2.2 Model Discretization Process

Observe that the problem (9) is formulated within the following unbounded spatial domain $Q_\infty^T = (0, \infty) \times (0, \infty) \times (0, T)$. As a result, numerical calculations are impossible. In order, to resolve the numerical problem (9), we choice a numerical bounded domain where we resolve (9) using the finite differences approximations. Therefore, we take into account the domains listed below:

$$\mathcal{D} = \Big\{(S, r) : S \in [S_l, S_r],\ S_l \neq 0, \quad and\ r \in [r_l, r_r],\ r_l \neq 0\Big\}. \tag{16}$$

Let us define the uniform grid of the domain $\mathcal{D}$: let $\Delta S = \frac{S_r - S_l}{N_S}$ and $\Delta r = \frac{r_r - r_l}{N_r}$. So, we can find the sequences $(S_i)_i$, $(r_j)_j$:

$$S_i = S_l + i\Delta S \quad \forall \quad i = 0, \ldots, N_S, \tag{17}$$

$$r_j = r_l + j\Delta r \quad \forall \quad j = 0, \ldots, N_r. \tag{18}$$

Considering the next estimations:

$$\vartheta(S_i, t^k, r_j) \approx \vartheta^k_{i,j},$$

$\forall\ i = 0, \ldots, N_S, k = 0, \ldots, N_t$ and $j = 0, \ldots, N_r$.

- *Using Riemann-Liouville derivative as in* [32] *Caputo time derivative discretization term* $({}^c D^\gamma_t \vartheta)$ *is written as:*

$$D^\gamma_t \vartheta(S, t, r) = \frac{1}{\Gamma(1-\gamma)} \frac{d}{dt} \int\limits_0^t \frac{\vartheta(S, \xi, r) - \vartheta(S, 0, r)}{(t-\xi)^\gamma} d\xi.$$

Let $\vartheta \in C^1([0, t])$, so we obtain:

$$D^\gamma_t \vartheta(S, t, r) = \frac{1}{\Gamma(1-\gamma)} \int\limits_0^t \frac{d\vartheta(S, \xi, r)}{d\xi} (t-\xi)^{-\gamma}\, d\xi = {}^c D^\gamma_t \vartheta(S, t, r). \tag{19}$$

As in [32], the following approximation is given at the point (S_i, t^{k+1}, r_j):

$$ {}^c D^\gamma_t \vartheta(S_i, t^{k+1}, r_j) \approx \frac{(\Delta t)^{-\gamma}}{\Gamma(2-\gamma)} \sum_{m=0}^{k} (\vartheta^{k+1-m}_{i,j} - \vartheta^{k-m}_{i,j}) b_m, \tag{20}$$

$$\approx \alpha_0 \sum_{m=0}^{k-1} (b_{m+1} - b_m) \vartheta^{k-m}_{i,j} - \alpha_0 \vartheta^0_{i,j} b_k + \alpha_0 \vartheta^{k+1}_{i,j}, \tag{21}$$

where $\alpha_0 = \frac{(\Delta t)^{-\gamma}}{\Gamma(2-\gamma)}$ and $b_m = (m+1)^{1-\gamma} - m^{1-\gamma}$.

- *At* (S_i, r_j, t^{k+1}), *the decomposition of the problem* (13), we get:

$$\frac{(\Delta t)^{-\gamma}}{\Gamma(2-\gamma)}\sum_{m=0}^{k}(\vartheta_{i,j}^{k+1-m}-\vartheta_{i,j}^{k-m})b_m =$$
$$\frac{\rho\sigma\eta S_i}{4\Delta S\Delta r}\left(\vartheta_{i+1,j+1}^{k+1}-\vartheta_{i+1,j-1}^{k+1}-\vartheta_{i-1,j+1}^{k+1}+\vartheta_{i-1,j-1}^{k+1}\right). \quad (22)$$

- *At* (S_i, r_j, t^{k+1}), *the decomposition of the problem* (14), we get:

$$\begin{aligned}
&\frac{(\Delta t)^{-\gamma}}{\Gamma(2-\gamma)}\sum_{m=0}^{k}(\vartheta_{i,j}^{k+1-m}-\vartheta_{i,j}^{k-m})b_m \\
&=\frac{r_jS_i}{\Delta s}\left(\vartheta_{i+1,j}^{k+1}-\vartheta_{i,j}^{k+1}\right)+\frac{\mu_1+\mu_2r_j}{\Delta r}\left(\vartheta_{i,j+1}^{k+1}-\vartheta_{i,j}^{k+1}\right)-r_j\vartheta_{i,j}^{k} \\
&=\frac{r_jS_i}{\Delta s}\left(\frac{1}{2}(\vartheta_{i+1,j+1}^{k+1}+\vartheta_{i+1,j-1}^{k+1})-\vartheta_{i,j}^{k+1}\right) \\
&+\frac{\mu_1+\mu_2r_j}{\Delta r}\left(\vartheta_{i,j+1}^{k+1}-\vartheta_{i,j}^{k+1}\right)-r_j\vartheta_{i,j}^{k}.
\end{aligned} \quad (23)$$

- *At* (S_i, r_j, t^{k+1}), *the decomposition of the problem* (15), we get:

$$\begin{aligned}
&\frac{(\Delta t)^{-\gamma}}{\Gamma(2-\gamma)}\sum_{m=0}^{k}(\vartheta_{i,j}^{k+1-m}-\vartheta_{i,j}^{k-m})b_m \\
&=\frac{\sigma^2S_i^2}{2(\Delta S)^2}\left(\vartheta_{i+1,j}^{k+1}-2\vartheta_{i,j}^{k+1}+\vartheta_{i-1,j}^{k+1}\right)+\frac{\eta^2}{2(\Delta r)^2}\left(\vartheta_{i,j+1}^{k+1}-2\vartheta_{i,j}^{k+1}+\vartheta_{i,j-1}^{k+1}\right) \\
&=\frac{\sigma^2S_i^2}{2(\Delta S)^2}\left(\vartheta_{i+1,j}^{k+1}-2\vartheta_{i,j}^{k+1}+\frac{1}{2}(\vartheta_{i-1,j+1}^{k+1}+\vartheta_{i-1,j-1}^{k+1})\right) \\
&+\frac{\eta^2}{2(\Delta r)^2}\left(\vartheta_{i,j+1}^{k+1}-2\vartheta_{i,j}^{k+1}+\vartheta_{i,j-1}^{k+1}\right).
\end{aligned} \quad (24)$$

Remark 1 In discretizing Eqs. (13), (14), and (15), we employed implicit time finite difference techniques. The benefit of these techniques resides in the unconditional stability.

2.3 The Solution's Stability

$\forall \ j = 1, \ldots, N_r - 1$ *and* $i = 1, \ldots, N_S - 1$, the spatial discretization of the Eq. (9) is given by:

$${}^cD_t^{\gamma}\vartheta_{i,j}=\frac{\sigma^2S_i^2}{2(\Delta S)^2}\left(\vartheta_{i+1,j}-2\vartheta_{i,j}+\frac{1}{2}(\vartheta_{i-1,j+1}+\vartheta_{i-1,j-1})\right)$$

$$
\begin{aligned}
&+\frac{\rho\sigma\eta S_i}{4\Delta S\Delta r}\Big(\vartheta_{i+1,j+1}-\vartheta_{i,j+1}-\vartheta_{i+1,j}+\vartheta_{i,j}\Big)\\
&+\frac{\eta^2}{2(\Delta r)^2}\Big(\vartheta_{i,j+1}-2\vartheta_{i,j}+\vartheta_{i,j-1}\Big)\\
&+\frac{r_jS_i}{\Delta s}\Big(\frac{1}{2}(\vartheta_{i+1,j+1}+\vartheta_{i+1,j-1})-\vartheta_{i,j}\Big)\\
&+\frac{\mu_1+\mu_2r_j}{\Delta r}\Big(\vartheta_{i,j+1}-\vartheta_{i,j}\Big)-r_j\vartheta_{i,j}, \qquad (25)
\end{aligned}
$$

where we denoted by $\vartheta_{i,j}=\vartheta_{i,j}(t)\ \forall\ t\in[0,T]$. So, we have this expression: $j=1,\ldots,N_r-1$ and $\forall i=1,\ldots,N_S-1$

$$
\begin{aligned}
{}^cD_t^\gamma\vartheta_{i,j}&=\alpha_i\vartheta_{i+1,j}-\varepsilon_{i,j}\vartheta_{i,j}+\overline{\omega}_i\vartheta_{i-1,j+1}+\omega_i\vartheta_{i-1,j-1}\\
&+\delta_{i,j}\vartheta_{i+1,j+1}+\overline{\delta}_{i,j}\vartheta_{i+1,j-1}+(a_0+f_j)\vartheta_{i,j+1}+a_0\vartheta_{i,j-1}, \qquad (26)
\end{aligned}
$$

where $a_0,\alpha_i,f_j,\beta_i,\gamma_{i,j},\varepsilon_{i,j},\omega_i$ and $\delta_{i,j}$ are real numbers and expressed as:

$$
\begin{aligned}
a_0&=\frac{\eta^2}{2(\Delta r)^2},\ \alpha_i=\frac{\sigma^2S_i^2}{2(\Delta S)^2},\ \beta_i=\frac{\rho\sigma\eta S_i}{4\Delta S\Delta r},\\
f_j&=\frac{\mu_1+\mu_2r_j}{\Delta r},\ \gamma_{i,j}=\frac{r_jS_i}{\Delta s}, \qquad (27)
\end{aligned}
$$

$$
\delta_{i,j}=\frac{1}{2}\gamma_{i,j}+\beta_i,\quad \omega_i=\frac{1}{2}\alpha_i+\beta_i,\quad \varepsilon_{i,j}=2a_0+2\alpha_i+\gamma_{i,j}+r_j+f_j, \qquad (28)
$$

we note that, if $u=a+b\in\mathbb{R}$ its conjugate $\overline{u}=a-b\in\mathbb{R}$.

Let $N_{Sr}=(N_S-1)\times(N_r-1)$. Let the vector $\mathbf{P}\in\mathbb{R}^{N_{Sr}}$ by:

$$
\begin{aligned}
\mathbf{P}(t)=[&\vartheta_{1,1},\vartheta_{1,2},\ldots,\vartheta_{1,N_r-1},\vartheta_{2,1},\vartheta_{2,2},\ldots,\vartheta_{2,N_r-1},\ldots,\\
&\vartheta_{N_S-1,1},\ldots,\vartheta_{N_S-1,N_r-1}]\in\mathbb{R}^{N_{Sr}}.
\end{aligned}
$$

So, we can write the system (26) as follows:

$$
{}^cD_t^\gamma\mathbf{P}(t)=A\mathbf{P}(t)+\mathbf{F}(t), \qquad (29)
$$

with I.C.:

$$
\mathbf{P}(0)=\mathbf{P}_0. \qquad (30)
$$

Remark 2 (1) $A = (a_{i,j})_{i,j} \in \mathbb{R}^{N_{Sr} \times N_{Sr}}$ has eight diagonals and is explained as follows:

◇The diagonal elements of A :

$$a_{i,i} \in \{-\varepsilon_{k,\ell}, \quad \ell = 1, \dots, N_r - 1, \quad k = 1, \dots, N_S - 1, \}. \tag{31}$$

◇The non-diagonal elements of A for $i \neq j$:

$$\begin{aligned} a_{i,j} \in \{&\alpha_k, \ \delta_{k,\ell}, \ \overline{\delta}_{k,\ell}, \ a_0, \ a_0 + f_\ell, \ \omega_k, \ \overline{\omega}_k, \\ &\ell = 1, \dots, N_r - 1 \, k = 1, \dots, N_S - 1, \}. \end{aligned} \tag{32}$$

(2) Let $\mathbf{F}(t) \in \mathbb{R}^{N_{Sr}}$ is a vector function exclusively determined by the trace of the solution $\vartheta(S, r, t)$ at the boundary $\Gamma_{\mathcal{D}}$ of the domain $\mathcal{D}$ given in (16) and the coefficients $\alpha_i, \delta_{i,j}, \overline{\delta}_{i,j}, a_0, f_j, \omega_i, \overline{\omega}_i$. So we can write $\mathbf{F}(t)$ as follows:

$$\begin{aligned} \mathbf{F}(t) = \Big[&a_0 \vartheta_{1,0} + \overline{\delta}_{1,1} \vartheta_{2,0} + \phi(1), \phi(2), \dots, \phi(j), \dots, \\ &\phi(N_r - 2), \phi(N_r - 1) + \delta_{1,N_r-1} \vartheta_{2,N_r} \\ &+ g_{N_r-1} \vartheta_{1,N_r}, \psi(2), \dots, \psi(i), \dots, \psi(N_S - 2), \theta(1) + \omega_{N_S-1} \vartheta_{N_S-2,0} \\ &+ a_0 \vartheta_{N_S-1,0}, \theta(2), \dots, \theta(j), \dots, \theta(N_r - 2), \theta(N_r - 1) \\ &+ \overline{\omega}_{N_S-1} \vartheta_{N_S-2,N_r} + g_{N_r-1} \vartheta_{N_S-1,N_r} \Big] \in \mathbb{R}^{N_{Sr}} \end{aligned} \tag{33}$$

where $g_j = a_0 + f_j$, the functions θ, ϕ and the vector $\psi \in \mathbb{R}^{(N_S-3)\times(N_r-1)}$ are given by:

$$\triangleright \ \phi(j) = \overline{\omega}_1 \vartheta_{0,j+1} + \omega_1 \vartheta_{0,j-1}, \quad j = 1, \dots, N_r - 1, \tag{34}$$

$$\triangleright \ \theta(j) = \alpha_{N_S-1} \vartheta_{N_S,j} + \delta_{N_S-1,j} \vartheta_{N_S,j+1} + \overline{\delta}_{N_S,j} \vartheta_{N_S,j-1}, \ j = 1, \dots, N_r - 1, \tag{35}$$

$$\begin{aligned} &\triangleright \ \psi(i) = \Big[\omega_i \vartheta_{i-1,0} + \overline{\delta}_{i,1} \vartheta_{i+1,0} + a_0 \vartheta_{i,0}, 0, \dots, 0, \\ &\overline{\omega}_i \vartheta_{i-1,N_r} + \delta_{i,N_r-1} \vartheta_{i+1,N_r} + g_{N_r-1} \vartheta_{i,N_r} \Big] \in \mathbb{R}^{N_r-1}, \quad i = 2, \dots, N_S - 2. \end{aligned} \tag{36}$$

where $\vartheta_{N_S,j}, \vartheta_{0,j}, \vartheta_{i,N_r}$ and $\vartheta_{i,0}$ depict the problem's distinct boundary conditions $\vartheta(S, r, t)|_{\Gamma_{\mathcal{D}}}$.

From [8, 12], we prove that the system (29)–(30) has an analytic solution:

$$\mathbf{P}(t) = \chi_{[0,t]} \big(t^{\gamma-1} E_{\gamma,\gamma}(At^\gamma) \big) * \chi_{[0,t]} \mathbf{F}(t) + E_{\gamma,1}(At^\gamma) \mathbf{P}_0, \tag{37}$$

where $\chi_{[0,t]}$ is the characteristic function of $[0, t]$, $E_{\gamma,1}(\cdot)$, $E_{\gamma,\gamma}(\cdot)$ are the Mittag-Leffler functions and the symbol $(*)$ means the convolution product.

Theorem 1 *For any* $\mathbf{P}_0 \in \mathbb{R}^{N_{Sr}}$ *such that* $\|\mathbf{P}_0\|_\infty \leq \varrho$, $(\varrho > 0)$, *the solution* $\mathbf{P}(t)$ *to the problem (29)–(30) is stable and complies with the stability identity:*

$$\|\mathbf{P}(t)\|_\infty \leq \varrho E_{\gamma,1}(\mathcal{M}t^\gamma) + \mathcal{E}t^\gamma E_{\gamma,\gamma}(\mathcal{M}t^\gamma), \quad \forall t \in [0, T], \tag{38}$$

where $\|A\|_\infty \leq \mathcal{M}$, $\mathcal{E} = \mathcal{E}(\frac{1}{\Delta S}, \frac{1}{\Delta r})$ *and* ϱ *are constants independent of* t. *Moreover, we get:*

$$\mathcal{K} \geq \|\mathbf{P}(t)\|_\infty, \quad \forall t \in [0, T], \tag{39}$$

where $\mathcal{K} = \mathcal{K}(\frac{1}{\Delta S}, \frac{1}{\Delta r})$ *is a constant independent of* t.

By examining the analytical solution's expression P(t) $\mathbf{P}(t)$ provided in (37), we can infer that:

$$\begin{aligned}
\|\mathbf{P}(t)\|_\infty &\leq \|\mathbf{P}_0\|_\infty \|E_{\gamma,1}(At^\gamma)\|_\infty + \|\chi_{[0,t]}\big(t^{\gamma-1}E_{\gamma,\gamma}(At^\gamma)\big) * \chi_{[0,t]}\mathbf{F}(t)\|_\infty, \\
&\leq \varrho \|E_{\gamma,1}(At^\gamma)\|_\infty + \|E_{\gamma,\gamma}(At^\gamma)\|_\infty \|\chi_{[0,t]}\big(t^{\gamma-1}\big) * \chi_{[0,t]}\mathbf{F}(t)\|_\infty, \\
&\leq \varrho \|E_{\gamma,1}(At^\gamma)\|_\infty + \|E_{\gamma,\gamma}(At^\gamma)\|_\infty \|\mathbf{F}(t)\|_\infty \left(\int_0^t s^{\gamma-1}\, ds \right),
\end{aligned} \tag{40}$$

where $\chi_{[0,t]}$ is the characteristic function of $[0, t]$. As a result of this,

$$\|\mathbf{P}(t)\|_\infty \leq \varrho \|E_{\gamma,1}(At^\gamma)\|_\infty + \frac{t^\gamma}{\gamma} \|E_{\gamma,\gamma}(At^\gamma)\|_\infty \|\mathbf{F}(t)\|_\infty. \tag{41}$$

We are aware of the fact that:

$$E_{\gamma,1}(At^\gamma) = \sum_{k\geq 0} \frac{A^k t^{\gamma k}}{\Gamma(\gamma k + 1)}, \tag{42}$$

$$E_{\gamma,\gamma}(At^\gamma) = \sum_{k\geq 0} \frac{A^k t^{\gamma k}}{\Gamma(\gamma k + \gamma)}. \tag{43}$$

As a result, we can conclude that:

$$\|E_{\gamma,1}(At^\gamma)\|_\infty \leq \sum_{k\geq 0} \frac{t^{\gamma k}}{\Gamma(\gamma k + 1)} \|A\|_\infty^k, \tag{44}$$

$$\|E_{\gamma,\gamma}(At^\gamma)\|_\infty \leq \sum_{k\geq 0} \frac{t^{\gamma k}}{\Gamma(\gamma k + \gamma)} \|A\|_\infty^k. \tag{45}$$

Consider the following:

$$\|A\|_\infty = \max_k \Big\{ \sum_{\ell=1}^{N_{Sr}} |a_{k,\ell}| \Big\}$$

From the previous results, we can prove that:

$$\begin{aligned}\|A\|_\infty &= \varepsilon_{i,i} + \left(\alpha_i + \delta_{i,j} + \overline{\delta}_{i,j} + a_0 + a_0 + f_j + \omega_i + \overline{\omega}_i\right)_{i\neq j},\\ &= 4\alpha_i + 2\gamma_{i,j} + 4a_0 + 2f_j + r_j,\\ &\forall j = 0, \ldots, N_r,\ i = 0, \ldots, N_S.\end{aligned} \tag{46}$$

Given the definitions of the coefficients a_0, α_i, $\delta_{i,j}$, $\overline{\delta}_{i,j}$, $a_0 + f_j$, ω_i, $\overline{\omega}_i$, and $\varepsilon_{i,i}$, it can be established that there exists a positive constant $\mathcal{M}$ that ensures:

$$\|A\|_\infty \leq \mathcal{M}. \tag{47}$$

As a result, employing the identities (44), (45) and (47), we obtain:

$$\|E_{\gamma,1}(At^\gamma)\|_\infty \leq E_{\gamma,1}(\mathcal{M}t^\gamma) \quad \text{and} \quad \|E_{\gamma,\gamma}(At^\gamma)\|_\infty \leq E_{\gamma,\gamma}(\mathcal{M}t^\gamma). \tag{48}$$

We can deduce that there exists a constant $\mathcal{E} = \mathcal{E}(\frac{1}{\Delta S}, \frac{1}{\Delta r})$ and independent of t in such a way that:

$$\|\mathbf{F}(t)\|_\infty \leq \mathcal{E}, \quad \forall t \geq 0. \tag{49}$$

In reality, the constant $\mathcal{E} = \mathcal{E}(\frac{1}{\Delta S}, \frac{1}{\Delta r})$ stems from the estimates derived for each coefficient a_0, α_i, $\delta_{i,j}$, $\overline{\delta}_{i,j}$, $a_0 + f_j$, ω_i, $\overline{\omega}_i$ where each one is dependent on $\frac{1}{\Delta S}$ and $\frac{1}{\Delta r}$, (refer to Remark 2). Ultimately, the identity (92) can be deduced from (41), (48), and (49). Given that the Mittag-Leffler functions $E_{\gamma,1}(\mathcal{M}t^\gamma)$, $E_{\gamma,\gamma}(\mathcal{M}t^\gamma)$ are bounded $\forall\ t\ \in [0, T]$, the identification can then be determined from (39). □

Definition 4 ([16]) $Y = (y_{i,j})_{i,j} \in \mathbb{R}^{n\times n}$ is referred to as Metzler if all its off-diagonal elements are positive, expressed as $0 \leq y_{i,j}\ \forall\ 1 \leq j \neq i \leq n$.

Theorem 2 *If $\rho_r = 0$, the matrix V is then Metzler.*
Let $\rho_r \in]-1, 1[\backslash\{0\}$ and $d = s_r - s_l$. If θ and the ratio $\frac{\Delta s}{\Delta r}$ of the spatial steps satisfy:

$$\theta \geq r_r \ \ \textit{and} \ \ \frac{\Delta s}{\Delta r} \leq \frac{1}{|\rho|}\min\{\frac{s_l}{\sigma_r}, \frac{2rd}{\sigma_r r_l N_s}\}, \tag{50}$$

the matrix V is then Metzler.

We need to research the signs of the coefficients $f_j\ \delta_{i,j}$, $\overline{\delta}_{i,j}$, $\alpha_{i,j} and \overline{\alpha}_{i,j}$ with respect to the values of ρ_r, θ_r. As a result, we must identify the following situations:

- When $v_r \leq \theta_r$, f_j are positive.
- When $0 < \rho_r < 1$: since $\alpha_{i,j}$, $\delta_{i,j}$ are positive, then $\overline{\alpha}_{i,j}$, $\overline{\delta}_{i,j}$ must be positive:

$$\overline{\delta}_{i,j} = \frac{1}{2}\gamma_i - \beta_{i,j} = \frac{v s_i}{2\Delta s} - \frac{\rho_r\sigma_r s_i r_j}{4\Delta s\Delta r} \geq 0, \tag{51}$$

$$\overline{\alpha}_{i,j} = \frac{1}{2}\tau_{i,j} - \beta_{i,j} = \frac{s_i^2 r_j}{4(\Delta s)^2} - \frac{\rho_r\sigma_r s_i r_j}{4\Delta s\Delta r} \geq 0. \tag{52}$$

As a result, we obtain:

$$\frac{\Delta s}{\Delta r} \le \frac{2r\Delta s}{\rho_r \sigma_r r_j}, \quad \forall j = 0, \ldots, N_r, \tag{53}$$

$$\frac{\Delta s}{\Delta r} \le \frac{s_i}{\rho_r \sigma_r}, \quad \forall i = 0, \ldots, N_s. \tag{54}$$

Or

$$\frac{\Delta s}{\Delta r} \le \frac{2rd}{\rho_r \sigma_r r_l N_s}, \quad \text{where } d = s_r - s_l, \tag{55}$$

$$\frac{\Delta s}{\Delta r} \le \frac{s_l}{\rho_r \sigma_r}. \tag{56}$$

Hence, we obtain:

$$\frac{\Delta s}{\Delta r} \le \frac{1}{\rho} \min\left\{\frac{2vd}{\sigma_r r_l N_s}, \frac{s_l}{\sigma_r}\right\}. \tag{57}$$

- When $\rho \in]-1, 0[$: since $\overline{\delta}_{i,j}$, $\overline{\alpha}_{i,j}$ are positive, then the terms $\delta_{i,j}$, $\alpha_{i,j}$ must be positive:

$$\delta_{i,j} = \frac{1}{2}\gamma_i + \beta_{i,j} = \frac{vs_i}{2\Delta s} + \frac{\rho_r \sigma_r s_i r_j}{4\Delta s \Delta r} \ge 0, \tag{58}$$

$$\alpha_i = \frac{1}{2}\tau_{i,j} + \beta_{i,j} = \frac{s_i^2 r_j}{4(\Delta s)^2} + \frac{\rho_r \sigma_r s_i r_j}{4\Delta s \Delta r} \ge 0. \tag{59}$$

Following the same procedure as described previously, we obtain at this stage of analysis

$$\frac{\Delta s}{\Delta r} \le -\frac{1}{\rho} \min\left\{\frac{2vd}{\sigma_r r_l N_s}, \frac{s_l}{\sigma_r}\right\}. \tag{60}$$

From (57) and (60), we get the inequalities (50).

□

Theorem 3 *Given any* $\mathbf{X}_0 \in \mathbb{R}^{N_{sv}}$ *with* $\varrho \ge \|\mathbf{X}_0\|_\infty$ *(*$\varrho > 0$*), the solution* $\mathbf{X}(t)$ *is stable and meets the stability identity:*

$$\|\mathbf{X}(t)\|_\infty \le \varrho E_{\gamma,1}(\mathcal{Y} t^\gamma) + \mathcal{E} t^\gamma E_{\gamma,\gamma}(\mathcal{Y} t^\gamma), \quad \forall t \in [0, T], \tag{61}$$

where $\|X\|_\infty \le \mathcal{Y}$, $\mathcal{E} = \mathcal{E}(\frac{1}{\Delta s}, \frac{1}{\Delta r})$ *and* ϱ *are constants independent of* t*. Furthermore:*

$$\mathcal{K} \ge \|\mathbf{X}(t)\|_\infty, \quad \forall t \in [0, T], \tag{62}$$

where $\mathcal{K} = \mathcal{K}(\frac{1}{\Delta s}, \frac{1}{\Delta r})$ *is a constant independent of* t.

Based on the analytical solution $\mathbf{X}(t)$, we reach the conclusion that:

$$\begin{aligned}
\|\mathbf{X}(t)\|_\infty &\leq \|\mathbf{X}_0\|_\infty \|E_{\gamma,1}(Vt^\gamma)\|_\infty + \|\chi_{[0,t]}\big(t^{\gamma-1}E_{\gamma,\gamma}(Vt^\gamma)\big) * \chi_{[0,t]}\mathbf{Y}(t)\|_\infty, \\
&\leq \varrho \|E_{\gamma,1}(Vt^\gamma)\|_\infty + \|E_{\gamma,\gamma}(Vt^\gamma)\|_\infty \|\chi_{[0,t]}\big(t^{\gamma-1}\big) * \chi_{[0,t]}\mathbf{Y}(t)\|_\infty, \\
&\leq \varrho \|E_{\gamma,1}(Vt^\gamma)\|_\infty + \|E_{\gamma,\gamma}(Vt^\gamma)\|_\infty \|\mathbf{Y}(t)\|_\infty \left(\int_0^t s^{\gamma-1}\, ds \right),
\end{aligned}$$

where $\chi_{[0,t]}$ is the characteristic function of $[0, t]$. Consequently:

$$\|\mathbf{X}(t)\|_\infty \leq \varrho \|E_{\gamma,1}(Vt^\gamma)\|_\infty + \frac{t^\gamma}{\gamma} \|E_{\gamma,\gamma}(Vt^\gamma)\|_\infty \|\mathbf{Y}(t)\|_\infty. \tag{63}$$

We are aware of the following:

$$\begin{aligned}
E_{\gamma,1}(Vt^\gamma) &= \sum_{k\geq 0} \frac{V^k t^{\gamma k}}{\Gamma(\gamma k + 1)}, \\
E_{\gamma,\gamma}(Vt^\gamma) &= \sum_{k\geq 0} \frac{V^k t^{\gamma k}}{\Gamma(\gamma k + \gamma)}.
\end{aligned}$$

As a result, we can conclude:

$$\|E_{\gamma,1}(Vt^\gamma)\|_\infty \leq \sum_{k\geq 0} \frac{t^{\gamma k}}{\Gamma(\gamma k + 1)} \|V\|_\infty^k, \tag{64}$$

$$\|E_{\gamma,\gamma}(Vt^\gamma)\|_\infty \leq \sum_{k\geq 0} \frac{t^{\gamma k}}{\Gamma(\gamma k + \gamma)} \|V\|_\infty^k, \tag{65}$$

where:

$$\|V\|_\infty = \max_k \left\{ \sum_{\ell=1}^{N_{sv}} |a_{k,\ell}\| \right\} \tag{66}$$

As a result, we can establish:

$$\begin{aligned}
\|V\|_\infty &= \epsilon_{i,i} + \big(\tau_{i,j} + \delta_{i,j} + a_j + a_j + f_j + \overline{\delta}_{i,j} + \alpha_{i,j} + \overline{\alpha}_{i,j}\big)_{i\neq j}, \\
&= 4\tau_{i,j} + 2\gamma_i + 4a_j + r + 2f_j, \quad \forall j = 0, \ldots, N_r, i = 0, \ldots, N_s.
\end{aligned}$$

Derived from the coefficients' definition $a_j, \tau_{i,j}, \gamma_i, f_j$, we state that there is a constant $0 < \mathcal{M}$ in such a way that:

$$\mathcal{Y} \geq \|V\|_\infty. \tag{67}$$

In the end, by (64), (65) and (67), we get:

$$\|E_{\gamma,1}(Vt^\gamma)\|_\infty \leq E_{\gamma,1}(\mathcal{Y}t^\gamma) \quad \text{and} \quad \|E_{\gamma,\gamma}(Vt^\gamma)\|_\infty \leq E_{\gamma,\gamma}(\mathcal{Y}t^\gamma). \tag{68}$$

Regarding the definition of $\mathbf{Y}(t)$, there is a constant $\mathcal{E} = \mathcal{E}(\frac{1}{\Delta s}, \frac{1}{\Delta r})$ that is independent of t in such a way that:

$$\|\mathbf{Y}(t)\|_\infty \leq \mathcal{E}, \quad \forall t \geq 0. \tag{69}$$

Eventually, (61) can be obtained from Eqs. (63), (68) and (69). Given the Mittag-Leffler functions $E_{\gamma,1}(\mathcal{Y}t^\gamma)$, $E_{\gamma,\gamma}(\mathcal{Y}t^\gamma)$ are bounded $\forall\, t \in [0, T]$, then we will be able to demonstrate (62). □

3 Time-Fractional Heston Model

The following section introduces the problem related to the evaluation of an European put option using the fractional Heston model:

$$\begin{aligned} &{}^cD_\tau^\gamma \zeta + \frac{1}{2} v s^2 \frac{\partial^2 \zeta}{\partial s^2} + \rho \sigma v s \frac{\partial^2 \zeta}{\partial s \partial v} + \frac{1}{2}\sigma^2 v \frac{\partial^2 \zeta}{\partial v^2} \\ &+ r s \frac{\partial \zeta}{\partial s} + \kappa(\theta - v)\frac{\partial \zeta}{\partial v} - r\zeta = 0, \end{aligned} \tag{70}$$

for all $(s, v, \tau) \in \mathcal{Q}_\infty^T = (0, \infty) \times (0, \infty) \times (0, T)$.
We take into account the following boundary conditions:

$$\zeta(s, v, T) = (K - s)^+, \tag{71}$$

$$\lim_{s \to 0} \zeta(s, v, \tau) = K, \qquad \lim_{s \to \infty} \zeta(s, v, \tau) = 0, \tag{72}$$

$$\lim_{v \to 0} \zeta(s, v, \tau) = \lim_{v \to \infty} \zeta(s, v, \tau) = (K - s)^+, \tag{73}$$

Let's restate the previous problem (70)–(73) using the newly introduced time variable:

$$t = T - \tau.$$

Consequently, we can straightforwardly infer:

$${}^cD_\tau^\gamma \zeta = -{}^cD_t^\gamma \zeta.$$

As a result, the updated problem using the new variable $(s, v, t) \in \mathcal{Q}_{\infty}^{T}$ is presented by

$$
\begin{aligned}
{}^{c}D_{t}^{\gamma}\zeta &= \frac{1}{2}vs^{2}\frac{\partial^{2}\zeta}{\partial s^{2}} + \rho\sigma vs\frac{\partial^{2}\zeta}{\partial s\partial v} + \frac{1}{2}\sigma^{2}v\frac{\partial^{2}\zeta}{\partial v^{2}} \\
&+rs\frac{\partial\zeta}{\partial s} + \kappa(\theta - v)\frac{\partial\zeta}{\partial v} - r\zeta,
\end{aligned} \tag{74}
$$

with the new boundary conditions as follows:

$$
\zeta(s, v, 0) = (K - s)^{+}, \tag{75}
$$

$$
\lim_{s\to 0}\zeta(s, v, t) = K, \qquad \lim_{s\to\infty}\zeta(s, v, t) = 0, \tag{76}
$$

$$
\lim_{v\to 0}\zeta(s, v, t) = \lim_{v\to\infty}\zeta(s, v, t) = (K - s)^{+}, \tag{77}
$$

3.1 Splitting Technique

In the subsequent steps, we propose the utilization of a splitting method [2]. In the discretization process, the splitting method enables the independent treatment of each operator through an appropriate numerical scheme. The following approach ensures the preservation of the numerical properties of each scheme employed for the respective operators.

We partition the interval $[0, T]$ on (N_{t+1}) evenly spaced points in the following manner:

$$
\Delta t = \frac{T}{N_t} \quad \text{where} \quad t^k = k\Delta t, \quad \text{for all} \quad k = 0, \ldots, N_t.
$$

As a result of this, we have:

$$
[0, T] = \bigcup_{k=0}^{N_t-1}[t^k, t^{k+1}].
$$

Take a look at the following approximation:

$$
\zeta(s, v, t^k) \approx \zeta^k(s, v),
$$

where $(s, v) \in (0, \infty) \times (0, \infty)$ and $\forall\, k = 0, \ldots, N_t$.

The splitting approach is based on the solution of (74) on each interval $[t^k, t^{k+1}]$, $\forall\, k = 0, \ldots, N_t - 1$, based upon the following steps:

(i) Solving:

$$
\begin{aligned}
&{}^{c}D_t^{\gamma}\zeta = \rho\sigma v s\frac{\partial^2\zeta}{\partial s\partial v}, \quad \text{in} \quad [t^k, t^{k+\frac{1}{3}}], \\
&\text{with I.C. at } t^k: \quad \zeta^k.
\end{aligned}
\tag{78}
$$

As a result, we find a solution at time step $t^{k+\frac{1}{3}}$ as indicated by $\zeta^{k+\frac{1}{3}}$.

(ii) Thus, we proceed to solve the equation:

$$
\begin{aligned}
&{}^{c}D_t^{\gamma}\zeta = rs\frac{\partial\zeta}{\partial s} + \kappa(\theta - v)\frac{\partial\zeta}{\partial v} - r\zeta, \quad \text{on} \quad [t^{k+\frac{1}{3}}, t^{k+\frac{2}{3}}], \\
&\text{with I.C. at } t^{k+\frac{1}{3}}: \quad \zeta^{k+\frac{1}{3}},
\end{aligned}
\tag{79}
$$

The solution at time $t^{k+\frac{2}{3}}$ is indicated by $\zeta^{k+\frac{2}{3}}$.

(iii) We solve the equation:

$$
\begin{aligned}
&{}^{c}D_t^{\gamma}\zeta = \frac{1}{2}vs^2\frac{\partial^2\zeta}{\partial s^2} + \frac{1}{2}\upsilon^2 v\frac{\partial^2\zeta}{\partial v^2}, \quad \text{on} \quad [t^{k+\frac{2}{3}}, t^k], \\
&\text{with I.C. at } t^{k+\frac{2}{3}}: \quad \zeta^{k+\frac{2}{3}}.
\end{aligned}
\tag{80}
$$

We persist in this approach until reaching the ultimate time point $T = t^{N_t}$ and we proceed to resolve simultaneously Eqs. (78), (79) and (80).

3.2 Model Discretization Process

In order to address the numerical problem presented by (74), it is imperative to designate a bounded numerical domain where we can resolve (74) through finite difference approximations. Consequently, we define this designated numerical domain:

$$
\mathcal{D} = \Big\{(s, v) : s \in [s_l, s_r], \ s_l \neq 0, \quad v \in [v_l, v_r], \ v_l \neq 0\Big\}.
\tag{81}
$$

We establish a uniform grid on the domain $\mathcal{D}$ in the following manner: let $\Delta s = \frac{s_r - s_l}{N_s}$ and $\Delta v = \frac{v_r - v_l}{N_v}$. Now, we can construct the sequences $(s_i)_i$ and $(v_j)_j$:

$$
\begin{aligned}
s_i &= s_l + i\Delta s \quad \forall \quad i = 0, \ldots, N_s, \\
v_j &= v_l + j\Delta v \quad \forall \quad j = 0, \ldots, N_v.
\end{aligned}
$$

Take into account the approximations presented below:

$$
\zeta(s_i, v_j, t^k) \approx \zeta_{i,j}^k,
$$

$\forall\, k = 0, \ldots, N_t,\ j = 0, \ldots, N_v$ and $i = 0, \ldots, N_s$.

- *Discretizing the Caputo time derivative term* (${}^c D_t^\gamma \zeta$): the concept involves employing the modified Riemann-Liouville derivative as in [32], in which it is defined by:

$$D_t^\gamma \zeta(s, v, t) = \frac{1}{\Gamma(1-\gamma)} \frac{d}{dt} \int_0^t \frac{\zeta(s, v, \xi) - \zeta(s, v, 0)}{(t-\xi)^\gamma} \, d\xi.$$

Let $\zeta \in C^1([0, t])$. Next, we have:

$$D_t^\gamma \zeta(s, v, t) = \frac{1}{\Gamma(1-\gamma)} \int_0^t \frac{d\zeta(s, v, \xi)}{d\xi} (t-\xi)^{-\gamma} \, d\xi = {}^c D_t^\gamma \zeta(s, v, t).$$

At (s_i, v_j, t^{k+1}), as identified in [32], we have this estimation:

$$\begin{aligned} {}^c D_t^\gamma \zeta(s_i, v_j, t^{k+1}) &\approx \frac{(\Delta t)^{-\gamma}}{\Gamma(2-\gamma)} \sum_{m=0}^{k} (\zeta_{i,j}^{k+1-m} - \zeta_{i,j}^{k-m}) b_m, \\ &\approx \alpha_0 \sum_{m=0}^{k-1} (b_{m+1} - b_m) \zeta_{i,j}^{k-m} - \alpha_0 \zeta_{i,j}^0 b_k + \alpha_0 \zeta_{i,j}^{k+1}, \end{aligned}$$

where $\alpha_0 = \frac{(\Delta t)^{-\gamma}}{\Gamma(2-\gamma)}$ and $b_m = (m+1)^{1-\gamma} - m^{1-\gamma}$.

- At (s_i, v_j, t^{k+1}), *the problem* (78) *will be discretized* as follows:

$$\begin{aligned} &\frac{(\Delta t)^{-\gamma}}{\Gamma(2-\gamma)} \sum_{m=0}^{k} (\zeta_{i,j}^{k+1-m} - \zeta_{i,j}^{k-m}) b_m \\ &= \frac{\rho \sigma s_i v_j}{\Delta s \Delta v} \left(\zeta_{i+1,j+1}^{k+1} - \zeta_{i,j+1}^{k+1} - \zeta_{i+1,j}^{k+1} + \zeta_{i,j}^{k+1} \right). \end{aligned} \tag{82}$$

- At (s_i, v_j, t^{k+1}) *the problem* (79) *will be discretized* as follows:

$$\begin{aligned} &\frac{(\Delta t)^{-\gamma}}{\Gamma(2-\gamma)} \sum_{m=0}^{k} (\zeta_{i,j}^{k+1-m} - \zeta_{i,j}^{k-m}) b_m \\ &= \frac{r s_i}{\Delta s} \left(\frac{1}{2} (\zeta_{i+1,j+1}^{k+1} + \zeta_{i+1,j-1}^{k+1}) - \zeta_{i,j}^{k+1} \right) \\ &+ \frac{\kappa(\theta - v_j)}{\Delta v} \left(\zeta_{i,j+1}^{k+1} - \zeta_{i,j}^{k+1} \right) - r \zeta_{i,j}^{k}. \end{aligned} \tag{83}$$

- At (s_i, v_j, t^{k+1}), *the problem* (80) *will be discretized* as follows:

$$\frac{(\Delta t)^{-\gamma}}{\Gamma(2-\gamma)}\sum_{m=0}^{k}(\zeta_{i,j}^{k+1-m} - \zeta_{i,j}^{k-m})b_m$$
$$= \frac{s_i^2 v_j}{2(\Delta s)^2}\Big(\zeta_{i+1,j}^{k+1} - 2\zeta_{i,j}^{k+1} + \frac{1}{2}(\zeta_{i-1,j+1}^{k+1} + \zeta_{i-1,j-1}^{k+1})\Big)$$
$$+\frac{\sigma^2 v_j}{2(\Delta v)^2}\Big(\zeta_{i,j+1}^{k+1} - 2\zeta_{i,j}^{k+1} + \zeta_{i,j-1}^{k+1}\Big). \tag{84}$$

Remark 3 For the discretization of (78), (79) and (80), we employed implicit finite difference time techniques. The primary benefit of these schemes is their unconditional stability.

3.3 Investigation of Solution Stability

The total spatial discretization of Eq. (74) is expressed: $j = 1, \ldots, N_v - 1$ and $\forall\, i = 1, \ldots, N_s - 1$

$$\begin{aligned}{}^c D_t^{\gamma} \zeta_{i,j} = {} & \frac{s_i^2 v_j}{2(\Delta s)^2}\Big(\zeta_{i+1,j} - 2\zeta_{i,j} + \frac{1}{2}(\zeta_{i-1,j+1} + \zeta_{i-1,j-1})\Big)\\
&+\frac{\rho\sigma s_i v_j}{4\Delta s\Delta v}\Big(\zeta_{i+1,j+1} - \zeta_{i,j+1} - \zeta_{i+1,j} + \zeta_{i,j}\Big)\\
&+\frac{\sigma^2 v_j}{2(\Delta v)^2}\Big(\zeta_{i,j+1} - 2\zeta_{i,j} + \zeta_{i,j-1}\Big)\\
&+\frac{r s_i}{\Delta s}\Big(\frac{1}{2}(\zeta_{i+1,j+1} + \zeta_{i+1,j-1}) - \zeta_{i,j}\Big)\\
&+\frac{k(\theta - v_j)}{\Delta v}\Big(\zeta_{i,j+1} - \zeta_{i,j}\Big) - r\zeta_{i,j},\end{aligned}$$

where we note by $\zeta_{i,j} = \zeta_{i,j}(t)\ \forall\ t \in [0, T]$. As a result, we get the subsequent expression: for all $j = 1, \ldots, N_v - 1$ and $i = 1, \ldots, N_s - 1$

$$\begin{aligned}{}^c D_t^{\gamma} \zeta_{i,j} = {} & \tau_{i,j}\zeta_{i+1,j} - \epsilon_{i,j}\zeta_{i,j} + \overline{\alpha}_{i,j}\zeta_{i-1,j+1} + \alpha_{i,j}\zeta_{i-1,j-1}\\
&+ \delta_{i,j}\zeta_{i+1,j+1} + \overline{\delta}_{i,j}\zeta_{i+1,j-1} + (a_j + f_j)\zeta_{i,j+1} + a_j\zeta_{i,j-1},\end{aligned} \tag{85}$$

where the coefficients $a_j, \tau_{i,j}, \beta_{i,j}, \gamma_i, f_j, \delta_{i,j}, \alpha_{i,j}$ and $\epsilon_{i,j}$ are real numbers and are defined by:

$$a_j = \frac{\sigma^2 v_j}{2(\Delta v)^2},\ \tau_{i,j} = \frac{s_i^2 v_j}{2(\Delta s)^2},\ \beta_{i,j} = \frac{\rho\sigma s_i v_j}{4\Delta s\Delta v},\ f_j = \frac{k(\theta - v_j)}{\Delta v},\ \gamma_i = \frac{r s_i}{\Delta s},$$
$$\alpha_{i,j} = \frac{1}{2}\tau_{i,j} + \beta_{i,j},\ \delta_{i,j} = \frac{1}{2}\gamma_i + \beta_{i,j},\ \epsilon_{i,j} = 2\tau_{i,j} + 2a_j + \gamma_i + f_j + r,$$

where if $q = x + y \in \mathbb{R}$, its conjugate $\overline{q} = x - y \in \mathbb{R}$.

Let $N_{sv} = (N_s - 1) \times (N_v - 1)$. We define $\mathbf{Q} \in \mathbb{R}^{N_{sv}}$ by:

$$\mathbf{Q}(t) = [\zeta_{1,1}, \zeta_{1,2}, \dots, \zeta_{1,N_v-1}, \zeta_{2,1}, \zeta_{2,2}, \dots, \zeta_{2,N_v-1}, \dots, \\ \zeta_{N_s-1,1}, \dots, \zeta_{N_s-1,N_v-1}] \in \mathbb{R}^{N_{sv}}.$$

As a result, the system (85) is written as:

$$^{c}D_t^{\gamma}\mathbf{Q}(t) = B\mathbf{Q}(t) + \mathbf{M}(t), \tag{86}$$

where the I.C.:

$$\mathbf{Q}(0) = \mathbf{Q}_0. \tag{87}$$

Theorem 4 *If $\rho = 0$, then the matrix B is Metzler.*

Let $d = s_r - s_l$ and $\rho \in]-1, 1[\setminus\{0\}$. If θ and the ratio $\frac{\Delta s}{\Delta v}$ of the spatial steps satisfy the identities:

$$\theta \geq v_r \quad and \quad \frac{\Delta s}{\Delta v} \leq \frac{1}{|\rho|} \min\left\{\frac{s_l}{\sigma}, \frac{2rd}{\sigma v_l N_s}\right\}, \tag{88}$$

then the matrix B is Metzler.

In the following we investigate the signs of the coefficients f_j, $\delta_{i,j}$, $\overline{\delta}_{i,j}$, $\alpha_{i,j}$, $\overline{\alpha}_{i,j}$ with respect to the values of ρ, θ. From this vantage point, it is useful to separate the following two scenarios:

- If $v_r \leq \theta$, the coefficients f_j are positive.
- If $0 < \rho < 1[$: since $\delta_{i,j}, \alpha_{i,j}$ are positive, then the terms $\overline{\delta}_{i,j}, \overline{\alpha}_{i,j}$ must be positive:

$$\overline{\delta}_{i,j} = \frac{1}{2}\gamma_i - \beta_{i,j} = \frac{rs_i}{2\Delta s} - \frac{\rho\sigma s_i v_j}{4\Delta s\Delta v} \geq 0,$$

$$\overline{\alpha}_{i,j} = \frac{1}{2}\tau_{i,j} - \beta_{i,j} = \frac{s_i^2 v_j}{4(\Delta s)^2} - \frac{\rho\sigma s_i v_j}{4\Delta s\Delta v} \geq 0.$$

As a result, we get:

$$\frac{\Delta s}{\Delta v} \leq \frac{2r\Delta s}{\rho\sigma v_j}, \quad \forall j = 0, \dots, N_v,$$

$$\frac{\Delta s}{\Delta v} \leq \frac{s_i}{\rho\sigma}, \quad \forall i = 0, \dots, N_s.$$

Or

$$\frac{\Delta s}{\Delta v} \leq \frac{2rd}{\rho\sigma v_l N_s}, \quad \text{where } d = s_r - s_l,$$
$$\frac{\Delta s}{\Delta v} \leq \frac{s_l}{\rho\sigma}.$$

As a result, we conclude:

$$\frac{\Delta s}{\Delta v} \leq \frac{1}{\rho} \min\left\{\frac{2rd}{\sigma v_l N_s}, \frac{s_l}{\sigma}\right\}. \tag{89}$$

- When $\rho \in]-1, 0[$: since $\overline{\delta}_{i,j}$, $\overline{\alpha}_{i,j}$ are positive, then the terms $\delta_{i,j}$, $\alpha_{i,j}$ must be positive:

$$\delta_{i,j} = \frac{1}{2}\gamma_i + \beta_{i,j} = \frac{rs_i}{2\Delta s} + \frac{\rho\sigma s_i v_j}{4\Delta s\Delta v} \geq 0,$$

$$\alpha_i = \frac{1}{2}\tau_{i,j} + \beta_{i,j} = \frac{s_i^2 v_j}{4(\Delta s)^2} + \frac{\rho\sigma s_i v_j}{4\Delta s\Delta v} \geq 0.$$

Continuing in the same manner as before, we obtain:

$$\frac{\Delta s}{\Delta v} \leq -\frac{1}{\rho} \min\{\frac{2rd}{\sigma v_l N_s}, \frac{s_l}{\sigma}\}. \tag{90}$$

Departing from (89) and (90) we deduce (88).

□

Remark 4 In reference to (75)–(77), we conclude that $\zeta(s, t, v)|_{\Gamma_D}$ and $\zeta(S, t, v)|_{t=0}$ are positive. Even more, the coefficients $\tau_{i,j}$, $\delta_{i,j}$, a_j, $\overline{\delta}_{i,j}$, $a_j + f_j$, $\alpha_{i,j}$, $\overline{\alpha}_{i,j}$ are positive under condition (88). Consequently, $\mathbf{M}(t)$ and $\mathbf{Q}_0$ are positive.

On the basis of [8, 12], we can conclude the system (86)–(87) has an analytic solution defined by:

$$\mathbf{Q}(t) = E_{\gamma,1}(Bt^\gamma)\mathbf{Q}_0 + \chi_{[0,t]}\big(t^{\gamma-1}E_{\gamma,\gamma}(Bt^\gamma)\big) * \chi_{[0,t]}\mathbf{M}(t), \tag{91}$$

Theorem 5 *For any* $\mathbf{Q}_0 \in \mathbb{R}^{N_{sv}}$ *such that* $\|\mathbf{Q}_0\|_\infty \leq \varrho$, $(\varrho > 0)$, *the solution* $\mathbf{Q}(t)$ *to the problem (86)–(87) is stable and satisfies the stability identity:*

$$\|\mathbf{Q}(t)\|_\infty \leq \varrho E_{\gamma,1}(\mathcal{M}t^\gamma) + \mathcal{E}t^\gamma E_{\gamma,\gamma}(\mathcal{M}t^\gamma), \quad \forall t \in [0, T], \tag{92}$$

where $\|A\|_\infty \leq \mathcal{M}$, $\mathcal{E} = \mathcal{E}(\frac{1}{\Delta s}, \frac{1}{\Delta v})$ *and* ϱ *are constants independent of* t. *Moreover, we have:*

$$\|\mathbf{Q}(t)\|_\infty \leq \mathcal{K}, \quad \forall t \in [0, T], \tag{93}$$

where $\mathcal{K} = \mathcal{K}(\frac{1}{\Delta s}, \frac{1}{\Delta v})$ *is a constant independent of* t.

For a more in-depth understanding, we encourage readers to consult the provided source [3] for additional information.

□

4 Conclusion

This study presents a novel approach, employing the splitting method, to address the pricing of European options arising from stochastic processes. The first scenario involves stochastic interest rates, specifically in the cases of the time-fractional Vasicek and Time-Fractional Cox-Ingersoll-Ross processes. Subsequently, we explore the situation where volatility is stochastic, focusing on the time-fractional Heston model. In all instances, we adopt implicit time finite difference method, chosen for their advantage of unconditional stability.

Competing Interests The authors have no conflicts of interest to declare that are relevant to the content of this chapter.

References

1. G. Amit, J. Singh, D. Kumar, Sushila, An efficient analytical approach for fractional equal width equations describing hydro-magnetic waves in cold plasma. Physica A. **524** (2019)
2. H. Arfaoui, Stabilization method for the Saint-Venant equations by boundary control. Trans. Inst. Measure. Control **42**(16), 3290–3302 (2020)
3. H. Arfaoui, M. Kharrat, Stability analysis for pricing options via time fractional Heston model. Filomat **37**(9), 2685–2697 (2023)
4. M. Benchohra, J.R. Graef, F.Z. Mostefai, Weak solutions for boundary-value problems with nonlinear fractional differential inclusions. Nonlinear Dyn. Syst. Theory **3**, 227–237 (2011)
5. A. Bensoussan, On the theory of option pricing. Acta. Appl. Math. **2**, 139–158 (1984)
6. F. Black, M.S. Scholes, The pricing of options and corporate liabilities. J. Polit. Econ. **81**, 279–296 (1973)
7. J.C. Cox, J.E. Ingersoll, S.A. Ross, LA theory of the term structure of interest rates. Econometrica **53**, 385–407 (1985)
8. V. Daftardar-Gejji, A. Babakhani, Analysis of a system of fractional differential equations. J. Math. Anal. Appl. **293**, 511–522 (2004)
9. V. Daftardar-Gejji, S. Bhalekar, Solving multiterm linear and nonlinear diffusion wave equations of fractional order by Adomian decomposition method. Appl. Math. Comput. **202**, 113–120 (2008)

10. B. Dumitru, Othman M. Pshtiwan, V.C. Miguel, R.O. Yenny, Some modifications in conformable fractional integral inequalities. Adv. Differ. Equ. **374**, 1–25 (2020)
11. A. Erdelyi, W. Magnus, F. Oberhettinger, F.G. Tricomi, *Higher Transcendental Functions*, Vol. III, ed. by E. Robert (Krieger Publishing Co., Inc.: Melbourne, Australia, 1981)
12. R. Garrappa, Exponential integrators for time-fractional partial differential equations. Eur. Phys. J. Spec. Top. **222**, 1915–1927 (2013)
13. M.A.M. Ghandehari, M. Ranjbar, European option pricing of fractional version of the Black-Scholes model: approach via expansion in series. Int. J. Nonlinear. Sci. **17** (2014)
14. G.H. Golub, L. Van, F. Charles, *Matrix Computations* (The Johns Hopkins University Press. Baltimore and London, 1996). ISBN 0-8018-5414-8
15. S.L. Heston, Closed form solution for options with stochastic volatility with application to bonds and currency options. Rev. Financ. Stud. **6**, 327–343 (1993)
16. T. Kaczorek, *Positive 1D and 2D Systems* (Springer, London, UK, 2002)
17. M. Kharrat, Computation of conditional expectation using Malliavin calculus under Vasicek model. Wulfenia J. **21**(4), 143–152 (2014)
18. M. Kharrat, Closed-form solution of European option under fractional Heston model. Nonlinear Dyn. Syst. Theory. **18**(2), 191–195 (2018)
19. M. Kharrat, Pricing American put Option under Fractional Heston Model. Pramana J. Phys. **3** (2021). https://doi.org/10.1007/s12043-020-02039-z
20. M. Kharrat, Pricing American put option under fractional model. Filomat **35**(10) (2021)
21. M. Kharrat, H. Arfaoui, A new stabled relaxation method for pricing European options under the time-fractional Vasicek model. Comput. Econ. **61**, 1745–1763 (2023) https://doi.org/10.1007/s10614-022-10264-4
22. M. Kharrat, Stability analysis for pricing European options regarding the interest rate generated by the time fractional Cox-Ingersoll-Ross processes. Methodol. Comput. Appl. Probab. **25**(50) (2023). https://doi.org/10.1007/s11009-023-10021-6
23. A.A. Kilbas, H.M. Srivastava, J.J. Trujillo, *Theory and Applications of fractional Differential Equations* (Elsevier Science B.V, Amsterdam, The Netherlands, 2006)
24. D. Kumar, J. Singh, D. Baleanu, On the analysis of vibration equation involving a fractional derivative with Mittag-Leffler law. Math. Method. Appl. Sci. **43**(1) (2019)
25. S. Kumar, A. Yildirim, Y. Khan, H. Jafari, K. Sayevand, L. Wei, Analytical solution of fractional Black-Scholes European option pricing equation by using Laplace transform. J. Fractional Calc. Appl. **2** (2012)
26. I. Podlubny, *Fractional Differential Equations Calculus* (Academic Press, New York, 1999)
27. C. Rafael, N.E. Vera, J. Lucas, F.V. Ferran, An ETD method for American options under the Heston Model. Comput. Model. Eng. Sci. **124**(2), (2020). https://doi.org/10.32604/cmes.2020.010208
28. H.M. Srivastava, V.P. Dubey, R. Kumar, J. Singh, D. Kumar, D. Baleanu, An efficient computational approach for a fractional-order biological population model with carrying capacity. Chaos. Solition. Fract. **138** (2020)
29. O. Vasicek, An equilibrium characterization of the term structure. J. Financ. Econ. **5**(2), 177–188 (1977)
30. Y. Xiaozhong, W. LifeiEmail, S. Shuzhen, Z. Xue, A universal difference method for time-space fractional Black-Scholes equation. Adv. Differ. Equ-Ny. **1** (2016)
31. J.M. Yu, Y.W. Luo, S.B. Zhou, X.R. Lin, Existence and uniqueness for nonlinear multi-variables fractional differential equations. Nonlinear Dyn. Syst. Theory **2**, 213–221 (2011)
32. H. Zhang, F. Liu, I. Turner, Q. Yang, Numerical solution of the time fractional Black-Scholes model governing European options. Comput. Math. Appl. **71**(16), 1772–1783 (2016)

Certain Generalization of Appell's Functions and Riemann–Liouville Fractional Derivative Operator and Their Applications

Saddam Husain and Nabiullah Khan

Abstract Here, the Riemann–Liouville fractional derivative operator concept and a new and interesting extended form of Appell's functions are discussed. We discovered new formulae for fractional derivatives of various well-known functions in terms of new extended Appell's hypergeometric functions of two variables and Lauricella hypergeometric functions of three variables, with a view toward the analytic properties and application of the new Riemann–Liouville-type fractional derivative operator. Additionally, we defined the Mellin transformations of that function. Next, we created generating functions for generalized extended hypergeometric functions to validate our new operator using an extended Riemann–Liouville fractional derivative operator and a new definition of extended Appell's functions.

1 Introduction and Preliminaries

Many generalizations and extensions of special functions and mathematical physics have been investigated in the past several years. In particular, special functions involving more than one variable provide a new variety of mathematical techniques for resolving huge classes of partial differential equations that are frequently encountered in physical problems. Most mathematical physics, special functions, and their generalizations are motivated by physical considerations. In which Riemann–Liouville fractional derivative operator plays an important role. The Riemann–Liouville fractional derivative operator offers a powerful mathematical tool for capturing the behavior of non-integer order systems, incorporating memory effects, and describing processes with fractal and self-similar properties. Its wide range of applications and its compatibility with existing mathematical frameworks make it a valuable tool for researchers and practitioners in various disciplines. Motivated by their importance in various fields, recently Özarslan et al. [16] studied the extended Appell's func-

S. Husain (✉)
Department of Mathematics and Statistics, Integral University, Lucknow 226026, India
e-mail: saddam@iul.ac.in

S. Husain · N. Khan
Department of Applied Mathematics, Aligarh Muslim University, Aligarh 202002, India

M. Kharrat et al. (eds.), *Modeling of Discrete and Continuous Systems*, Forum for Interdisciplinary Mathematics, https://doi.org/10.1007/978-981-97-8715-9_10

tion using the extended beta function given by Chaudhary et al. [4]. He defined the Riemann–Liouville fractional derivative operator and found the formula of some well-known functions. Furthermore, he defined their Millen transform and some generating relations for extended hypergeometric functions via a generalized fractional derivative operator. In 2022, Sadek [19–21] gave various ideas related to fractional derivative operators and related problems like fractional backward differentiation formulas for solving fractional differential matrix equations, controllability, and observability for fractal linear dynamical systems and stability of conformable linear infinite-dimensional systems and stability of conformable linear infinite-dimensional systems. After that, in 2023, Sadek et al. [22, 23] gave another idea for a new type of fractional derivatives involving exponential cotangent function in their kernels and also discussed their applications and a new numerical approach to solving the fractional differential Riccati equations (FDRC) numerically. Khan et al. [7] studied the extended beta function with the help of the Mittag-Leffler function in two parameters and discussed their application. Inspired by the work mentioned earlier, in this paper, we will study the further generalization of Appell's functions, and the Riemann–Liouville fractional derivative operator using extended beta functions given by khan et al. [7].

Throughout the paper, the set of all real numbers is represented by the symbol $\mathbb{R}$. The set of natural numbers is represented by the symbol $\mathbb{N}$. The set of complex numbers by the symbol $\mathbb{C}$. The set of positive real numbers is represented by the symbol $\mathbb{R}^+$ and the set of non-negative real numbers by the symbol $\mathbb{R}_0^+$. Due to importance of special functions many authors (see more details in [6–12, 15, 17]) have recently developed extensions to the Euler beta function, gamma function, Gauss hypergeometric function, confluent hypergeometric function, and many other functions. Below, we mentioned some well-known helpful definitions as we were required to discuss our main problem as follows:

Definition 1 The Euler beta function is defined as [2, 4]:

$$\mathcal{B}(\kappa_1, \kappa_2) = \int_0^1 \tau^{\kappa_1 - 1} (1 - \tau)^{\kappa_2 - 1} d\tau,$$

$$(\Re(\kappa_1) > 0,\ \Re(\kappa_2) > 0).$$

The classical Gauss hypergeometric function $F(\kappa_1, \kappa_2; \kappa_3; z)$ and confluent hypergeometric function $\Phi(\kappa_2; \kappa_3; z)$ are defined as [18]

$$F(\kappa_1, \kappa_2; \kappa_3; z) = \frac{1}{\mathcal{B}(\kappa_2, \kappa_3 - \kappa_2)} \int_0^1 \tau^{\kappa_2 - 1} (1 - \tau)^{\kappa_3 - \kappa_2 - 1} (1 - z\tau)^{-\kappa_1}\, d\tau, \quad (1)$$

$$(\ |\arg(1 - z)| < \pi;\ \Re(\kappa_3) > \Re(\kappa_2) > 0\ ),$$

$$\Phi\,(\kappa_2;\kappa_3;z) = \frac{1}{\mathcal{B}(\kappa_2,\kappa_3-\kappa_2)} \int_0^1 \tau^{\kappa_2-1}\,(1-\tau)^{\kappa_3-\kappa_2-1}\, exp\,(z\tau)\,d\tau, \qquad (2)$$

$$(\;\Re(\kappa_3) > \Re(\kappa_2) > 0\;).$$

By using the series expansion of $(1-z\tau)^{-\kappa_1}$ and $\exp(z\tau)$ in (1) and (2), respectively, the hypergeometric and confluent hypergeometric functions are written in terms of beta function as

$$F(\kappa_1,\kappa_2;\kappa_3;z) = \sum_{n=0}^{\infty} (\kappa_1)_n \frac{\mathcal{B}(\kappa_2+n,\kappa_3-\kappa_2)}{\mathcal{B}(\kappa_2,\kappa_3-\kappa_2)} \frac{z^n}{n!}, \qquad (3)$$

$$(\;|z|<1,\;\Re(\kappa_3) > \Re(\kappa_2) > 0),$$

and

$$\Phi\,(\kappa_2;\kappa_3;z) = \sum_{n=0}^{\infty} \frac{\mathcal{B}(\kappa_2+n,\kappa_3-\kappa_2)}{\mathcal{B}(\kappa_2,\kappa_3-\kappa_2)} \frac{\omega^n}{n!}, \qquad (4)$$

$$(|z|<1,\;\Re(\kappa_3) > \Re(\kappa_2) > 0).$$

A new extended beta function in terms of extended Mittag-Leffler function given by Khan et al. [7] is defined as follows:

$$\mathcal{B}^{\wp,\eta,\nu}_{\alpha,\beta}(\kappa_1,\kappa_2) = \int_0^1 \tau^{\kappa_1-1}\,(1-\tau)^{\kappa_2-1}\,E_{\alpha,\beta}\left(-\frac{\wp}{\tau^{\eta}(1-\tau)^{\nu}}\right)d\tau, \qquad (5)$$

$$(\Re(\kappa_1) > 0,\;\Re(\kappa_2) > 0;\,\alpha,\beta,\eta,\nu \in \mathbb{R}^+_0;\;\wp \geq 0).$$

Here, $E_{\alpha,\beta}(.)$ is known as two-parameter Mittag-Leffler function [13, 14] defined as follows:

$$E_{\alpha,\beta}\,(z) = \sum_{n=0}^{\infty} \frac{z^n}{\Gamma(\alpha n+\beta)}, \qquad (\alpha,\beta \in \mathbb{R}^+_0, z \in \mathbb{C}).$$

With the help of new extended beta functions (5), the extended Gauss hypergeometric and confluent hypergeometric functions and their integral representation are defined as follows:

$$F_{\alpha,\beta}^{\wp,\eta,\nu}(\kappa_1,\kappa_2;\kappa_3;z)=\sum_{n=0}^{\infty}(a)_n\,\frac{\mathcal{B}_{\alpha,\beta}^{\wp,\eta,\nu}(\kappa_2+n,\kappa_3-\kappa_2)}{\mathcal{B}(\kappa_2,\kappa_3-\kappa_2)}\frac{z^n}{n!},\tag{6}$$

$$(\wp\geq 0,\ |z|<1,\ \alpha,\beta,\eta,\nu\in\mathbb{R}_0^+,\ \Re(\kappa_3)>\Re(\kappa_2)>0),$$

$$\Phi_{\alpha,\beta}^{\wp,\eta,\nu}(\kappa_2;\kappa_3;z)=\sum_{n=0}^{\infty}\frac{\mathcal{B}_{\alpha,\beta}^{\wp,\eta,\nu}(\kappa_2+n,\kappa_3-\kappa_2)}{\mathcal{B}(\kappa_2,\kappa_3-\kappa_2)}\frac{z^n}{n!},\tag{7}$$

$$(\wp\geq 0,\ \alpha,\beta,\eta,\nu\in\mathbb{R}_0^+,\ \Re(\kappa_3)>\Re(\kappa_2)>0).$$

$$\begin{aligned}F_{\alpha,\beta}^{\wp,\eta,\nu}(\kappa_1,\kappa_2;\kappa_3;z)=&\frac{1}{\mathcal{B}(\kappa_2,\kappa_3-\kappa_2)}\\&\times\int_0^1\tau^{\kappa_2-1}(1-\tau)^{\kappa_3-\kappa_2-1}(1-z\tau)^{-\kappa_1}E_{\alpha,\beta}\left(-\frac{\wp}{\tau^{\eta}(1-\tau)^{\nu}}\right)d\tau,\end{aligned}\tag{8}$$

$$(\wp\in\mathbb{R}_0^+,\alpha,\beta,\eta,\nu\in\mathbb{R}_0^+\ \textit{and}\ |\arg(1-z)|<\pi;\ \Re(\kappa_3)>\Re(\kappa_2)>0),$$

and

$$\begin{aligned}\Phi_{\alpha,\beta}^{\wp,\eta,\nu}(\kappa_2;\kappa_3;z)=&\frac{1}{\mathcal{B}(\kappa_2,\kappa_3-\kappa_2)}\\&\times\int_0^1\tau^{\kappa_2-1}(1-\tau)^{\kappa_3-\kappa_2-1}e^{z\tau}E_{\alpha,\beta}\left(-\frac{\wp}{\tau^{\eta}(1-\tau)^{\nu}}\right)d\tau,\end{aligned}\tag{9}$$

$$(\wp\in\mathbb{R}_0^+,\alpha,\beta,\eta,\nu\in\mathbb{R}_0^+\ \textit{and}\ \Re(\kappa_3)>\Re(\kappa_2)>0).$$

This research aims to investigate the extended Riemann–Liouville fractional derivative operator, determine the formula of some known functions through the use of extended Appell's functions, and discuss the generating functions of generalized extended hypergeometric functions through the application of a new generalized extended fractional derivative operator. Here is a summary of our paper.

In Sect. 2, we define a generalization of two variables' extended Appell's function $F^{\wp,\eta,\nu}_{1;\alpha,\beta}(\kappa_1,\kappa_2,\kappa_3;\kappa_4;x,y)$, $F^{\wp,\eta,\nu}_{2;\alpha,\beta}(\kappa_1,\kappa_2,\kappa_3;\kappa_4,\kappa_5;x,y)$ and a generalized extended Lauricella's hypergeometric functions of three variables $F^{3,\alpha,\beta,\eta,\nu}_{\mathcal{D},\wp}(\kappa_1,\kappa_2,\kappa_3,\kappa_4;\kappa_5;x,y,z)$. Also, we study integral transforms of generalized extended Appell's function of two variables. In Sect. 3, we define generalized extended Riemann–Liouville fractional derivative operator with the help of extended beta function and find fractional derivative formula of some known functions. In Sect. 4, we introduce the Mellin transform of extended Riemann–Liouville fractional derivative. Finally, in Sect. 5, we derive the generating function of generalized extended hypergeometric functions by using extended fractional derivative operator.

2 An Extended Appell's Functions

Let us define the further extensions of Appell's functions of two variables $F^{\wp,\eta,\nu}_{1;\alpha,\beta}(\kappa_1,\kappa_2,\kappa_3;\kappa_4;x,y)$, $F^{\wp,\eta,\nu}_{2;\alpha,\beta}(\kappa_1,\kappa_2,\kappa_3;\kappa_4,\kappa_5;x,y)$ as well as Lauricella's hypergeometric function of three variables $F^{3,\alpha,\beta,\eta,\nu}_{\mathcal{D},\wp}(\kappa_1,\kappa_2,\kappa_3,\kappa_4;\kappa_5;x,y,z)$ with the help of new extended beta functions (5) as follows:

$$F^{\wp,\eta,\nu}_{1;\alpha,\beta}(\kappa_1,\kappa_2,\kappa_3;\kappa_4;x,y)=\sum_{n,m=0}^{\infty}\frac{\mathcal{B}^{\wp,\eta,\nu}_{\alpha,\beta}(\kappa_1+m+n,\kappa_4-\kappa_1)}{\mathcal{B}(\kappa_1,\kappa_4-\kappa_1)}(\kappa_2)_n(\kappa_3)_m\frac{x^n}{n!}\frac{y^m}{m!},\tag{10}$$

$$(\max\{|x|,|y|\}<1\ ;\ \alpha,\beta,\vartheta,\nu\in\mathbb{R}^+\ and\ \wp\geq 0).$$

$$F^{\wp,\eta,\nu}_{2;\alpha,\beta,}(\kappa_1,\kappa_2,\kappa_3;\kappa_4,\kappa_5;x,y)$$

$$=\sum_{n,m=0}^{\infty}\frac{(\kappa_1)_{m+n}\mathcal{B}^{\wp,\eta,\nu}_{\alpha,\beta}(\kappa_2+n,\kappa_4-\kappa_2)\mathcal{B}^{\wp,\eta,\nu}_{\alpha,\beta}(\kappa_3+m,\kappa_5-\kappa_3)}{\mathcal{B}(\kappa_2,\kappa_4-\kappa_2)\mathcal{B}(\kappa_3;\kappa_5-\kappa_3)}\frac{x^n}{n!}\frac{y^m}{m!},\tag{11}$$

$$(|x|+|y|<1;\quad \alpha,\beta,\vartheta,\nu,\in\mathbb{R}^+\ and\ \wp\geq 0).$$

$$F^{3,\alpha,\beta,\eta,\nu}_{\mathcal{D},\wp}(\kappa_1,\kappa_2,\kappa_3,\kappa_4;\kappa_5;x,y,z)$$

$$=\sum_{n,m,r=0}^{\infty}\frac{\mathcal{B}^{\wp,\eta,\nu}_{\alpha,\beta}(\kappa_1+m+n+r,\kappa_5-\kappa_1)(\kappa_2)_m(\kappa_3)_n(\kappa_4)_r}{\mathcal{B}(\kappa_1,\kappa_5-\kappa_1)}\frac{x^n}{n!}\frac{y^m}{m!}\frac{z^r}{r!},\tag{12}$$

$$(\sqrt{|x|}+\sqrt{|y|}+\sqrt{|z|}<1\ ;\ \alpha,\beta,\vartheta,\nu\in\mathbb{R}_0^+\ and\ \wp\geq 0).$$

Remark 1 If $\alpha = \beta = \nu = \eta = 1$ in (10), (11) and (12) then we get the well-known result defined by Özarslan et al. (see [16] Eq. 3, 4, 5).

3 Integral Representation

In this section, we discuss the integral representation of $F_{1;\alpha,\beta}^{\wp,\eta,\nu}(\kappa_1, \kappa_2, \kappa_3; \kappa_4; x, y)$ and $F_{2;\alpha,\beta}^{\wp,\eta,\nu}(\kappa_1, \kappa_2, \kappa_3; \kappa_4, \kappa_5; x, y)$ as follows:

Theorem 1 *The following integral representation of* $F_{1;\alpha,\beta}^{\wp,\eta,\nu}(\kappa_1, \kappa_2, \kappa_3; \kappa_4; x, y)$ *holds true:*

$$F_{1;\alpha,\beta}^{\wp,\eta,\nu}(\kappa_1, \kappa_2, \kappa_3; \kappa_4; x, y)$$

$$= \frac{\Gamma(\kappa_4)}{\Gamma(\kappa_1)\Gamma(\kappa_4 - \kappa_1)} \int_0^1 \tau^{\kappa_1 - 1}(1-\tau)^{\kappa_4 - \kappa_1 - 1}(1 - x\tau)^{-\kappa_2}(1 - y\tau)^{-\kappa_3} E_{\alpha,\beta}\left(\frac{-\wp}{\tau^{\eta}(1-\tau)^{\nu}}\right) d\tau, \tag{13}$$

$$(\wp \geq 0\ ,\ |arg(1-x)| < \pi,\ \ |arg(1-y)| < \pi;\ \ \Re(\kappa_4) > \Re(\kappa_1) > 0)$$

$$(\Re(\kappa_2) > 0,\ \ \Re(\kappa_3) > 0;\ \ \eta, \nu, \alpha, \beta \in \mathbb{R}_0^+).$$

Proof Let us assume that if $|x| < 1,\ |y| < 1,\ \Re(\kappa_2) > 0$ *and* $\Re(\kappa_3) > 0$. By using the binomial expansion of $(1 - x\tau)^{-\kappa_2}$ and $(1 - y\tau)^{-\kappa_3}$ and considering the fact that the series involved are uniformly convergent and the given integral

$$\int_0^1 \tau^{\kappa_1 + m + n - 1}(1-\tau)^{\kappa_4 - \kappa_1 - 1} E_{\alpha,\beta}\left(\frac{-\wp}{\tau^{\eta}(1-\tau)^{\nu}}\right) d\tau,$$

is absolutely convergent for $m, n \in \mathbb{N}_0, \Re(\kappa_4) > \Re(\kappa_1) > 0;\ \ \eta, \nu, \alpha, \beta \in \mathbb{R}^+$ and $\wp \geq 0$ because of the fact that

$$\left| \int_0^1 \tau^{\kappa_1 + m + n - 1}(1-\tau)^{\kappa_4 - \kappa_1 - 1} E_{\alpha,\beta}\left(\frac{-\wp}{\tau^{\eta}(1-\tau)^{\nu}}\right) d\tau \right| \leq \int_0^1 \tau^{\kappa_1 + m + n - 1}(1-\tau)^{\kappa_4 - \kappa_1 - 1} d\tau,$$

we have by interchanging the order of summation and integration that

$$\int_0^1 \tau^{\kappa_1-1}(1-\tau)^{\kappa_4-\kappa_1-1}(1-x\tau)^{-\kappa_2}(1-y\tau)^{-\kappa_3} E_{\alpha,\beta}\left(\frac{-\wp}{\tau^\eta(1-\tau)^\nu}\right) d\tau$$

$$= \int_0^1 \tau^{\kappa_1-1}(1-\tau)^{\kappa_4-\kappa_1-1} E_{\alpha,\beta}\left(\frac{-\wp}{\tau^\eta(1-\tau)^\nu}\right) \sum_{n=0}^{\infty}(\kappa_2)_n \frac{(x\tau)^n}{n!} \sum_{m=0}^{\infty}(\kappa_3)_m \frac{(y\tau)^m}{m!} d\tau$$

$$= \sum_{n=0}^{\infty}\sum_{m=0}^{\infty}(\kappa_2)_n(\kappa_3)_m \frac{x^n}{n!}\frac{y^m}{m!} \int_0^1 \tau^{\kappa_1+m+n-1}(1-\tau)^{\kappa_4-\kappa_1-1} E_{\alpha,\beta}\left(\frac{-\wp}{\tau^\eta(1-\tau)^\nu}\right) d\tau$$

$$= \sum_{n,m=0}^{\infty} \mathcal{B}_{\alpha,\beta}^{\wp,\eta,\nu}(\kappa_1+m+n, \kappa_4-\kappa_1)(\kappa_2)_n(\kappa_3)_m \frac{x^n}{n!}\frac{y^m}{m!}$$

$$= \mathcal{B}(\kappa_1, \kappa_4-\kappa_1) \sum_{n,m=0}^{\infty} \frac{\mathcal{B}_{\alpha,\beta}^{\wp,\eta,\nu}(\kappa_1+m+n, \kappa_4-\kappa_1)}{\mathcal{B}(\kappa_1, \kappa_4-\kappa_1)}(\kappa_2)_n(\kappa_3)_m \frac{x^n}{n!}\frac{y^m}{m!}$$

$$= \frac{\Gamma(\kappa_1)\Gamma(\kappa_4-\kappa_1)}{\Gamma(\kappa_4)} F_{1;\alpha,\beta}^{\wp,\eta,\nu}(\kappa_1, \kappa_2, \kappa_3; \kappa_4; x, y).$$

□

Theorem 2 *The following integral representation of* $F_{2;\alpha,\beta}^{p,\eta,\nu}(\kappa_1, \kappa_2, \kappa_3; \kappa_4, \kappa_5; x, y)$ *holds true:*

$$F_{2;\alpha,\beta,}^{\wp,\eta,\nu,}(\kappa_1, \kappa_2, \kappa_3; \kappa_4, \kappa_5; x, y) = \frac{1}{\mathcal{B}(\kappa_2, \kappa_4-\kappa_2)\mathcal{B}(\kappa_3, \kappa_5-\kappa_3)} \times \int_0^1\int_0^1 \frac{\tau^{\kappa_2-1}(1-\tau)^{\kappa_4-\kappa_2-1}s^{\kappa_3-1}(1-s)^{\kappa_5-\kappa_3-1}}{(1-x\tau-ys)^{\kappa_1}} \times E_{\alpha,\beta}\left(\frac{-\wp}{s^\eta(1-s)^\nu}\right) E_{\alpha,\beta}\left(\frac{-\wp}{s^\eta(1-s)^\nu}\right) d\tau\, ds, \tag{14}$$

($\wp \geq 0$; $|x|+|y|<1$; $\Re(\kappa_4) > \Re(\kappa_2) > 0$; $\Re(\kappa_5) > \Re(\kappa_3) > 0$; $\Re(\kappa_1) > 0$; $\alpha, \beta, \eta, \nu \in \mathbb{R}^+$).

Proof Suppose that $|x|+|y|<1$ and $\Re(\kappa_1) > 0$ expanding $(1-x\tau-ys)^{\kappa_1}$, we get

$$\int_0^1\int_0^1 \frac{\tau^{\kappa_2-1}(1-\tau)^{\kappa_4-\kappa_2-1}s^{\kappa_3-1}(1-s)^{\kappa_5-\kappa_3-1}}{(1-x\tau-ys)^{\kappa_1}} E_{\alpha,\beta}\left(\frac{-\wp}{s^\eta(1-s)^\nu}\right) E_{\alpha,\beta}\left(\frac{-\wp}{s^\eta(1-s)^\nu}\right) d\tau\, ds$$

$$=\int_0^1\int_0^1 \tau^{\kappa_2-1}(1-\tau)^{\kappa_4-\kappa_2-1}E_{\alpha,\beta}\left(\frac{-\wp}{s^\eta(1-s)^\nu}\right)s^{\kappa_3-1}(1-s)^{\kappa_5-\kappa_3-1}E_{\alpha,\beta}\left(\frac{-\wp}{s^\eta(1-s)^\nu}\right)$$
$$\times\sum_{N=0}^{\infty}\frac{(\kappa_1)_N(x\tau+ys)^N}{N!}d\tau\, ds,$$

we know that the summation formula,

$$\left\{\sum_{N=0}^{\infty}f(N)\frac{(x+y)^N}{N!}=\sum_{n=0}^{\infty}\sum_{m=0}^{\infty}f(m+n)\frac{x^n y^m}{n!m!}\right\}.$$

We obtain

$$=\sum_{n=0}^{\infty}\sum_{m=0}^{\infty}(\kappa_1)_{m+n}\int_0^1\int_0^1 \tau^{\kappa_2-1}(1-\tau)^{\kappa_4-\kappa_2-1}E_{\alpha,\beta}\left(\frac{-\wp}{s^\eta(1-s)^\nu}\right)$$
$$\times s^{\kappa_3-1}(1-s)^{\kappa_5-\kappa_3-1}E_{\alpha,\beta}\left(\frac{-\wp}{s^\eta(1-s)^\nu}\right)\frac{(x\tau)^n}{n!}\frac{(ys)^m}{m!}dtds$$
$$=\sum_{n=0}^{\infty}\sum_{m=0}^{\infty}(\kappa_1)_{m+n}\int_0^1 \tau^{\kappa_2+n-1}(1-\tau)^{\kappa_4-\kappa_2-1}E_{\alpha,\beta}\left(\frac{-\wp}{s^\eta(1-s)^\nu}\right)d\tau$$
$$\times\int_0^1 s^{\kappa_3+m-1}(1-s)^{\kappa_5-\kappa_3-1}E_{\alpha,\beta}\left(\frac{-\wp}{s^\eta(1-s)^\nu}\right)ds\,\frac{x^n y^m}{n!m!}$$
$$=\sum_{n=0}^{\infty}\sum_{m=0}^{\infty}(\kappa_1)_{m+n}\mathcal{B}_{\alpha,\beta}^{\wp,\eta,\nu}(\kappa_2+n,\kappa_4-\kappa_2)\mathcal{B}_{\alpha,\beta}^{\wp,\eta,\nu}(\kappa_3+m,\kappa_5-\kappa_3)\frac{x^n y^m}{n!m!}$$
$$=\mathcal{B}(\kappa_2,\kappa_4-\kappa_2)\mathcal{B}(\kappa_3,\kappa_5-\kappa_3)F_{2;\alpha,\beta}^{\wp,\eta,\nu,}(\kappa_1,\kappa_2,\kappa_3;\kappa_4,\kappa_5;x,y).$$

Hence, we get the desired result (14). □

4 An Extended Riemann–Liouville Fractional Derivative

The well-known classical Riemann–Liouville fractional derivative of order ϑ defined as

$$\mathcal{D}_z^\vartheta\{f(z)\}=\frac{1}{\Gamma(-\vartheta)}\int_0^z f(\tau)(z-\tau)^{-\vartheta-1}d\tau,\quad(\Re(\vartheta)<0),$$

where the line on complex τ plane that goes from 0 to z and for $j-1<\Re(\vartheta)<j$ such that $j=1,2,3,....,$ we have

$$\mathcal{D}_z^{\vartheta}\{f(z)\} = \frac{d^j}{dz^j}\mathcal{D}_z^{\vartheta-j}\{f(z)\}$$
$$= \frac{d^j}{dz^j}\left\{\frac{1}{\Gamma(j-\vartheta)}\int_0^z f(\tau)(z-1)^{-\vartheta+j-1}d\tau\right\}.$$

The study of fractional calculus has become a popular research subject due to its numerous applications in diverse fields of science and engineering, including fluid flow, electrical networks, and probability theory. Srivastava et al. [24] and several authors [3, 5] provide a well-organized summary of the research on the topic of fractional calculus and its applications, and this is a valuable contribution. Srivastava and Manocha [25] also provide an explanation for the application of fractional derivatives in generating functions theory.

After introducing new parameters, we proposed an extended Riemann–Liouville fractional derivative operator using the Mittag-Leffler function as a kernel operator defined as follows:

$$\mathcal{D}_{z;\alpha,\beta}^{\vartheta,p,\eta,\nu}\{f(z)\} = \frac{1}{\Gamma(-\vartheta)}\int_0^z f(\tau)(z-\tau)^{-\vartheta-1}E_{\alpha,\beta}\left(\frac{-\wp z^{\eta+\nu}}{\tau^n(z-\tau)^\nu}\right)d\tau, \tag{15}$$

$$(\Re(\vartheta) < 0;\ \Re(p) > 0;\ p = 0;\ \eta, \nu, \alpha, \beta \in \mathbb{R}_0^+),$$

and for $j - 1 < \Re(\vartheta) < j$ such that $j = 1, 2, 3, ...$

Here, an extended Riemann–Liouville fractional derivative operator $\mathcal{D}_{z;\alpha,\beta}^{\vartheta,p,\eta,\nu}$ has an important semi-group property (or convolution property), i.e., $\mathcal{D}_{z;\alpha,\beta}^{\vartheta,p,\eta,\nu} * \mathcal{D}_{z;\alpha,\beta}^{\zeta,p,\eta,\nu} = \mathcal{D}_{z;\alpha,\beta}^{\vartheta+\zeta,p,\eta,\nu}$ for arbitrary $\Re(\vartheta) < 0$ and $\Re(\zeta) < 0$.

$$\mathcal{D}_{z;\alpha,\beta}^{\vartheta,p,\eta,\nu}\{f(z)\} = \frac{d^j}{dz^j}\mathcal{D}_z^{\vartheta-j}\{f(z)\}$$
$$= \frac{d^j}{dz^j}\left\{\frac{1}{\Gamma(j-\vartheta)}\int_0^z f(\tau)(z-1)^{-\vartheta+j-1}E_{\alpha,\beta}\left(\frac{-\wp z^{\eta+\nu}}{\tau^n(z-\tau)^\nu}\right)d\tau\right\},$$

where the integration path is a line in the complex t-plane from 0 to z.

Remark 2 The classical Riemann–Liouville fractional derivative operator is obtained when $\wp = \alpha = \beta = \vartheta = \eta = 0$ and if $\alpha = \beta = \vartheta = \eta = 1$ in (15) then we obtained the known result (see [16], Eq. 6).

5 Applications

The generalized Riemann–Liouville fractional derivative operator has been used to compute the fractional derivative formulae of a few fundamental functions in the following theorems.

Theorem 3 *The following relation holds true:*

$$\mathcal{D}_{z,\alpha,\beta}^{\vartheta,\wp,\eta,\nu}(z^{\varrho}) = \frac{z^{\varrho-\vartheta}}{\Gamma(-\vartheta)}\mathcal{B}_{\alpha,\beta}^{\wp,\eta,\nu}(\varrho+1,-\vartheta), \tag{16}$$

$$(\Re(\varrho) > -1;\ \Re(\vartheta) < 0;\ \alpha,\beta,\eta,\nu \in \mathbb{R}^{+};\ \wp \geq 0).$$

Proof By using (15), we have

$$\mathcal{D}_{z,\alpha,\beta}^{\vartheta,\wp,\eta,\nu}(z^{\varrho}) = \frac{1}{\Gamma(-\vartheta)}\int_0^z \tau^{\varrho}(z-\tau)^{-\vartheta-1}E_{\alpha,\beta}\left(\frac{-\wp z^{\eta+\nu}}{\tau^{\eta}(z-\tau)^{\nu}}\right)d\tau,$$

by putting $\tau = zv$, $d\tau = zdv$ in above equation, we get

$$=\frac{1}{\Gamma(-\vartheta)}\int_0^z (zv)^{\varrho}(z-zv)^{-\varrho-1}E_{\alpha,\beta}\left(\frac{-\wp z^{\eta+\nu}}{z^{\eta+\nu}v^{n}(1-v)^{\nu}}\right)zdv$$

$$=\frac{z^{\varrho+\vartheta}}{\Gamma(-\vartheta)}\int_0^1 v^{\varrho}(1-v)^{-\vartheta-1}E_{\alpha,\beta}\left(\frac{-p}{v^{\eta}(1-v)^{\nu}}\right)du$$

$$=\frac{z^{\varrho-\vartheta}}{\Gamma(-\vartheta)}\mathcal{B}_{\alpha,\beta}^{\wp,\eta,\nu}(\varrho+1,-\vartheta).$$

Thus, we obtain what we want. □

Theorem 4 *For $\Re(\varrho) > 0$; $\Re(\kappa_1) > 0$; $\Re(\vartheta) < 0$ and $|z| < 1$, the following relation holds true:*

$$\mathcal{D}_{z;\alpha,\beta}^{\varrho-\vartheta,\wp,\eta,\nu}\left\{z^{\varrho-1}(1-z)^{-\kappa_1}\right\} = \frac{\Gamma(\varrho)}{\Gamma(\vartheta)}z^{\vartheta-1}F_{\alpha,\beta}^{\wp,\eta,\nu}(\kappa_1,\varrho;\vartheta;z). \tag{17}$$

Proof Using (15), we have

$$\mathcal{D}_{z;\alpha,\beta}^{\varrho-\vartheta,\wp,\eta,\nu}\left\{z^{\varrho-1}(1-z)^{-\kappa_1}\right\}$$

$$=\frac{1}{\Gamma(\vartheta-\varrho)}\int_0^z \tau^{\varrho-1}(1-\tau)^{-\kappa_1}(z-\tau)^{\vartheta-\varrho-1}E_{\alpha,\beta}\left(\frac{-\wp z^{\eta+\nu}}{\tau^{\eta}(z-\tau)^{\nu}}\right)d\tau,$$

$$=\frac{z^{\vartheta-\varrho-1}}{\Gamma(\vartheta-\varrho)}\int_0^z \tau^{\varrho-1}(1-\tau)^{-\kappa_1}(1-\frac{\tau}{z})^{\vartheta-\varrho-1}E_{\alpha,\beta}\left(\frac{-\wp z^{\eta+\nu}}{\tau^{n}(z-\tau)^{\nu}}\right)d\tau.$$

By inserting $\tau = zv,\ d\tau = zdv$ in the above equation, we have

$$=\frac{z^{\vartheta-\varrho-1}}{\Gamma(\vartheta-\varrho)}\int_0^1 (zv)^{\varrho-1}(1-zv)^{-\kappa_1}(1-v)^{\vartheta-\varrho-1}E_{\alpha,\beta}\left(\frac{-\wp z^{\eta+\nu}}{z^{\eta+\nu}v^{\eta}(1-v)^{\nu}}\right)zdv,$$

$$=\frac{z^{\vartheta-1}}{\Gamma(\vartheta-\varrho)}\int_0^1 v^{\varrho-1}(1-v)^{\vartheta-\varrho-1}(1-zv)^{-\kappa_1}E_{\alpha,\beta}\left(\frac{-p}{v^{\eta}(1-v)^{\nu}}\right)dv.$$

Using (8) in the above expression, we get

$$\begin{aligned}\mathcal{D}^{\varrho-\vartheta,\wp,\eta,\nu}_{z;\alpha,\beta}\left\{z^{\varrho-1}(1-z)^{-\kappa_1}\right\} &=\frac{z^{\vartheta-1}}{\Gamma(\vartheta-\varrho)}\mathcal{B}(\varrho,\vartheta-\varrho)F^{\wp,\eta,\nu}_{\alpha,\beta}(\kappa_1,\varrho;\vartheta;z)\\ &=\frac{z^{\vartheta-1}}{\Gamma(\vartheta-\varrho)}\frac{\Gamma(\varrho)\Gamma(\vartheta-\varrho)}{\Gamma(\vartheta)}F^{\wp,\eta,\nu}_{\alpha,\beta}(\kappa_1,\varrho;\vartheta;z)\\ &=\frac{\Gamma(\varrho)}{\Gamma(\vartheta)}z^{\vartheta-1}F^{\wp,\eta,\nu}_{\alpha,\beta}(\kappa_1,\varrho;\vartheta;z).\end{aligned}$$

Thus, we arrive at the desired result. □

Theorem 5 *For* $\Re(\varrho) > \Re(\vartheta) > 0$, $\Re(\gamma) > 0$, $\Re(\delta) > 0$, $|\kappa_1 z| < 1$, *and* $|\kappa_2 z| < 1$, *the following relation holds true:*

$$\mathcal{D}^{\varrho-\vartheta,\wp,\eta,\nu}_{z;\alpha,\beta}\left\{z^{\varrho-1}(1-\kappa_1 z)^{-\gamma}(1-\kappa_2 z)^{-\delta}\right\}=z^{\vartheta-1}\frac{\Gamma(\varrho)}{\Gamma(\vartheta)}F^{\wp,\eta,\nu}_{1;\alpha,\beta}(\varrho,\gamma,\delta,\epsilon;\vartheta;\kappa_1 z,\kappa_2 z). \quad (18)$$

Proof By using (15), we have

$$\mathcal{D}^{\varrho-\vartheta,\wp,\eta,\nu}_{z;\alpha,\beta}\left\{z^{\varrho-1}(1-\kappa_1 z)^{-\gamma}(1-\kappa_2 z)^{-\delta}\right\}$$

$$=\frac{1}{\Gamma(\vartheta-\varrho)}\int_0^z \tau^{\varrho-1}(1-\kappa_1\tau)^{-\gamma}(1-\kappa_2\tau)^{-\delta}E_{\alpha,\beta}\left(\frac{-\wp z^{\eta+\nu}}{\tau^{n}(z-\tau)^{\nu}}\right)(z-\tau)^{\vartheta-\varrho-1}d\tau,$$

$$=\frac{z^{\vartheta-\varrho-1}}{\Gamma(\vartheta-\varrho)}\int_0^z \tau^{\varrho-1}(1-\kappa_1\tau)^{-\gamma}(1-\kappa_2\tau)^{-\delta}(1-\frac{\tau}{z})^{\vartheta-\varrho-1}E_{\alpha,\beta}\left(\frac{-\wp z^{\eta+\nu}}{\tau^{\eta}(z-\tau)^{\nu}}\right)d\tau.$$

Putting $\tau = zv, d\tau = zdv$,

$$=\frac{z^{\vartheta-\varrho-1}}{\Gamma(\vartheta-\varrho)}\int_0^z (zv)^{\varrho-1}(1-\kappa_1 vz)^{-\gamma}(1-\kappa_2 vz)^{-\delta}(1-v)^{\vartheta-\varrho-1}E_{\alpha,\beta}\left(\frac{-\wp z^{\eta+\nu}}{(zv)^{\eta}(1-zv)^{\nu}}\right)zdv$$

$$=\frac{z^{\vartheta-1}}{\Gamma(\vartheta-\varrho)}\int_0^z (v)^{\varrho-1}(1-\kappa_1 vz)^{-\gamma}(1-\kappa_2 vz)^{-\delta}(1-v)^{\vartheta-\varrho-1}E_{\alpha,\beta}\left(\frac{-\wp}{v^{\eta}(1-v)^{\nu}}\right)dv.$$

Applying definition (13) in the above expression, we get

$$\begin{aligned}\mathcal{D}^{\varrho-\vartheta,\wp,\eta,\nu}_{z;\alpha,\beta}\left\{z^{\varrho-1}(1-\kappa_1 z)^{-\gamma}(1-\kappa_2 z)^{-\delta}\right\} =&\frac{z^{\vartheta-1}}{\Gamma(\vartheta-\varrho)}\frac{\Gamma(\varrho)\Gamma(\vartheta-\varrho)}{\Gamma(\vartheta)}\\ &\times F^{\wp,\eta,\nu}_{1;\alpha,\beta}(\varrho,\gamma,\delta;\vartheta;\kappa_1 z,\kappa_2 z)\\ =&z^{\vartheta-1}\frac{\Gamma(\varrho)}{\Gamma(\vartheta)}F^{\wp,\eta,\nu}_{1;\alpha,\beta}(\varrho,\gamma,\delta;\vartheta;\kappa_1 z,\kappa_2 z).\end{aligned}$$

We complete the proof of this theorem. □

Theorem 6 *For* $\Re(\varrho) > \Re(\vartheta) > 0$, $\Re(\gamma) > 0$, $\Re(\delta) > 0$, $\Re(\epsilon) > 0$, $|\kappa_1 z| < 1$, $|\kappa_2 z| < 1$, *and* $|\kappa_3 z| < 1$, *the following relation holds true:*

$$\begin{aligned}&\mathcal{D}^{\varrho-\vartheta,\wp,\eta,\nu}_{z;\alpha,\beta}\left\{z^{\varrho-1}(1-\kappa_1 z)^{-\gamma}(1-\kappa_2 z)^{-\delta}(1-\kappa_3 z)^{-\epsilon}\right\}\\ &=z^{\vartheta-1}\frac{\Gamma(\varrho)}{\Gamma(\vartheta)}F^{3,\eta,\nu}_{D,p,\alpha,\beta}(\varrho,\gamma,\delta,\epsilon;\vartheta;\kappa_1 z,\kappa_2 z,\kappa_3 z),\end{aligned} \tag{19}$$

$$(\alpha,\beta,\eta,\nu\in\mathbb{R}^+ \text{ and } \wp\geq 0).$$

Proof With the help of (18), we get

$$\mathcal{D}^{\varrho-\vartheta,\wp,\eta,\nu}_{z;\alpha,\beta}\left\{z^{\varrho-1}(1-\kappa_1 z)^{-\gamma}(1-\kappa_2 z)^{-\delta}(1-\kappa_3 z)^{-\epsilon}\right\}$$

$$=\frac{z^{\vartheta-1}}{\Gamma(\vartheta-\varrho)}\sum_{m,n,r=0}^{\infty}\frac{(\gamma)_m(\delta)_n(\epsilon)_r}{m!n!r!}\kappa_1^m\,\kappa_2^n\,\kappa_3^r\,\mathcal{B}^{\wp,\eta,\nu}_{\alpha,\beta}(\varrho+m+n+r,\vartheta-\varrho)z^{m+n+r}$$

$$=\frac{\mathcal{B}(\varrho,\vartheta-\varrho)}{\Gamma(\vartheta-\varrho)}z^{\vartheta-1}\sum_{m,n,r=0}^{\infty}\frac{\mathcal{B}^{\wp,\eta,\nu}_{\alpha,\beta}(\varrho+m+n+r,\vartheta-\varrho)}{\mathcal{B}(\varrho,\vartheta-\varrho)}\frac{(\gamma)_m(\delta)_n(\epsilon)_r}{m!n!r!}$$

$$\times\kappa_1^m\kappa_2^n\kappa_3^r(\kappa_1 z)^m(\kappa_2 z)^n(\kappa_3 z)^r$$

$$=z^{\vartheta-1}\frac{\Gamma(\varrho)}{\Gamma(\vartheta)}F^{3,\eta,\nu}_{D,\wp,\alpha,\beta}(\varrho,\gamma,\delta,\epsilon;\vartheta;\kappa_1 z,\kappa_2 z,\kappa_3 z).$$

Thus, we obtain what we want. □

Theorem 7 *For* $\Re(\varrho) > \Re(\vartheta) > 0$, $\Re(\kappa_1) > 0$, $\Re(\kappa_2) > 0$, $\Re(\gamma) > 0$, *and* $\left|\frac{x}{1-z}\right| < 1$, *the following relation holds true:*

$$\mathcal{D}_{z;\alpha,\beta}^{\varrho-\vartheta,\wp,\eta,\nu}\left\{z^{\varrho-1}(1-z)^{-\kappa_1}F_{\alpha,\beta}^{\wp,\eta,\nu}(\kappa_1,\kappa_2;\gamma;\frac{x}{1-z})\right\}$$

$$=\frac{1}{\mathcal{B}(\kappa_2,\gamma-\kappa_2)\Gamma(\vartheta-\varrho)}z^{\vartheta-1}F_{2;\alpha,\beta}^{\wp,\eta,\nu}(\kappa_1,\kappa_2,\varrho;\gamma,\vartheta;x,z), \tag{20}$$

$$(\alpha,\beta,\eta,\nu\in\mathbb{R}^+ \ \ and\ \ \wp\geq 0).$$

Proof By using the relation (15) and (11), we obtain

$$\mathcal{D}_{z;\alpha,\beta}^{\varrho-\vartheta,\wp,\eta,\nu}\left\{z^{\varrho-1}(1-z)^{-\kappa_1}F_{\alpha,\beta}^{\wp,\eta,\nu}(\kappa_1,\kappa_2;\gamma;\frac{x}{1-z})\right\}$$

$$=\mathcal{D}_{z;\alpha,\beta}^{\varrho-\vartheta,\wp,\eta,\nu}\left\{z^{\varrho-1}(1-z)^{-\kappa_1}\frac{1}{\mathcal{B}(\kappa_2,\gamma-\kappa_2)}\sum_{n=0}^{\infty}\frac{(\kappa_1)_n\mathcal{B}_{\alpha,\beta}^{\wp,\eta,\nu}(\kappa_2+n,\gamma-\kappa_2)}{n!}\left(\frac{x}{1-z}\right)^n\right\}$$

$$=\frac{1}{\mathcal{B}(b,\gamma-b)}D_{z;\alpha,\beta}^{\varrho-\vartheta,\wp,\eta,\nu}\left\{z^{\varrho-1}\sum_{n=0}^{\infty}(\kappa_1)_n\mathcal{B}_{\alpha,\beta}^{\wp,\eta,\nu}(\kappa_2+n,\gamma-\kappa_2)\frac{x^n}{n!}(1-z)^{-\kappa_1-n}\right\}$$

$$=\frac{1}{\mathcal{B}(\kappa_2,\gamma-\kappa_2)}\sum_{n=0}^{\infty}(\kappa_1)_n\mathcal{B}_{\alpha,\beta}^{\wp,\eta,\nu}(\kappa_2+n,\gamma-\kappa_2)\frac{x^n}{n!}D_{z;\alpha,\beta}^{\varrho-\vartheta,p,\eta,\nu}\left\{z^{\varrho-1}(1-z)^{-\kappa_1-n}\right\}$$

$$=\frac{1}{\mathcal{B}(\kappa_2,\gamma-\kappa_2)}\sum_{n=0}^{\infty}(\kappa_1)_n\mathcal{B}_{\alpha,\beta}^{\wp,\eta,\nu}(\kappa_2+n,\gamma-\kappa_2)\frac{x^n}{n!}\frac{z^{\vartheta-1}}{\Gamma(\vartheta-\varrho)}\mathcal{B}(\varrho,\vartheta-\varrho)$$
$$\times F_{\alpha,\beta}^{\wp,\eta,\nu}(\kappa_1+n,\varrho;\vartheta;z)$$

$$=\frac{z^{\vartheta-1}}{\mathcal{B}(\kappa_2,\gamma-\kappa_2)\Gamma(\vartheta-\varrho)}\sum_{n=0}^{\infty}(\kappa_1)_n\mathcal{B}_{\alpha,\beta}^{\wp,\eta,\nu}(\kappa_2+n,\gamma-\kappa_2)\frac{x^n}{n!}\mathcal{B}(\varrho,\vartheta-\varrho)$$
$$\times F_{\alpha,\beta}^{\wp,\eta,\nu}(\kappa_1+n,\varrho;\vartheta;z)$$

$$=\frac{z^{\vartheta-1}}{\mathcal{B}(\kappa_2,\gamma-\kappa_2)\Gamma(\vartheta-\varrho)}$$
$$\times\sum_{n=0}^{\infty}(\kappa_1)_n\mathcal{B}_{\alpha,\beta}^{\wp,\eta,\nu}(\kappa_2+n,\gamma-\kappa_2)\frac{x^n}{n!}\mathcal{B}(\varrho,\vartheta-\varrho)\sum_{m=0}^{\infty}(\kappa_1+n)_m\frac{\mathcal{B}_{\alpha,\beta}^{\wp,\eta,\nu}(\varrho+m,\vartheta-\varrho)}{\mathcal{B}(\varrho,\vartheta-\varrho)}\frac{z^m}{m!}$$

$$=\frac{z^{\vartheta-1}}{\mathcal{B}(\kappa_2,\gamma-\kappa_2)\Gamma(\vartheta-\varrho)}\sum_{n,m=0}^{\infty}(\kappa_1)_n(\kappa_1+n)_m\mathcal{B}_{\alpha,\beta}^{\wp,\eta,\nu}(\kappa_2+n,\gamma-\kappa_2)\mathcal{B}_{\alpha,\beta}^{\wp,\eta,\nu}(\varrho+m,\vartheta-\varrho)$$
$$\times\frac{x^nz^m}{n!m!}$$

$$=\frac{z^{\vartheta-1}}{\mathcal{B}(\kappa_2,\gamma-\kappa_2)\Gamma(\vartheta-\varrho)}\sum_{n,m=0}^{\infty}(\kappa_1)_{n+m}\mathcal{B}_{\alpha,\beta}^{\wp,\eta,\nu}(\kappa_2+n,\gamma-\kappa_2)\mathcal{B}_{\alpha,\beta}^{\wp,\eta,\nu}(\varrho+m,\vartheta-\varrho)\frac{x^nz^m}{n!m!}$$

$$=\frac{1}{\mathcal{B}(\kappa_2,\gamma-\kappa_2)}\sum_{n,m=0}^{\infty}\mathcal{B}_{\alpha,\beta}^{\wp,\eta,\nu}(\kappa_2+n,\gamma-\kappa_2)\frac{x^n}{n!}\frac{(\kappa_1)_{n+m}}{m!}\frac{\mathcal{B}_{\alpha,\beta}^{\wp,\eta,\nu}(\varrho+m,\vartheta-\varrho)}{\Gamma(\vartheta-\varrho)}z^{\vartheta+m-1}$$

$$=\frac{1}{\mathcal{B}(\kappa_2,\gamma-\kappa_2)\Gamma(\vartheta-\varrho)}z^{\vartheta-1}F_{2;\alpha,\beta}^{\wp,\eta,\nu}(\kappa_1,\kappa_2,\varrho;\gamma,\vartheta;x,z).$$

□

Thus, we obtain what we want.

Example 1 Let $f(z)=\sum_{n=0}^{\infty}a_n z^n$ be a power series expansion of an analytic function f(z) in the disk $|z|<\rho$, then following relation holds:

$$\mathcal{D}_{z;\alpha,\beta}^{\vartheta,\wp,\eta,\nu}\left\{z^{\varrho-1}f(z)\right\}=\mathcal{D}_{z;\alpha,\beta}^{\vartheta,\wp,\eta,\nu}\left\{z^{\varrho-1}\sum_{n=0}^{\infty}a_n z^n\right\}$$

$$=\frac{z^{\varrho-\vartheta-1}}{\Gamma(-\vartheta)}\sum_{n-0}^{\infty}a_n\mathcal{B}_{\alpha,\beta}^{\wp,\eta,\nu}(\varrho+n,-\vartheta)z^n,$$

$$(\Re(\varrho)>0,\ \ \Re(\vartheta)<0,\ \ \alpha,\beta,\eta,\nu\in\mathbb{R}^+\ \ and\ \ \wp\geq 0).$$

Proof We have

$$\mathcal{D}_{z;\alpha,\beta}^{\vartheta,\wp,\eta,\nu}\left\{z^{\varrho-1}f(z)\right\}=\mathcal{D}_{z;\alpha,\beta}^{\vartheta,\wp,\eta,\nu}\left\{z^{\varrho-1}\sum_{n=0}^{\infty}a_n z^n\right\}$$

$$=\sum_{n=0}^{\infty}a_n\mathcal{D}_{z;\alpha,\beta}^{\vartheta,\wp,\eta,\nu}\left\{z^{\varrho+n-1}\right\}$$

$$=\frac{z^{\varrho-\vartheta-1}}{\Gamma(-\vartheta)}\sum_{n=0}^{\infty}a_n\mathcal{B}_{\alpha,\beta}^{\wp,\eta,\nu}(\varrho+n,-\vartheta)z^n$$

$$=\frac{1}{\Gamma(-\vartheta)}\int_0^z t^{\varrho-1}\sum_{n=0}^{\infty}a_n\tau^{\eta}(z-\tau)^{-\vartheta-1}E_{\alpha,\beta}\left(\frac{-\wp z^{\eta+\nu}}{\tau^{\eta}(z-\tau)^{\nu}}\right)d\tau$$

$$=\frac{z^{-\vartheta-1}}{\Gamma(-\vartheta)}\int_0^z\tau^{\varrho-1}\sum_{n=0}^{\infty}a_n\tau^{\eta}(1-\frac{\tau}{z})^{-\vartheta-1}E_{\alpha,\beta}\left(\frac{-\wp}{(\frac{\tau}{z})^{\eta}(1-\frac{\tau}{z})^{\nu}}\right)d\tau,$$

if we choose $\frac{\tau}{z}=u$ then $zdu=d\tau$, putting this value in above expression, we have

$$=\frac{z^{\vartheta-1}}{\Gamma(-\vartheta)}\int_0^1 (uz)^{\varrho-1}\sum_{n=0}^{\infty} a_n(uz)^n(1-u)^{-\vartheta-1}E_{\alpha,\beta}\left(\frac{-\wp}{u^{\eta}(1-u)^{\nu}}\right)zdu$$

$$=\frac{z^{\varrho-\vartheta-1}}{\Gamma(-\vartheta)}\int_0^1 (u)^{\varrho-1}(1-u)^{-\vartheta-1}E_{\alpha,\beta}\left(\frac{-\wp}{u^{\eta}(1-u)^{\nu}}\right)\sum_{n=0}^{\infty} a_n(uz)^n du.$$

By using uniformly convergent condition

$$=\sum_{n=0}^{\infty} a_n\frac{z^{\varrho+n-\vartheta-1}}{\Gamma(-\vartheta)}\int_0^1 u^{\varrho+n-1}(1-u)^{-\vartheta-1}E_{\alpha,\beta}\left(\frac{-\wp}{u^{\eta}(1-u)^{\nu}}\right)du$$

$$=\sum_{n=0}^{\infty} a_n\frac{z^{\varrho+n-\vartheta-1}}{\Gamma(-\vartheta)}\mathcal{B}_{\alpha,\beta}^{\wp,\eta,\nu}(\varrho+n,-\vartheta)$$

$$=\frac{z^{\varrho-\vartheta-1}}{\Gamma(-\vartheta)}\sum_{n=0}^{\infty} a_n\mathcal{B}_{\alpha,\beta}^{\wp,\eta,\nu}(\varrho+n,-\vartheta)z^n.$$

Hence proved. □

6 Mellin Transforms

This section deals with significant Mellin transform and extended Riemann–Liouville fractional derivative operator, which have been defined as follows:

Theorem 8 *The Mellin transform of an extended Riemann–Liouville fractional derivative operator is defined in* (15) *as follows:*

$$\mathcal{M}\left[\mathcal{D}_{z,\alpha,\beta}^{\vartheta,\wp,\eta,\nu}\{z^{\varrho}:s\}\right]=\frac{\Gamma_{(s)}^{\alpha,\beta}}{\Gamma(-\vartheta)}\mathcal{B}(\varrho+\eta s+1,\nu s-\vartheta)z^{\varrho-\vartheta},\tag{21}$$

$$(\Re(\varrho)>-1,\ \Re(\vartheta)<0,\ \Re(s)>0,\ \alpha,\beta,\eta,\nu\in\mathbb{R}^{+}\ \textit{and}\ \wp\geq 0).$$

Proof By using definition of Mellin transform and (15), we have

$$\mathcal{M}\left[\mathcal{D}_{z,\alpha,\beta}^{\vartheta,\wp,\eta,\nu}\{z^{\varrho}:s\}\right]$$

$$=\int_0^\infty \mathcal{P}^{s-1}\mathcal{D}^{\vartheta,\wp,\eta,\nu}_{z;\alpha,\beta}(z^{\varrho})d\mathcal{P}$$

$$=\frac{1}{\Gamma(-\vartheta)}\int_0^\infty \mathcal{P}^{s-1}\int_0^z \tau^{\varrho}(z-\tau)^{-\vartheta-1}E_{\alpha,\beta}\left(\frac{-\wp z^{\eta+\nu}}{\tau^{\eta}(z-\tau)^{\nu}}\right)d\tau d\mathcal{P}$$

$$=\frac{z^{-\vartheta-1}}{\Gamma(-\vartheta)}\int_0^\infty \mathcal{P}^{s-1}\int_0^z \tau^{\varrho}(1-\frac{\tau}{z})^{-\vartheta-1}E_{\alpha,\beta}\left(\frac{-\wp z^{\eta+\nu}}{\tau^{\eta}(z-\tau)^{\nu}}\right)d\tau d\mathcal{P}$$

$$=\frac{z^{-\vartheta-1}}{\Gamma(-\vartheta)}\int_0^\infty \mathcal{P}^{s-1}\int_0^1 u^{\varrho}z^{\varrho}(1-u)^{-\vartheta-1}E_{\alpha,\beta}\left(\frac{-\wp}{u^{\eta}(1-u)^{\nu}}\right)z\,du\,d\mathcal{P},$$

since

$$\int_0^1 u^{\varrho}(1-u)^{-\vartheta-1}E_{\alpha,\beta}\left(\frac{-\wp}{u^{\eta}(1-u)^{\nu}}\right)du,$$

and

$$\int_0^\infty \mathcal{P}^{s-1}E_{\alpha,\beta}\left(\frac{-\wp}{u^{\eta}(1-u)^{\nu}}\right)d\mathcal{P},$$

are absolutely convergent, the order of integration will be interchanged and we have

$$=\frac{z^{-\vartheta-1}}{\Gamma(-\vartheta)}\int_0^1 u^{\varrho}(1-u)^{-\vartheta-1}\int_0^\infty \mathcal{P}^{s-1}E_{\alpha,\beta}\left(\frac{-\wp}{u^{\eta}(1-u)^{\nu}}\right)d\mathcal{P}du.$$

If we put $\gamma=\frac{\mathcal{P}}{u^{\eta}(1-u)^{\nu}}$ implies $u^{\eta}(1-u)^{\nu}d\gamma=d\mathcal{P}$, then we see that

$$=\frac{z^{-\vartheta-1}}{\Gamma(-\vartheta)}\int_0^1 u^{\varrho}(1-u)^{-\vartheta-1}\int_0^\infty (\gamma u^{\eta}(1-u)^{\nu})^{s-1}\,E_{\alpha,\beta}(-\gamma)u^{\eta}(1-u)^{\nu}d\gamma du$$

$$=\frac{z^{\varrho-\vartheta}}{\Gamma(-\vartheta)}\int_0^1 u^{\varrho+\eta s}(1-u)^{\nu s-\vartheta-1}\int_0^\infty \gamma^{s-1}E_{\alpha,\beta}(-\gamma)d\gamma du$$

$$=\frac{\Gamma^{\alpha,\beta}_{(s)}}{\Gamma(-\vartheta)}z^{\varrho-\vartheta}\int_0^1 u^{\varrho+\eta s}(1-u)^{\nu s-\vartheta-1}du$$

$$=\frac{\Gamma^{\alpha,\beta}_{(s)}}{\Gamma(-\vartheta)}\mathcal{B}(\varrho+\eta s+1,\nu s-\vartheta)z^{\varrho-\vartheta}.$$

Thus, we complete the proof of this theorem. □

Theorem 9 *The following relation holds true:*

$$\mathcal{M}\left[\mathcal{D}_{z,\alpha,\beta}^{\vartheta,\wp,\eta,\nu}\left\{(1-z)^{-\kappa_1}\right\}:s\right]=\frac{\Gamma_{(s)}^{\alpha,\beta}}{\Gamma(-\vartheta)}z^{-\vartheta}\sum_{n=0}^{\infty}\frac{(\kappa_1)_n}{n!}z^n\mathcal{B}(n+\eta s+1,\nu s-\vartheta), \tag{22}$$

$$(\Re(\kappa_1>0,\ \Re(s)>0),\ \Re(\vartheta)<0,\ \alpha,\beta,\eta,\nu\in\mathbb{R}^+,\ \wp\geq 0\ and\ |z|<1).$$

Proof By using binomial expansion of $(1-z)^{-\kappa_1}$ and (21), we get

$$\begin{aligned}\mathcal{M}\left[\mathcal{D}_{z,\alpha,\beta}^{\vartheta,\wp,\eta,\nu}\left\{(1-z)^{-\kappa_1}\right\}:s\right]&=\sum_{n=0}^{\infty}\frac{(\kappa_1)_n}{n!}\mathcal{M}\left[\mathcal{D}_{z,\alpha,\beta}^{\vartheta,\wp,\eta,\nu}\left\{z^n\right\}:s\right]\\&=\frac{\Gamma_{(s)}^{\alpha,\beta}}{\Gamma(-\vartheta)}\sum_{n=0}^{\infty}\frac{(\kappa_1)_n}{n!}\mathcal{B}(n+\eta s+1,\nu s-\vartheta)z^{n-\vartheta}\\&=\frac{\Gamma_{(s)}^{\alpha,\beta}}{\Gamma(-\vartheta)}z^{-\vartheta}\sum_{n=0}^{\infty}\frac{(\kappa_1)_n}{n!}z^n\mathcal{B}(n+\eta s+1,\nu s-\vartheta).\end{aligned}$$

Hence, we achieve the desired result (22). □

7 Generating Functions for an Extended Hypergeometric Functions

In this part, we will explore using extended Appell's functions and the fractional derivative operator to construct generating functions for extended hypergeometric functions.

Theorem 10 *The following relation of extended hypergeometric function holds true:*

$$\sum_{n=0}^{\infty}\frac{(\varrho)_n}{n!}F_{\alpha,\beta}^{\wp,\eta,\nu}(\varrho+n,\kappa_1;\kappa_2;x)\tau^n=(1-\tau)^{-\varrho}F_{\alpha,\beta}^{\wp,\eta,\nu}(\varrho,\kappa_1,\kappa_2;\frac{x}{1-\tau}), \tag{23}$$

$$(\ |x|<\min\{1,|1-\tau|\};\Re(\varrho)>0,\Re(\kappa_2)>\Re(\kappa_1)>0\ ).$$

Proof By using well-known identity such as

$$[(1-x)-\tau]^{-\varrho}=(1-\tau)^{-\varrho}\left(1-\frac{x}{1-\tau}\right)^{-\varrho},$$

and expand the left-hand side of above expression for $|\tau|<|1-x|$, we have

$$\sum_{n=0}^{\infty}\frac{(\varrho)_n}{n!}(1-x)^{-\varrho}\left(\frac{\tau}{1-x}\right)^n=(1-\tau)^{-\varrho}\left(1-\frac{x}{1-\tau}\right)^{-\varrho}.$$

Now, multiply both sides of above expression by x^{κ_1-1} and applying the extended fractional derivative operator $\mathcal{D}_{x,\alpha,\beta}^{\kappa_1-\kappa_2,\wp,\eta,\nu}$, we get

$$\mathcal{D}_{x,\alpha,\beta}^{\kappa_1-\kappa_2,\wp,\eta,\nu}\left\{\sum_{n=0}^{\infty}\frac{(\varrho)_n}{n!}(1-x)^{-\varrho}\left(\frac{\tau}{1-x}\right)^n x^{\kappa_1-1}\right\}$$

$$=(1-\tau)^{-\varrho}\mathcal{D}_{x,\alpha,\beta}^{\kappa_1-\kappa_2,\wp,\eta,\nu}\left\{x^{\kappa_1-1}\left(1-\frac{x}{1-\tau}\right)^{-\varrho}\right\}.$$

Interchanging the order which is valid for $\Re(\kappa_1)>0$ and $|t|<|1-x|$, we get

$$\sum_{n=0}^{\infty}\frac{(\varrho)_n}{n!}\tau^\eta\mathcal{D}_{x,\alpha,\beta}^{\kappa_1-\kappa_2,\wp,\eta,\nu}\left\{(1-x)^{-(\varrho+n)}x^{a-1}\right\}=(1-\tau)^{-\varrho}\mathcal{D}_{x,\alpha,\beta}^{\kappa_1-\kappa_2,\wp,\eta,\nu}\left\{\left(x^{\kappa_1-1}(1-\frac{x}{1-\tau}\right)^{-\varrho}\right\}.$$

By using (17) in the above expression we achieve the result (23). □

Theorem 11 *For the new extended hypergeometric function, the following relation holds true:*

$$\sum_{n=0}^{\infty}\frac{(\varrho)_n}{n!}F_{\alpha,\beta}^{\wp,\eta,\nu}(\rho-n,\kappa_1;\kappa_2;x)\tau^\eta=(1-\tau)^{-\varrho}F_{1;\alpha,\beta}^{\wp,\eta,\nu}\left(\kappa_1,\rho,\varrho;\kappa_2,x,\frac{x\tau}{1-\tau}\right). \tag{24}$$

Proof Using the following identities:

$$[1-(1-x)\tau]^{-\varrho}=(1-\tau)^{-\varrho}\left(1+\frac{x\tau}{1-\tau}\right)^{-\varrho}.$$

By using Binomial expansion to expanding the left-hand side of above relation for $|\tau|<|1-x|$, we have

$$\sum_{n=0}^{\infty}\frac{(\varrho)_n}{n!}(1-x)^n\tau^n=(1-\tau)^{-\varrho}\left(1-\frac{-x\tau}{1-\tau}\right)^{-\varrho}.$$

Now, multiplying both sides by $x^{\kappa_1-1}(1-x)^{-\rho}$ and applying the extended fractional derivative operator $\mathcal{D}_{x,\alpha,\beta}^{\kappa_1-\kappa_2,\wp,\eta,\nu}$, we have

$$\mathcal{D}_{x,\alpha,\beta}^{\kappa_1-\kappa_2,\wp,\eta,\nu}\left\{\sum_{n=0}^{\infty}\frac{(\varrho)_n}{n!}x^{\kappa_1-1}(1-x)^{-\rho+n}\tau^n\right\}$$

$$= (1-\tau)^{-\varrho}\mathcal{D}_{x,\alpha,\beta}^{\kappa_1-\kappa_2,\wp,\eta,\nu}\left\{x^{\kappa_1-1}(1-x)^{-\rho}\left(1-\frac{-x\tau}{1-\tau}\right)^{-\varrho}\right\}.$$

Interchanging the order which is valid for $\Re(\kappa_1) > 0$ and $|xt| < |1-x|$, we get

$$\sum_{n=0}^{\infty}\frac{(\varrho)_n}{n!}\mathcal{D}_{x,\alpha,\beta}^{\kappa_1-\kappa_2,\wp,\eta,\nu}\left\{x^{\kappa_1-1}(1-x)^{-\rho+n}\tau^n\right\}$$

$$= (1-\tau)^{-\varrho}\mathcal{D}_{x,\alpha,\beta}^{\kappa_1-\kappa_2,\wp,\eta,\nu}\left\{x^{\kappa_1-1}(1-x)^{-\rho}\left(1-\frac{-x\tau}{1-\tau}\right)^{-\varrho}\right\}.$$

By using (16) and (17) in the above expression, we achieve the result (24). □

Theorem 12 *For the new extended hypergeometric function, the following relation holds true:*

$$\sum_{n=0}^{\infty}\frac{(\varrho)_n}{n!}\,F_{\alpha,\beta}^{\wp,\eta,\nu}(-n,\delta;\zeta;y)\,F_{\alpha,\beta}^{\wp,\eta,\nu}(\varrho+n,\kappa_1;\kappa_2;x)\,\tau^n$$

$$= (1-\tau)^{-\varrho}F_{2;\alpha,\beta}^{\wp,\eta,\nu}\left(\varrho,\kappa_1,\delta;\kappa_2,\zeta;\frac{x}{1-\tau},\frac{-y\tau}{1-\tau}\right). \quad (25)$$

Proof In the expression (23), by replacing t by $(1-y)\tau$, multiplying in the resulting equality by $y^{\delta-1}$, and using extended fractional derivative operator $D_{y,\alpha,\beta}^{\delta-\zeta,p,\eta,\nu}$, we have

$$\mathcal{D}_{y,\alpha,\beta}^{\delta-\zeta,\wp,\eta,\nu}\left\{\sum_{n=0}^{\infty}\frac{(\varrho)_n}{n!}y^{\delta-1}F_{\alpha,\beta}^{\wp,\eta,\nu}(\varrho+n,\kappa_1;\kappa_2;x)(1-y)^n\tau^n\right\}$$

$$= \mathcal{D}_{y,\alpha,\beta}^{\delta-\zeta,\wp,\eta,\nu}\left\{(1-(1-y)\tau)^{-\varrho}\,y^{\delta-1}\,F_{\alpha,\beta}^{\wp,\eta,\nu}\left(\varrho,\kappa_1,\kappa_2;\frac{x}{1-(1-y)\tau}\right)\right\}.$$

Now, interchanging the order, which is valid for $|x| < 1$, $\left|\frac{1-x}{1-y}\tau\right| < 1$ and $\left|\frac{x}{1-\tau}\right| + \left|\frac{y\tau}{1-\tau}\right| < 1$, we get

$$\sum_{n=0}^{\infty}\frac{(\varrho)_n}{n!}\mathcal{D}_{y,\alpha,\beta}^{\delta-\zeta,\wp,\eta,\nu}\left\{y^{\delta-1}(1-y)^n\right\}F_{\alpha,\beta}^{\wp,\eta,\nu}(\varrho+n,\kappa_1;\kappa_2;x)t^n$$

$$= \mathcal{D}_{y,\alpha,\beta}^{\delta-\zeta,\wp,\eta,\nu}\left\{y^{\delta-1}\left(1-\frac{-y\tau}{1-\tau}\right)^{-\varrho}F_{\alpha,\beta}^{\wp,\eta,\nu}\left(\varrho,\kappa_1,\kappa_2;\frac{\frac{x}{1-\tau}}{1-\frac{-y\tau}{1-\tau}}\right)\right\}.$$

By using (16) and (19), we get the desired result. □

8 Conclusions

In this article, extended Appell's functions, Riemann–Liouville fractional derivative operators, and their Mellin transform were derived. We also studied some applications of Riemann–Liouville fractional derivative operators and with their help, we concluded some generating functions of extended hypergeometric functions. In order to justify our new operator, we know that fractional calculus is used in various engineering and science domains. With the results obtained in this article, there are several uses and applications in various fields. We come to another conclusion for further research.

For future direction, recently Ghayasuddin et al. [1] defined a new beta function using three-parameter Mittag-Leffler function and with the help of this beta functions, we derive a new generalization of Appell's functions of two- and three-variable as well as Riemann–Liouville fractional derivative operators as follows:

$$F^{\wp}_{1;\alpha,\beta,\gamma}(\kappa_1,\kappa_2,\kappa_3;\kappa_4;x,y)=\sum_{n,m=0}^{\infty}\frac{\mathcal{B}^{\wp}_{\alpha,\beta,\gamma}(\kappa_1+m+n,\kappa_4-\kappa_1)}{\mathcal{B}(\kappa_1,\kappa_4-\kappa_1)}(\kappa_2)_n(\kappa_3)_m\frac{x^n}{n!}\frac{y^m}{m!},$$

$$(\max\{|x|,|y|\}<1\ ;\ \alpha,\beta,\gamma\in\mathbb{R}^+\ and\ \wp\geq 0).$$

$$F^{\wp}_{2;\alpha,\beta,\gamma}(\kappa_1,\kappa_2,\kappa_3;\kappa_4,\kappa_5;x,y)$$

$$=\sum_{n,m=0}^{\infty}\frac{(\kappa_1)_{m+n}\mathcal{B}^{\wp}_{\alpha,\beta,\gamma}(\kappa_2+n,\kappa_4-\kappa_2)\mathcal{B}^{\wp}_{\alpha,\beta,\gamma}(\kappa_3+m,\kappa_5-\kappa_3)}{\mathcal{B}(\kappa_2,\kappa_4-\kappa_2)\mathcal{B}(\kappa_3;\kappa_5-\kappa_3)}\frac{x^n}{n!}\frac{y^m}{m!},$$

$$(|x|+|y|<1;\quad \alpha,\beta,\gamma\in\mathbb{R}^+\ and\ \wp\geq 0).$$

$$F^{3,\alpha,\beta,\gamma}_{\mathcal{D},\wp}(\kappa_1,\kappa_2,\kappa_3,\kappa_4;\kappa_5;x,y,z)$$

$$=\sum_{n,m,r=0}^{\infty}\frac{\mathcal{B}^{\wp}_{\alpha,\beta,\gamma}(\kappa_1+m+n+r,\kappa_5-\kappa_1)(\kappa_2)_m(\kappa_3)_n(\kappa_4)_r}{\mathcal{B}(\kappa_1,\kappa_5-\kappa_1)}\frac{x^n}{n!}\frac{y^m}{m!}\frac{z^r}{r!},$$

$$(\sqrt{|x|}+\sqrt{|y|}+\sqrt{|z|}<1\ ;\ \alpha,\beta,\gamma\in\mathbb{R}^+_0\ and\ \wp\geq 0).$$

and

$$\mathcal{D}^{\vartheta,\wp}_{z;\alpha,\beta,\gamma}\{f(z)\}=\frac{1}{\Gamma(-\vartheta)}\int_0^z f(\tau)(z-\tau)^{-\vartheta-1}E^{\gamma}_{\alpha,\beta}\left(\frac{-\wp z^2}{\tau(z-\tau)}\right)dt$$

$$(\Re(\vartheta)<0;\ \ \wp\geq 0;\alpha,\beta,\gamma\in\mathbb{R}^+),$$

for $j-1<\Re(\vartheta)<j$ such that $j=1,2,3,....$

$$\mathcal{D}^{\vartheta,\wp}_{z;\alpha,\beta,\gamma}\{f(z)\}=\frac{d^j}{dz^j}\mathcal{D}^{\vartheta-j}_{z}\{f(z)\}$$
$$=\frac{d^j}{dz^j}\left\{\frac{1}{\Gamma(j-\vartheta)}\int_0^z f(\tau)(z-1)^{-\vartheta+j-1}E^{\gamma}_{\alpha,\beta}\left(\frac{-\wp z^2}{\tau(z-\tau)}\right)d\tau\right\},$$

and study their properties and applications.

Acknowledgements The author Saddam Husain would like to thank Integral University, Lucknow, India for providing the manuscript number (MCN): IU/R&D/2024-MCN0002557 for the present research work.

Conflict of Interest: The author declare no conflict of interest.

References

1. M. Ali, M. Ghayasuddin, A note on extended beta, Gauss and confluent hypergeometric functions. Ital. J. Pure Appl. Math. **46**(2) (2020)
2. G.E. Andrews, R. Askey, R. Roy, *Special Functions* (Cambridge University Press, Cambridge, 1999)
3. S.I. Butt, M. Nadeem, G. Farid, On Caputo fractional derivatives via exponential s-convex functions. Turk. J. Sci. **5**(2), 140–146 (2020)
4. M.A. Chaudhry, A. Qadir, M. Rafique, S.M. Zubair, Extension of Euler's beta function. J. Comput. Appl. Math. **78**, 19–32 (1997)
5. L. Chen, G. Farid, S.I. Butt, S.B. Akbar, Boundedness of fractional integral operators containing Mittag-Leffler functions. Turk. J. Ineq **4**(1), 14–24 (2020)
6. M. Ghayasuddin, N.U. Khan, A Study of extended beta, Gauss and confluent hypergeometric functions. Int. J. Appl. Math. **33**(1), 1–13 (2020)
7. N.U. Khan, S. Husain, A note on extended beta function involving generalized Mittag-Leffler function and its applications. TWMS J. Appl. Eng. Math. **12**(1), 71–81 (2022)
8. N.U. Khan, S. Husain, T. Usman, S. Araci, Results concerning the analysis of multi-index Whittaker function. J. Math. (2022)
9. N.U. Khan, S. Husain, M.I. Khan, Analysis of an extended Whittaker function and its properties. Honam Math. J. **45**(2), 184–197 (2023)
10. N.U. Khan, S. Husain, O. Khan, A novel kind of new Beta logarithmic function and their properties. Hacettepe J. Math. Stat. **52**(4) (2023)
11. N. Khan, R. Sk, S. Husain, M.A. Shah, The study of generalized Hurwitz-Lerch zeta function and fractional kinetic equations. Res. Math. **11**(1) (2024), https://doi.org/10.1080/27684830.2024.2331419
12. D.M. Lee, A.K. Rathie, R.K. Parmar, Y.S. Kim, Generalization of extended beta function, Hypergeometric and Confluent Hypergeometric functions. Honam Math. J. **33**(2), 187–206 (2011)
13. G.M. Mittag-Leffler, Sur la nouvelle function $E_\alpha(x)$. C. R. Acad. Sci. Paris **137**, 554–558 (1903)
14. G.M. Mittag-Leffler, Sur la répresentation analytique d'une branche uniforme d'une fonction monogène. Acta Math. **29**, 101–181 (1905)

15. E. Özergin, M.A. Özarslan, A. Altın, Extension of gamma, beta and hypergeometric function. J. Compt. Appl. Math. **235**(16), 4601–4610 (2011)
16. E. Özergin, M.A.Özarslan, Some generating relations for extended hypergeometric functions via generalized fractional derivative operator. Math. Comput. Modell. **52**(9–10), 1825–1833 (2010)
17. R.K. Parmar, *A New Generalization of Gamma, Beta, Hypergeometric and Confluent Hypergeometric Functions* (Le Matematiche, LXVIII, 2013), pp. 33–52
18. E.D. Rainville, *Special Functions* (The Macmillan Company, New York, 1960)
19. L. Sadek, Fractional BDF methods for solving fractional differential matrix equations. Int. J. Appl. Comput. Math. **8**(5), 238 (2022)
20. L. Sadek, Controllability and observability for fractal linear dynamical systems. J. Vibr. Control **29**(19–20), 4730–4740 (2022)
21. L. Sadek, Stability of conformable linear infinite-dimensional systems. Int. J. Dyn. Control **11**(3), 1276–1284 (2023)
22. L. Sadek, A. S. Bataineh, H. Talibi Alaoui, I. Hashim, The novel Mittag-Leffler-Galerkin method: application to a Riccati differential equation of fractional order. Fractal Fractional (MDPI), **7**(4), 302 (2023)
23. L. Sadek, A Cotangent fractional derivative with the application. Fractal Fractional (MDPI), **7**(6), 444 (2023)
24. H.M. Srivastava, R.K. Saxena, Operators of fractional integration and their applications. Appl. Math. Comput. **118**(1), 1–52 (2001)
25. H.M. Srivastava, H.L. Manocha, *A Treatise on Generating Functions* (Halsted Press, Ellis Horwood Limited, Chichester; John Wiley and Sons, New York, 1984)

A Study of Integral Transforms Associated with Srivastava–Daoust Multivariable Hypergeometric Functions

Abdelmajid Belafhal, Fatima Khannous, and Talha Usman

Abstract A large number of integral transforms involving the product of some special functions which play a key role in the domain of mathematical physics draws the attention of many authors. In this paper, we present the evolution of several integral transforms involving the product of Whittaker and Bessel functions with the weight $e^{-p\rho^2+2q\rho^{2\gamma}}$, which are expressed in terms of Srivastava–Daoust multivariable hypergeometric functions. Some properties of these functions and more results as special cases of our main results are also deduced by specializing the suitable values of the indexes of the Whittaker function.

1 Introduction and Preliminaries

Recently, many authors have elaborated on interesting integral transforms involving the product of special functions with different weights. The Bessel function, introduced in the integrand of these integrals, plays an important role since it is introduced once the cylindrical coordinates are used in the mathematical physics field [1–9]. In the present work, we established some integral transform with the weight $e^{-p\rho^2+2q\rho^{2\gamma}}$. These integrals are evaluated in terms of the Srivastava–Daoust multivariable hypergeometric functions. The obtained integral transforms play an interesting role in several fields of physics like plasma physics, neutrons physics, and radio physics.

We recall here some basic definitions. In the first, we consider the particular multiple hypergeometric functions series known as the generalized Lauricella series in several variables defined by Srivastava and Karlsson [10], Sahai and Verma [11].

A. Belafhal · F. Khannous
LPNAMME, Laser Physics Group, Department of Physics, Faculty of Sciences, Chouaïb Doukkali University, P. B 20, 24000 El Jadida, Morocco

T. Usman (✉)
Preparatory Studies Center, Mathematics and Computing Skills Unit, University of Technology and Applied Sciences, P.O. Box: 484, 411 Sur, Oman
e-mail: talhausman.maths@gmail.com

M. Kharrat et al. (eds.), *Modeling of Discrete and Continuous Systems*, Forum for Interdisciplinary Mathematics, https://doi.org/10.1007/978-981-97-8715-9_11

$$F_{SD}^{(5)}\left(\begin{matrix} r & : a, b, c, d, & e & : a'', b'', c'', d'', & e'' ; \\ s & : a', b', c', d', e' & & : a''', b''', c''', d''', e'''; \end{matrix} t, u, x, y, z\right) = \sum_{h,k,l,m,n=0}^{+\infty} \Omega\left(h, k, l, m, n\right) \frac{t^h}{h!} \frac{u^k}{k!} \frac{x^l}{l!} \frac{y^m}{m!} \frac{z^n}{n!}, \tag{1}$$

$$F_{SD}^{(4)}\left(\begin{matrix} r & : a, b, c, & d & : a'', b'', c'', & d'' ; \\ s & : a', b', c', d' & & : a''', b''', c''', d'''; \end{matrix} u, x, y, z\right) = \sum_{k,l,m,n=0}^{+\infty} \Omega\left(k, l, m, n\right) \frac{u^k}{k!} \frac{x^l}{l!} \frac{y^m}{m!} \frac{z^n}{n!}, \tag{2}$$

and

$$F_{SD}^{(3)}\left(\begin{matrix} r : a, b, & c : a'', b'', & c'' ; \\ s : a', b', c' & : a''', b''', c''' ; \end{matrix} x, y, z\right) = \sum_{l,m,n=0}^{+\infty} \Omega\left(l, m, n\right) \frac{x^l}{l!} \frac{y^m}{m!} \frac{z^n}{n!}, \tag{3}$$

where

$$\Omega\left(h, k, l, m, n\right) = \frac{(r)_{ah+bk+cl+dm+en}}{(s)_{a'h+b'k+c'l+d'm+e'n}} \frac{\left(a''\right)_h \left(b''\right)_k \left(c''\right)_l \left(d''\right)_m \left(e''\right)_n}{\left(a'''\right)_h \left(b'''\right)_k \left(c'''\right)_l \left(d'''\right)_m \left(e'''\right)_n}, \tag{4}$$

$$\Omega\left(k, l, m, n\right) = \frac{(r)_{ak+bl+cm+dn}}{(s)_{a'k+b'l+c'm+d'n}} \frac{\left(a''\right)_k \left(b''\right)_l \left(c''\right)_m \left(d''\right)_n}{\left(a'''\right)_k \left(b'''\right)_l \left(c'''\right)_m \left(d'''\right)_n}, \tag{5}$$

and

$$\Omega\left(l, m, n\right) = \frac{(r)_{al+bm+cn}}{(s)_{a'l+b'm+c'n}} \frac{\left(a''\right)_l \left(b''\right)_m \left(c''\right)_n}{\left(a'''\right)_l \left(b'''\right)_m \left(c'''\right)_n}. \tag{6}$$

In the above definitions, and with $\lambda + \chi \in \mathbb{C}\backslash Z_0^-$, the Pochhammer symbol is defined in terms of the gamma function Γ by Srivastava and Karlsson [10]

$$(\lambda)_\chi = \frac{\Gamma\left(\lambda + \chi\right)}{\Gamma\left(\lambda\right)}. \tag{7}$$

To generalize (1), (2), and (3), we give the Srivastava–Daoust multivariable hypergeometric functions expressed as follows [10, 11]:

$$F\begin{matrix} A : B^{(1)}; .....; B^{(k)} \\ C : D^{(1)}; .....; D^{(k)} \end{matrix}\begin{pmatrix} x_1 \\ \vdots \\ x_k \end{pmatrix}$$
$$= F_{SD}^{(k)}\left(\begin{matrix} [(a) : \theta^{(1)}, ..., \theta^{(k)}] : [(b^{(1)}) : \phi^{(1)}]; ...; [(b^{(k)}) : \phi^{(k)}]; \\ [(c) : \psi^{(1)}, ..., \psi^{(k)}] : [(d^{(1)}) : \delta^{(1)}]; ...; [(d^{(k)}) : \delta^{(k)}]; \end{matrix} x_1, ..., x_k\right)$$
$$= \sum_{s_1,...,s_{k=0}}^{+\infty} \Omega(s_1, ..., s_k) \prod_{i=0}^{k} \frac{x_i^{s_i}}{s_i!}, \tag{8}$$

where

$$\Omega(s_1, ..., s_k) = \frac{\prod_{j=1}^{A} (a_j)_{s_1\theta_j^{(1)}+...+s_k\theta_j^{(k)}} \prod_{j=1}^{B^{(1)}} \left(b_j^{(1)}\right)_{s_1\phi_j^{(1)}} \cdots \prod_{j=1}^{B^{(k)}} \left(b_j^{(k)}\right)_{s_k\phi_j^{(k)}}}{\prod_{j=1}^{C} (c_j)_{s_1\psi_j^{(1)}+...+s_k\psi_j^{(k)}} \prod_{j=1}^{D^{(1)}} \left(d_j^{(1)}\right)_{s_1\delta_j^{(1)}} \cdots \prod_{j=1}^{D^{(k)}} \left(d_j^{(k)}\right)_{s_k\delta_j^{(k)}}}. \tag{9}$$

In this last equation, the real and positive coefficients are given by $\theta_j^{(t)}$, $j = 1, ..., A$; $\phi_j^{(t)}$; $j = 1, ..., B^{(t)}$; $\psi_j^{(t)}$; $j = 1, ..., C$; $\delta_j^{(t)}$; $j = 1, ..., D^{(t)}$; $\forall t \in \{1, ..., k\}$, with (a) abbreviates the array of A parameters $a_1, ..., a_A$, $\left(b^{(t)}\right)$ abbreviates the array of $B^{(t)}$ parameters $b_j^{(t)}$, $j = 1, ..., B^{(t)}$; $\forall t \in \{1, ..., k\}$, with similar interpretations for (c) and $\left(d^{(t)}\right)$, $t = 1, ..., k$; etc. for the precise conditions under with the multiple hypergeometric series in (8) converges absolutely.

Also recalling the hypergeometric function ${}_pF_q$ defined by Srivastava and Karlsson [10]

$${}_pF_q\left(\alpha_1, ..., \alpha_p; \beta_1, ..., \beta_q; z\right) = \sum_{n=0}^{+\infty} \frac{(\alpha_1)_n ... (\alpha_p)_n}{(\beta_1)_n ... (\beta_q)_n} \frac{z^n}{n!}, \tag{10}$$

with $\beta_j \neq 0, -1, -2, ...; j = 1, 2, ..., q$.

We give also the series representation of the Bessel function of the first kind of order ν, for each $z \in \mathbb{C}\setminus(-\infty, 0)$, as follows [12]:

$$J_\nu(z) = \sum_{k=0}^{+\infty} \frac{(-1)^k}{k!\Gamma(\nu + k + 1)} \left(\frac{z}{2}\right)^{\nu+2k}, \tag{11}$$

and for the Whittaker function defined by Buchholz [13]

$$M_{\varepsilon,\eta}(z) = z^{\eta+1/2} e^{-z/2} {}_1F_1(\eta - \varepsilon + 1/2; 2\eta + 1; z)$$
$$= z^{\eta+1/2} e^{-z/2} \sum_{n=0,\infty} \frac{(\eta - \varepsilon + 1/2)_n}{(2\eta + 1)_n} \frac{z^n}{n!}. \tag{12}$$

Note that this last function can be replaced by taking some special values of ε and η by the exponential function, the hyperbolic sine function, the modified Bessel function $I_\nu(z)$, generalized Laguerre polynomial L_n^m, Hermite polynomial H_n, incomplete gamma function $\gamma(2\eta, z)$, error function $erf(z)$, Tricomi confluent hypergeometric function $U(a, b, z)$, and so on.

The present paper is organized as follows. In Sect. 2, we elaborate on five theorems to establish our main results. Also, three lemmas are established by finding some properties of the considered series. From our results, we also deduce two important properties of the Srivastava–Daoust functions $F_{SD}^{(4)}$ and $F_{SD}^{(5)}$. We elaborated in the last section, the particular cases in the form of corollaries of our main results, several integral transform whose integrands are associated with exponential, sinh, Bessel, modified Bessel, incomplete gamma, erf and Tricomi functions or Hermite, and generalized Laguerre polynomials.

2 Main Results

Here, we compute some closed forms of integral transforms involving the product of some special functions with the weight $e^{-p\rho^2+2q\rho^{2\gamma}}$. The integral transforms are expressed in terms of the multivariable hypergeometric functions as given in the following theorems:

Theorem 1 *Let* $\mathrm{Re}\, p >, \forall\gamma \neq 0, \mathrm{Re}\,\mu > -1$ *and* $\max\{|x|, |y|, |z|\} < 1$, *then*

$$\int_0^R \rho^\mu J_\nu(\alpha\rho)\, e^{-p\rho^2+2q\rho^{2\gamma}}\, d\rho$$
$$= \frac{(\alpha R/2)^\nu}{\nu!\,(\mu+\nu+1)} R^{\mu+1} e^{-pR^2} F_{SD}^{(3)}\left(\begin{matrix} \frac{\mu+\nu+1}{2} & : 1, \gamma, - & : -, -, & 1\ ; \\ \frac{\mu+\nu+3}{2} & : 1, \gamma, 1 & : \nu+1, -, -; \end{matrix}\ x, y, z\right) \tag{13}$$

where

$$x = -R^2\alpha^2/4\ ,\ y = 2qR^{2\gamma}\ \textit{and}\ z = pR^2. \tag{14}$$

Proof Denoting the left-hand side of (13) by A and using the definition of the Bessel function of the first kind given by (11), one finds

$$A = \frac{(\alpha/2)^\nu}{\nu!} \sum_{l=0}^{+\infty} \frac{(-\alpha^2/4)^l}{l!(\nu+1)_l} A_l, \tag{15}$$

where

$$A_l = \int_0^R \rho^{2l+\mu+\nu} e^{-p\rho^2+2q\rho^{2\gamma}} d\rho. \tag{16}$$

By introducing the expansion of the exponential function, (16) can be written as

$$A_l = \sum_{m=0}^{+\infty} \frac{(2q)^m}{m!} A_m, \tag{17}$$

where

$$A_m = \int_0^R \rho^{2l+2\gamma m+\mu+\nu} e^{-p\rho^2} d\rho. \tag{18}$$

By the use of the following identities [14]

$$\gamma(\varepsilon, x) = \frac{x^\varepsilon}{\varepsilon} {}_1F_1(\varepsilon; \varepsilon+1; -x), \tag{19}$$

with Re $\varepsilon > 0$,

$$\int_0^u x^{\chi-1} e^{-\varepsilon x} dx = \frac{1}{\varepsilon^\chi} \gamma(\chi, \varepsilon u), \tag{20}$$

with Re $\chi > 0$, and

$${}_1F_1(\varepsilon; \gamma; z) = e^z {}_1F_1(\gamma - \varepsilon; \gamma; -z) \tag{21}$$

(15) becomes

$$A = \frac{(\alpha R/2)^\nu}{\nu!(\mu+\nu+1)} R^{\mu+1} e^{-pR^2} \sum_{l,m,n=0}^{+\infty} \Omega(l,m,n) \frac{x^l}{l!} \frac{y^m}{m!} \frac{z^n}{n!}, \tag{22}$$

where

$$\Omega(l,m,n) = \frac{\left(\frac{\mu+\nu+1}{2}\right)_{l+\gamma m}}{\left(\frac{\mu+\nu+3}{2}\right)_{l+\gamma m+n}} \frac{(1)_n}{(\nu+1)_l}, \tag{23}$$

with

$$x = -R^2\alpha^2/4\ ,\ y = 2qR^{2\gamma} \text{ and } z = pR^2. \tag{24}$$

The last summation of (22) is the Srivastava–Daoust multivariable hypergeometric function $F_{SD}^{(3)}$ given by (3) and the required result (13) easily follows. This completes the proof of Theorem 1.

Theorem 2 *Let* $\operatorname{Re} p > 0,\ \forall\gamma \neq 0,\ \operatorname{Re}\mu > -1$ *and* $\max\{|u|, |x|, |y|, |z|\} < 1$, *then*

$$\int_0^R \rho^{\mu} J_{\nu}(\alpha\rho) J_{\lambda}(\chi\rho) e^{-p\rho^2+2q\rho^{2\gamma}} d\rho$$

$$= \frac{(\alpha R/2)^{\nu}(\chi R/2)^{\lambda}}{\nu!\lambda!(\mu+\nu+\lambda+1)} R^{\mu+1} e^{-pR^2} \times F_{SD}^{(4)}\left(\begin{matrix}\frac{\mu+\nu+\lambda+1}{2} & :1,1,\gamma,- & :-,-,-, & 1 & ; \\ \frac{\mu+\nu+\lambda+3}{2} & :1,1,\gamma,1, & :\lambda+1,\nu+1,-,- & ; \end{matrix} u,x,y,z\right), \tag{25}$$

where

$$u = -\chi^2 R^2/4\,,\ x = -R^2\alpha^2/4\,,\ y = 2qR^{2\gamma}\ and\ z = pR^2. \tag{26}$$

Proof To prove (25), starting with the expansion of Bessel function J_{λ} given by (11) in the integral and by denoting the left-hand side of (25) by B, one can write

$$B = \frac{(\chi/2)^{\lambda}}{\lambda!} \sum_{k=0}^{+\infty} \frac{(-\chi^2/4)^k}{k!(\lambda+1)_k} B_k, \tag{27}$$

where

$$B_k = \int_0^R \rho^{\mu+\lambda+2k} J_{\nu}(\alpha\rho) e^{-p\rho^2+2q\rho^{2\gamma}} d\rho. \tag{28}$$

With the help of Theorem 1 and the identity (7), (27) can be expressed as

$$B = \frac{(\alpha R/2)^{\nu}(\chi R/2)^{\lambda}}{\nu!\lambda!(\mu+\nu+\lambda+1)} R^{\mu+1} e^{-pR^2} \sum_{k,l,m,n=0}^{+\infty} \Omega(k,l,m,n) \frac{u^k}{k!}\frac{x^l}{l!}\frac{y^m}{m!}\frac{z^n}{n!}, \tag{29}$$

where

$$\Omega(k,l,m,n) = \frac{\left(\frac{\mu+\nu+\lambda+1}{2}\right)_{k+l+\gamma m}}{\left(\frac{\mu+\nu+\lambda+3}{2}\right)_{k+l+\gamma m+n}} \frac{(1)_n}{(\lambda+1)_k(\nu+1)_l}, \tag{30}$$

and u, x, y, and z are given by (26).

Finally making the resulting expression deduced from (29) and (30) in terms of the Srivastava–Daoust multivariable hypergeometric function $F_{SD}^{(4)}$, the required result (25) is obtained. This completes the proof of Theorem 2.

Theorem 3 *Let* $\operatorname{Re} p > 0,\ \gamma \neq 0$ *and* $\max\{|x|, |y|, |z|\} < 1$, *then*

$$\int_0^R \rho^{\mu} M_{\varepsilon,\eta}\left(2\beta\rho^2\right) e^{-p\rho^2+2q\rho^{2\gamma}} d\rho$$
$$= \frac{(2\beta)^{\eta+1/2} R^{\mu+2\eta+2}}{(\mu+2\eta+2)} e^{-(p+\beta)R^2} \tag{31}$$
$$\times F_{SD}^{(3)}\left(\begin{matrix}\mu/2\ +\eta+1 : 1,\gamma, - : 1/2\ +\eta-\varepsilon, -, 1; \\ \mu/2\ +\eta+2 : 1,\gamma,\ 1\ : 2\eta+1, -,\ \ -\ \ ;\end{matrix} x, y, z\right),$$

where
$$x = 2\beta R^2,\ y = 2qR^{2\gamma}\ and\ z = (\rho+\beta)\,R^2. \tag{32}$$

Proof Denoting the left-hand side of (31) by C and using the expansions of the exponential function and of the Whittaker function given by (12), we can write

$$C = (2\beta)^{\eta+1/2} \sum_{l,m=0}^{+\infty} \frac{(1/2\ +\eta-\varepsilon)_l}{(2\eta+1)_l} \frac{(2\beta)^l}{l!} \frac{(2q)^m}{m!} C_{lm}, \tag{33}$$

where
$$C_{lm} = \int_0^R \rho^{\mu+2\eta+2l+2\gamma m+1} e^{-(p+\beta)\rho^2} d\rho. \tag{34}$$

With the help of (19), (20), and (21), it is easy to find the following identity:

$$\int_0^{\sqrt{u}} \rho^{2\chi-1} e^{-\varepsilon\rho^2} d\rho = \frac{u^{\chi}}{2\chi} e^{-\varepsilon u} \sum_{n=0}^{+\infty} \frac{(1)_n}{(\chi+1)_n} \frac{(\varepsilon\mu)^n}{n!}, \tag{35}$$

and consequently C becomes

$$C = \frac{(2\beta)^{\eta+1/2} R^{\mu+2\eta+2}}{(\mu+2\eta+2)} e^{-(p+\beta)R^2} \sum_{l,m,n=0}^{+\infty} \Omega(l,m,n) \frac{x^l}{l!} \frac{y^m}{m!} \frac{z^n}{n!}, \tag{36}$$

where
$$\Omega(l,m,n) = \frac{(\mu/2\ +\eta+1)_{l+\gamma m}}{(\mu/2\ +\eta+2)_{l+\gamma m+n}} \frac{(1/2\ +\eta-\varepsilon)_l (1)_n}{(2\eta+1)_l}, \tag{37}$$

with x, y, and z are given by (32).

This completes the proof of Theorem 3 by the use of definition of $F_{SD}^{(3)}$ given by (3).

Theorem 4 *Let* Re $p > 0$, Re $\mu > -1$ *and* $\max\{|u|, |x|, |y|, |z|\} < 1$, *then*

$$\int_0^R \rho^{\mu} M_{\varepsilon,\eta}\left(2\beta\rho^2\right) J_{\nu}(\alpha\rho)\, e^{-p\rho^2+2q\rho^{2\gamma}}\, d\rho$$
$$= \frac{(2\beta)^{\eta+1/2}\,(\alpha R/2\,)^{\nu} R^{\mu+2\eta+2}}{\nu!\,(\mu+\nu+2\eta+2)} e^{-(p+\beta)R^2}$$
$$\times F_{SD}^{(4)}\left(\begin{matrix}\frac{\mu+\nu}{2}+\eta+1:1,1,\gamma,-:1/2\,+\eta-\varepsilon,-,-,1;\\ \frac{\mu+\nu}{2}+\eta+2:1,1,\gamma,1:2\eta+1,\nu+1,-,-;\end{matrix}\,u,x,y,z\right), \tag{38}$$

where

$$u = 2\beta R^2, x = -R^2\alpha^2/4\ , y = 2qR^{2\gamma}\, and\, z = (p+\beta)\, R^2. \tag{39}$$

Proof By using the expression of the Whittaker function given by (12) and denoting the left-hand side of (38) by D, we get

$$D = (2\beta)^{\eta+1/2} \sum_{k=0}^{+\infty} \frac{(1/2\,+\eta-\varepsilon)_k}{(2\eta+1)_k} \frac{(2\beta)^k}{k!} D_k, \tag{40}$$

where

$$D_k = \int_0^R \rho^{\mu+2\eta+2k+1} J_{\nu}(\alpha\rho)\, e^{-(p+\beta)\rho^2+2q\rho^{2\gamma}}\, d\rho. \tag{41}$$

Applying Theorem 1, and after tedious algebraic calculations, (40) can be rearranged as

$$D = \frac{(2\beta)^{\eta+1/2}\,(\alpha R/2\,)^{\nu} R^{\mu+2\eta+2}}{\nu!\,(\mu+\nu+2\eta+2)} e^{-(p+\beta)R^2} \sum_{k,l,m,n=0}^{+\infty} \Omega\,(k,l,m,n)\, \frac{u^k}{k!}\frac{x^l}{l!}\frac{y^m}{m!}\frac{z^n}{n!}, \tag{42}$$

where

$$\Omega\,(k,l,m,n) = \frac{\left(\frac{\mu+\nu}{2}+\eta+1\right)_{k+l+\gamma m}}{\left(\frac{\mu+\nu}{2}+\eta+2\right)_{k+l+\gamma m+n}} \frac{(1/2+\eta-\varepsilon\,)_k(1)_n}{(2\eta+1)_k(\nu+1)_l}, \tag{43}$$

and u, x, y, and z are given by (39).

It is easy to see from (42) and the definition of $F_{SD}^{(4)}$ given by (2) that (38) is proved. This completes the proof of Theorem 4.

Theorem 5 *Let* Re $\mu > -1$, Re $p > 0$, $\gamma \neq 0$ *and* max $\{|t|\,, |u|\,, |x|\,, |y|\,, |z|\} < 1$, *then*

$$\int_0^R \rho^{\mu} M_{\varepsilon,\eta}\left(2\beta\rho^2\right) J_\nu(\alpha\rho) J_\lambda(\chi\rho) e^{-p\rho^2+2q\rho^{2\gamma}} d\rho$$

$$= \frac{(2\beta)^{\eta+1/2}(\alpha R/2)^{\nu}(\chi R/2)^{\lambda} R^{\mu+2\eta+2}}{\nu!\lambda!(\mu+\nu+\lambda+2\eta+2)} e^{-(p+\beta)R^2}$$

$$\times F_{SD}^{(5)}\left(\begin{matrix} \frac{\mu+\nu+\lambda}{2}+\eta+1 : 1,1,1,\gamma,- : -, 1/2+\eta-\varepsilon, -, -, 1 ; \\ \frac{\mu+\nu+\lambda}{2}+\eta+2 : 1,1,1,\gamma, 1 : \lambda+1, 2\eta+1, \nu+1, -, -; \end{matrix} t,u,x,y,z\right), \tag{44}$$

where

$$t=-\chi^2R^2/4,\ u=2\beta R^2,\ x=-R^2\alpha^2/4,\ y=2qR^{2\gamma}\ and\ z=(p+\beta)R^2. \tag{45}$$

Proof Let E denote the left-hand side of (44) and using the expression of $J_\nu(\chi\rho)$ given by (11), one can write

$$E = \frac{(\chi/2)^{\lambda}}{\lambda!}\sum_{h=0}^{+\infty}\frac{\left(-\chi^2/4\right)^h}{h!(\lambda+1)_h}E_h, \tag{46}$$

where

$$E_h = \int_0^R \rho^{\mu+\lambda+2h} M_{\varepsilon,\eta}\left(2\beta\rho^2\right) J_\nu(\alpha\rho) e^{-p\rho^2+2q\rho^{2\gamma}} d\rho. \tag{47}$$

Applying Theorem 4 to this last integral and using the identity (7), one finds

$$E = \frac{(2\beta)^{\eta+1/2}(\alpha R/2)^{\nu}(\chi R/2)^{\lambda} R^{\mu+2\eta+2}}{\nu!\lambda!(\mu+\nu+\lambda+2\eta+2)} e^{-(p+\beta)R^2} \times \sum_{h,k,l,m,n=0}^{+\infty}\Omega(h,k,l,m,n)\frac{t^h}{h!}\frac{u^k}{k!}\frac{x^l}{l!}\frac{y^m}{m!}\frac{z^n}{n!}, \tag{48}$$

where

$$\Omega(h,k,l,m,n) = \frac{\left(\frac{\mu+\nu+\lambda}{2}+\eta+1\right)_{h+k+l+\gamma m}}{\left(\frac{\mu+\nu+\lambda}{2}+\eta+2\right)_{h+k+l+\gamma m+n}}\frac{(1/2+\eta-\varepsilon)_k(1)_n}{(\lambda+1)_h(2\eta+1)_k(\nu+1)_l}, \tag{49}$$

and t, u, x, y, and z are given by (45).

The summation in (48) is equal to $F_{SD}^{(5)}$ given (1) and finally (44) is proved. This completes the proof of Theorem 5.

From our results that we have just introduced, we deduce also two important properties of the Srivastava–Daoust functions $F_{SD}^{(4)}$ and $F_{SD}^{(5)}$ appearing in Theorems 4 and 5, respectively. These properties are presented in the following two lemmas.

Lemma 1 *The undermentioned property holds true:*

$$F_{SD}^{(4)}\left(\begin{matrix}\frac{\mu+\nu}{2}+\eta+1\ :1,1,\gamma,-:1/2\ +\eta-\varepsilon,-,-,1\ ;\\ \frac{\mu+\nu}{2}+\eta+2:1,1,\gamma,\ 1\ :2\eta+1,\nu+1,-,\ \ -\ ;\end{matrix}u,x,y,z\right)$$
$$=F_{SD}^{(4)}\left(\begin{matrix}\frac{\mu+\nu}{2}+\eta+1\ :1,1,\gamma,-:-,1/2\ +\eta-\varepsilon,-,1\ ;\\ \frac{\mu+\nu}{2}+\eta+2:1,1,\gamma,\ 1\ :\nu+1,2\eta+1,-,\ \ -\ ;\end{matrix}x,u,y,z\right), \tag{50}$$

Proof. To prove Theorem 4 and to evaluate the integral D, we have used the expression of the Whittaker function in the integral of (38). Here, we will evaluate the integral of this identity by using the expansion of the Bessel function J_ν. So, by denoting the left-hand side of (38) by F and using (11), one finds

$$F=\frac{(\alpha/2\,)^{\nu}}{\nu!}\sum_{k=0}^{+\infty}\frac{\left(-\alpha^{2}/4\,\right)^{k}}{k!(\nu+1)_{k}}F_{k}, \tag{51}$$

where

$$F_{k}=\int_{0}^{R}\rho^{\mu+\nu+2k}M_{\varepsilon,\eta}\left(2\beta\rho^{2}\right)e^{-p\rho^{2}+2q\rho^{2\gamma}}d\rho. \tag{52}$$

With the help of Theorem 3, (51) becomes

$$F=\frac{(2\beta)^{\eta+1/2}\,(\alpha R/2\,)^{\nu}R^{\mu+2\eta+2}}{\nu!\,(\mu+\nu+2\eta+2)}e^{-(p+\beta)R^{2}}$$
$$\times\sum_{k,l,m,n=0}^{+\infty}\frac{\left(\frac{\mu+\nu}{2}+\eta+1\right)_{k+l+\gamma m}}{\left(\frac{\mu+\nu}{2}+\eta+2\right)_{k+l+\gamma m+n}}\frac{(1/2+\eta-\varepsilon\,)_{l}(1)_{n}}{(\nu+1)_{k}(2\eta+1)_{l}}\frac{x^{k}}{k!}\frac{u^{l}}{l!}\frac{y^{m}}{m!}\frac{z^{n}}{n!}, \tag{53}$$

which can be written as

$$F=\frac{(2\beta)^{\eta+1/2}\,(\alpha R/2\,)^{\nu}R^{\mu+2\eta+2}}{\nu!\,(\mu+\nu+2\eta+2)}e^{-(p+\beta)R^{2}}$$
$$\times F_{SD}^{(4)}\left(\begin{matrix}\frac{\mu+\nu}{2}+\eta+1:1,1,\gamma,-:-,1/2\ +\eta-\varepsilon,-,1;\\ \frac{\mu+\nu}{2}+\eta+2:1,1,\gamma,1:\nu+1,2\eta+1,-,\ \ -\ ;\end{matrix}x,u,y,z\right), \tag{54}$$

where u, x, y, and z are given by (39).

By equating the integrals D and F, it is easy to prove the property given by (50).

Lemma 2 *The undermentioned property holds true:*

$$\begin{aligned} &F_{SD}^{(5)}\left(\begin{matrix}\frac{\mu+\nu+\lambda}{2}+\eta+1: 1,1,1,\gamma,- : -,\frac{1}{2}+\eta-\varepsilon,-,-, \quad 1 \; ; \\ \frac{\mu+\nu+\lambda}{2}+\eta+2 : 1,1,1,\gamma,\ 1 : \lambda+1, 2\eta+1, \nu+1, -, \quad - \; ;\end{matrix} t,u,x,y,z\right) \\ &= F_{SD}^{(5)}\left(\begin{matrix}\frac{\mu+\nu+\lambda}{2}+\eta+1: 1,1,1,\gamma,- : \frac{1}{2}+\eta-\varepsilon,-,-, \quad 1 \; ; \\ \frac{\mu+\nu+\lambda}{2}+\eta+2 : 1,1,1,\gamma,\ 1 : 2\eta+1, \lambda+1, \nu+1, \quad - \; ;\end{matrix} u,t,x,y,z\right) \end{aligned} \tag{55}$$

Proof To evaluate the integral (44), we have used the expansion of the Bessel function. Now, we will calculate the integral E of (44) by expending the Whittaker function. So, by denoting the hand side of the above equation by G, one finds where

$$G = (2\beta)^{\eta+1/2} \sum_{h=0}^{+\infty} \frac{\left(\frac{1}{2}+\eta-\varepsilon\right)_h}{(2\eta+1)_h} \frac{(2\beta)^h}{h!} G_h, \tag{56}$$

where

$$G_h = \int_0^R \rho^{\mu+2\eta+2h+1} J_\nu(\alpha\rho)\, J_\lambda(\chi\rho)\, e^{-(p+\beta)\rho^2+2q\rho^{2\gamma}}\, d\rho. \tag{57}$$

This last integral can be evaluated with the help of Theorem 2 which gives

$$\begin{aligned} G = &\frac{(2\beta)^{\eta+1/2}\,(\alpha R/2\,)^{\nu}(\chi\, R/2\,)^{\lambda}\, R^{\mu+2\eta+2}}{\nu!\lambda!\,(\mu+\nu+\lambda+2\eta+2)} e^{-(p+\beta)R^2} \\ &\times \sum_{h,k,l,m,n}^{+\infty} \frac{\left(\frac{\mu+\nu+\lambda}{2}+\eta+1\right)_{h+k+l+\gamma m}}{\left(\frac{\mu+\nu+\lambda}{2}+\eta+2\right)_{h+k+l+\gamma m+n}} \frac{(1/2+\eta-\varepsilon\,)_h (1)_n}{(2\eta+1)_h(\lambda+1)_k(\nu+1)_l} \frac{u^h}{h!}\frac{t^k}{k!}\frac{x^l}{l!}\frac{y^m}{m!}\frac{z^n}{n!}, \end{aligned} \tag{58}$$

where t, u, x, y, and z are given by (45).

Finally, by equating E and G, it is easy to prove (55) by using (1).

3 Special Cases

These cases will be treated by taking special values to the parameters ε and η of the Whittaker function $M_{\varepsilon,\eta}$, to the parameter ν of the Bessel function J_ν and the conditions of the used of theorems must be verified. The variables of $F_{SD}^{(3)}$, $F_{SD}^{(4)}$, and $F_{SD}^{(5)}$ used in this section are already defined in the previous theorems.

3.1 Case of ε and $\eta = -\varepsilon - 1/2$

The Whittaker function is given in terms of the exponential function as $M_{\varepsilon,-\varepsilon-1/2}(z) = e^{z/2}/z^{\varepsilon}$.

Corollary 1 *The undermentioned integral transform holds true:*

$$\begin{aligned}&\int_0^R \rho^{\mu-2\varepsilon} e^{-(p-\beta)\rho^2+2q\rho^{2\gamma}} d\rho \\ &= \frac{R^{\mu-2\varepsilon+1}}{(\mu-2\varepsilon+1)} e^{-(p+\beta)R^2} F_{SD}^{(3)}\left(\begin{matrix}\frac{\mu+1}{2}-\varepsilon : 1, \gamma, - : -, -, \ 1 \ ; \\ \frac{\mu+3}{2}-\varepsilon : 1, \gamma, 1 : -, -, -;\end{matrix} x, y, z\right).\end{aligned} \tag{59}$$

This corollary is established easily by applying Theorem 3 with the considered particular parameters ε and η.

Corollary 2 *The undermentioned integral transform holds true:*

$$\begin{aligned}&\int_0^R \rho^{\mu-2\varepsilon} J_\nu(\alpha\rho) e^{-(p-\beta)\rho^2+2q\rho^{2\gamma}} \\ &= \frac{\left(^{\alpha R}/_2\right)^\nu}{\nu!} \frac{R^{\mu-2\varepsilon+1}}{(\mu+\nu-2\varepsilon+1)} e^{-(p+\beta)R^2} \\ &\times F_{SD}^{(4)}\left(\begin{matrix}\frac{\mu+\nu+1}{2}-\varepsilon \ : 1, 1, \gamma, - : -, -, -, \ 1 \ ; \\ \frac{\mu+\nu+1}{2}-\varepsilon+1 : 1, 1, \gamma, 1 \ : -, \nu+1, -, -;\end{matrix} u, x, y, z\right).\end{aligned} \tag{60}$$

By applying Theorem 4 and with the particular values of ε and η, the corollary is proved.

Corollary 3 *The undermentioned integral transform holds true:*

$$\begin{aligned}&\int_0^R \rho^{\mu-2\varepsilon} J_\nu(\alpha\rho) J_\lambda(\chi\rho) e^{-(p-\beta)\rho^2+2q\rho^{2\gamma}} d\rho \\ &= \frac{(\alpha R/2)^\nu (\chi R/2)^\lambda R^{\mu-2\varepsilon+1}}{\nu!\lambda!(\mu+\nu+\lambda-2\varepsilon+1)} e^{-(p+\beta)R^2} \\ &\times F_{SD}^{(5)}\left(\begin{matrix}\frac{\mu+\nu+\lambda+1}{2}-\varepsilon \ : 1, 1, 1, \gamma, - \ : -, -, -, -, \ \ 1 \ \ ; \\ \frac{\mu+\nu+\lambda+3}{2}-\varepsilon \ : 1, 1, 1, \gamma, \ 1 \ : \lambda+1, -, \nu+1, -, -;\end{matrix} t, u, x, y, z\right).\end{aligned} \tag{61}$$

From Theorem 5, one can deduce this corollary by taking the considered particular indexes of the Whittaker function.

3.2 *Case of $\varepsilon = 0$ and $\eta = 1/2$*

The Whittaker is linked to: $M_{0,1/2}\ (2z) = 2\sinh(z)$.

Corollary 4 *The undermentioned integral transform holds true:*

$$\int_0^R \rho^{\mu} \sinh\left(\beta\rho^2\right) e^{-p\rho^2+2q\rho^{2\gamma}}\, d\rho = \frac{\beta R^{\mu+3}}{(\mu+3)} e^{-(p+\beta)R^2} F_{SD}^{(3)}\left(\begin{matrix} \frac{\mu+3}{2} : 1, \gamma, - : 1, -, 1; \\ \frac{\mu+5}{2} : 1, \gamma, 1 : 2, -, -; \end{matrix} x, y, z\right). \tag{62}$$

In this case, we apply Theorem 3 and the corollary is established easily.

Corollary 5 *The undermentioned integral transform holds true:*

$$\int_0^R \rho^{\mu} \sinh\left(\beta\rho^2\right) J_{\nu}(\alpha\rho)\, e^{-p\rho^2+2q\rho^{2\gamma}}\, d\rho = \frac{\beta(\alpha R/2\,)^{\nu} R^{\mu+3}}{\nu!\,(\mu+\nu+3)} e^{-(p+\beta)R^2} \times F_{SD}^{(4)}\left(\begin{matrix} \frac{\mu+\nu+3}{2} : 1, 1, \gamma, - : 1, -, -, 1; \\ \frac{\mu+\nu+5}{2} : 1, 1, \gamma, 1 : 2, \nu+1, -, -; \end{matrix} u, x, y, z\right). \tag{63}$$

This corollary can be proved directly from Theorem 4 by setting $\varepsilon = 0$ and $\eta = 1/2$.

Corollary 6 *The undermentioned integral transform holds true:*

$$\int_0^R \rho^{\mu} \sinh\left(\beta\rho^2\right) J_{\nu}(\alpha\rho)\ J_{\lambda}(\chi\rho)\, e^{-p\rho^2+2q\rho^{2\gamma}}\, d\rho = \frac{\beta(\alpha R/2\,)^{\nu}(\chi R/2\,)^{\lambda} R^{\mu+3}}{\nu!\lambda!\,(\mu+\nu+\lambda+3)} \times e^{-(p+\beta)R^2} F_{SD}^{(5)}\left(\begin{matrix} \frac{\mu+\nu+\lambda+3}{2} : 1, 1, 1, \gamma, - : -, 1, -, -, \quad 1 \quad ; \\ \frac{\mu+\nu+\lambda+5}{2} : 1, 1, 1, \gamma, \ 1 \ : \lambda+1, 2, \nu+1, -, -; \end{matrix} t, u, x, y, z\right). \tag{64}$$

By setting $\varepsilon = 0$ and $\eta = {}^1/_2$, Theorem 5 yields this corollary.

3.3 *Case of $\varepsilon = 0$ and η*

By setting $\varepsilon = 0$ and η, the Whittaker function is defined by: $M_{0,\eta}\left(2\beta\rho^2\right) = 2^{2\eta+1/2}\,\eta!\sqrt{\beta}\rho I_{\eta}\left(\beta\rho^2\right)$ with I_{η} is the modified Bessel function of order η.

Corollary 7 *The undermentioned integral transform holds true:*

$$\int_0^R \rho^{\mu+1} I_\eta\left(\beta\rho^2\right) e^{-p\rho^2+2q\rho^{2\gamma}} d\rho = \frac{\left(\beta R^2/2\right)^\eta R^{\mu+2}}{\eta!\,(\mu+2\eta+2)} e^{-(p+\beta)R^2} \times F_{SD}^{(3)}\left(\begin{matrix}\frac{\mu}{2}+\eta+1 : 1,\gamma,- : \eta+1/2\,,-,1; \\ \frac{\mu}{2}+\eta+2 : 1,\gamma,\;1\;: 2\eta+1,-,-;\end{matrix} x,y,z\right), \quad (65)$$

Theorem 3 yields easily this corollary by taking $\varepsilon = 0$ *and* η.

Corollary 8 *The undermentioned integral transform holds true:*

$$\int_0^R \rho^{\mu+1} I_\eta\left(\beta\rho^2\right) J_\nu(\alpha\rho)\, e^{-p\rho^2+2q\rho^{2\gamma}} d\rho = \frac{\left(\beta R^2/2\right)^\eta (\alpha R/2)^\nu R^{\mu+2}}{\eta!\nu!\,(\mu+\nu+2\eta+2)} e^{-(p+\beta)R^2} \times F_{SD}^{(4)}\left(\begin{matrix}\frac{\mu+\nu}{2}+\eta+1 : 1,1,\gamma,- : \eta+\frac{1}{2},-,-,\;1\;; \\ \frac{\mu+\nu}{2}+\eta+2 : 1,1,\gamma,\;1\;: 2\eta+1,\nu+1,-,-;\end{matrix} u,x,y,z\right). \quad (66)$$

By using Theorem 4 and taking the indexes of the Whittaker function $\varepsilon = 0$ *and* η, *the corollary is proved.*

Corollary 9 *The undermentioned integral transform holds true:*

$$\int_0^R \rho^{\mu+1} I_\eta\left(\beta\rho^2\right) J_\nu(\alpha\rho)\, J_\lambda(\chi\rho)\, e^{-p\rho^2+2q\rho} d\rho$$

$$= \frac{\left(\beta R^2/2\right)^\eta (\alpha R/2)^\nu (\chi R/2)^\lambda R^{\mu+2}}{\eta!\nu!\lambda!\,(\mu+\nu+\lambda+2\eta+2)} e^{-(p+\beta)} \times F_{SD}^{(5)}\left(\begin{matrix}\frac{\mu+\nu+\lambda}{2}+\eta+1 : 1,1,1,\gamma,- : -,\eta+\frac{1}{2},-,-,\;1\;; \\ \frac{\mu+\nu+\lambda}{2}+\eta+2 : 1,1,1,\gamma,1 : \lambda+1,2\eta+1,\nu+1,-,-;\end{matrix} t,u,x,y,z\right). \quad (67)$$

By taking $\varepsilon = 0$ and η as particular indexes of $M_{\varepsilon,\eta}$ and with the help of Theorem 5, one can establish (66).

3.4 *Case of* $\varepsilon = \frac{p'+1}{2} + s$ *and* $\eta = \frac{p'}{2}$

In this case, the Whittaker function is related to the generalized Laguerre polynomial by $M_{\varepsilon,\eta}\left(2\beta\rho^2\right) = \frac{s!}{(p'+1)_s}(2\beta)^{\frac{p'+1}{2}}\rho^{p'+1}e^{-\beta\rho^2}L_s^{(p')}\left(2\beta\rho^2\right)$.

Corollary 10 *Replacing the Whittaker function by the above identity and applying Theorem 3, one obtains*

$$\int_0^R \rho^{\mu+p'+1} L_s^{(p')}\left(2\beta\rho^2\right) e^{-(p+\beta)\rho^2+2q\rho^{2\gamma}} d\rho = \frac{\left(p'+1\right)_s}{s!} \frac{1}{(\mu+p'+2)} R^{\mu+p'+2}$$
$$\times e^{-(p+\beta)R^2} F_{SD}^{(3)}\left(\begin{matrix} \frac{\mu+p'}{2}+1 : 1, \gamma, - : -s, -, \; 1 \; ; \\ \frac{\mu+p'}{2}+2 : 1, \gamma, \; 1 \; : p'+1, -, -; \end{matrix} x, y, z\right). \tag{68}$$

Corollary 11 *The undermentioned integral transform holds true:*

$$\int_0^R \rho^{\mu+p'+1} L_s^{(p')}\left(2\beta\rho^2\right) J_\nu(\alpha\rho) e^{-(p+\beta)\rho^2+2q\rho^{2\gamma}} d\rho = \frac{\left(p'+1\right)_s}{s!} \frac{\left({}^{\alpha R}/_2\right)^\nu R^{\mu+p'+2}}{\nu!\left(\mu+\nu+p'+2\right)}$$
$$\times e^{-(p+\beta)R^2} F_{SD}^{(4)}\left(\begin{matrix} \frac{\mu+\nu+p'}{2}+1 : 1, 1, \gamma, - : -s, -, -, \; 1 \; ; \\ \frac{\mu+\nu+p'}{2}+2 : 1, 1, \gamma, \; 1 \; : p'+1, \nu+1, -, -; \end{matrix} u, x, y, z\right). \tag{69}$$

By setting $\varepsilon = \frac{p'+1}{2} + s$ *and* $\eta = \frac{p'}{2}$, *and by using Theorem 4, the corollary is derived.*

Corollary 12 *The undermentioned integral transform holds true:*

$$\int_0^R \rho^{\mu+p'+1} L_s^{(p')}\left(2\beta\rho^2\right) J_\nu(\alpha\rho) J_\lambda(\chi\rho) e^{-(p+\beta)\rho^2+2q\rho^{2\gamma}} d\rho$$
$$= \frac{\left(p'+1\right)_s}{s!} \frac{\left({}^{\alpha R}/_2\right)^\nu \left({}^{\chi R}/_2\right)^\lambda R^{\mu+p'+2}}{\nu!\lambda!\left(\mu+\nu+\lambda+p'+2\right)} e^{-(p+\beta)R^2}$$
$$\times F_{SD}^{(5)}\left(\begin{matrix} \frac{\mu+\nu+\lambda+p'}{2}+1 : 1, 1, 1, \gamma, - : -, -s, -, -, \; 1 \; ; \\ \frac{\mu+\nu+\lambda+p'}{2}+2 : 1, 1, 1, \gamma, 1 : \lambda+1, p'+1, \nu+1, -, -; \end{matrix} t, u, x, y, z\right). \tag{70}$$

By using Theorem 5 and setting the particular indexes of the Whittaker function this corollary is established easily.

3.5 *Case of* $\varepsilon = p' + 1/4$ *and* $\eta = -1/4$

If we set these parameters, $M_{\varepsilon,\eta}$ is expressed in terms of the Hermite polynomial of the order $2p'$, as: $M_{p'+1/4,-1/4}\left(z^2\right) = \frac{(-1)^{p'} p'!}{(2p')!} e^{-z^2/2} \sqrt{z} H_{2p'}(z)$.

Corollary 13 *The undermentioned integral transform holds true:*

$$\int_0^R \rho^{\mu+1/2} H_{2p'}\left(\sqrt{2\beta}\rho\right) e^{-(p+\beta)\rho^2+2q\rho^{2\gamma}} d\rho$$

$$= \frac{(-1)^{p'}(2p')! R^{\mu+3/2}}{p'!(\mu+3/2)} e^{-(p+\beta)R^2} F_{SD}^{(3)}\left(\begin{matrix}\frac{\mu}{2}+3/4 & : 1,\gamma,- : -p',-,1; \\ \frac{\mu}{2}+7/4 & : 1,\gamma,1 : 1/2,-,-;\end{matrix} x,y,z\right). \tag{71}$$

With the help of Theorem 3, the corollary is derived by taking $\varepsilon = p' + 1/4$ *and* $\eta = -1/4$.

Corollary 14 *The undermentioned integral transform holds true:*

$$\int_0^R \rho^{\mu+1/2} H_{2p'}\left(\sqrt{2\beta}\rho\right) J_\nu(\alpha\rho) e^{-(p+\beta)\rho^2+2q\rho^{2\gamma}} = \frac{(-1)^{p'}(2p')!(\alpha R/2)^\nu R^{\mu+3/2}}{p'!\nu!(\mu+\nu+3/2)}$$

$$\times e^{-(p+\beta)R^2} F_{SD}^{(4)}\left(\begin{matrix}\frac{\mu+\nu}{2}+3/4 & : 1,1,\gamma,- : -p',-,-,1; \\ \frac{\mu+\nu}{2}+7/4 & : 1,1,\gamma,1 : 1/2,\nu+1,-,-;\end{matrix} x,y,z\right). \tag{72}$$

With the above indexes of $M_{p'+1/4,-1/4}$ *and by use of Theorem 4, (71) is deduced.*

Corollary 15 *The undermentioned integral transform holds true:*

$$\int_0^R \rho^{\mu+1/2} H_{2p'}\left(\sqrt{2\beta}\rho\right) J_\nu(\alpha\rho) J_\lambda(\chi\rho) e^{-(p+\beta)\rho^2+2q\rho^{2\gamma}} d\rho$$

$$= \frac{(-1)^{p'}(2p')!(\alpha R/2)^\nu}{p'!\nu!} \frac{(\chi R/2)^\lambda R^{\mu+3/2}}{\lambda!(\mu+\nu+\lambda+3/2)} e^{-(p+\beta)R^2}$$

$$\times F_{SD}^{(5)}\left(\begin{matrix}\frac{\mu+\nu+\lambda}{2}+\frac{3}{4} & : 1,1,1,\gamma,- : -,-p',-,-, & 1 & ; \\ \frac{\mu+\nu+\lambda}{2}+\frac{7}{4} & : 1,1,1,\gamma, 1 : \lambda+1,1/2,\nu+1,-,-; & & \end{matrix} t,u,x,y,z\right). \tag{73}$$

As the above corollary and using Theorem 5, (72) is derived.

3.6 *Case of* $\varepsilon = p' + 3/4$ *and* $\eta = 1/4$

In this case, the Whittaker function is linked to the Hermite polynomial of order $(2p'+1)$ by the identity: $M_{p'+3/4,1/4}\left(z^2\right) = \frac{(-1)^{p'}p'!}{2(2p'+1)!}e^{-z^2/2}\sqrt{z}H_{2p'+1}(z)$.

Corollary 16 *Taking* $\varepsilon = p' + 3/4$ *and* $\eta = 1/4$ *in Theorem 3, we get*

$$\int_0^R \rho^{\mu+1/2}\, H_{2p'+1}\left(\sqrt{2\beta}\rho\right) e^{-(p+\beta)\rho^2+2q\rho^{2\gamma}}\, d\rho = \frac{(-1)^{p'}\,(2p'+1)!}{p'!}\frac{\sqrt{8\beta}\,R^{\mu+5/2}}{(\mu+5/2\,)}$$

$$\times\, e^{-(p+\beta)R^2}\, F_{SD}^{(3)}\left(\begin{matrix}\frac{\mu}{2}+5/4\ :1,\gamma,-:-p',-,1;\\ \frac{\mu}{2}+9/4\ :1,\gamma,1:3/2\,,-,-;\end{matrix}\,x,y,z\right). \tag{74}$$

Corollary 17 *The use of the considered indexes of $M_{\varepsilon,\eta}$ and Theorem 4 yields*

$$\int_0^R \rho^{\mu+1/2}\, H_{2p'+1}\left(\sqrt{2\beta}\rho\right) J_\nu(\alpha\rho)\, e^{-(p+\beta)\rho^2+2q\rho^{2\gamma}}$$

$$= \frac{(-1)^{p'}\,(2p'+1)!}{p'!}\frac{\sqrt{8\beta}(\alpha R/2\,)^\nu R^{\mu+5/2}}{\nu!\,(\mu+\nu+5/2\,)}\, e^{-(p+\beta)R^2} \tag{75}$$

$$\times\, F_{SD}^{(4)}\left(\begin{matrix}\frac{\mu+\nu}{2}+5/4\ :1,1,\gamma,-:-p',-,-,\ \ 1\ ;\\ \frac{\mu+\nu}{2}+9/4\ :1,1,\gamma,1\ :3/2\,,\nu+1,-,-;\end{matrix}\,u,x,y,z\right).$$

Corollary 18 *With the help of Theorem 5, the following formula is derived:*

$$\int_0^R \rho^{\mu+1/2}\, H_{2p'+1}\left(\sqrt{2\beta}\rho\right) J_\nu(\alpha\rho)\, J_\lambda(\chi\rho)\, e^{-(p+\beta)\rho^2+2q\rho^{2\gamma}}\, d\rho$$

$$= \frac{(-1)^{p'}\,(2p'+1)!\sqrt{8\beta}}{p'!}\frac{(\alpha R/2\,)^\nu(\chi R/2\,)^\lambda R^{\mu+5/2}}{\nu!\lambda!\,(\mu+\nu+\lambda+5/2\,)} e^{-(p+\beta)R^2}$$

$$\times\, F_{SD}^{(5)}\left(\begin{matrix}\frac{\mu+\nu+\lambda}{2}+\frac{5}{4}\ :1,1,1,\gamma,-:-,-p',-,-,\ \ 1\ \ ;\\ \frac{\mu+\nu+\lambda}{2}+\frac{9}{4}\ :1,1,1,\gamma,1:\lambda+1,3/2\,,\nu+1,-,-;\end{matrix}\,t,u,x,y,z\right). \tag{76}$$

3.7 Case of $\varepsilon = \eta - 1/2$ and η

In this case, the Whittaker function is expressed in terms of the incomplete gamma function $\gamma\,(2\eta, z)$ as $M_{\eta-1/2\,,\eta}\,(z) = 2\eta.z^{1/2\,-\eta}e^{z/2}\,\gamma\,(2\eta, z)$.

Corollary 19 *The undermentioned integral transform holds true:*

$$\int_0^R \rho^{\mu+1--2\eta}\gamma\left(2\eta, 2\beta\rho^2\right)e^{-(p-\beta)\rho^2+2q\rho^{2\gamma'}}\, d\rho$$

$$= \frac{(2\beta)^{2\eta}}{2\eta}\frac{R^{\mu+2\eta+2}}{(\mu+2\eta+2)}e^{-(p+\beta)R^2}\, F_{SD}^{(3)}\left(\begin{matrix}\frac{\mu}{2}+\eta+1\ :1,\gamma',-:1,-,\ \ 1\ \ ;\\ \frac{\mu}{2}+\eta+2\ \ :1,\gamma',1\ :2\eta+1,-,-;\end{matrix}\,x,y,z\right). \tag{77}$$

This result is established directly by applying Theorem 3 and using the particular indexes of the Whittaker function.

Corollary 20 *By letting* $\varepsilon = \eta - 1/2$ *, we get a particular case of Theorem 4 asserted by the following identity:*

$$\int_0^R \rho^{\mu+1--2\eta} \gamma\left(2\eta, 2\beta\rho^2\right) J_\nu(\alpha\rho)\, e^{-(p-\beta)\rho^2+2q\rho^{2\gamma'}} = \frac{(2\beta)^{2\eta}(\alpha R/2)^\nu R^{\mu+\eta+2}}{2\eta\nu!\,(\mu+\nu+2\eta+2)}$$
$$\times\, e^{-(p+\beta)R^2} F_{SD}^{(4)}\left(\begin{array}{l} \frac{\mu+\nu}{2}+\eta+1 : 1, 1, \gamma', - : 1, -, -, \quad 1 \quad ; \\ \frac{\mu+\nu}{2}+\eta+2 : 1, 1, \gamma', \; 1 \;\; : 2\eta+1, \nu+1, -, 1; \end{array} u, x, y, z\right). \tag{78}$$

Corollary 21 *As the above corollary and by using Theorem 5, one finds*

$$\int_0^R \rho^{\mu+1--2\eta} \gamma\left(2\eta, 2\beta\rho^2\right) J_\nu(\alpha\rho)\, J_\lambda(\chi\rho)\, e^{-(p-\beta)\rho^2+2q\rho^{2\gamma'}} d\rho$$
$$= \frac{(2\beta)^{2\eta}}{2\eta\nu!} \frac{(\alpha R/2)^\nu (\chi R/2)^\lambda R^{\mu+2\eta+2}}{\nu!\lambda!\,(\mu+\nu+\lambda+2\eta+2)} e^{-(p+\beta)R^2}$$
$$\times\, F_{SD}^{(5)}\left(\begin{array}{l} \frac{\mu+\nu+\lambda}{2}+\eta+1 : 1, 1, 1, \gamma', - : -, 1, -, -, \quad 1 \quad ; \\ \frac{\mu+\nu+\lambda}{2}+\eta+2 : 1, 1, 1, \gamma', \; 1 \;\; : \lambda+1, 2\eta+1, \nu+1, -, -; \end{array} t, u, x, y, z\right). \tag{79}$$

3.8 Case of $\varepsilon = -1/4$ and $\eta = 1/4$

The relation between $M_{-1/4,1/4}$ and the erf function is given, in this case, by $M_{-1/4,1/4}\left(z^2\right) = \frac{\sqrt{\pi}}{2} e^{z^2/2} \sqrt{z}.erf\,(z)$.

Corollary 22 *Using Theorem 3 and employing the above relation between* $M_{-1/4,1/4}$ *and erf functions, we get*

$$\int_0^R \rho^{\mu+1/2} erf\left(\sqrt{2\beta}\rho\right) e^{-(p-\beta)\rho^2+2q\rho^{2\gamma}} d\rho$$
$$= \sqrt{\frac{8\beta}{\pi}} \frac{R^{\mu+5/2}}{(\mu+5/2)} e^{-(p+\beta)R^2} F_{SD}^{(3)}\left(\begin{array}{l} \frac{\mu}{2}+5/4 \;\; : 1, \gamma, - : 1, -, \quad 1 \;; \\ \frac{\mu}{2}+9/4 \;\; : 1, \gamma, 1 : 3/2\,, -, -; \end{array} x, y, z\right). \tag{80}$$

Corollary 23 *For* $\varepsilon = -1/4$ *and* $\eta = 1/4$ *, Theorem 4 yields*

$$\int_0^R \rho^{\mu+1/2}\, erf\left(\sqrt{2\beta}\rho\right) J_\nu(\alpha\rho)\, e^{-(p-\beta)\rho^2+2q\rho^{2\gamma}} = \frac{\sqrt{8\beta/\pi}\,(\alpha R/2)^\nu R^{\mu+5/2}}{\nu!\,(\mu+\nu+5/2)}$$

$$\times e^{-(p+\beta)R^2} F_{SD}^{(4)}\left(\begin{matrix}\frac{\mu+\nu}{2}+5/4 \ : 1,1,\gamma,- : 1,-,-,1; \\ \frac{\mu+\nu}{2}+9/4 \ : 1,1,\gamma,1 : 3/2\,,\nu+1,-,-;\end{matrix} u,x,y,z\right). \tag{81}$$

Corollary 24 *The undermentioned integral transform holds true:*

$$\int_0^R \rho^{\mu+1/2}\, erf\left(\sqrt{2\beta}\rho\right) J_\nu(\alpha\rho)\, J_\lambda(\chi\rho)\, e^{-(p-\beta)\rho^2+2q\rho^{2\gamma}}\, d\rho$$

$$= \frac{\sqrt{8\beta/\pi}\,(\alpha R/2)^\nu(\chi R/2)^\lambda R^{\mu+5/2}}{\nu!\lambda!\,(\mu+\nu+\lambda+5/2)} e^{-(p+\beta)R^2}$$

$$\times F_{SD}^{(5)}\left(\begin{matrix}\frac{\mu+\nu+\lambda}{2}+\frac{5}{4} : 1,1,1,\gamma,- : -,1,-,-, \quad 1 \quad ; \\ \frac{\mu+\nu+\lambda}{2}+\frac{9}{4} : 1,1,1,\gamma,1 : \lambda+1,3/2\,,\nu+1,-,-;\end{matrix} t,u,x,y,z\right). \tag{82}$$

Applying Theorem 5 with $\varepsilon = -1/4$ and $\eta = 1/4$, one finds easily (81).

3.9 Case of $\varepsilon = -s/2$ and $\eta = 1/4$

The Whittaker function is given in terms of the Tricomi confluent hypergeometric function $U(a,b,z)$ by

$$M_{-s/2,1/4}\left(2\beta\rho^2\right) = \frac{2^{\frac{s+1}{2}}\Gamma(\frac{s}{2}+1/4)\beta^{1/4}}{4\sqrt{\pi}}\rho^{1/2}\left[U\left(s,-2\sqrt{\beta}\rho\right) - U\left(s,2\sqrt{\beta}\rho\right)\right].$$

Corollary 25 *Setting $\varepsilon = -s/2$ and $\eta = 1/4$ and using Theorem 3, we get*

$$\int_0^R \rho^{\mu+1/2}\left[U\left(s,-2\sqrt{\beta}\rho\right) - U\left(s,2\sqrt{\beta}\rho\right)\right] e^{-p\rho^2+2q\rho^{2\gamma}}\, d\rho$$

$$= \frac{\sqrt{\pi\beta}\,R^{\mu+5/2}}{2^{\frac{s}{2}-9/4}\,\Gamma\left(\frac{s}{2}+1/4\right)(\mu+5/2)} e^{-(p+\beta)R^2} \tag{83}$$

$$\times F_{SD}^{(3)}\left(\begin{matrix}\frac{\mu}{2}+5/4 \ : 1,\gamma,- : 3/4+\frac{s}{2},-,1; \\ \frac{\mu}{2}+9/4 \ : 1,\gamma,1 : 3/2\,,-, \quad - \quad ;\end{matrix} x,y,z\right).$$

Corollary 26 *As the above corollary, Theorem 4 yields the following transformation:*

$$\int_0^R \rho^{\mu+1/2}\left[U\left(s,-2\sqrt{\beta}\rho\right)-U\left(s,2\sqrt{\beta}\rho\right)\right]J_\nu(\alpha\rho)\,e^{-p\rho^2+2q\rho^{2\gamma}}$$
$$=\frac{\sqrt{\pi\beta}(\alpha R/2)^\nu R^{\mu+5/2}}{2^{\frac{s}{2}-9/4}\Gamma\left(\frac{s}{2}+1/4\right)\nu!(\mu+\nu+5/2)}e^{-(p+\beta)R^2}$$
$$\times F_{SD}^{(4)}\left(\begin{matrix}\frac{\mu+\nu}{2}+\frac{5}{4}:1,1,\gamma,-:\frac{s}{2}+\frac{3}{4},-,-, & 1 & ;\\ \frac{\mu+\nu}{2}+\frac{9}{4}:1,1,\gamma,\ 1\ :3/2\ ,\nu+1,-,-; & & \end{matrix}u,x,y,z\right). \tag{84}$$

Corollary 27 *The undermentioned integral transform holds true:*

$$\int_0^R \rho^{\mu+1/2}\left[U\left(s,-2\sqrt{\beta}\rho\right)-U\left(s,2\sqrt{\beta}\rho\right)\right]J_\nu(\alpha\rho)\,J_\lambda(\chi\rho)\,e^{-p\rho^2+2q\rho^{2\gamma}}d\rho$$
$$=\frac{\sqrt{\pi\beta}(\alpha R/2)^\nu(\chi R/2)^\lambda R^{\mu+5/2}}{2^{\frac{s}{2}-9/4}\Gamma\left(\frac{s}{2}+1/4\right)\nu!\lambda!(\mu+\nu+\lambda+5/2)}e^{-(p+\beta)R^2}$$
$$\times F_{SD}^{(5)}\left(\begin{matrix}\frac{\mu+\nu+\lambda}{2}+\frac{5}{4}:1,1,1,\gamma,-:-,\frac{s}{2}+\frac{3}{4},-,-, & 1 & ;\\ \frac{\mu+\nu+\lambda}{2}+\frac{9}{4}:1,1,1,\gamma,1:\lambda+1,3/2\ ,\nu+1,-, & - & ;\end{matrix}t,u,x,y,z\right). \tag{85}$$

This corollary can be derived directly from Theorem 5 by setting $\varepsilon=-s/2$ *and* $\eta=1/4$.

3.10 *Case of* $\mu=-\nu=1/2$

In this case, we use the following identity [14]:

$$J_{-1/2}(z)=\sqrt{\frac{2}{\pi z}}\cos(z). \tag{86}$$

Corollary 28 *By the use of Theorems 1, 2, 4, and 5, we arrive at the following transformations:*

$$\int_0^R \cos(\alpha\rho)\,e^{-p\rho^2+2q\rho^{2\gamma}}d\rho=R.e^{-pR^2}F_{SD}^{(3)}\left(\begin{matrix}1/2\ :1,\gamma,-:-,-, & 1 & ;\\ 3/2\ :1,\gamma,1\ :1/2\ ,-,-; & & \end{matrix}x,y,z\right) \tag{87}$$

where x and y are given by (14),

$$\int_0^R \cos(\alpha\rho)\, J_\lambda(\chi\rho)\, e^{-p\rho^2+2q\rho^{2\gamma}}\, d\rho$$
$$= \frac{R(\chi R/2)^\lambda}{\Gamma(\lambda+2)} e^{-pR^2} F_{SD}^{(4)}\left(\begin{matrix} \frac{\lambda+1}{2} : 1, 1, \gamma, - : -, -, -, \quad 1 \ ; \\ \frac{\lambda+3}{2} : 1, 1, \gamma, 1 : \lambda+1, 1/2, -, - \ ; \end{matrix} u, x, y, z\right), \tag{88}$$

where u, x, y, and z are given by (26)

$$\int_0^R M_{\varepsilon,\eta}\left(2\beta\rho^2\right) \cos(\alpha\rho)\, e^{-p\rho^2+2q\rho^{2\gamma}}\, d\rho$$
$$= R\frac{\left(2\beta R^2\right)^{\eta+1/2}}{2(\eta+1)} e^{-(p+\beta)R^2} F_{SD}^{(4)}\left(\begin{matrix} \eta+1 : 1, 1, \gamma, - : 1/2 + \eta - \varepsilon, -, -, \ 1 \ ; \\ \eta+2 : 1, 1, \gamma, 1 : 2\eta+1, 1/2, -, - \quad ; \end{matrix} u, x, y, z\right), \tag{89}$$

where x, y, and z are given by (39) and

$$\int_0^R M_{\varepsilon,\eta}\left(2\beta\rho^2\right) \cos(\alpha\rho)\, J_\lambda(\chi\rho) e^{-p\rho^2+2q\rho^{2\gamma}}\, d\rho = \frac{R\left(2R^2\beta\right)^{\eta+1/2} \left({}^{\chi R}/{}_2\right)^\lambda}{\lambda!(\lambda+2\eta+2)} e^{-(p+\beta)R^2}$$
$$\times F_{SD}^{(5)}\left(\begin{matrix} \frac{\lambda}{2}+\eta+1 : 1, 1, 1, \gamma, - : -, 1/2 + \eta - \varepsilon, -, -, 1; \\ \frac{\lambda}{2}+\eta+2 : 1, 1, 1, \gamma, 1 : \lambda+1, 2\eta+1, 1/2, -, -; \end{matrix} t, u, x, y, z\right), \tag{90}$$

where t, u, x, y, and z are given by (45).

3.11 Case of $\mu = \nu = 1/2$

We use in this case the well-known identity [14]

$$J_{1/2}(z) = \sqrt{\frac{2}{\pi z}} \sin(z). \tag{91}$$

Corollary 29 *As the above corollary, the following integral transforms hold true:*

$$\int_0^R \sin(\alpha\rho)\, e^{-p\rho^2+2q\rho^{2\gamma}}\, d\rho = \frac{\alpha R^2}{2} e^{-pR^2} F_{SD}^{(3)}\left(\begin{matrix} 1 : 1, \gamma, - : -, -, \quad 1 \ ; \\ 2 : 1, \gamma, 1 : 3/2, -, -; \end{matrix} x, y, z\right), \tag{92}$$

where x, y, and z are given by (14)

$$\int_0^R \sin(\alpha\rho)\, J_\lambda(\chi\rho)\, e^{-p\rho^2+2q\rho^{2\gamma}}\, d\rho$$
$$= \alpha R^2 \frac{(\chi R/2)^\lambda}{\lambda!\,(\lambda+2)} e^{-pR^2} F_{SD}^{(4)}\left(\begin{matrix} \lambda/2+1 : 1, 1, \gamma, - : -, -, -, \quad 1 \ ; \\ \lambda/2+2 : 1, 1, \gamma, 1 : \lambda+1, 3/2, -, - \ ; \end{matrix} u, x, y, z\right), \tag{93}$$

where u, x, y, and z are given by (26)

$$\int_0^R M_{\varepsilon,\eta}\left(2\beta\rho^2\right)\sin(\alpha\rho)\, e^{-p\rho^2+2q\rho^{2\gamma}}\, d\rho = \frac{\alpha R^2}{(2\eta+3)}\left(2\beta R^2\right)^{\eta+1/2} e^{-(p+\beta)R^2} \times F_{SD}^{(4)}\left(\begin{matrix}\eta+3/2 & :1,1,\gamma,- & :1/2+\eta-\varepsilon,-,-, & 1; \\ \eta+5/2 & :1,1,\gamma,1 & :2\eta+1,3/2, & -,- \;\; ; \end{matrix} u,x,y,z\right), \tag{94}$$

with u, x, y, and z are given by (39), and

$$\int_0^R M_{\varepsilon,\eta}\left(2\beta\rho^2\right)\sin(\alpha\rho)\, J_\lambda(\chi\rho)\, e^{-p\rho^2+2q\rho^{2\gamma}}\, d\rho$$
$$= \alpha(2\beta)^{\eta+1/2}\frac{(\chi R/2)^\lambda R^{2\eta+3}}{\lambda!\,(\lambda+2\eta+3)} e^{-(p+\beta)R^2} \times F_{SD}^{(5)}\left(\begin{matrix}\frac{\lambda+3}{2}+\eta : 1,1,1,\gamma,- : -,1/2+\eta-\varepsilon,-,-, \;\; 1; \\ \frac{\lambda+5}{2}+\eta : 1,1,1,\gamma,1 : \lambda+1,2\eta+1,3/2,-,- ; \end{matrix} t,u,x,y,z\right), \tag{95}$$

where t, u, x, y, and z are given by (45).

4 Conclusion

Motivated by multiple applications in the diverse field of mathematical physics such as neutron physics, plasma physics, and radio physics. In this research, we have derived several new integral transforms with a weight $e^{-p\rho^2+2q\rho^{2\gamma}}$ involving some special functions. These integrals are expressed in terms of Srivastava–Daoust multivariable hypergeometric functions. From our main results, we have established several interesting integral transforms in the form of corollaries, which can be used in diverse fields.

References

1. A. Belafhal, E.M. El Halba, T. Usman, An Integral transform involving the product of Bessel functions and Whittaker Function and its application. Int. J. Appl. Comput. Math. **6**, 1–11 (2020)
2. A. Belafhal, E.M. El Halba, T. Usman, Certain integral transforms and their applications in propagation of Laguerre-Gaussian Schell-model beams. Contemp. Math. **4**(4), 637–646 (2023)
3. A. Belafhal, E.M. El Halba, T. Usman, A note on some representations of Appell and Horn functions. Adv. Stud. Contemp. Math. **30**, 5–16 (2020)
4. A. Belafhal, E.M. El Halba, T. Usman, An integral transform and its application in the propagation of Lorentz-Gaussian beams. Commun. Math. **29**, 483–491 (2021)
5. A. Belafhal, H. Benzehoua, T. Usman, Certain integral transforms and their application to generate new Laser waves: Exton-Gaussian beams. Adv. Math. Models Appl. **6**, 206–217 (2021)
6. A. Belafhal, H. Benzehoua, A. Balhamri, T. Usman, An advanced method for evaluating Lommel integral and its application in marine environment. J. Comput. Appl. Math. **416**, 114600-1-11 (2022)
7. A. Belafhal, S. Hennani, A note on some integrals used in laser field involving the product of Bessel functions. Phys. Chem. News **61**, 59–62 (2011)
8. A. Belafhal, N. Nossir, Essakali L. Dalil, T. Usman, Integral transforms involving the product of Humbert and Bessel functions and its application. AIMS Math. **5**, 1260–1274 (2020)
9. A. Belafhal, S. Chib, F. Khannous, T. Usman, Evaluation of integral transforms using special functions with applications to biological tissues. Comput. Appl. Math. **40**, 1–23 (2021)
10. H.M. Srivastava, P.W. Karlsson, *Multiple Gaussian Hypergeometric Series* (John Wiley and Sons, New York, 1985)
11. V. Sahai, A. Verma, Recursion formulas for Srivastava's general triple hypergeometric function. Asian-Eur. J. Math. **9**(3), 1650063 (2016)
12. N. Watson, *Treatise on the Theory of Bessel Functions*, 2nd edn. (Cambridge University Press, Cambridge, 1944)
13. H. Buchholz, *The Confluent Hypergeometric Function* (Springer, 1969)
14. I.S. Gradshteyn, I.M. Ryzhik, *Table of Integrals, Series, and Products*, 5th edn. (Academic Press, New York, 1994)

A Measure Second-Order Maximal Monotone Differential Inclusion

D. Azzam-Laouir

Abstract In this work, we are principally concerned with the existence of right-continuous with bounded variation maps, which solve a second-order differential problem driven by a family of maximal monotone operators. The considered operators depend on both the time and the state variables.

Keywords Bounded variation · Differential measure · Lipschitz-continuous · Maximal monotone operator · Vladimirov distance · Right-continuous · Second order

2010 MSC 34A60 · 28A25 · 28C20

1 Introduction

Let $a > 0$ and $\mathcal{E}$ be a separable real Hilbert space. Let us take, for $(s, y) \in [0, a] \times \mathcal{E}$, a maximal monotone operator (m.m.op) of $\mathcal{E}$, that one denotes by $\mathfrak{B}(s, y)$, such that the set-valued map $(s, y) \mapsto \mathfrak{B}(s, y)$ is right-continuous with bounded variation (RCBV) on $[0, a]$, and Lipschitz-continuous on $\mathcal{E}$, where we mean the existence of a nondecreasing function $\vartheta : [0, a] \to [0, \infty[$, which is right-continuous on $[0, a[$, and a positive constant κ such that for $0 \le s_1 \le s_2 \le a,\ \ y_1, y_2 \in \mathcal{E}$,

$$dis(\mathfrak{B}(s_1, y_1), \mathfrak{B}(s_2, y_2)) \le d\vartheta(]s_1, s_2]) + \kappa\|y_1 - y_2\| = \vartheta(s_2) - \vartheta(s_1) + \kappa\|y_1 - y_2\|,$$

here $dis(\cdot, \cdot)$ denotes the Vladimirov's pseudo-distance between m.m.ops introduced in [51]; see relation (8). Let us also take a map $\mathcal{G} : [0, a] \times \mathcal{E} \times \mathcal{E} \longrightarrow \mathcal{E}$ of Carathéodory type.

Our goal is the study of the existence of an RCBV map $w : [0, a] \to \mathcal{E}$ and a Lipschitz-continuous map $z : [0, a] \to \mathcal{E}$ solving the following second-order measure differential problem:

D. Azzam-Laouir (✉)
LAOTI Laboratory, Faculty of Exact Sciences and Informatics, University of Jijel, Jijel 18000, Algeria
e-mail: dalilalaouir@univ-jijel.dz

M. Kharrat et al. (eds.), *Modeling of Discrete and Continuous Systems*, Forum for Interdisciplinary Mathematics, https://doi.org/10.1007/978-981-97-8715-9_12

$$(P_{\mathcal{G}})\begin{cases} z(0)=z_0,\, w(0)=w_0\in D(\mathfrak{B}(0,z_0));\\ w(s)\in D(\mathfrak{B}(s,z(s)))\quad \forall s\in[0,a];\\ \dfrac{dz}{d\lambda}(s)=w(s)\quad \lambda-a.e.\, s\in[0,a];\\ -\dfrac{dw}{d\nu}(s)\in \mathfrak{B}(s,z(s))w(s)+\mathcal{G}(s,z(s),w(s))\dfrac{d\lambda}{d\nu}(s)\quad d\nu-a.e.\, s\in[0,a], \end{cases}$$

where λ is the Lebesgue-measure on $[0, a]$, $d\vartheta$ the Stieltjes-measure associated with ϑ and $\nu := \lambda + d\vartheta$, while $\frac{d\lambda}{d\nu}$ is the density of λ with respect to the measure ν.

It is worth to mention that in the literature exists a great number of papers that have treated the first-order differential inclusion (FODI) with m.m.ops depending on the time as well as on the state, with different types of variation: Lipschitz-continuous, absolutely continuous (AC), and continuous with bounded variation (CBV). We refer the interested reader to [13–16, 22, 26, 32–34, 43, 47, 50, 51] and their references. As regards the RCBV time-dependence we cite references [10, 17, 33].

To our current knowledge, the second-order differential inclusion (SODI) with m.m.ops satisfying absolutely continuous variation was carried out in the recent works [21, 30, 45]. However, we are not aware of any result dealing with (SODI) with m.m.ops where the variation is RCBV. This fact justifies our motivation in this contribution, which forms a prolongation to the second order of RCBV results about (FODI) cited above.

A second motivation in the treatment of (DI) with m.m.ops is the named "sweeping process" (SP). This process is a differential inclusion driven by the normal cone to nonempty, closed, and convex sets, which depend on the time (moving sets). It was introduced by J.J. Moreau and largely studied in a series of pioneering works [35–37] thanks to their applications to several mechanical problems. It turns out that this normal cone is a m.m.op, so that (DI) with m.m.ops is a generalization of (SP).

Existence of RCBV solutions to the (FOSP) was treated in many promising papers; we cite, for example, [1, 5, 19, 20, 29, 31, 38, 41, 48, 49]. However, there are few works dealing with the RCBV (SOSP) in comparison with those concerning AC and CBV solutions; see [2, 3, 7–9, 11, 12, 23, 24, 27, 28, 42, 46].

Actually, regarding the more general case of RCBV (SOSP), we are aware of references [4, 6, 41]. In [4], using an adapted catching-up algorithm or Moreau algorithm, the authors gave an existence result for the second-order sweeping process $(\mathcal{SP})$ (see Corollary 2 of our main theorem), where the sets considered in the normal cone are r-prox regular, and the external force $\mathcal{G}$ is a closed convex set-valued map satisfying a time-dependent growth condition. Still with r-prox regular sets in the normal cone and set-valued perturbations where the growth condition is only supposed on their elements of minimal norm, Nacry et al. introduced, in [41], a very interesting new mixed first-order sweeping process in the Hilbert space $\mathcal{E}^2$. They have adapted a mixed catching-up algorithm to prove the existence of RCBV solutions, from which they have derived existence for both first-order and second-order sweeping processes. The same problem was considered in [6], where the sets in the normal cone are in the wider class of subsmooth sets.

2 Notations and Preliminaries

All over the manuscript, $\langle \cdot, \cdot \rangle$ is the inner product of $\mathcal{E}$ and $\| \cdot \|$ the induced norm. For $r > 0$, $r\overline{B}_{\mathcal{E}}$ is the closed ball of $\mathcal{E}$ centered at 0 with radius r. $\mathcal{B}([0, a])$ is the Borel-tribe on $[0, a]$, and $Id_{\mathcal{E}}$ is the identity mapping of $\mathcal{E}$, and χ_Q is the characteristic function of a set Q. For $\mathcal{S} \subset \mathcal{E}$, $\overline{co}(\mathcal{S})$ is the closed convex hull of $\mathcal{S}$, and by separation theorem:

$$\overline{co}(\mathcal{S}) = \Big\{\bar{x} \in \mathcal{E} : \ \langle \bar{x}, e\rangle \leq \sup_{\bar{y} \in \mathcal{S}} \langle \bar{y}, e\rangle \ \forall e \in \mathcal{S}\Big\}. \tag{1}$$

Let $p \in [1, +\infty[$ and let μ be a positive measure on $[0, a]$, by $L^p([0, a], \mathcal{E}; \mu)$ we denote the Banach space of classes of p-μ-integrable mappings.

Now we introduce definition and useful properties of mappings with bounded variation; see [31, 35, 38–40] for more features.

Let $f : [0, a] \longrightarrow \mathcal{E}$. The variation of f in $[0, a]$ is the positive extended real number defined by

$$var(f; [0, a]) := \sup \sum_{j=1}^{n} \| f(s_j) - f(s_{j-1})\|, \tag{2}$$

here the sup is considered in $[0, +\infty]$ over all the finite subdivisions of $[0, a]$ by the points $s_0 < s_1 < \cdots < s_n$ (n is arbitrary). We say that f is with bounded variation if, and only if, $var(f; [0, a]) < +\infty$.

For $s \in [0, a[$, the right limit of f at the point s, if it exists, is given by

$$f^+(s) = \lim_{t \downarrow s} f(t),$$

and if $s = a$, $f^+(s) = f(s)$. We say that f is right-continuous (RC) at the point s if, $f^+(s) = f(s)$, and f is RCBV on $[0, a]$, if it is right-continuous with bounded variation on $[0, a]$.

Now, let μ be a positive Radon measure on $[0, a]$ and $\hat{\mu}$ be a measure on $[0, a]$ with values in $\mathcal{E}$ and a density $\frac{d\hat{\mu}}{d\mu} \in L^1_{loc}([0, a], \mathcal{E}; \mu)$. Then

$$\frac{d\hat{\mu}}{d\mu}(s) = \lim_{\epsilon \downarrow 0} \frac{d\hat{\mu}([s, s+\epsilon])}{d\mu([s, s+\epsilon])} = \lim_{\epsilon \downarrow 0} \frac{d\hat{\mu}([s-\epsilon, s])}{d\mu([s-\epsilon, s])} \quad d\mu - a.e. \, s \in [0, a]. \tag{3}$$

The measure $\hat{\mu}$ is absolutely continuous w.r.t μ iff $\hat{\mu} = \frac{d\hat{\mu}}{d\mu}\mu$, that is, $\frac{d\hat{\mu}}{d\mu}$ is a density of $\hat{\mu}$ w.r.t μ, and a mapping $g : [0, a] \longrightarrow \mathcal{E}$ is $\hat{\mu}$-integrable iff the mapping $s \mapsto g(s)\frac{d\hat{\mu}}{d\mu}(s)$ is μ-integrable, and we have

$$\int_{[0,a]} g(s) d\hat{\mu}(s) = \int_{[0,a]} g(s) \frac{d\hat{\mu}}{d\mu}(s) d\mu(s). \tag{4}$$

If $f : [0, a] \longrightarrow \mathcal{E}$ is BV and df is its differential measure, then we have

$$f^+(\tau) - f^+(s) = \int_{]s,\tau]} df \quad \forall 0 \le s \le \tau \le a,$$

so that if f is RCBV, then

$$f(\tau) - f(s) = \int_{]s,\tau]} df \quad \forall 0 \le s \le \tau \le a. \tag{5}$$

Reversely, if one has a mapping $g \in L^1([0, a], \mathcal{E}; \mu)$ such that $f(s) = f(0) + \int_{]0,s]} g\, d\mu$ for all $s \in [0, a]$, then g is RCBV and $df = g\, d\mu$, that is, g is a density of the vector measure df w.r.t the measure μ. Applying (3), one obtains

$$g(s) = \frac{df}{d\mu}(s) = \lim_{\epsilon \downarrow 0} \frac{df([s, s+\epsilon])}{d\mu([s, s+\epsilon])} = \lim_{\epsilon \downarrow 0} \frac{df([s-\epsilon, s])}{d\mu([s-\epsilon, s])}. \tag{6}$$

If $\mu(\{s\}) > 0$, (6) ensures that

$$g(s) = \frac{df}{d\mu}(s) = \frac{df(\{s\})}{d\mu(\{s\})} \quad \text{and} \quad \frac{d\lambda}{d\mu}(s) = 0. \tag{7}$$

Next, we present the terminology and the needed properties of m.m.ops. For the details, we refer back to [18, 25, 52].

Let $\mathfrak{B} : \mathcal{E} \rightrightarrows \mathcal{E}$ be a set-valued map, and let $D(\mathfrak{B})$ and $Gr(\mathfrak{B})$ its domain and graph. We say that the operator $\mathfrak{B}$ is monotone if $\langle z_1 - z_2, \varsigma_1 - \varsigma_2 \rangle \ge 0$ for all $(\varsigma_j, z_j) \in Gr(\mathfrak{B})$ $(j = 1, 2)$, and the monotone operator $\mathfrak{B}$ is said to be maximal, if its graph is maximal for inclusion of the graphs of monotone operators of $\mathcal{E}$.

If $\mathfrak{B}$ is a m.m.op of $\mathcal{E}$, then for every $\varsigma \in D(\mathfrak{B})$, the set $\mathfrak{B}(\varsigma)$ is nonempty closed and convex. Its element of minimal norm will be denoted, in the remainder of the paper, by $\mathfrak{B}^0(\varsigma)$.

For $\eta > 0$, we denote by $J_\eta^{\mathfrak{B}} := (Id_{\mathcal{E}} + \eta\mathfrak{B})^{-1}$ the resolvent of $\mathfrak{B}$. The operator $J_\eta^{\mathfrak{B}}$, defined on all of $\mathcal{E}$, is single-valued and satisfies: $J_\eta^{\mathfrak{B}}(\varsigma) \in D(\mathfrak{B})$, for all $\varsigma \in \mathcal{E}$.

Let $\mathfrak{B}_1 : D(\mathfrak{B}_1) \rightrightarrows \mathcal{E}$ and $\mathfrak{B}_2 : D(\mathfrak{B}_2) \rightrightarrows \mathcal{E}$ be two m.m.ops. Then, $dis(\mathfrak{B}_1, \mathfrak{B}_2)$ is the Vladimirov's pseudo-distance between $\mathfrak{B}_1$ and $\mathfrak{B}_2$, defined as follows:

$$dis(\mathfrak{B}_1, \mathfrak{B}_2) = \sup\left\{ \frac{\langle z_1 - z_2, \varsigma_2 - \varsigma_1 \rangle}{1 + \|z_1\| + \|z_2\|} : (\varsigma_j, z_j) \in Gr(\mathfrak{B}_j),\ j = 1, 2 \right\}. \tag{8}$$

We also need to recall some lemmas that are crucial for our investigation. We refer to [33, 51] for their demonstrations.

Lemma 1 *Let* $\mathfrak{B}$ *be a m.m.op of* $\mathcal{E}$. *Let* $\varsigma \in \overline{D(\mathfrak{B})}$ *and* $\bar{\varsigma} \in \mathcal{E}$ *satisfying*

$$\langle \mathfrak{B}^0(\varpi) - \bar{\varsigma}, \varpi - \varsigma \rangle \geq 0 \ \ \forall \varpi \in D(\mathfrak{B}),$$

then $\varsigma \in D(\mathfrak{B})$ *and* $\bar{\varsigma} \in \mathfrak{B}(\varsigma)$.

Lemma 2 *Let* $\mathfrak{B}$, $\mathfrak{B}_1$ *and* $\mathfrak{B}_2$ *be m.m.ops of* $\mathcal{E}$. *Then*
1) for $\eta > 0$ *and* $\varsigma \in D(\mathfrak{B}_1)$

$$\|\varsigma - J_\eta^{\mathfrak{B}_2}(\varsigma)\| \leq \eta\|\mathfrak{B}_1^0(\varsigma)\| + dis(\mathfrak{B}_1, \mathfrak{B}_2) + \sqrt{\eta\big(1 + \|\mathfrak{B}_1^0(\varsigma)\|\big)dis(\mathfrak{B}_1, \mathfrak{B}_2)}.$$

2) For $\eta > 0$ *and* $\varsigma, \bar{\varsigma} \in \mathcal{E}$

$$\|J_\eta^{\mathfrak{B}}(\varsigma) - J_\eta^{\mathfrak{B}}(\bar{\varsigma})\| \leq \|\varsigma - \bar{\varsigma}\|.$$

Lemma 3 *Let* $\mathfrak{B}_n$ $(n \in \mathbb{N})$ *and* $\mathfrak{B}$ *be m.m.ops of* $\mathcal{E}$ *such that* $dis(\mathfrak{B}_n, \mathfrak{B}) \to 0$. *Suppose also that* $\varsigma_n \in D(\mathfrak{B}_n)$ *with* $\varsigma_n \to \varsigma$ *and* $z_n \in \mathfrak{B}_n(\varsigma_n)$ *with* $z_n \to z$ *weakly for some* $\varsigma, z \in \mathcal{E}$. *Then* $\varsigma \in D(\mathfrak{B})$ *and* $z \in \mathfrak{B}(\varsigma)$.

Lemma 4 *Let* $\mathfrak{B}_n$ $(n \in \mathbb{N})$ *and* $\mathfrak{B}$ *be m.m.ops of* $\mathcal{E}$ *such that* $dis(\mathfrak{B}_n, \mathfrak{B}) \to 0$, *and for all* $n \in \mathbb{N}$ *and* $\varsigma \in D(\mathfrak{B}_n)$, $\|\mathfrak{B}_n^0(\varsigma)\| \leq m(1 + \|\varsigma\|)$ *for some positive constant* m. *Then for every* $\zeta \in D(\mathfrak{B})$, *one has the existence of a sequence* $(\zeta_n)_n$ *such that*

$$\zeta_n \in D(\mathfrak{B}_n), \quad \zeta_n \to \zeta \quad \text{and} \quad \mathfrak{B}_n^0(\zeta_n) \to \mathfrak{B}^0(\zeta).$$

Finally, let us recall the following discrete Gronwall's lemma.

Lemma 5 *Let* $(a_j)_j$, $(b_j)_j$, $(c_j)_j$, *and* $(x_j)_j$ *be sequences of positive numbers verifying:* $x_{j+1} \leq a_j + b_j\big(x_0 + x_1 + \cdots + x_{j-1}\big) + (1 + c_j)x_j$ *for* $j \geq 0$. *Then*

$$x_j \leq \left(x_0 + \sum_{k=0}^{j-1} a_k\right) \exp\left(\sum_{k=0}^{j-1} \big(k\, b_k + c_k\big)\right) \ \ \textit{for } j \geq 1.$$

We start by listing our needed assumptions.
(a_1) There exist a nondecreasing function $\vartheta : [0, a] \longrightarrow [0, +\infty[$, which is RC on $[0, a[$ with $\vartheta(0) = 0$ and $\vartheta(a) < +\infty$, and a positive constant κ such that

$$dis(\mathfrak{B}(s_1, y_1), \mathfrak{B}(s_2, y_2)) \leq d\vartheta(]s_1, s_2]) + \kappa\|y_1 - y_2\| \ \text{ for } \ 0 \leq s_1 \leq s_2 \leq a, \ \ y_1, y_2 \in \mathcal{E}.$$

(a_2) Fore some positive real constant c,

$$\|\mathfrak{B}^0(s, y)\bar{y}\| \leq c(1 + \|y\| + \|\bar{y}\|) \ \text{ for } \ s \in [0, a], \ \ y \in \mathcal{E}, \ \ \bar{y} \in D(\mathfrak{B}(s, y)).$$

(a_3) For every bounded subset A of $\mathcal{E}$, and for any $r \geq 0$, $D(\mathfrak{B}([0, a] \times A)) \cap r\overline{B}_{\mathcal{E}}$ is relatively compact.

(a_4) The map $s \mapsto \mathcal{G}(s, \cdot, \cdot)$ is Borel-measurable, i.e., $\mathcal{B}([0, a])$-measurable.
(a_5) There is a positive function $k(\cdot) \in L^1([0, a], \mathbb{R}; \lambda)$ such that

$$\|\mathcal{G}(s, x_1, x_2) - \mathcal{G}(s, y_1, y_2)\| \leq k(s)(\|x_1 - y_1\| + \|x_2 - y_2\|) \quad \forall s \in [0, a], \ \forall (x_1, x_2), (y_1, y_2) \in \mathcal{E} \times \mathcal{E}.$$

(a_6) Fore some positive real constant l,

$$\|\mathcal{G}(s, y, \bar{y})\| \leq l(1 + \|y\| + \|\bar{y}\|) \quad \forall (s, y, \bar{y}) \in [0, a] \times \mathcal{E} \times \mathcal{E}.$$

3 Main Result

This section is devoted to the statement and the proof of our main result. We will use some ideas from [4].

Theorem 1 *Let for every* $(s, y) \in [0, a] \times \mathcal{E}$, $\mathfrak{B}(s, y) : D(\mathfrak{B}(s, y)) \rightrightarrows \mathcal{E}$ *be a m.m.op satisfying* (a_1), (a_2), *and* (a_3). *Let* $\mathcal{G} : [0, a] \times \mathcal{E} \times \mathcal{E} \longrightarrow \mathcal{E}$ *such that* (a_4), (a_5), *and* (a_6) *are satisfied. Then for any* $(z_0, w_0) \in \mathcal{E} \times D(\mathfrak{B}(0, z_0))$, *there exist an RCBV map* $w : [0, a] \to \mathcal{E}$ *and a Lipschitz-continuous map* $z : [0, a] \to \mathcal{E}$ *such that the couple* (z, w) *is solution of problem* $(P_{\mathcal{G}})$, *and one has the estimates:*

$$\|w(\tau) - w(s)\| \leq b\big((\vartheta(\tau) - \vartheta(s)) + (\tau - s)\big) \quad \forall 0 \leq s \leq \tau \leq a,$$

$$\|z(\tau) - z(s)\| \leq R(\tau - s) \quad \forall 0 \leq s \leq \tau \leq a,$$

for some positive constants b *and* R.

Proof As in reference [29], we take a sequence $(\epsilon_n)_{n \geq 1} \subset]0, 1]$ which decreases to zero, and we choose a partition of $[0, a]$ by the points $0 = s_0^n < s_1^n < \cdots < s_{\iota_n}^n = a$ (with $\lim_{n \to \infty} \iota_n = +\infty$), for which we have

$$\nu(]s_j^n, s_{j+1}^n[) = |s_{j+1}^n - s_j^n| + d\vartheta(]s_j^n, s_{j+1}^n[) \leq \epsilon_n \quad \text{for } j = 0, \ldots, \iota_n - 1, \tag{9}$$

for instance, $\epsilon_n = \frac{a}{\iota_n}$. For each $j \in \{0, \ldots, \iota_n - 1\}$, let us set

$$\delta_{j+1}^n = d\vartheta(]s_j^n, s_{j+1}^n]), \quad \eta_{j+1}^n = s_{j+1}^n - s_j^n, \quad \beta_{j+1}^n = \nu(]s_j^n, s_{j+1}^n]). \tag{10}$$

For every $n \in \mathbb{N}$, put $z_0^n = z_0$, $w_0^n = w_0 \in D(\mathfrak{B}(0, z_0))$, and defined by recurrence, sequences $(w_j^n)_{0 \leq j \leq \iota_n - 1}$, $(z_j^n)_{0 \leq j \leq \iota_n - 1}$ as follows:

$$z_{j+1}^n = z_j^n + \eta_{j+1}^n w_j^n, \tag{11}$$

and

$$w_{j+1}^n = J_{j+1}^n\Big(w_j^n - \int_{s_j^n}^{s_{j+1}^n} \mathcal{G}(\tau, z_j^n, w_j^n)d\lambda(\tau)\Big), \tag{12}$$

where $J_{j+1}^n := J_{\beta_{j+1}^n}^{\mathfrak{B}(s_{j+1}^n, z_{j+1}^n)} = \big(Id_{\mathcal{E}} + \beta_{j+1}^n \mathfrak{B}(s_{j+1}^n, z_{j+1}^n)\big)^{-1}$. From the properties of the resolvent, we have that $w_{j+1}^n \in D(\mathfrak{B}(s_{j+1}^n, z_{j+1}^n))$ and

$$-\frac{1}{\beta_{j+1}^n}\Big(w_{j+1}^n - w_j^n + \int_{s_j^n}^{s_{j+1}^n} \mathcal{G}(\tau, z_j^n, w_j^n)d\lambda(\tau)\Big) \in \mathfrak{B}(s_{j+1}^n, z_{j+1}^n)w_{j+1}^n. \tag{13}$$

For $s \in [s_j^n, s_{j+1}^n[$, $j = 0, \ldots, \iota_n - 1$, set

$$w_n(s) = w_j^n + \frac{\nu(]s_j^n, s])}{\nu(]s_j^n, s_{j+1}^n])}\Big(w_{j+1}^n - w_j^n + \int_{s_j^n}^{s_{j+1}^n} \mathcal{G}(\tau, z_j^n, w_j^n)d\lambda(\tau)\Big) - \int_{s_j^n}^{s} \mathcal{G}(\tau, z_j^n, w_j^n)d\lambda(\tau), \tag{14}$$

$$z_n(s) = z_j^n + \frac{s - s_j^n}{\eta_{j+1}^n}(z_{j+1}^n - z_j^n), \tag{15}$$

and $w_n(a) = w_{\iota_n}^n$, $z_n(a) = z_{\iota_n}^n$. So that $w_n(s_j^n) = w_j^n$, $z_n(s_j^n) = z_j^n$, z_n is absolutely continuous on $[0, a]$ and w_n is an RCBV mapping on $[0, a]$.

For the clarity of the rest of the proof, we will share it into three parts.

Part 1. We show here that $(w_n)_n$, $(z_n)_n$ are bounded in variation and in norm. Involving Lemma 2, one has from (12), (a_1), (a_2), and (a_6), for $j = 0, 1, \ldots, \iota_n - 1$ (keeping in mind that $\sqrt{pq} \le \frac{p}{2} + q$ for positive real numbers p, q),

$$\begin{aligned}
\|w_{j+1}^n - w_j^n\| &\le \Big\|J_{j+1}^n\Big(w_j^n - \int_{s_j^n}^{s_{j+1}^n} \mathcal{G}(\tau, z_j^n, w_j^n)d\lambda(\tau)\Big) - J_{j+1}^n(w_j^n)\Big\| + \|J_{j+1}^n(w_j^n) - w_j^n\| \\
&\le \int_{s_j^n}^{s_{j+1}^n} \|\mathcal{G}(\tau, z_j^n, w_j^n)\|d\lambda(\tau) + \beta_{j+1}^n\|\mathfrak{B}^0(s_j^n, z_j^n)w_j^n\| + dis\big(\mathfrak{B}(s_{j+1}^n, z_{j+1}^n), \mathfrak{B}(s_j^n, z_j^n)\big) \\
&\quad + \sqrt{\beta_{j+1}^n(1 + \|\mathfrak{B}^0(s_j^n, z_j^n)w_j^n\|)dis\big(\mathfrak{B}(s_{j+1}^n, z_{j+1}^n), \mathfrak{B}(s_j^n, z_j^n)\big)} \\
&\le l(1 + \|z_j^n\| + \|w_j^n\|)\eta_{j+1}^n + \big(1 + c(1 + \|z_j^n\| + \|w_j^n\|)\big)\beta_{j+1}^n + \kappa\|z_{j+1}^n - z_j^n\| \\
&\quad + \sqrt{\beta_{j+1}^n\big(1 + c(1 + \|z_j^n\| + \|w_j^n\|)\big)(\beta_{j+1}^n + \kappa\|z_{j+1}^n - z_j^n\|)} \\
&\le l(1 + \|z_j^n\| + \|w_j^n\|)\beta_{j+1}^n + \big(1 + c(1 + \|z_j^n\| + \|w_j^n\|)\big)\beta_{j+1}^n + \kappa\|z_{j+1}^n - z_j^n\| \\
&\quad + \beta_{j+1}^n\frac{1}{2}\big(1 + c(1 + \|z_j^n\| + \|w_j^n\|)\big) + \beta_{j+1}^n + \kappa\|z_{j+1}^n - z_j^n\|,
\end{aligned}$$

that is,

$$\begin{aligned}\|w_{j+1}^n - w_j^n\| &\le (l + 1 + \frac{3}{2}(1+c))(1 + \|z_j^n\| + \|w_j^n\|)\beta_{j+1}^n + 2\kappa\|z_{j+1}^n - z_j^n\| \\ &=: l_1(1 + \|z_j^n\| + \|w_j^n\|)\beta_{j+1}^n + \kappa_1\|z_{j+1}^n - z_j^n\|, \qquad (16)\end{aligned}$$

which entails

$$\|w_{j+1}^n\| \le \|w_j^n\| + l_1(1 + \|z_j^n\| + \|w_j^n\|)\beta_{j+1}^n + \kappa_1\|z_{j+1}^n - z_j^n\|. \qquad (17)$$

On the other hand, we have from (11),

$$\|z_{j+1}^n - z_j^n\| = \eta_{j+1}^n\|w_j^n\| \qquad (18)$$

and (using (9))

$$\begin{aligned}\|z_j^n\| &= \|z_{j-1}^n + \eta_j^n w_{j-1}^n\| = \|z_0 + \eta_1^n w_0^n + \eta_2^n w_1^n + \cdots + \eta_j^n w_{j-1}^n\| \\ &\le \|z_0\| + \epsilon_n(\|w_0^n\| + \|w_1^n\| + \cdots + \|w_{j-1}^n\|). \qquad (19)\end{aligned}$$

Replace now relations (18), (19) in (17), one gets (using (9)),

$$\begin{aligned}\|w_{j+1}^n\| &\le \|w_j^n\| + l_1\Big(1 + \|z_0\| + \epsilon_n(\|w_0^n\| + \|w_1^n\| + \cdots + \|w_{j-1}^n\|) + \|w_j^n\|\Big)\beta_{j+1}^n + \kappa_1\|w_j^n\|\epsilon_n \\ &\le l_1(1 + \|z_0\|)\beta_{j+1}^n + l_1\epsilon_n\beta_{j+1}^n(\|w_0^n\| + \|w_1^n\| + \cdots + \|w_{j-1}^n\|) \\ &+ (1 + (l_1\beta_{j+1}^n + \kappa_1\epsilon_n))\|w_j^n\|.\end{aligned}$$

The application of Lemma 5 gives us for all $j \le \iota_n$,

$$\begin{aligned}\|w_j^n\| &\le \Big(\|w_0\| + \sum_{i=0}^{j-1} l_1(1 + \|z_0\|)\beta_{i+1}^n\Big)\exp\Big(\sum_{i=0}^{j-1}\big(i\, l_1\epsilon_n\beta_{i+1}^n + (l_1\beta_{i+1}^n + \kappa_1\epsilon_n)\big)\Big) \\ &\le \Big(\|w_0\| + l_1(1 + \|z_0\|)\nu(]0,a])\Big)\exp\Big(l_1\frac{a}{\iota_n}\iota_n\sum_{i=0}^{j-1}\beta_{i+1}^n + l_1\nu(]0,a]) + \kappa_1 a\Big) \\ &\le \Big(\|w_0\| + l_1(1 + \|z_0\|)\nu(]0,a])\Big)\exp\Big(l_1 a\nu(]0,a]) + l_1\nu(]0,a]) + a\kappa_1\Big) =: l_2.\end{aligned}$$

Using this last estimate in (19), one obtains for all $1 \le j \le \iota_n$,

$$\|z_j^n\| \le \|z_0\| + \epsilon_n\iota_n l_2 = \|z_0\| + a\, l_2,$$

and replacing in (18) and in (16), it comes that

$$\|z_{j+1}^n - z_j^n\| \le l_2\eta_{j+1}^n,$$

and

$$\|w_{j+1}^n - w_j^n\| \le l_1(1 + \|z_0\| + al_2 + l_2)\beta_{j+1}^n + \kappa_1 l_2 \beta_{j+1}^n =: l_3 \beta_{j+1}^n.$$

Consequently, for all $1 \le j \le \iota_n$,

$$\|z_j^n\| \le R, \qquad \|w_j^n\| \le R \tag{20}$$

and for all $0 \le j < \iota_n - 1$,

$$\|z_{j+1}^n - z_j^n\| \le R(s_{j+1}^n - s_j^n), \quad \|w_{j+1}^n - w_j^n\| \le R\,\nu(]s_j^n, s_{j+1}^n]), \tag{21}$$

so that

$$\sup_{n\in\mathbb{N}} var(z_n) \le a\,R \quad \text{and} \quad \sup_{n\in\mathbb{N}} var(w_n) \le R\,\nu(]0,a]),$$

where $R = \max(l_2, \|z_0\| + a\,l_2, l_3)$. Now, for $s \in [s_j^n, s_{j+1}^n[$, we have from (9), (14), (20), (21), and (a_6),

$$\|w_n(s) - w_j^n\| \le R\epsilon_n + 2l(1 + 2R)\epsilon_n \le R + 2l(1 + 2R) =: R_1,$$

and from (9), (15), (20), and (21), we get

$$\|z_n(s) - z_j^n\| \le R.$$

These last relations give us, for all $n \in \mathbb{N}$,

$$\|w_n(s)\| \le R + R_1 =: R_2 \quad \text{and} \quad \|z_n(s)\| \le 2R \quad \forall s \in [0, a],$$

that is,

$$\sup_n \|w_n\| := \sup_n \big(\sup_{s\in[0,a]} \|w_n(s)\|\big) \le R_2 \quad \text{and} \quad \sup_n \|z_n\| \le 2R. \tag{22}$$

Consequently, $(z_n)_n$, $(w_n)_n$ are bounded in norm and in variation.

Part 2. One shows, in this second part, that $(z_n)_n$, $(w_n)_n$ and $(\frac{dw_n}{d\nu})_n$ are convergent. One defines the step-functions $\sigma_n,\ \varrho_n : [0, a] \longrightarrow [0, a]$ by

$$\sigma_n(s) = s_j^n, \quad \varrho_n(s) = s_{j+1}^n \quad \text{for } s \in]s_j^n, s_{j+1}^n], \ j = 0, 1, \ldots, \iota_n - 1, \quad \text{and } \sigma_n(0) = \varrho_n(0) = 0.$$

Obviously,

$$\sigma_n(s) \uparrow s \quad \text{and} \quad \varrho_n(s) \downarrow s \quad \text{as } n \to \infty.$$

Set, for all $s \in [0, a]$,

$$\Upsilon_n(s) = \sum_{j=0}^{\iota_n - 1} \frac{1}{\nu(]s_j^n, s_{j+1}^n])} \Big(w_{j+1}^n - w_j^n + \int_{s_j^n}^{s_{j+1}^n} \mathcal{G}(\tau, z_j^n, w_j^n) d\lambda(\tau) \Big) \chi_{]s_j^n, s_{j+1}^n]}(s).$$

So, from (14),

$$w_n(s) = w_0 + \int_{]0,s]} \Upsilon_n(\tau) d\nu(\tau) - \int_{]0,s]} \mathcal{G}(\tau, z_n(\sigma_n(\tau)), w_n(\sigma_n(\tau))) d\lambda(\tau).$$

As $\frac{d\lambda}{d\nu}$ is a density of λ w.r.t ν, then by (4), one gets for every $s \in [0, a]$,

$$w_n(s) = w_0 + \int_{]0,s]} \Big(\Upsilon_n(\tau) - \mathcal{G}(\tau, z_n(\sigma_n(\tau)), w_n(\sigma_n(\tau))) \frac{d\lambda}{d\nu}(\tau) \Big) d\nu(\tau),$$

which means that $\frac{dw_n}{d\nu}$ is a density of the vector measure dw_n w.r.t ν and for ν-almost all $s \in [0, a]$,

$$\frac{dw_n}{d\nu}(s) = \Upsilon_n(s) - \mathcal{G}(s, z_n(\sigma_n(s)), w_n(\sigma_n(s))) \frac{d\lambda}{d\nu}(s),$$

whence coming back to the definition of Υ_n, one has for ν-almost all $s \in [0, a]$,

$$\begin{aligned} &\frac{dw_n}{d\nu}(s) + \mathcal{G}(s, z_n(\sigma_n(s)), w_n(\sigma_n(s))) \frac{d\lambda}{d\nu}(s) \\ &= \sum_{j=0}^{\iota_n - 1} \frac{1}{\nu(]s_j^n, s_{j+1}^n])} \Big(w_{j+1}^n - w_j^n + \int_{s_j^n}^{s_{j+1}^n} \mathcal{G}(\tau, z_j^n, w_j^n) d\lambda(\tau) \Big) \chi_{]s_j^n, s_{j+1}^n]}(s). \end{aligned} \tag{23}$$

Referring to (a_6), (20), and (23), one gets

$$\Big\| \frac{dw_n}{d\nu}(s) + \mathcal{G}(s, z_n(\sigma_n(s)), w_n(\sigma_n(s))) \frac{d\lambda}{d\nu}(s) \Big\| \leq R + l(1 + 2R) =: R_3 \quad \nu - a.e. s \in [0, a]. \tag{24}$$

Moreover, since $\nu = d\vartheta + \lambda$, it is obvious that $\frac{d\lambda}{d\nu}(s) \leq 1$. Then, by (a_6) and (20), it results that

$$\Big\| \mathcal{G}(s, z_n(\sigma_n(s)), w_n(\sigma_n(s))) \frac{d\lambda}{d\nu}(s) \Big\| \leq l(1 + 2R) =: R_4 \quad \nu - a.e. s \in [0, a], \tag{25}$$

by consequent, we may infer the existence of a ν-null Borel subset $J \subset [0, a]$, such that

$$\Big\| \frac{dw_n}{d\nu}(s) \Big\| \leq R_3 + R_4 =: R_5 \quad \forall s \in [0, a] \setminus J. \tag{26}$$

Now, remark by (13) and (23), that for each $n \in \mathbb{N}$, there is a ν-null Borel subset $J_n \subset [0, a]$, such that

$$\frac{dw_n}{d\nu}(s) + \mathcal{G}(s, z_n(\sigma_n(s)), w_n(\sigma_n(s)))\frac{d\lambda}{d\nu}(s) \in -\mathfrak{B}(\varrho_n(s), z_n(\varrho_n(s)))w_n(\varrho_n(s)) \quad \forall s \in [0, a] \setminus J_n, \tag{27}$$

with

$$w_n(\varrho_n(s)) \in D\big(\mathfrak{B}(\varrho_n(s), z_n(\varrho_n(s)))\big) \quad \forall s \in [0, a]. \tag{28}$$

Our aim next is to show the uniform convergence on $[0, a]$ of $(w_n)_n$. On the one side, one has from (5), for all $s, t \in I$ $(s \leq t)$,

$$w_n(t) = w_n(s) + \int_{]s,t]} \frac{dw_n}{d\nu}(\tau)\, d\nu(\tau),$$

whence, taking into account relation (26), one obtains

$$\|w_n(t) - w_n(s)\| \leq R_5 \nu(]s, t]), \tag{29}$$

which clearly shows that $(w_n)_n$ is equi-right-continuous with bounded variation. On the other side, involving (22) and (28), we see that for any $s \in [0, a]$, the sequence $\big(w_n(\varrho_n(s))\big)_n$ is included in the set $D([0, a] \times R\overline{B}_{\mathcal{E}}) \cap R\overline{B}_{\mathcal{E}}$, so that it is relatively compact (see (a_3)). From (29), one has for all $s \in [0, a]$, (since $\varrho_n(s) \downarrow s$)

$$\|w_n(\varrho_n(s)) - w_n(s)\| \leq R_5\, \nu(]s, \varrho_n(s)]) \to 0 \text{ as } n \to \infty, \tag{30}$$

concluding that $(w_n(s))_n$ is relatively compact as well. The application of Helly–Banach theorem (see [44]) gives us the existence of an RCBV mapping w such that $(w_n(s))_n$ converges strongly to $w(s)$ for any $s \in [0, a]$. Moreover, by (29), one gets

$$\|w(t) - w(s)\| \leq R_5\, \nu(]s, t]), \tag{31}$$

and from (30)

$$\|w_n(\varrho_n(s)) - w(s)\| \leq \|w_n(\varrho_n(s)) - w_n(s)\| + \|w_n(s) - w(s)\| \to 0 \text{ as } n \to \infty. \tag{32}$$

Regarding the convergence of $(z_n)_n$, let us define for all $s \in [0, a]$,

$$\alpha_n(s) = \sum_{j=0}^{\iota_n - 1} \frac{1}{s_{j+1}^n - s_j^n}(z_{j+1}^n - z_j^n)\chi_{]s_j^n, s_{j+1}^n]}(s),$$

and observe that from (11), $\alpha_n(s) = w_n(\sigma_n(s))$ for λ-a.e. $s \in [0, a]$. Whence, relation (15) implies that for any $s \in [0, a]$,

$$z_n(s) = z_0 + \int_{]0,s]} \alpha_n(\tau) d\lambda(\tau) = z_0 + \int_{]0,s]} w_n(\sigma_n(\tau)) d\lambda(\tau), \tag{33}$$

and from (22),

$$\|z_n(t) - z_n(s)\| \leq R|t-s| \quad \forall s, t \in [0, a]. \tag{34}$$

From (29) a second time, one obtains for all $s \in [0, a[$,

$$\|w_n(\sigma_n(s)) - w_n(s)\| \leq R_5\, \nu(]\sigma_n(s), s]) \leq R_5 \epsilon_n \to 0 \text{ as } n \to \infty. \tag{35}$$

Coming back to (22) and (35) and the strong convergence in $\mathcal{E}$ of $(w_n(s))_n$ to $w(s)$, one obtains by Lebesgue dominated convergence theorem, that for all $s \in [0, a]$,

$$\lim_{n\to\infty} z_n(s) = z_0 + \int_{]0,s]} w(\tau) d\lambda(\tau) =: z(s), \tag{36}$$

that is $(z_n(s))_n$ converges strongly to $z(s)$ for all $s \in [0, a]$, and for λ-almost every $s \in [0, a]$,

$$\frac{dz}{d\lambda}(s) = w(s), \tag{37}$$

and from (34),

$$\|z(t) - z(s)\| \leq R|t-s| \quad \forall s, t \in [0, a], \tag{38}$$

that is, z is Lipschitz-continuous on $[0, a]$.

Now, observe that $(\frac{dw_n}{d\nu})_n$ is a bounded sequence in $L^2([0, a], \mathcal{E}; \nu)$ due to relation (26), so that it converges weakly in $L^2([0, a], \mathcal{E}; \nu)$ to some mapping $u \in L^2([0, a], \mathcal{E}; \nu)$. In particular, for all $s \in [0, a]$

$$\int_{]0,s]} \frac{dw_n}{d\nu}(\tau)\, d\nu(\tau) \longrightarrow \int_{]0,s]} u(\tau)\, d\nu(\tau) \quad \text{weakly in } \mathcal{E}.$$

Since dw_n has $\frac{dw_n}{d\nu}$ as a density with respect to the measure ν, it comes that for all $s \in [0, a]$,

$$w_n(s) = w_0 + \int_{]0,s]} \frac{dw_n}{d\nu}(\tau)\, d\nu(\tau)$$

and since for any $s \in [0, a]$, $(w_n(s))_n$ converges strongly in $\mathcal{E}$ and then weakly to $w(s)$, one deduces from preceding facts that

$$w(s) = w_0 + \int_{]0,s]} u(\tau)\, d\nu(\tau) \quad \forall s \in [0, a].$$

This shows that $\frac{dw}{d\nu}(s) = u(s)$ for ν-a.e. $s \in [0, a]$. Consequently, $(\frac{dw_n}{d\nu})_n$ converges weakly in $L^2([0, a], \mathcal{E}; \nu)$ to $\frac{dw}{d\nu}$.

Part 3. We show, in this last part, that the limit couple (z, w) is the required solution.

First, observe that from (28), $w_n(\varrho_n(s)) \in D\big(\mathfrak{B}(\varrho_n(s), z_n(\varrho_n(s)))\big)$ for all $s \in [0, a]$. Also, from (a_1), we have that for all $s \in [0, a]$,

$$dis\big(\mathfrak{B}(\varrho_n(s), z_n(\varrho_n(s))), \mathfrak{B}(s, z(s))\big) \le d\vartheta(]s, \varrho_n(s)]) + \kappa \|z_n(\varrho_n(s)) - z(s)\|,$$

but it is clear that (using (34) and (9))

$$\|z_n(\varrho_n(s)) - z(s)\| \le \|z_n(\varrho_n(s)) - z_n(s)\| + \|z_n(s) - z(s)\| \le R\epsilon_n + \|z_n(s) - z(s)\|,$$

so, from (36), it comes that

$$\lim_{n\to\infty} dis\big(\mathfrak{B}(\varrho_n(s), z_n(\varrho_n(s))), \mathfrak{B}(s, z(s))\big) = 0. \tag{39}$$

On the other side, from (22) and (a_2), we see that $\big(\mathfrak{B}^0\big(\varrho_n(s), z_n(\varrho_n(s))\big)w_n(\varrho_n(s))\big)_n$ is a weakly relatively compact sequence in $\mathcal{E}$. Using this fact, (32) and (39), one deduces, thanks to Lemma 3, that $w(s) \in D(\mathfrak{B}(s, z(s)))$ for all $s \in [0, a]$.

Now, reverting to the weak convergence in $L^2([0, a], \mathcal{E}; \nu)$ of the sequence $(\frac{dw_n}{d\nu})_n$ to $\frac{dw}{d\nu}$, the application of Mazur's theorem, asserts existence of a sequence that one denotes by $(\xi_n)_n$ such that for each $n \in \mathbb{N}$, $\xi_n \in co\{\frac{dw_q}{d\nu} : q \ge n\}$, and $(\xi_n)_n$ converges strongly in $L^2([0, a], \mathcal{E}; \nu)$ to $\frac{dw}{d\nu}$. This fact guarantees the existence of a subsequence $(\xi_{n_i})_i$, such that $(\xi_{n_i}(s))_i$ converges to $\frac{dw}{d\nu}(s)$ for ν-almost every $s \in [0, a]$. That is, there is a ν-null Borel subset $J' \subset [0, a]$, and for $s \in [0, a] \setminus J'$,

$$\xi_{n_i}(s) \longrightarrow \frac{dw}{d\nu}(s) \in \bigcap_n \overline{co}\{\frac{dw_q}{d\nu} : q \ge n\}.$$

It results, by (1), that for any fixed $x^* \in \mathcal{E}$,

$$\big\langle \frac{dw}{d\nu}(s), x^* \big\rangle \le \limsup_{n\to\infty} \big\langle \frac{dw_n}{d\nu}(s), x^* \big\rangle. \tag{40}$$

Next, observe that from (a_5) and (29), for all $s \in [0, a]$, one has

$$\begin{aligned}
&\left\|\mathcal{G}\big(s, z_n(\sigma_n(s)), w_n(\sigma_n(s))\big)\frac{d\lambda}{d\nu}(s) - \mathcal{G}(s, z(s), w(s))\frac{d\lambda}{d\nu}(s)\right\| \\
&\leq k(s)\big(\|z_n(\sigma_n(s)) - z(s)\| + \|w_n(\sigma_n(s)) - w(s)\|\big)\frac{d\lambda}{d\nu}(s) \\
&\leq k(s)\big(\|z_n(\sigma_n(s)) - z(s)\| + \|w_n(\sigma_n(s)) - w_n(s)\| + \|w_n(s) - w(s)\|\big)\frac{d\lambda}{d\nu}(s) \\
&\leq k(s)\big(\|z_n(\sigma_n(s)) - z(s)\| + R_5\nu(]\sigma_n(s), s]) + \|w_n(s) - w(s)\|\big)\frac{d\lambda}{d\nu}(s),
\end{aligned}$$

whence by the strong convergence of $(w_n(s))_n$ to $w(s)$, and of $(z_n(s))_n$ to $z(s)$ and relation (7), it results that for all $s \in [0, a]$,

$$\mathcal{G}\big(s, z_n(\sigma_n(s)), w_n(\sigma_n(s))\big)\frac{d\lambda}{d\nu}(s) \longrightarrow \mathcal{G}(s, z(s), w(s))\frac{d\lambda}{d\nu}(s) \quad \text{as } n \to \infty. \tag{41}$$

Finally, to show that (z, w) is the searched solution, we are in need to use Lemma 1. Taking into account that $w(s) \in D(\mathfrak{B}(s, z(s)))$ for all $s \in [0, a]$, one has to prove that for ν-almost all fixed $s \in [0, a]$ and for any $y \in D(\mathfrak{B}(s, z(s)))$

$$\left\langle \mathfrak{B}^0(s, z(s))y + \frac{dw}{d\nu}(s) + \mathcal{G}(s, z(s), w(s))\frac{d\lambda}{d\nu}(s), y - w(s)\right\rangle \geq 0.$$

Indeed, let $s \in [0, a]$. By (a_2), using Lemma 4, we may ensure the existence of a sequence $(\zeta_n)_n$, such that

$$\zeta_n \in D\big(\mathfrak{B}(\varrho_n(s), z_n(\varrho_n(z)))\big), \quad \zeta_n \longrightarrow y \text{ and } \mathfrak{B}^0\big(\varrho_n(s), z_n(\varrho_n(s))\big)\zeta_n \longrightarrow \mathfrak{B}^0(s, z(s))y. \tag{42}$$

Since $\mathfrak{B}(\varrho_n(s), z_n(\varrho_n(s)))$ is monotone, referring to (27), one has for each $n \in \mathbb{N}$ and $s \in [0, a] \setminus J_n$,

$$\left\langle w_n(\varrho_n(s)) - \zeta_n, \frac{dw_n}{d\nu}(s) + \mathcal{G}(s, z_n(\sigma_n(s)), w_n(\sigma_n(s)))\frac{d\lambda}{d\nu}(s) + \mathfrak{B}^0(\varrho_n(s), z_n(\varrho_n(s)))\zeta_n\right\rangle \leq 0. \tag{43}$$

Next, let $s \in [0, a] \setminus \big(J \cup J' \cup (\bigcup_n J_n)\big)$. From (40), one has

$$\left\langle\frac{dw}{d\nu}(s), w(s) - y\right\rangle \leq \limsup_{n\to\infty}\left\langle\frac{dw_n}{d\nu}(s), w(s) - y\right\rangle,$$

and since from (26) and (43), one has for each $n \in \mathbb{N}$,

$$\Big\langle\frac{dw_n}{d\nu}(s), w(s)-y\Big\rangle \le \Big\langle\frac{dw_n}{d\nu}(s), w(s)-w_n(\varrho_n(s))\Big\rangle + \Big\langle\frac{dw_n}{d\nu}(s), w_n(\varrho_n(s))-\zeta_n\Big\rangle$$
$$+\Big\langle\frac{dw_n}{d\nu}(s), \zeta_n - y\Big\rangle$$
$$\le R_5\big(\|w(s)-w_n(\varrho_n(s))\| + \|\zeta_n - y\|\big) + \Big\langle\frac{dw_n}{d\nu}(s), w_n(\varrho_n(s))-\zeta_n\Big\rangle$$
$$\le R_5\big(\|w(s)-w_n(\varrho_n(s))\| + \|\zeta_n - y\|\big)$$
$$+\Big\langle\mathcal{G}\big(s, z_n(\sigma_n(s)), w_n(\sigma_n(s))\big)\frac{d\lambda}{d\nu}(s) + \mathfrak{B}^0(\varrho_n(s), z_n(\varrho_n(s)))\zeta_n, \zeta_n - w_n(\varrho_n(s))\Big\rangle$$

one concludes, using (32), (41), and (42), that

$$\Big\langle\frac{dw}{d\nu}(s), w(s)-y\Big\rangle \le \Big\langle\mathcal{G}(s, z(s), w(s))\frac{d\lambda}{d\nu}(s) + \mathfrak{B}^0(s, z(s))y, y - w(s)\Big\rangle, \quad (44)$$

which, with the fact that $w(s) \in D(\mathfrak{B}(s, z(s)))$, $z(0) = z_0$, $w(0) = w_0$, and (37), signifies that (z, w) is a solution to problem $(P_{\mathcal{G}})$ satisfying the estimates in (38) and (31).

Remark 1 If the operators that one has considered depend only on the time and satisfy:
(a_1') there exists a nondecreasing function $\vartheta : [0, a] \longrightarrow [0, +\infty[$, which is RC on $[0, a[$ with $\vartheta(0) = 0$ and $\vartheta(a) < +\infty$ such that

$$dis(\mathfrak{B}(\tau), \mathfrak{B}(s)) \le d\vartheta(]s, \tau]) = \vartheta(\tau) - \vartheta(s) \ \text{ for } \ 0 \le s \le \tau \le a.$$

(a_2') There is a positive constant c such that

$$\|\mathfrak{B}^0(s)y\| \le c(1 + \|y\|) \ \text{ for } \ s \in [0, a], \ y \in D(\mathfrak{B}(s)).$$

(a_3') For any $s \in [0, a]$ and any $r \ge 0$, $D(\mathfrak{B}(s)) \cap r\overline{B}_{\mathcal{E}}$ is relatively compact.
One has the following corollary.

Corollary 1 *Let for every $s \in [0, a]$, $\mathfrak{B}(s) : D(\mathfrak{B}(s)) \rightrightarrows \mathcal{E}$ be a m.m.op satisfying (a_1'), (a_2'), and (a_3'). Let $\mathcal{G} : [0, a] \times \mathcal{E} \times \mathcal{E} \longrightarrow \mathcal{E}$ be such that (a_4), (a_5) and (a_6) are satisfied. Then for any $(z_0, w_0) \in \mathcal{E} \times D(\mathfrak{B}(0))$, there is an RCBV map $w : [0, a] \longrightarrow \mathcal{E}$ and a Lipschitz-continuous map $z : [0, a] \to \mathcal{E}$ such that the couple (z, w) is a solution of the problem*

$$\begin{cases} z(0) = z_0, \ w(0) = w_0; \\ w(s) \in D(\mathfrak{B}(s)) \quad \forall s \in [0, a]; \\ \dfrac{dz}{d\lambda}(s) = w(s) \quad \lambda - a.e. \ s \in [0, a]; \\ -\dfrac{dw}{d\nu}(s) \in \mathfrak{B}(s)w(s) + \mathcal{G}(s, z(s), w(s))\dfrac{d\lambda}{d\nu}(s) \quad d\nu - a.e. \ s \in [0, a]. \end{cases}$$

Remark 2 As it was indicated in the introduction, the normal cone, in the sense of convex analysis, of a nonempty, closed, and convex set K, that we denote by $N_K(\cdot)$ and which is defined, for all $y_0 \in K$, by

$$N_K(y_0) = \{y^* \in \mathcal{E} : \langle y^*, z - y_0 \rangle \leq 0 \quad \forall z \in K\},$$

and $N_K(y_0) = \emptyset$ if $y_0 \notin K$ is a m.m.op containing 0, i.e., $N_K^0(y_0) = 0$. It is also clear that $D(N_K(\cdot)) = K$. Furthermore, we have that (see [51])

$$dis(N_{K_1}, N_{K_2}) = d_H(K_1, K_2)$$

for any nonempty, closed, and convex sets $K_1, K_2 \subset \mathcal{E}$, where d_H denotes the Hausdorff distance.

So that we can derive from our main theorem the following corollary, which concerns existence of RCBV solutions to (SOSP).

Indeed, let us take $\mathfrak{B}(\tau, y) = N_{\mathcal{S}(\tau,y)}$, the normal cone to the closed convex set $\mathcal{S}(\tau, y)$, and consider the following assumptions on $\mathcal{S}$:
(a_1'') there exists a nondecreasing function $\vartheta : [0, a] \longrightarrow [0, +\infty[$, which is RC on $[0, a[$ with $\vartheta(0) = 0$ and $\vartheta(a) < +\infty$ and a positive real constant κ such that

$$|d_{\mathcal{S}(\tau_1,y_1)}(z) - d_{\mathcal{S}(\tau_2,y_2)}(z)| \leq d\vartheta(]\tau_1, \tau_2]) + \kappa\|y_1 - y_2\| \text{ for } 0 \leq \tau_1 \leq \tau_2 \leq a,\ y_1, y_2, z \in \mathcal{E};$$

(a_2'') For any bounded subset A of $\mathcal{E}$ and for any $r \geq 0$, $\mathcal{S}([0, a] \times A) \cap r\overline{B}_{\mathcal{E}}$ is relatively compact.

Corollary 2 *Let $\mathcal{S} : [0, a] \times \mathcal{E} \rightrightarrows \mathcal{E}$ be a nonempty, convex, and closed valued multi-mapping, verifying (a_1'') and (a_2''). Let $\mathcal{G} : [0, a] \times \mathcal{E} \times \mathcal{E} \longrightarrow \mathcal{E}$ be such that (a_4), (a_5), and (a_6) are satisfied. Then for any $(z_0, w_0) \in \mathcal{E} \times \mathcal{S}(0, z_0)$, there is an RCBV map $w : [0, a] \to \mathcal{E}$ and a Lipschitz-continuous map $z : [0, a] \to \mathcal{E}$ such that the couple (z, w) is solution of the measure problem*

$$(\mathcal{SP}) \begin{cases} z(0) = z_0,\ w(0) = w_0; \\ w(\tau) \in \mathcal{S}(\tau, z(\tau)) \quad \forall \tau \in [0, a]; \\ \dfrac{dz}{d\lambda}(\tau) = w(\tau) \quad \lambda - a.e.\ \tau \in [0, a]; \\ -\dfrac{dw}{d\nu}(\tau) \in N_{\mathcal{S}(\tau,z(\tau))}(w(\tau)) + \mathcal{G}(\tau, z(\tau), w(\tau))\dfrac{d\lambda}{d\nu}(\tau) \quad d\nu - a.e.\ \tau \in [0, a]. \end{cases}$$

Remark 3 It is worth to mention that this last corollary is a particular case of the results obtained in [4, 6, 41] for the RCBV second-order sweeping process, since the sets taken into references [4, 41] are r-prox regulars (the class of prox regular sets contains strictly the class of convex sets), and in [6] the sets are in the wider class of subsmooth sets. Moreover, the perturbations considered there are set-valued.

Competing Interests: The author has no conflicts of interest to declare that are relevant to the content of this chapter.

References

1. S. Adly, T. Haddad, L. Thibault, Convex sweeping process in the framework of measure differential inclusions and evolution variational inequalities. Math. Program, **148** (1-2, Ser. B), 5–47 (2014)
2. S. Adly, B.K. Le, Unbounded second-order state-dependent Moreau's sweeping processes in Hilbert spaces. J. Optim. Theory Appl. **169**, 407–423 (2016)
3. S. Adly, B.K. Le, Unbounded state-dependent sweeping process with perturbations in uniformly convex and q-uniformly smooth Banach spaces. Numer. Algebra Control Optim. **8**, 81–95 (2018)
4. S. Adly, F. Nacry, An existence result for discontinuous second-order nonconvex state-dependent sweeping process. Appl. Math. Optim. **79**, 515–546 (2019)
5. S. Adly, F. Nacry, L. Thibault, Discontinuous sweeping process with prox-regular sets, ESAIM: control. Optim. Calc. Var. **23**(4), 1293–1329 (2017)
6. M. Aissous, F. Nacry, V.A.T. Nguyen, First and second order state-dependent bounded subsmooth sweeping processes. Linear Nonlinear Anal. **6**(3), 447–472 (2020)
7. F. Aliouane, D. Azzam-Laouir, A second order differential inclusion with proximal normal cone in Banach spaces. Topol. Methods Nonlinear Anal. **44**(1), 143–160 (2014)
8. F. Aliouane, D. Azzam-Laouir, Second-order sweeping process with a Lipschitz perturbation. J. Math. Anal. Appl. **452**, 729–746 (2017)
9. F. Aliouane, D. Azzam-Laouir, C. Castaing, M.D.P. Monteiro Marques, Second-order time and state sweeping process in Hilbert space. J. Optim. Theory Appl. **182**, 153–188 (2019)
10. D. Azzam-Laouir, A measure differential inclusion with time-dependent maximal monotone operators. Topol. Methods Nonlinear Anal. **59**, 941–956 (2022)
11. D. Azzam-Laouir, Mixed semicontinuous perturbation of a second order nonconvex sweeping process. Electron. J. Qual. Theory of Differ. Equ. **37**, 1–9 (2008)
12. D. Azzam-Laouir, S. Izza, Existence of solutions for second-order perturbed nonconvex sweeping process. Comput. Math. Appl. **62**, 1736–1744 (2011)
13. D. Azzam-Laouir, W. Belhoula, C. Castaing, M.D.P. Monteiro Marques, Perturbed evolution problems with absolutely continuous variation in time and applications, J. Fixed Point Theory Appl. **21**, 40 (2019)
14. D. Azzam-Laouir, W. Belhoula, C. Castaing, M.D.P. Monteiro Marques, Multi-valued perturbation to evolution problems involving time dependent maximal monotone operators. Evol. Equ. Control Theory **9**(1), 219–254 (2020)
15. D. Azzam-Laouir, I. Boutana-Harid, Mixed semicontinuous perturbation to an evolution problem with time-dependent maximal monotone operator. J. Nonlinear Convex Anal. **20**(1), 39–52 (2018)
16. D. Azzam-Laouir, C. Castaing, M.D.P. Monteiro Marques, Perturbed evolution problems with continuous bounded variation in time and applications. Set-Valued Var. Anal. **26**(3), 693–728 (2018)
17. D. Azzam-Laouir, C. Castaing, M.D.P. Monteiro Marques, BV right-continuous solutions of differential inclusions involving time dependent maximal monotone operators. ArXiv: 2103.01113v1, 1–43 (2021)
18. V. Barbu, *Nonlinear Semigroups and Differential Equations in Banach Spaces* (Noordhoff International Publishing Leyden, 1976)
19. H. Benabdellah, C. Castaing, M.A. Gamal Ibrahim, *BV Solutions of Multivalued Differential Equations on Closed Moving Sets in Banach Spaces, Banach Center Publications*, vol. 32 (Institute of Mathematics, Polish academy of Sciences, Warszawa, 1995)
20. H. Benabdellah, C. Castaing, A. Salvadori, A. Syam, Nonconvex sweeping process. J. Appl. Anal. **2**(2), 17–40 (1996)
21. M. Benguessoum, D. Azzam-Laouir, On first and second-order perturbed differential inclusions governed my maximal monotone operators. Accepted for publication in Topol. Methods Nonlinear Anal. (2024)

22. M. Benguessoum, D. Azzam-Laouir, C. Castaing, On a time and state dependent maximal monotone operator coupled with a sweeping process with perturbations. Set-Valued Var. Anal. **29**, 191–219 (2021)
23. M. Bounkhel, Existence results for second order convex sweeping processes in p-uniformly smooth and q-uniformly convex Banach spaces. Electron. J. Qual. Theory of Differ. Equ. **27**, 1–10 (2012)
24. M. Bounkhel, D. Laouir-Azzam, Existence results for second-order nonconvex sweeping processes. Set-Valued Var. Anal. **12**(3), 291–318 (2004)
25. H. Brézis, *Opérateurs Maximaux Monotones* (North Holland Publishing Company, Amsterdam-London, 1973)
26. K. Camlibel, L. Ianelli, A. Tanwani, Convergence of proximal solutions for evolution inclusions with time-dependent maximal monotone operators. Math. Program. Ser. A **194**(1–2), 1017–1059 (2022)
27. C. Castaing, Quelques problèmes d'évolution du second ordre. Sem. Anal. Convexe. Montpellier, Exposé 5 (1988)
28. C. Castaing, A.G. Ibrahim, M.F. Yarou, Some contributions to nonconvex sweeping process. J. Nonlinear Convex Anal. **10**(1), 1–20 (2009)
29. C. Castaing, M.D.P. Monteiro Marques, Evolution problems associated with nonconvex closed moving sets with bounded variation. Portugal. Math., **53**, 73–87 (1996)
30. C. Castaing, M.D.P. Monteiro Marques, P. Raynaud de Fitte, Second order evolution problems with time-dependent maximal monotone operator and applications. Adv. Math. Econ. **22**, 25–77 (2018)
31. J.F. Edmond, L. Thibault, BV solutions of non convex sweeping process differential inclusion with perturbation. J. Diff. Equ. **22**, 135–179 (2006)
32. N. Kenmochi, Solvability of nonlinear evolution equations with time-dependent constraints and applications. Bull. Fac. Educ. Chiba Univ. **30**, 1–81 (1981)
33. M. Kunze, M.D.P. Monteiro Marques, BV solutions to evolution problems with time-dependent domains. Set-Valued Anal. **5**, 57–72 (1997)
34. B.K. Le, Well-posedness and nonsmooth Lyapunov pairs for state-dependent maximal monotone differential inclusions. Optimization **69**(6), 1187–1217 (2020)
35. M.D.P. Monteiro Marques, Differential inclusions nonsmooth mechanical problems, shocks and dry friction. Prog. Nonlinear Differ. Equ. Appl. Birkhauser **9** (1993)
36. J.J. Moreau, Rafle par un convexe variable I. Sem. Anal. Convexe Montpellier, Exposé 15 (1971)
37. J.J. Moreau, Evolution problem associated with a moving convex set in a Hilbert space. J. Differiential. Equ. **26**, 347–374 (1977)
38. J.J. Moreau, Bounded variation in time, in topics in non-smooth mechanics, in *Birkhäuser Verlag*, ed. by J.J. Moreau, P.D. Panagiotopoulos, G. Strang (Basel-Boston-Berlin, 1988)
39. J.J. Moreau, Sur les mesures différentielles des fonctions vectorielles à variation bornée. Sem. Anal. Convexe. Montpellier, Exposé 17 (1975)
40. J.J. Moreau, M. Valadier, A chain rule involving vector functions of bounded variations. J. Funct. Anal. **74**(2), 333–345 (1987)
41. F. Nacry, J. Noel, L. Thibault, On first and second order state-dependent prox-regular sweeping process. Pure Appl. Funct. Anal. **6**(6), 1453–1493 (2021)
42. J. Noel, Second-order general perturbed sweeping process differential inclusion. J. Fixed Point Theory Appl. **20**, 133 (2018)
43. N.H. Pavel, Nonlinear evolution operators and semigroups. Lecture Notes in Mathematics, vol. 1260. (Springer, New York, 1987)
44. J.E. Porter, Helly selection principle for functions of bounded P-variation. Rocky Mt. J. Math. **35**(2), 675–679 (2005)
45. S. Saidi, On a second-order functional evolution problem with time and state dependent maximal monotone operators. Evol. Equ. Control Theory **11**(4), 1001–1035 (2022)
46. M. Schatzman, A class of nonlinear differential equations of second order in time. Nonlinear Anal. Theory Methods Appl. **2**(3), 355–373 (1978)

47. F. Selamnia, D. Azzam-Laouir, M.D.P. Monteiro Marques, Evolution problems involving state-dependent maximal monotone operators. Appl. Anal. **10**(1), 297–313 (2022)
48. L. Thibault, Moreau sweeping process with bounded retraction. J. Convex Anal. **23**(4), 1051–1098 (2016)
49. A.A. Tolstonogov, BV solutions of a convex sweeping process with a composed perturbation. Evol. Equ. Control Theory **11**(2), 537–557 (2022)
50. A.A. Tolstonogov, BV continuous solutions of an evolution inclusion with maximal monotone operator and non-convex valued perturbation, Existence theorem. Set-Valued Var. Anal. **29**, 29–60 (2021)
51. A.A. Vladimirov, Nonstationnary dissipative evolution equation in Hilbert space. Nonlinear Anal. **17**, 499–518 (1991)
52. I.I. Vrabie, Compactness methods for nonlinear evolution equations, in *Pitman Monographs and Surveys in Pure and Applied mathematics, Longman Scientific and Technical*, vol. 32 (John Wiley and Sons, Inc. New York, 1987)

Certain Integrals Involving Generalized Laguerre Polynomials and Bessel Function and Their Application

Abdelmajid Belafhal, El Mostafa El Halba, and Talha Usman

Abstract The principal aim of the present work is to evaluate a new class of integrals involving the product of generalized Laguerre polynomials $L_n^{(\alpha)}$ and Bessel functions J_ν which deduce some new results in terms of Kummer confluent hypergeometric function. Further, we discuss some special cases of our main results and deduce some well-known results. Some particular cases of these integrals involving Hermite and Laguerre polynomials are also derived. Finally, the results obtained are applied in the problem of propagation of a Laguerre–Bessel–Gaussian beam through a spiral phase plate.

1 Introduction and Preliminaries

Earlier, several integrals involving the product of associated Laguerre polynomials and Bessel function with an argument and an exponential function $e^{-\sigma x^2}$ are evaluated [1, 17]. The propagation of laser beams through a turbulent atmosphere or oceanic turbulence or an ABCD optical system is very important [3–11] plays very important role in laser physics and linear or non-linear optics. Recently, some Schell-model beams propagating through paraxial optical systems in turbulent atmosphere have been studied [12, 15]. Many theories presented in these papers require the evaluation of infinite integrals involving the product of generalized Laguerre polynomials (GLP) and Bessel functions which has seven parameters and is a particular Laplace-type integral. In this work, we will evaluate two integrals $I_{n,\alpha}^{\nu}$ and $I_{n,\alpha}^{\nu,\chi}$ used in the above fields and investigate some special cases.

The following definitions are essential for the present investigation: The classical Gauss's hypergeometric function is defined by

A. Belafhal · E. M. El Halba
LPNAMME, Laser Physics Group, Department of Physics, Faculty of Sciences, Chouaïb Doukkali University, P. B 20, 24000 El Jadida, Morocco

T. Usman (✉)
Preparatory Studies Center, Mathematics and Computing Skills Unit, University of Technology and Applied Sciences, P.O. Box: 484, 411 Sur, Oman
e-mail: talhausman.maths@gmail.com

M. Kharrat et al. (eds.), *Modeling of Discrete and Continuous Systems*, Forum for Interdisciplinary Mathematics, https://doi.org/10.1007/978-981-97-8715-9_13

$$ {}_2F_1\left[\begin{matrix} a, b; \\ c; \end{matrix} x\right] = \sum_{n=0}^{\infty} \frac{(a)_n (b)_n}{(c)_n} \frac{x^n}{n!} = {}_2F_1(a, b; c; x), \tag{1} $$

where $(a)_n$ $(a \in \mathbb{C})$ is the well-known Pochhammer symbol [2, 16].

The Kummer confluent hypergeometric function ${}_1F_1$ [2, 16] is defined by the series representation

$$ {}_1F_1(a; b; x) = \sum_{k=0}^{\infty} \frac{(a)_k}{(b)_k} \frac{x^k}{k!}. \tag{2} $$

The Bessel function of the first kind of order ν [14, 18] is defined by

$$ J_\nu(x) = \sum_{k=0}^{\infty} \frac{(-1)^k \left(\frac{x}{2}\right)^{\nu+2k}}{k!\Gamma(\nu+k+1)}; \quad \forall x \in \mathbb{C} \setminus (-\infty, 0). \tag{3} $$

It is well known that

$$ J_{\frac{1}{2}}(x) = \sqrt{\frac{2}{\pi x}} \sin x $$

and

$$ J_{\frac{-1}{2}}(x) = \sqrt{\frac{2}{\pi x}} \cos x. $$

The generalized Laguerre polynomials $L_n^{(\alpha)}(x)$ [16, 19] are defined as

$$ L_n^{(\alpha)}(x) = \frac{(1+\alpha)_n}{n!} {}_1F_1(-n; 1+\alpha; x). \tag{4} $$

From (4), it follows that

$$ L_n^{(\alpha)}(x) = \sum_{k=0}^{n} \frac{(-1)^k (1+\alpha)_n x^n}{k!(n-k)!(1+\alpha)_k}. \tag{5} $$

We know that Hermite polynomials $H_n(x)$ are a special case of GLP given by

$$ H_{2n}(x) = (-1)^n 2^{2n} n! L_n^{\frac{-1}{2}}(x^2) \tag{6} $$

and

$$ H_{2n+1}(x) = (-1)^n 2^{2n+1} n! x L_n^{\frac{1}{2}}(x^2), \tag{7} $$

where $H_n(x)$ are the Hermite polynomials [16].

The GLP have the following relationship with the Whittaker functions $M_{k,\mu}(x)$ and $W_{k,\mu}(x)$:

$$ L_n^{(\alpha)}(x) = \frac{(1+\alpha)_n}{n!} x^{\frac{1}{2}-\frac{\alpha}{2}} e^{\frac{x}{2}} M_{n+\frac{1}{2}+\frac{\alpha}{2}, \frac{\alpha}{2}}(x) \tag{8} $$

and

$$L_n^{(\alpha)}(x) = \frac{(-1)^n}{n!} x^{\frac{1}{2}-\frac{\alpha}{2}} e^{\frac{x}{2}} W_{n+\frac{1}{2}+\frac{\alpha}{2}, \frac{\alpha}{2}}(x), \tag{9}$$

where $M_{k,\mu}(x)$ and $W_{k,\mu}(x)$ [20, 21] are given, respectively, by

$$M_{k,\mu}(x) = x^{\mu+\frac{1}{2}} e^{-\frac{x}{2}} {}_1F_1\left(\frac{1}{2} + \mu - k,\ 2\mu + 1;\ x\right) \tag{10}$$

and

$$W_{k,\mu}(x) = x^{\mu+\frac{1}{2}} e^{-\frac{x}{2}}\, U\left(\frac{1}{2} + \mu - k,\ 2\mu + 1;\ x\right). \tag{11}$$

2 Evaluation of $I_{n,\alpha}^{\nu}$

Here, we evaluate certain integrals involving the product of generalized Laguerre polynomials (GLP) and Bessel function which are expressed in terms of Kummer confluent hypergeometric function as given in Theorem 1.

Theorem 1 *The undermentioned formula holds true:*

$$\begin{aligned}\int_0^{\infty} x^{2\gamma+1} e^{-\sigma x^2} L_n^{(\alpha)}(\beta x^2) J_\nu(\mu x) dx &= \frac{\mu^\nu \Gamma(n+\alpha+1)}{2^{\nu+1}\sigma^{\left(\frac{\nu}{2}+\gamma+1\right)}} \\ \times e^{-\frac{\mu^2}{4\sigma}} \sum_{k=0}^{n} \frac{\Gamma(\gamma+k+1-\frac{\nu}{2})(\frac{-\beta}{\sigma})^k}{k!\Gamma(n-k+1)\Gamma(\alpha+k+1)} & L_{k+\gamma-\frac{\nu}{2}}^{(\nu)}\left(\frac{\mu^2}{4\sigma}\right).\end{aligned} \tag{12}$$

The condition of the validity of the above result is $\Re(\sigma) > 0$ *and* $\Re(2\gamma + \nu + 1) > -1$, *and is well verified in our study.*

Proof In order to derive (12), we use the following identity:

$$\int_0^{\infty} x^{\mu'} e^{-\sigma x^2} J_\nu(\mu x) dx = \frac{\mu^\nu \Gamma\left(\frac{\mu'+\nu+1}{2}\right)}{2^{\nu+1}\sigma^{\left(\frac{\mu'+\nu+1}{2}\right)}\nu!} {}_1F_1\left(\frac{\mu'+\nu+1}{2}; \nu+1; -\frac{\mu^2}{4\sigma}\right), \tag{13}$$

with $\Re(\sigma) > 0$, $\Re(\mu' + \nu) > -1$, $\Gamma(.)$ is the gamma function [16] and the GLP is defined by

$$L_n^{(\alpha)}(\beta x^2) = \sum_{k=0}^{n} \frac{\Gamma(n+\alpha+1)(-\beta x^2)^k}{k!\Gamma(n-k+1)\Gamma(\alpha+k+1)}. \tag{14}$$

So, the integral $I_{n,\alpha}^{\nu}$ becomes

$$I_{n,\alpha}^{\nu}=\sum_{k=0}^{n}\frac{\Gamma(n+\alpha+1)(-\beta)^k}{k!\Gamma(n-k+1)\Gamma(\alpha+k+1)}I_k, \tag{15}$$

where I_k is given by

$$I_k=\frac{\mu^{\nu}\Gamma\left(\frac{\nu}{2}+\gamma+k+1\right)}{2^{\nu+1}\sigma^{\left(\frac{\nu}{2}+\gamma+k+1\right)}\nu!}\,{}_1F_1\left(\frac{\nu}{2}+\gamma+k+1;\nu+1;-\frac{\mu^2}{4\sigma}\right). \tag{16}$$

By using the following transformation formula [14]:

$${}_1F_1(a;b;z)=e^{z}\,{}_1F_1(b-a;b;-z), \tag{17}$$

the integral $I_{n,\alpha}^{\nu}$ (15) becomes

$$\int_0^{\infty}x^{2\gamma+1}e^{-\sigma x^2}L_n^{(\alpha)}(\beta x^2)J_{\nu}(\mu x)dx=\frac{\mu^{\nu}\Gamma\left(n+\alpha+1\right)}{2^{\nu+1}\sigma^{\left(\frac{\nu}{2}+\gamma+1\right)}\nu!}e^{-\frac{\mu^2}{4\sigma}}$$

$$\times\sum_{k=0}^{n}\frac{\Gamma(\frac{\nu}{2}+\gamma+k+1)(\frac{-\beta}{\sigma})^k}{k!\Gamma(n-k+1)\Gamma(\alpha+k+1)}\,{}_1F_1\left(\frac{\nu}{2}-\gamma-k;\nu+1;\frac{\mu^2}{4\sigma}\right). \tag{18}$$

Note that (18) containing a terminating series representation of the Kummer function given by

$${}_1F_1\left(\frac{\nu}{2}-\gamma-k;\nu+1;\frac{\mu^2}{4\sigma}\right)=\sum_{m=0}^{\infty}\frac{[-(\gamma+k-\frac{\nu}{2})]_m}{(\nu+1)_m}\frac{\left(\frac{\mu^2}{4\sigma}\right)^m}{m!}. \tag{19}$$

Now with the help of the relation

$$(-\alpha)_m=\begin{cases}(-1)^m\frac{\alpha!}{(\alpha-m)!}, & 0\le m\le\alpha\\ 0, & m>\alpha,\end{cases} \tag{20}$$

(19) can be rewritten as

$${}_1F_1\left(-(\gamma+k-\frac{\nu}{2});\nu+1;\frac{\mu^2}{4\sigma}\right)=\sum_{m=0}^{\gamma+k-\frac{\nu}{2}}\frac{(-1)^m(\gamma+k-\frac{\nu}{2})!}{(\gamma+k-\frac{\nu}{2}-m)!(\nu+1)_m}\frac{\left(\frac{\mu^2}{4\sigma}\right)^m}{m!}. \tag{21}$$

Thus, if a numerator parameter is a zero or a negative integer, the Kummer series terminates in view of the identity (20), and by using the relation between the Kummer function and the GLP in (4) and in view of (18), the required result follows.

Here, we consider some special cases in view of Theorem 1. We are interested here to evaluate the integral involving the product of Hermite polynomials with Bessel functions of the first kind of arbitrary order given under.

Corollary 1 *The undermentioned formula holds true:*

$$\int_0^\infty x^{2\gamma+1} e^{-\sigma x^2} H_{2n}(\sqrt{\beta}x) J_\nu(\mu x) dx = \frac{(-1)^n n! \mu^\nu \Gamma\left(n+\frac{1}{2}\right)}{2^{\nu+1-2n} \sigma^{\left(\frac{\nu}{2}+\gamma+1\right)}} \times e^{-\frac{\mu^2}{4\sigma}} \sum_{k=0}^{n} \frac{\Gamma(\gamma+k+1-\frac{\nu}{2})(\frac{-\beta}{\sigma})^k}{k!\Gamma(n-k+1)\Gamma(k+\frac{1}{2})} L_{k+\gamma-\frac{\nu}{2}}^{(\nu)}\left(\frac{\mu^2}{4\sigma}\right), \tag{22}$$

and

$$\int_0^\infty x^{2\gamma} e^{-\sigma x^2} H_{2n+1}(\sqrt{\beta}x) J_\nu(\mu x) dx = \frac{(-1)^n n! \sqrt{\beta} \mu^\nu \Gamma\left(n+\frac{3}{2}\right)}{2^{\nu-2n} \sigma^{\left(\frac{\nu}{2}+\gamma+1\right)}} \times e^{-\frac{\mu^2}{4\sigma}} \sum_{k=0}^{n} \frac{\Gamma(\gamma+k-\frac{\nu}{2})(\frac{-\beta}{\sigma})^k}{k!\Gamma(n-k+1)\Gamma(k+\frac{3}{2})} L_{k+\gamma-\frac{\nu}{2}}^{(\nu)}\left(\frac{\mu^2}{4\sigma}\right). \tag{23}$$

Using the relations in (6), (7) *and in view of* (12), *the result in* (22) *and* (23) *can easily be obtained after some simplifications.*

Corollary 2 *The undermentioned formula holds true:*

$$\int_0^\infty x^{2\gamma-\alpha+2} e^{(\frac{\beta}{2}-\sigma)x^2} M_{n+\frac{\alpha+1}{2},\mu}(\beta x^2)\, J_\nu(\mu x) dx = \frac{n! \mu^\nu \Gamma(n+\alpha+1)}{(1+\alpha)_n \beta^{\frac{1-\alpha}{2}} 2^{\nu+1} \sigma^{\left(\frac{\nu}{2}+\gamma+1\right)}} \times e^{-\frac{\mu^2}{4\sigma}} \sum_{k=0}^{n} \frac{\Gamma(\gamma+k+1-\frac{\nu}{2})(\frac{-\beta}{\sigma})^k}{k!\Gamma(n-k+1)\Gamma(k+\alpha+1)} L_{k+\gamma-\frac{\nu}{2}}^{(\nu)}\left(\frac{\mu^2}{4\sigma}\right), \tag{24}$$

and

$$\int_0^\infty x^{2\gamma-\alpha+2} e^{(\frac{\beta}{2}-\sigma)x^2} W_{n+\frac{\alpha+1}{2},\mu}(\beta x^2)\, J_\nu(\mu x) dx = \frac{n! \mu^\nu \Gamma(n+\alpha+1)}{(-1)^n \beta^{\frac{1-\alpha}{2}} 2^{\nu+1} \sigma^{\left(\frac{\nu}{2}+\gamma+1\right)}} \times e^{-\frac{\mu^2}{4\sigma}} \sum_{k=0}^{n} \frac{\Gamma(\gamma+k+1-\frac{\nu}{2})(\frac{-\beta}{\sigma})^k}{k!\Gamma(n-k+1)\Gamma(k+\alpha+1)} L_{k+\gamma-\frac{\nu}{2}}^{(\nu)}\left(\frac{\mu^2}{4\sigma}\right). \tag{25}$$

Using the relations in (8), (9) *and in view of* (12), *the result in* (24) *and* (25) *can easily be obtained after some simplifications.*

Corollary 3 *If we take* $\nu = 0$ *in (12), we arrive at the undermentioned formula:*

$$\int_0^\infty x^{2\gamma+1} e^{-\sigma x^2} L_n^{(\alpha)}(\beta x^2) J_0(\mu x) dx = \frac{\Gamma(n+\alpha+1)}{2\sigma^{(\gamma+1)}} \times e^{-\frac{\mu^2}{4\sigma}} \sum_{k=0}^{n} \frac{\Gamma(\gamma+k+1)(\frac{-\beta}{\sigma})^k}{k!\Gamma(n-k+1)\Gamma(\alpha+k+1)} L_{k+\gamma}\left(\frac{\mu^2}{4\sigma}\right), \tag{26}$$

where $\Re(\sigma) > 0$, $\Re(2\gamma + 1) > -1$ *and* $L_{k+\gamma}$ *is the Laguerre polynomial of degree* $k + \gamma$.

Corollary 4 *If we take* $\nu = \alpha = 0$ *in (12), we obtain the undermentioned formula:*

$$\int_0^\infty x^{2\gamma+1} e^{-\sigma x^2} L_n(\beta x^2) J_0(\mu x) dx = \frac{n!}{2\sigma^{(\gamma+1)}} \times e^{-\frac{\mu^2}{4\sigma}} \sum_{k=0}^{n} \frac{\Gamma(\gamma + k + 1)(\frac{-\beta}{\sigma})^k}{\Gamma(n - k + 1)(k!)^2} L_{k+\gamma}\left(\frac{\mu^2}{4\sigma}\right), \tag{27}$$

where $\Re(\sigma) > 0$ *and* $\Re(2\gamma + 1) > -1$.

Corollary 5 *If we place* $\gamma = \alpha = \nu = 0$ *in (4), we get*

$$\int_0^\infty x e^{-\sigma x^2} L_n(\beta x^2) J_0(\mu x) dx = \frac{n!}{2\sigma} e^{-\frac{\mu^2}{4\sigma}} \sum_{k=0}^{n} \frac{(\frac{-\beta}{\sigma})^k}{(n - k)!k!} L_k\left(\frac{\mu^2}{4\sigma}\right). \tag{28}$$

Corollary 6 *The undermentioned integral formula holds true:*

$$\int_0^\infty x e^{-\sigma x^2} L_n(\beta x^2) J_0(\mu x) dx = \frac{1}{2\sigma}\left(1 - \frac{\beta}{\sigma}\right)^n e^{-\frac{\mu^2}{4\sigma}} L_n\left[\frac{\beta\mu^2}{4\sigma(\beta - \sigma)}\right]. \tag{29}$$

It should be noted that by using the following identity [19]:

$${}_1F_1\left(-\rho; \alpha + 1; \frac{sz}{1+s}\right) = \frac{1}{(1+s)^\rho} \sum_{m=0}^{\infty} \frac{(-s)^m(-\rho)_m}{(\alpha + 1)_m} L_m^{(\alpha)}(z), \tag{30}$$

and (4), one finds that (29) is identical to Eq. 6.651 of [14], which is an important result used in the study of Laguerre–Gaussian beams.

3 Evaluation of $I_{n,\alpha}^{\nu,\chi}$

Here, we evaluate certain integrals involving the GLP and the product of two Bessel functions by two different methods given, respectively, by Theorems 2 and 3. We show that this other integral plays an important role in laser physics.

Theorem 2 *The undermentioned formula holds true:*

$$\int_0^\infty x^{2\gamma+1}e^{-\sigma x^2}L_n^{(\alpha)}(\beta x^2)J_\nu(\mu x)J_\chi(\lambda x)dx$$
$$=\frac{\mu^\nu\lambda^\chi}{2^{\nu+\chi+1}\sigma^{\left(\frac{\nu+\chi}{2}+\gamma+1\right)}\nu!}\sum_{k=0}^{n}\frac{\Gamma(n+\alpha+1)(\frac{-\beta}{\sigma})^k}{k!\Gamma(n-k+1)\Gamma(\alpha+k+1)}$$
$$\times\sum_{m=0}^{+\infty}\frac{\Gamma(m+\gamma+k+\frac{\nu+\chi}{2}+1)(\frac{-\lambda^2}{4\sigma})^m}{\Gamma(m+\chi+1)m!}{}_2F_1\left(-m,-\chi-m;\nu+1;\frac{\mu^2}{\lambda^2}\right). \tag{31}$$

Proof This integral can be evaluated by using the relation in (see [14, Eq. 6.633]). We have indeed

$$\int_0^\infty x^{2\gamma'+1}e^{-\sigma x^2}J_\nu(\mu x)J_\chi(\lambda x)dx=\frac{\mu^\nu\lambda^\chi}{2^{\nu+\chi+1}\sigma^{\left(\frac{\nu+\chi}{2}+\gamma'+1\right)}\nu!}$$
$$\times\sum_{m=0}^{+\infty}\frac{\Gamma(m+\frac{\nu+\chi}{2}+\gamma'+1)(\frac{-\lambda^2}{4\sigma})^m}{m!\Gamma(m+\chi+1)}{}_2F_1\left(-m,-\chi-m;\nu+1;\frac{\mu^2}{\lambda^2}\right),$$

with $\Re(\sigma)>0$ and $\Re(2\gamma'+\chi+\nu)>-2$; $\lambda>0$ and $\mu>0$ and the expansion of GLP is given by (14).

Corollary 7 *On assigning $\nu=\chi=0$ in (31), we obtain*

$$\int_0^\infty x^{2\gamma+1}e^{-\sigma x^2}L_n^{(\alpha)}(\beta x^2)J_0(\mu x)J_0(\lambda x)dx=\frac{1}{2\sigma^{(\gamma+1)}}$$
$$\times\sum_{k=0}^{n}\frac{\Gamma(n+\alpha+1)(\frac{-\beta}{\sigma})^k}{k!\Gamma(n-k+1)\Gamma(\alpha+k+1)}\sum_{m=0}^{+\infty}(m+1)_{\gamma+k}\frac{(\frac{-\lambda^2}{4\sigma})^m}{m!}{}_2F_1\left(-m,-m;1;\frac{\mu^2}{\lambda^2}\right), \tag{32}$$

where $\Re(\sigma)>0$, $\lambda>0$, and $\mu>0$.

Corollary 8 *On assigning $\alpha=\nu=\chi=0$ in (31), we arrive at the undermentioned formula:*

$$\int_0^\infty x^{2\gamma+1}e^{-\sigma x^2}L_n(\beta x^2)J_0(\mu x)J_0(\lambda x)dx=\frac{1}{2\sigma^{(\gamma+1)}}$$
$$\times\sum_{k=0}^{n}\frac{n!(\frac{-\beta}{\sigma})^k}{(k!)^2\Gamma(n-k+1)}\sum_{m=0}^{+\infty}(m+1)_{\gamma+k}\frac{(\frac{-\lambda^2}{4\sigma})^m}{m!}{}_2F_1\left(-m,-m;1;\frac{\mu^2}{\lambda^2}\right), \tag{33}$$

where $\Re(\sigma)>0$, $\lambda>0$, and $\mu>0$.

Corollary 9 *For $\gamma=\alpha=\nu=\chi=0$ in (31), we get the undermentioned formula:*

$$\int_0^{\infty} xe^{-\sigma x^2} L_n(\beta x^2) J_0(\mu x) J_0(\lambda x) dx = \frac{1}{2\sigma} \times \sum_{k=0}^{n} \frac{n!(\frac{-\beta}{\sigma})^k}{(k!)^2 \Gamma(n-k+1)} \sum_{m=0}^{+\infty} (m+1)_k \frac{(\frac{-\lambda^2}{4\sigma})^m}{m!} {}_2F_1\left(-m, -m; 1; \frac{\mu^2}{\lambda^2}\right), \tag{34}$$

where $\Re(\sigma) > 0$, $\lambda > 0$, *and* $\mu > 0$.

Now, we evaluate the integral $I_{n,\alpha}^{\nu,\chi}$ with the help of the expansion of Bessel function given in (3) in the form:

$$J_\chi(\lambda x) = \sum_{m=0}^{+\infty} \frac{(-1)^m (\lambda x/2)^{\chi+2m}}{m!\Gamma(\chi+m+1)}. \tag{35}$$

Theorem 3 *The undermentioned formula holds true:*

$$\int_0^{\infty} x^{2\gamma+1} e^{-\sigma x^2} L_n^{(\alpha)}(\beta x^2) J_\nu(\mu x) J_\chi(\lambda x) dx = \frac{\mu^\nu \lambda^\chi}{2^{\nu+\chi+1} \sigma^{(\frac{\nu+\chi}{2}+\gamma+1)}} \times e^{-\mu^2/4\sigma} \sum_{k=0}^{n} \frac{\Gamma(n+\alpha+1)(-\beta/\sigma)^k}{k!\Gamma(n-k+1)\Gamma(\alpha+k+1)} \times \sum_{m=0}^{+\infty} \frac{\Gamma(m+\gamma+k+\frac{\chi-\nu}{2}+1)(-\lambda^2/4\sigma)^m}{m!\Gamma(m+\chi+1)} L_{m+\gamma+k+\frac{\chi-\nu}{2}}^{(\nu)}\left(\mu^2/4\sigma\right). \tag{36}$$

It is equally easy to write the formula

$$I_{n,\alpha}^{\nu,\chi} = \sum_{m=0}^{+\infty} \frac{(-1)^m (\lambda/2)^{\chi+2m}}{m!\Gamma(\chi+m+1)} \int_0^{\infty} x^{2\gamma+2m+\chi+1} e^{-\sigma x^2} L_n^{(\alpha)}(\beta x^2) J_\nu(\mu x) dx, \tag{37}$$

and by using (12) and (35), one obtains

$$\int_0^{\infty} x^{2\gamma+1} e^{-\sigma x^2} L_n^{(\alpha)}(\beta x^2) J_\nu(\mu x) J_\chi(\lambda x) dx = \frac{\mu^\nu \lambda^\chi}{2^{\nu+\chi+1} \sigma^{(\frac{\nu+\chi}{2}+\gamma+1)}} \times e^{-\lambda^2/4\sigma} \sum_{k=0}^{n} \frac{\Gamma(n+\alpha+1)(-\beta/\sigma)^k}{k!\Gamma(n-k+1)\Gamma(\alpha+k+1)} \times \sum_{m=0}^{+\infty} \frac{\Gamma(m+\gamma+k+\frac{\nu-\chi}{2}+1)(-\mu^2/4\sigma)^m}{m!\Gamma(m+\chi+1)} L_{m+\gamma+k+\frac{\nu-\chi}{2}}^{(\chi)}\left(\lambda^2/4\sigma\right). \tag{38}$$

Now, we see that (31) and (36) are equivalent and we can find the second equation from the first by interchanging μ *by* λ, χ *by* ν, *and* $\chi - \nu$ *by* $\nu - \chi$.

Corollary 10 *On setting $\chi = \nu = 0$ in (36), we get the following special case:*

$$\int_0^\infty x^{2\gamma+1} e^{-\sigma x^2} L_n^{(\alpha)}(\beta x^2) J_0(\mu x) J_0(\lambda x) dx = \frac{1}{2\sigma^{(\gamma+1)}} \times e^{-\mu^2/4\sigma} \sum_{k=0}^{n} \frac{\Gamma(n+\alpha+1)(-\beta/\sigma)^k}{k!\Gamma(n-k+1)\Gamma(\alpha+k+1)} \times \sum_{m=0}^{+\infty} (m+1)_{k+\gamma} \frac{(-\lambda^2/4\sigma)^m}{m!} L_{m+\gamma+k}\left(\mu^2/4\sigma\right). \tag{39}$$

Corollary 11 *If we take $\alpha = \chi = \nu = 0$ in Eq. (36), we find*

$$\int_0^\infty x^{2\gamma+1} e^{-\sigma x^2} L_n(\beta x^2) J_0(\mu x) J_0(\lambda x) dx = \frac{1}{2\sigma^{(\gamma+1)}} \times e^{-\mu^2/4\sigma} \sum_{k=0}^{n} \binom{n}{k} \frac{(-\beta/\sigma)^k}{k!} \sum_{m=0}^{+\infty} (m+1)_{k+\gamma} \frac{(-\lambda^2/4\sigma)^m}{m!} L_{m+k+\gamma}\left(\mu^2/4\sigma\right). \tag{40}$$

Corollary 12 *The undermentioned formula holds true:*

$$\int_0^\infty x e^{-\sigma x^2} L_n(\beta x^2) J_0(\mu x) J_0(\lambda x) dx = \frac{1}{2\sigma} e^{-\mu^2/4\sigma} \times \sum_{k=0}^{n} \binom{n}{k} \frac{(-\beta/\sigma)^k}{k!} \sum_{m=0}^{+\infty} (m+1)_k \frac{(-\lambda^2/4\sigma)^m}{m!} L_{m+k}\left(\mu^2/4\sigma\right). \tag{41}$$

With $\gamma = 0$ and in view of (40), one can straightforwardly find (41).

It is difficult to show the equivalence between (31) and (36). For the comparison of these equations, we illustrate the two equations in Fig. 1. This last shows the compilation, for particular parameters β, $\gamma = 2$, $\sigma = 3.6$, and $\lambda = 1.5$, of the two expressions which are identical. From Fig. 2, we illustrate $I_{n,\alpha}^{\chi,\nu}$ as a function of the distance $z = \mu^2/4\sigma$ for different values of the parameters β, γ, n, and λ.

We can deduce from simulations the equivalence between the two equations and we can deduce the undermentioned corollary:

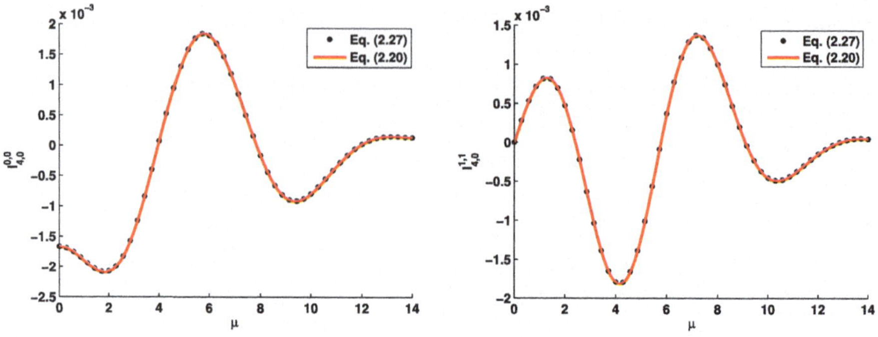

Fig. 1 Illustration of $I_{n,\alpha}^{\chi,\nu}$ as a function of the parameter μ calculated from (31) and (36) with $\beta = 2$, $\gamma = 2$, $\sigma = 3.6$, and $\lambda = 1.5$

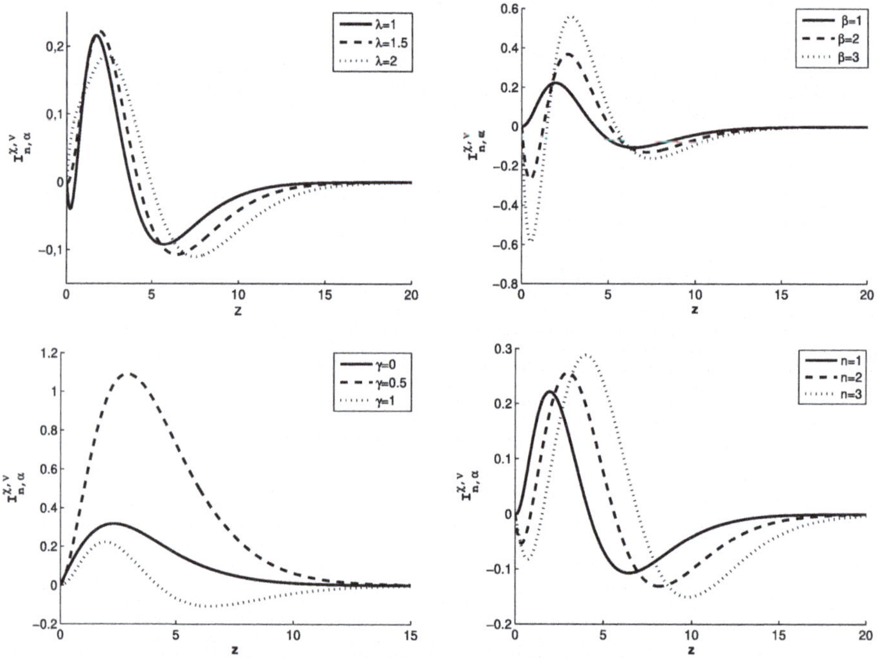

Fig. 2 Variation of $I_{n,\alpha}^{\chi,\nu}$ as a function of the distance z calculated from (31) and (36) for different values of β, γ, n, and λ

Corollary 13 *The undermentioned formula holds true:*

$$\frac{1}{\nu!}\sum_{k=0}^{n}\frac{\Gamma(n+\alpha+1)(\frac{-\beta}{\sigma})^k}{k!\Gamma(n-k+1)\Gamma(\alpha+k+1)}$$
$$\times\sum_{m=0}^{+\infty}\frac{\Gamma(m+\gamma+k+\frac{\nu+\chi}{2}+1)(\frac{-\lambda^2}{4\sigma})^m}{\Gamma(m+\chi+1)m!}\,{}_2F_1\left(-m,-\chi-m;\nu+1;\frac{\mu^2}{\lambda^2}\right)$$
$$=e^{-\mu^2/4\sigma}\sum_{k=0}^{n}\frac{\Gamma(n+\alpha+1)(-\beta/\sigma)^k}{k!\Gamma(n-k+1)\Gamma(\alpha+k+1)} \tag{42}$$
$$\times\sum_{m=0}^{+\infty}\frac{\Gamma(m+\gamma+k+\frac{\chi-\nu}{2}+1)(-\lambda^2/4\sigma)^m}{m!\Gamma(m+\chi+1)}\,L^{(\nu)}_{m+\gamma+k+\frac{\chi-\nu}{2}}\left(\mu^2/4\sigma\right).$$

4 Application

Here, we apply our main result to investigate the propagation of a Laguerre–Bessel–Gaussian beam through a spiral phase plate. This field is defined by

$$E(\rho,z=0)=L_n^{(\alpha)}(\beta\rho^2)J_\nu(\mu\rho)e^{-\sigma\rho^2}, \tag{43}$$

and the output field can be obtained from the Huygens–Fresnel diffraction Collins integral [13]:

$$E(r,z)=\int_0^\infty \rho^{\gamma+1}e^{-\sigma\rho^2}L_n^{(\alpha)}(\beta\rho^2)J_\nu(\mu\rho)J_\chi(\lambda\rho)d\rho. \tag{44}$$

By replacing $\gamma=0$ in (31) and in (36), we find the following expressions:

$$E(r,z)=\frac{\mu^\nu\lambda^\chi}{2^{\nu+\chi+1}\sigma^{(\frac{\nu+\chi}{2}+\gamma+1)}\nu!}\sum_{k=0}^{n}\frac{\Gamma(n+\alpha+1)(\frac{-\beta}{\sigma})^k}{k!\Gamma(n-k+1)\Gamma(\alpha+k+1)} \tag{45}$$
$$\times\sum_{m=0}^{+\infty}\frac{\Gamma(m+\gamma+k+\frac{\nu+\chi}{2}+1)(\frac{-\lambda^2}{4\sigma})^m}{\Gamma(m+\chi+1)m!}\,{}_2F_1\left(-m,-\chi-m;\nu+1;\frac{\mu^2}{\lambda^2}\right),$$

and

$$E(r,z)=\frac{\mu^{\nu}\lambda^{\chi}}{2^{\nu+\chi+1}\sigma^{\left(\frac{\nu+\chi}{2}+\gamma+1\right)}}e^{-\mu^2/4\sigma}\sum_{k=0}^{n}\frac{\Gamma(n+\alpha+1)(-\beta/\sigma)^k}{k!\Gamma(n-k+1)\Gamma(\alpha+k+1)}$$

$$\times\sum_{m=0}^{+\infty}\frac{\Gamma(m+\gamma+k+\frac{\chi-\nu}{2}+1)(-\lambda^2/4\sigma)^m}{m!\Gamma(m+\chi+1)}L^{(\nu)}_{m+\gamma+k+\frac{\chi-\nu}{2}}\left(\mu^2/4\sigma\right). \tag{46}$$

Equations (45) and (46) give the propagation of the considered beam's family in the space.

5 Conclusion

In the present paper, we have investigated two general and interesting formulas for the integral involving the product of Bessel functions with generalized Laguerre polynomials. The results deduced in this paper are general in advantage and likely to discover certain applications in the theory of special functions. On the other hand, the main results have importance as from which we have obtained some important integral formulas for special values of the parameters which are potentially useful for the laser specialists.

References

1. R.S. Alassar, H.A. Mavromatis, S.A. Sofianos, A new integral involving the product of Bessel functions and associated Laguerre polynomials. Acta Appl. Math. **100**, 263–267 (2008)
2. G.E. Andrews, R. Askey, R. Roy, *Special Functions* (Cambridge University Press, Cambridge, 1999)
3. A. Belafhal, E.M. El Halba, T. Usman, An Integral transform involving the product of Bessel functions and Whittaker Function and its application. Int. J. Appl. Comput. Math. **6**, 1–11 (2020)
4. A. Belafhal, E.M. El Halba, T. Usman, Certain integral transforms and their applications in propagation of Laguerre-Gaussian Schell-model beams. Contemp. Math. **4**(4), 637–646 (2023)
5. A. Belafhal, E.M. El Halba, T. Usman, A note on some representations of Appell and Horn functions. Adv. Stud. Contemp. Math. **30**, 5–16 (2020)
6. A. Belafhal, E.M. El Halba, T. Usman, An integral transform and its application in the propagation of Lorentz-Gaussian beams. Commun. Math. **29**, 483–491 (2021)
7. A. Belafhal, H. Benzehoua, T. Usman, Certain integral transforms and their application to generate new Laser waves: Exton-Gaussian beams. Adv. Math. Models Appl. **6**, 206–217 (2021)
8. A. Belafhal, H. Benzehoua, A. Balhamri, T. Usman, An advanced method for evaluating Lommel integral and its application in marine environment. J. Comput. Appl. Math. **416**, 114600-1-11 (2022)

9. A. Belafhal, S. Hennani, A note on some integrals used in laser field involving the product of Bessel functions. Phys. Chem. News **61**, 59–62 (2011)
10. A. Belafhal, N. Nossir, Essakali L. Dalil, T. Usman, Integral transforms involving the product of Humbert and Bessel functions and its application. AIMS Math. **5**, 1260–1274 (2020)
11. A. Belafhal, S. Chib, F. Khannous, T. Usman, Evaluation of integral transforms using special functions with applications to biological tissues. Comput. Appl. Math. **40**, 1–23 (2021)
12. J. Cang, P. Xiu, X. Liu, Propagation of Laguerre-Gaussian and Bessel-Gaussian Schell-model beams through paraxial optical systems in turbulent atmosphere. Opt. Laser Technol. **54**, 35–41 (2013)
13. S.A. Collins, Lens-system diffraction integral written in terms of matrix optics. J. Opt. Soc. Am. **60**, 1168–1177 (1970)
14. I.S. Gradshteyn, I.M. Ryzhik, *Table of Integrals, Series, and Products*, 5th edn. (Academic Press, New York, 1994)
15. F. Khannous, A. Belafhal, A new atmospheric spectral model for the marine environment. Optik-Int. J. Light Electron Opt. **153**, 86–94 (2018)
16. E.D. Rainville, *Special Functions* (Macmillan Company, New York, 1960) (Reprinted by Chelsea Publishing Company, Bronx, New York, 1971)
17. A.K. Shukla, A. Kumar, I.A. Salehbhai, The integral expression involving the family of Laguerre polynomials and Bessel function. Commun. Korean Math. Soc. **27**, 721–732 (2012)
18. G.N. Watson, *A Treatise on the Theory of Bessel Functions*, 2nd edn. (Cambridge University Press, Cambridge, 1944)
19. E.J. Weniger, On the analyticity of Laguerre series. J. Phys. A: Math. Theor. **41**, 425207–425257 (2008)
20. E.T. Whittaker, An expression of certain known functions as generalized hypergeometric functions. Bull. Amer. Math. Soc. **10**, 125–134 (1903)
21. E.T. Whittaker, G.N. Watson, *A Course of Modern Analysis*, 4th edn. (Cambridge University Press, Cambridge, Cambridge Mathematical Library, 1996)

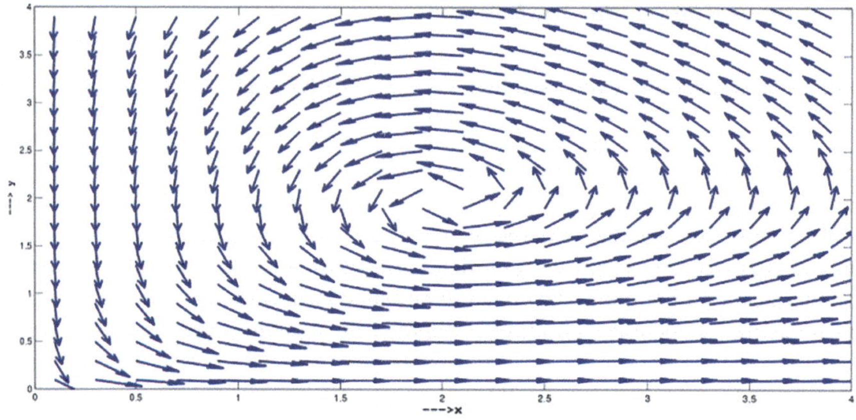

Fig. 1 Vector field of Lotka-Volterra model (Refer Sect. 5.2)

by Ronald Fisher. Solving the above equation, we get $ln(x) = kt + lnc$ and finally the solution of Malthusian growth model can be written as $x(t) = x_0e^{kt}$.

- Inspired by Malthusian model, mathematician Pierre Francois Verhulst proposed a new model for self-limiting growth in 1838 as follows: $\dot{N} = rN(1 - \frac{N}{K})$. Here K is the carrying capacity i.e. the maximum limit of individual applicable for environment's support. In other words, pressure on environmental resource will tend to zero in per-capita grow rate when the carrying capacity K is achieved by the population. This model captures the population growth at a proportional rate with respect to the current population. The solution of the model can be written in the form

$$N(t) = \frac{N_0Ke^{rt}}{K + N_0(e^{rt} - 1)} \tag{8}$$

The presence of excess individuals in same form can be expressed in this simplified form $\dot{N} = rN(1 - N)$. Since discrete maps perfectly capture population growth or decrease from Kth step to $(K + 1)$th step, logistic model can be reframed into discrete case. As an interesting example of population dynamics, logistic maps will be discussed later.

- Blumberg growth model is another interesting model in population dynamics, modified version of Verhulst growth model. He introduced the notion of shape parameter and observed that the inflexibility of the inflection point as the major limitation of the logistic curve and further observed that modification of natural growth rate term as a time dependent polynomial overcome this limitation. In 1968, he published the model in the form

$$\frac{dx}{dt} = rx^{\alpha}\left(1 - \frac{x}{K}\right)^{\gamma} \tag{9}$$

where α, γ are shape parameters, r is the Malthusian parameter [8]. This is also known as hyper-logistic equation. This popular model is widely used in biostatistics, computer virus propagation. This model has been used for social system modeling of co-operatively breeding cichild fish *Neolamprologus pulcher*, used for experimental research on complement-mediated hemolysis.

Fixed Point:

Definition 1 A fixed point or an equilibrium point (steady point or critical point) for a continuous-time model denotes the location of stagnation of flow. In other words, fixed point in phase space is invariant under dynamic flow. If $\dot{x} = 0$, this implies that the velocity is zero and therefore no flow is present in the system.

For example, if we consider $\frac{dx}{dt} = x^2 - kx$, then $x = 0, k$ are the fixed points of the system. Similarly, the equations involved with Duffing oscillator having no damping coefficient can be expressed as

$$\frac{dx_1}{dt} = x_2, \frac{dx_2}{dt} = -x_1 + x_1^3 \tag{10}$$

Then fixed points are $(0, 0), (\pm 1, 0)$.

Nature of Fixed Points:

First we will investigate the nature of fixed points via two-dimensional linear systems. Let us consider the system

$$\begin{aligned} \frac{dx}{dt} &= px + qy \\ \frac{dy}{dt} &= rx + sy \end{aligned} \tag{11}$$

Then we can rewrite the system in the matrix form $\frac{dX}{dt} = MX$ where $X = \begin{pmatrix} x \\ y \end{pmatrix}, M = \begin{pmatrix} p & q \\ r & s \end{pmatrix}$. Substituting X(t) by $e^{\lambda t}\alpha$ in the matrix form, we have $\lambda e^{\lambda t}\alpha = Me^{\lambda t}\alpha$. So, $M\alpha = \lambda\alpha$ will imply that λ can be an eigenvalue of the coefficient matrix and α is the corresponding eigenvector. Here λ measures the growth or change of the dynamic system while α is a fixed vector. Now we have to solve $\det(M - \lambda I) = 0$. Then

$$\lambda^2 - \lambda(p + s) + (ps - qr) = 0 \implies \lambda^2 - \lambda\rho + \delta = 0$$

where ρ, δ denote the trace and the determinant value of the coefficient matrix M respectively. Solving this qudratic in λ, we have $\lambda = \frac{\rho \pm \sqrt{\rho^2 - 4\delta}}{2}$.

Hence the nature of fixed points will be decided by the sign of discriminant $D = \sqrt{\rho^2 - 4\delta}$. The complete summary is available in the following Table 1.

Table 1 Nature of roots

Sign of D	Types of roots	Nature of fixed points
$D > 0$	Real roots	Saddle point (opposite sign) Node (same sign)
$D = 0$	Equal roots	Node
$D < 0$	Imaginary roots	Spiral (complex conjugate) or center

Linearization about the Fixed Point:

One Dimensional Case:

Consider the equation $\frac{dx}{dt} = f(x)$ and suppose x^* is a fixed point. So, $f(x^*) = 0$. Through the linearization about fixed point x^* will capture the qualitative features of a nonlinear system. We will create some disturbance (perturbation) away from x^* and will investigate the decay or growth of perturbation. Let $\delta x = x - x^*$ be a small perturbation (or a trajectory $x(t)$ is displaced by δx from the fixed point x^*). So,

$$\begin{aligned}\frac{d}{dt}(\delta x) &= \frac{d}{dt}(x - x^*)\\ &= \dot{x}\\ &= f(x)\\ &= f(x^* + \delta x)\\ &= f(x^*) + \delta x f'(x^*) + O(\delta x^2)\\ &= \delta x f'(x^*)\end{aligned}$$

Here x^* is a fixed point, so its derivative vanishes. Then expanding the function $f(x)$ in Taylor's series, we will get $O(\delta x^2)$. But we can neglect those quadratically small terms since $f'(x^*) \neq 0$. Then if $f'(x^*) > 0$, the perturbation grows exponentially, otherwise decays when $f'(x^*) < 0$. Hence the notion of stability of a fixed point comes into the picture.

Two Dimensional Case:

Let us consider the two-dimensional dynamical system $\dot{x} = f(x, y)$, $\dot{y} = g(x, y)$. Suppose $X = (x^*, y^*)$ be the equilibrium point of the system so that $f(x^*, y^*) = g(x^*, y^*) = 0$. In order to investigate the strength (stability) of this equilibrium point, we will provide infinitesimal perturbation to it as follows: $\xi = x - x^*$, $\zeta = y - y^*$. Then

$$\begin{aligned}\dot{\xi} &= \dot{x}\\ &= f(x, y)\\ &= f(x^* + \xi, y^* + \zeta)\\ &= f(x^*, y^*) + \xi \left(\frac{\partial f}{\partial x}\right)_{(x^*, y^*)} + \zeta \left(\frac{\partial f}{\partial y}\right)_{(x^*, y^*)} + \text{higher order terms}\end{aligned}$$

and similarly,

$$\begin{aligned}\dot{\zeta} &= \dot{y}\\ &= g(x, y)\\ &= g(x^* + \xi, y^* + \zeta)\\ &= g(x^*, y^*) + \xi \left(\frac{\partial g}{\partial x}\right)_{(x^*, y^*)} + \zeta \left(\frac{\partial g}{\partial y}\right)_{(x^*, y^*)} + \text{higher order terms}\end{aligned}$$

Now ignoring small values of higher order terms, we can rewrite the system in the matrix form as

$$\begin{pmatrix}\dot{\xi}\\ \dot{\zeta}\end{pmatrix} = \begin{pmatrix}\frac{\partial f}{\partial x}{}_{(x^*, y^*)} & \frac{\partial f}{\partial y}{}_{(x^*, y^*)}\\ \frac{\partial g}{\partial x}{}_{(x^*, y^*)} & \frac{\partial g}{\partial y}{}_{(x^*, y^*)}\end{pmatrix} \begin{pmatrix}\xi\\ \zeta\end{pmatrix}$$

Here the coefficient matrix $\begin{pmatrix}\frac{\partial f}{\partial x}{}_{(x^*, y^*)} & \frac{\partial f}{\partial y}{}_{(x^*, y^*)}\\ \frac{\partial g}{\partial x}{}_{(x^*, y^*)} & \frac{\partial g}{\partial y}{}_{(x^*, y^*)}\end{pmatrix}$ is called the Jacobian matrix at the equilibrium point and denoted as $J_{(x^*, y^*)}$. Depending on the sign of eigenvalues of this matrix, we can conclude the nature of fixed point and the stability can be decided.

Stability of Fixed Point:

The concept of stability is very old. In this regard, we will quote the famous statement by Landau and Lifshitz:

> The flows that occur in Nature must not only obey the equations of fluid dynamics, but also be stable.

Apart from initial approaches by Archimedes and Aristotle, the precise definition have been given by three legendary personalities—Lagrange, Poincare, Lyapunov. Aristotle emphasized on the qualitative research based on dynamic motion after a perturbation, while Archimedes observed the geometric scenario after perturbation in terms of statics. The stability of the solar system was the primary motivation of the researchers. Euler used the perturbation theory for the stability of planetary motion. Through the continuous development by Torricelli, Lagrange, Poisson, Poincaré,

Lyapunov, nowadays stability plays a vital role in the field of economics, engineering, biology, quantum mechanics etc. To check the nature of fixed points, we have already used the term perturbation. Basically there are mainly three kinds of perturbations including perturbations of initial values and of the parameters. Stability of a fixed point, an important concept is measured by direction of flow close to it. Small perturbation should not create high fluctuation of trajectories in case of stable fixed point. Heat equation is a common example of stable partial differential equation (Figs. 2 and 3).

Let us classify different cases of stability of fixed points. If both the eigenvalues are real but positive, then clearly $|\xi(t)|$, $|\zeta(t)|$ go to infinity as time tending to infinity. So, the trajectories will diverge after starting from the neighbourhood of the fixed point. This gives the idea of unstability. According to equal or unequal magnitudes of eigenvalues, we can select unstable star or node (Figs. 4 and 5). Similarly, for both negative, real choices of eigenvalues, the trajectory will approach to it exponentially in long course of time, beginning from the neighbourhood of the fixed point. Finally complex cases come into the scenario. If eigenvalues are complex conjugates, then depending upon the sign of the real part, we get stable or unstable spirals. This means that either the trajectories wind around the fixed point several times along spiral

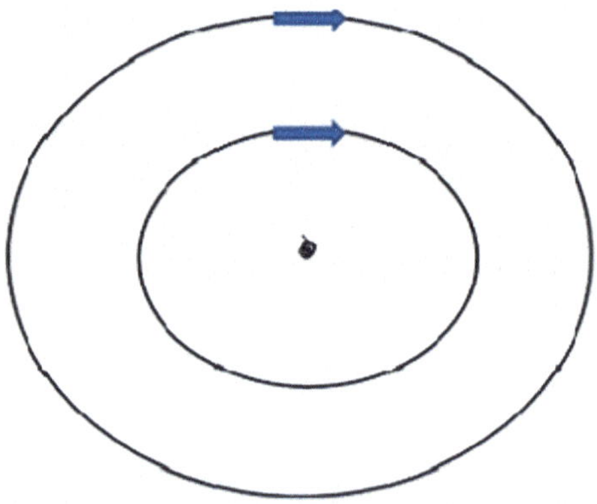

Fig. 2 For the case of purely imaginary eigenvalues, the trajectories don't reach the fixed point as $t \to \pm\infty$ and the stable fixed point is known as center

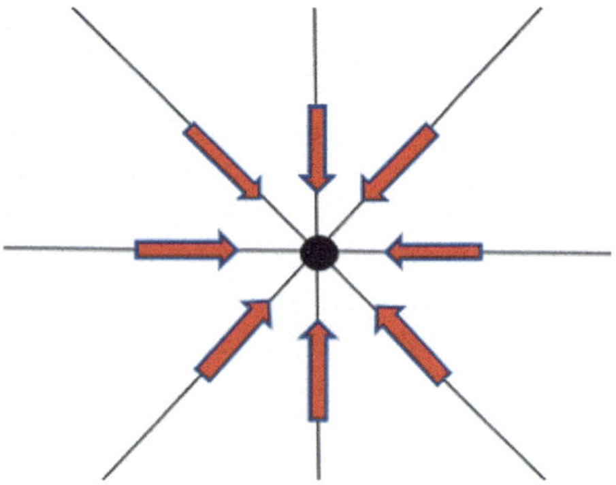

Fig. 3 For real and identical eigenvalues of the Jacobian matrix, the fixed point will be a star. Trajectories about a star are straight lines

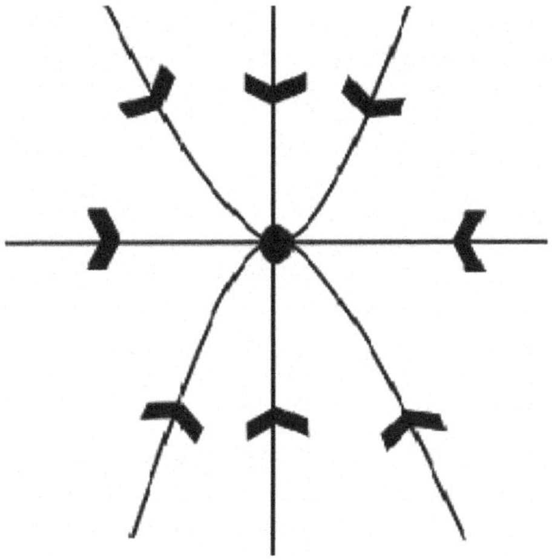

Fig. 4 If eigenvalues are real but negative, then the corresponding fixed point will be a stable node

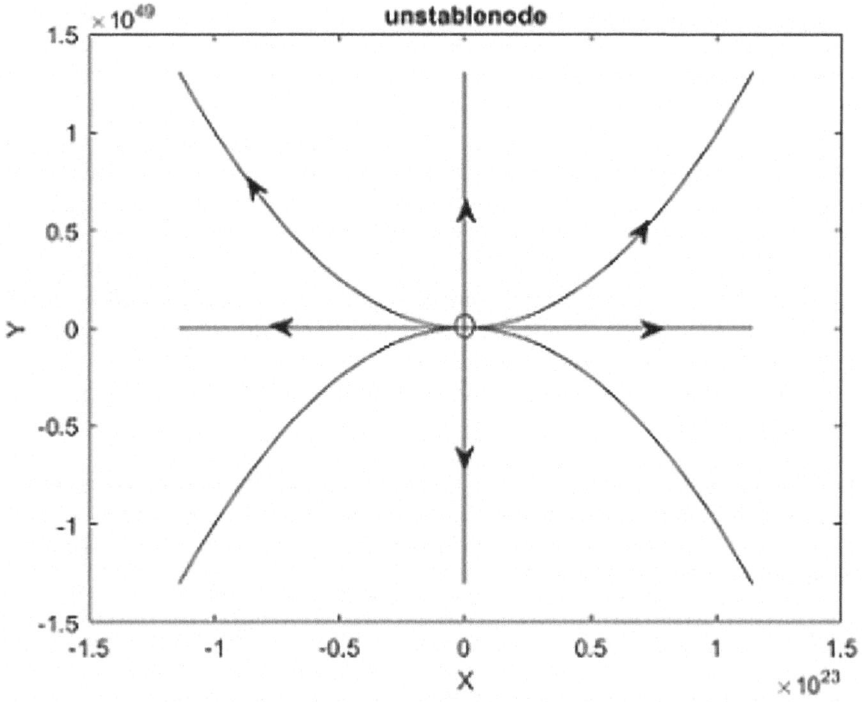

Fig. 5 If eigenvalues are real and positive, then the corresponding fixed point is an unstable node

Table 2 Stability of fixed points

Case	Types of Eigenvalues	Nature of fixed point
1.	Real and positive	Unstable star or node
2.	Real and negative	Stable star or node
3.	Purely imaginary	Center
4.	Opposite signs	Saddle
5.	Complex conjugates	Stable or unstable spiral or focus

directions or move away from it. For the case of purely imaginary eigenvalues, the trajectories don't reach the fixed point as $t \to \pm\infty$ and that neutrally stable fixed point is known as the center (Fig. 2). Here the perturbation periodically changes over time (Table 2).

Bifurcation:

Bifurcation is taken into consideration when small change in parameter values causes qualitative transform in the feature of system. This idea was first described by Henri Poincare in 1885. We can investigate bifurcation from two perspective—local and global. Local bifurcation depicts the qualitative change in the small neighbourhood of a fixed point. While global fluctuation in the state space is related to global bifurcation where the control parameter varies. Global bifurcation can't be restricted to a tiny neighbourhood and it creates fluctuation in the topology of the trajectories in the phase space. Collision of limit cycle with fixed points depicts the global scenario. Homoclinic and heteroclinic bifurcations are two common examples. For the continuous-time model, we can observe the change of stability of fixed points in any one of the following cases: Hopf, Saddle-Node, Transcritical and Pitchfork. Here the first type is considered as dynamic bifurcation while other three belong to the class of static bifurcation.

Definition 2 For the system $\dot{x} = f(x; M)$, with respect to the scalar control parameter α, static bifurcation of fixed points of the system takes place at the point (x_0, α^*) if (a) $f(x_0, \alpha^*) = 0$ and (b) the Jacobian matrix at (x_0, α^*) has single zero eigenvalue with other values having nonzero real parts.

Saddle-Node Bifurcation is such case when two fixed points move towards each other, coalesce and finally mutual annihilation takes place. This bifurcation is often referred as fold bifurcation or turning point bifurcation. The approach can be easily expressed by the following example: $\dot{x} = r - x^2$ where r is the parameter. Then two fixed points $x^* = \pm\sqrt{r}$ exist for $r \geq 0$ but no one exists for $r < 0$. Now the derivative at the fixed point is $-2x^* < 0$, so $x^* = \sqrt{r}$ is stable while $x^* = -\sqrt{r}$ is unstable.

Fig. 6 For $r > 0$, $f'(0) > 0$, $f'(r) = -r < 0$ and so $x^* = 0$ is unstable but $x^* = r$ is stable. On the other hand, if $r < 0$, then $f'(0) < 0$, $f'(r) = -r > 0$ and so $x^* = 0$ becomes stable and $x^* = r$ is unstable

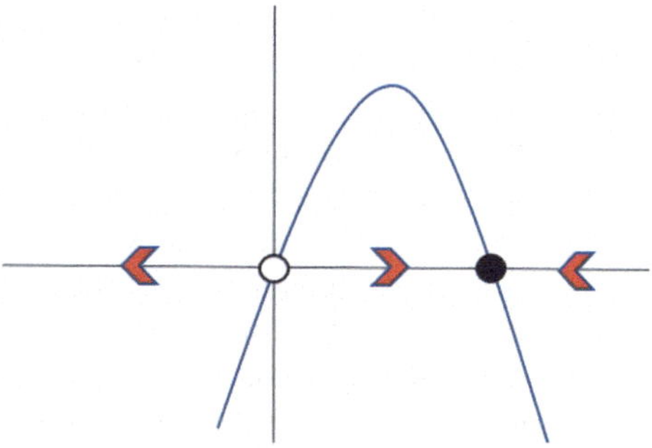

Transcritical Bifurcation is another case where the fixed point exists for every choice of the parameter although the modification in parameter creates the change in stability. The fixed point can't be destroyed. For clarification of the phenomenon, let's consider the example $\dot{x} = f(x) = rx - x^2$. Clearly $x^* = 0, r$ are the fixed points (Figs. 6 and 7).

In case of symmetric pairs, we can consider *Pitchfork Bifurcation*. Here the fixed points occur and disappear in symmetric pairs. Pitchfork bifurcations have two classes—subcritical and supercritical. In general, supercritical bifurcation is expressed in the form $\dot{x} = rx - x^3$ where r is the parameter. Note that for $r > 0$, the origin is only unstable while two stable fixed points $x^* = \pm\sqrt{r}$ are located at the symmetric positions with respect to the origin. For $r < 0$, the origin is only stable fixed point. But we can observe the effect of critical slowing down for $r = 0$. What does it mean by 'critical slowing down'? From $r < 0$ to $r = 0$, the origin is the

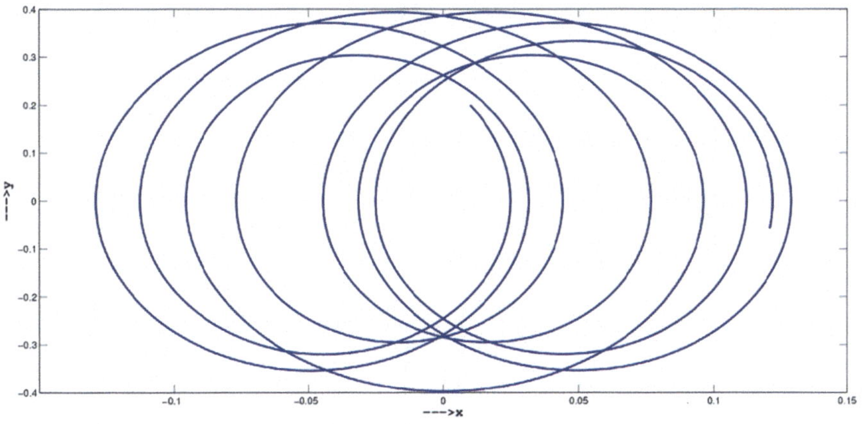

Fig. 7 Phase portrait of periodic forced pendulum

stable one but there is a very slow decay of the solution with respect to time. And finally the origin becomes unstable for $r > 0$.

Hopf Bifurcation is another special case where limit cycle is formed. Basically limit cycle is the isolated closed trajectory. Around 1890, Poincaré introduced the notion of periodic solution. Later Andronov and others improvised the idea in 1930. And Hopf described the idea for arbitrary dimension in 1942. So, this is also known as Poincaré–Andronov–Hopf bifurcation. As a part of dynamic bifurcation, this idea explores the change of stability of fixed points via modification of control parameter. As the parameter crosses the critical value, a limit cycle forms or disappears locally from the equilibria (Fig. 8).

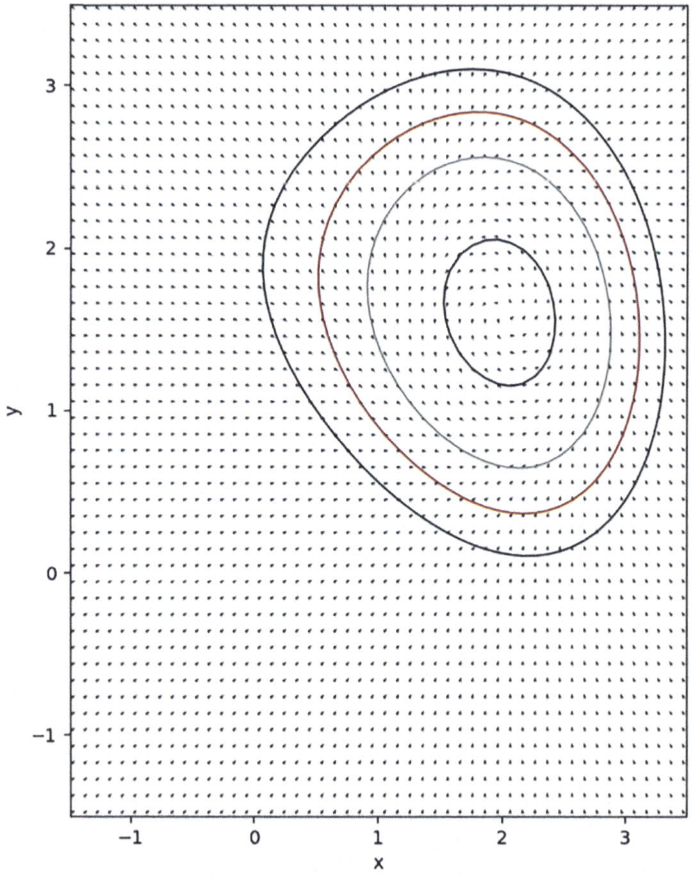

Fig. 8 This represents an example of limit cycle in Hopf bifurcation. As a summary, when the conjugate eigenpairs of linearised flow at the fixed or equilibrium point becomes purely imaginary, we can consider the issue of limit cycle and Hopf bifurcation. Limit cycle is basically an isolated trajectory. For subcritical case, limit cycle is unstable whereas limit cycle is stable for supercritical case

Let us consider two-dimensional continuous system $\dot{x} = f(x, y; \xi), \dot{y} = g(x, y; \xi)$ where $\xi \in \mathbf{R}$ is the only parameter. Suppose (x^*, y^*) is the fixed point which may depend on the values of ξ and the complex conjugate eigenvalues are given by $\lambda_i(\xi) = \alpha(\xi) \pm i\beta(\xi)$. For specific value of $\xi = \xi^*$, we will investigate the criteria for Hopf bifurcation. We need to check the following conditions:

I. $\alpha(\xi^*) = 0, \beta(\xi^*) = \rho(\text{say}) \neq 0$

II. $\frac{d\alpha(\xi)}{d\xi}|_{\xi=\xi^*} = k(\text{say}) \neq 0$

III. $k = \frac{1}{16}(f_{xxx} + g_{xxy} + f_{xyy}g_{yyy}) + \frac{1}{16\rho}\left\{f_{xy}(f_{xx} + f_{yy}) - g_{xy}(g_{xx} + g_{yy}) - f_{xx}g_{xx} + f_{yy}g_{yy}\right\} \neq 0$

IV. $f_{\xi x} + g_{\xi y} \neq 0.$

The second condition, known as Transversality depicts that the eigenvalues cross the imaginary axis having nonzero speed. Now, if $k < 0$, for $\xi > \xi^*$ the fixed point is called stable. Also local birth or destroy of periodic solutions will depend on 3rd and 4th conditions. For $\xi < \xi^*$, if $k.(f_{\xi x} + g_{\xi y}) > 0$ holds, then periodic solutions exist. Similarly, for $\xi > \xi^*$, if $k.(f_{\xi x} + g_{\xi y}) < 0$ holds, then periodicity occurs.

2.2 *Discrete System*

Discrete dynamical system is such one where time is treated as discrete variable i.e. values of variables for all times t can be captured through the information of variables at a initial time instance t_0. Discrete jump from one state to another is represented by recursive equations, iterated maps or difference equations in real or complex variables. General form is given by

$$x_{n+1} = f(x_n, n), \; x(0) = x_0 \tag{12}$$

where time n can assume only integral values. Usually a discrete evolution is displayed by a system of algebraic maps f. This concept of measuring physical quantities after finite time span (in terms of hour/week/month) is highly useful for the problems of rainfall, economic time series, radioactive decay, electronics.

2.2.1 Examples of Discrete Models

- Logistic Map: Renowned biologist Robert May proposed this one dimensional discrete map in search of bacteria growth in 1976. Map is defined by this recursive equation $x_{n+1} = \lambda x_n(1 - x_n)$ where λ is the only control parameter. The input x_n can assume values in $(0,\ 1)$. This model depicts the complex chaos originated from simple dynamical systems (Fig. 9).
- Tent Map: This is similar kind of discrete map having single parameter μ. The map is described by the relation

$$x_{n+1} = \begin{cases} \mu x_n, \; x_n < \frac{1}{2} \\ \mu(1 - x_n), \; x_n \geq \frac{1}{2} \end{cases}$$

- Henon Map: As the simplified case of Lorenz model, Michel Hénon proposed this two dimensional map. This discrete map is very popular in use of chaotic analysis. Maps are defined by the following relations:

$$x_{n+1} = 1 - a{x_n}^2 + y_n, \; y_{n+1} = bx_n$$

With the change of two control parameters, the sequence either diverges or converges to a strange attractor (Fig. 10).
- Arnold Cat Map: Vladimir Arnold, student of Kolmogorov solved Hilbert's 13th problem at the age of 20. He had remarkable contribution on the stability of dynamical systems. Arnold proposed a two-dimensional map incorporating approach of linear algebra in 1960. Using an image of a cat, he applied stress, stretch etc. keeping the volume fixed. After multiple iterations, original image was restored. This cat map is the transformation of two-dimensional torus T^2 into itself. This map is defined as follows:

$$\Gamma(X) = \begin{pmatrix} 2 & 1 \\ 1 & 1 \end{pmatrix} X \; (mod\ 1)$$

where $X = \begin{pmatrix} x \\ y \end{pmatrix}$ be the vector. The coefficient matrix shows the stretch of one unit domain in single direction and two units in other.
- LPA Model: Larvae-Pupae-Adult (LPA) model is very famous model in mathematical ecology. To analyze population of dynamics of flour beetles, *Tribolium castaneum* is most convenient for laboraty work. This popular model is used to describe the transition of 3 stages (larvae, pupae, adult) of flour beetles and non-linear aspects of cannibalism. The model is interpreted as the system of three nonlinear difference equations as follows:

$$\begin{aligned} L_{i+1} &= bA_i e^{-k_l L_i - k_a A_i} \\ P_{i+1} &= (1 - \mu_l)L_i \\ A_{i+1} &= P_i e^{-k_{pa} A_i} + (1 - \mu_a)A_i \end{aligned}$$

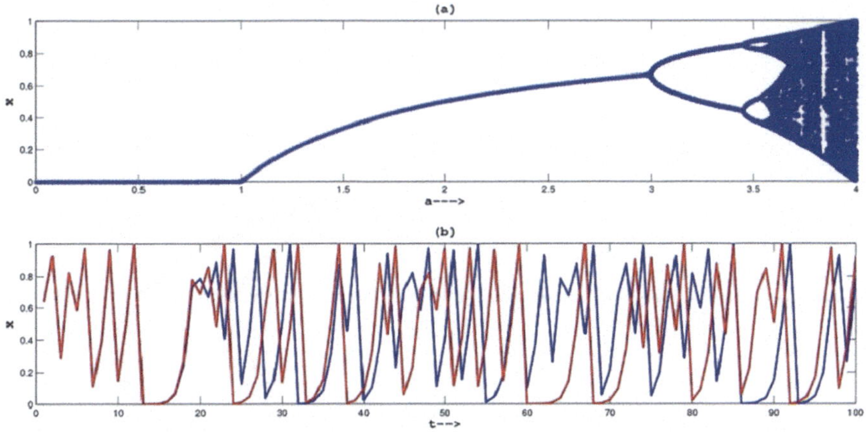

Fig. 9 **a** Represents the bifurcation diagram and **b** depicts sensitive dependence of Logistic Map. In general, value of λ, close to 3.8 presents the chaotic behaviour

Here L_i, P_i, A_i denote the the number of larvae, number of individuals in pupae stage and the number of adults at time i respectively. b is the rate of egg production by adults, k_l, k_a are respective cannibalism coefficients of eggs by larvae and adults. Two weeks is the general time for state transition. We can consider μ_l as the natural death rate of larvae, μ_a as the death rate of adults and k_{pa} as the level of cannibalism, pupae must achieve to become adult.

Fixed Point:

In discrete case, fixed points x^* are simply the solutions of $f(x) = x$. Clearly the trajectory can't go out of x. For example, if we consider the logistic map $x_{n+1} = \lambda x_n(1 - x_n)$, $0 \leq \lambda \leq 4$, the fixed points are the roots of $x^* = \lambda x^*(1 - x^*)$ i.e. points will be 0, $(1 - \frac{1}{\lambda})$. Similarly for the map $x_{n+1} = \sin x_n$, $x^* = 0$ is a fixed point.

Types of Fixed Points:

Let's consider the map $x_{n+1} = f(x_n)$ and x^* be the fixed point. Then introducing small perturbation ξ on x^*, just like continuous case, we get $\xi_{n+1} = (Jf)_{x^*}\xi_n$ where Jf is the matrix of first partial derivatives computed at x^*. Next, we can use linear transformation $\xi = P\eta$ to get the form $P\eta_{n+1} = (Jf)_{x^*}P\eta_n$. But what is P? Depending on nature of eigenvalues, we can construct the matrix P. Suppose all eigenvalues are complex. Then we can choose the eigenvectors $p_i's$ corresponding to eigenvalues λ_i's as the columns of P and accordingly, $P^{-1}(Jf)P$ will be diagonal.Instead of $(Jf)_{x^*}$, we can use the matrix A. If P is nonsingular, then we

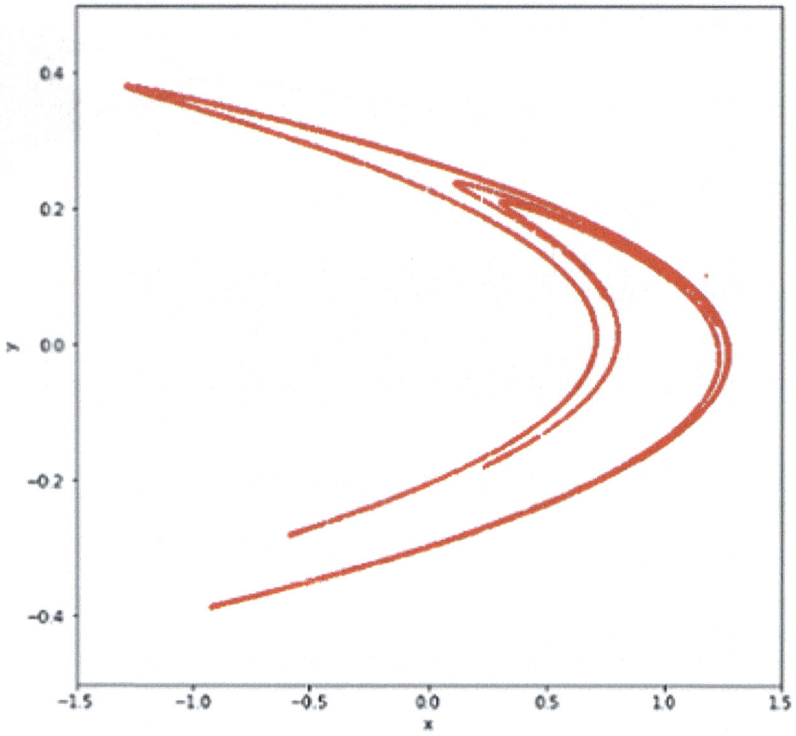

Fig. 10 Figure represents the discrete Henon map. Generally $a = 1.4,\ b = 0.3$ shows the chaotic behaviour

can have $\eta_{n+1} = P^{-1}AP\eta_n = J\eta_n$ where J is the Jordan canonical form. For all distinct eigenvalues, we can consider μ_i's as the diagonal entries of the Jordan form, which is eventually diagonal. Therefore, we have the relation $\eta_{n+1}^{(k)} = \mu_k \eta_n^{(k)}$. Now $\eta_n^{(k)}$ goes to infinity if $|\mu_k| > 1$ as k tends to infinity, $\eta_n^{(k)} = \eta_0^{(k)}$ for $\mu_k = 1$. Here we can summarise the nature of fixed point. A fixed point is called hyperbolic if all the corresponding eigenvalues lie completely inside the unit circle in the complex plane or outside the circle. If one or more eigenvalues lie on the unit circle, then the fixed point is known as nonhyperbolic fixed point (Table 3).

Table 3 Types of fixed points and location of eigenvalues

Type of fixed point	Location of eigenvalues with respect to the unit circle	Sub-category
Hyperbolic	Some values lie within circle Others lie outside the circle	Saddle point
Hyperbolic	All values lie outside circle	Source point
Hyperbolic	All values lie within circle	Sink point
Nonhyperbolic	One or more lie on the circle	–

Bifurcation:

Basic idea of bifurcation in a discrete system will not differ from the continuous-time system. Let's start with an example of one-dimensional map: $x_{n+1} = (1+\lambda)x_n - x_n^2$ where λ is the single parameter. Clearly λ is the nontrivial fixed point and 0 is the trivial one. Note that for $0 < \lambda < 2$, the nontrivial point $x^* = \lambda$ is stable but it becomes unstable for $\lambda < 0$. Similarly, for $-2 < \lambda < 0$, the trivial point $x^* = 0$ is stable but it becomes unstable for $\lambda > 0$.

Hopf Bifurcation in discrete case is exactly similar to that in continuous-time model. Let us consider two-dimensional discrete system

$$x_{n+1} = f(x_n, y_n; \xi),\ y_{n+1} = g(x_n, y_n; \xi) \tag{13}$$

where $\xi \in \mathbf{R}$ is the only parameter. Suppose (x^*, y^*) is the fixed point i.e. $x^* = f(x^*, y^*; \xi)$, $y^* = g(x^*, y^*; \xi)$ for all values of ξ and the complex conjugate eigenvalues are given by $\lambda_i(\xi) = \alpha(\xi) \pm i\beta(\xi)$. For specific value of $\xi = \xi^*$, Hopf bifurcation takes place at (x^*, y^*, ξ^*) if the following conditions hold:

a. $|\lambda_1(\xi^*)| = 1$

b. $\left(\frac{\partial \lambda_1(\xi)}{\partial \xi}\right)_{\xi=\xi^*} > 0$ where $\overline{\lambda_1}(\xi) = \lambda_2(\xi)$.

2.3 *Typical Features of Nonlinear System*

It's time to discuss unique features of nonlinear differential equations, differ from linear equations. For linear differential equation, every dependent variable and the involved derivative occurs in first degree only without having any product of dependent variable and/or corresponding derivatives. On the other hand, for nonlinear partial differential equations, unknown functions or derivatives are multiplied by each other. We can observe the presence of trigonometric functions, exponential functions and other transcendental functions in nonlinear equations. For example, $y\frac{dy}{dx} = \sqrt{x}(\frac{dy}{dx})^2 + k$ represents a nonlinear differential equation.

- Existence and Uniqueness of Solution: The reader may ask that does every initial value problem (IVP) possess one solution? If yes, then is the solution unique? Here, the notion of existence and uniqueness of solution comes to the picture. Rudolf Lipschitz was considered as the father figure to attempt the question of existence and uniqueness. After his work in 1876, Giuseppe Peano deduced the idea of non-unique solution around 1890 if function is continuous in x and y. Continuity of the function f defined from $\mathbb{R}^n$ to $\mathbb{R}^n$ for the equation $\frac{d}{dt}(x(t)) = f(x(t)),\ x(t_0) = x_0$ doesn't ensure the uniqueness of the solution. For example, the equation $\dot{x}(t) = 5|x(t)|^{\frac{4}{5}},\ x(0) = 0$ has different solutions $x(t) = 0,\ x(t) = t^5$.
- General solution: We can solve the first order linear ordinary differential equation easily because by applying general techniques for the equation having constant coefficients, we can obtain solution. Even for second and higher order linear ODE, we can obtain complimentary function as well as the particular integral. While a first order nonlinear ordinary differential equation can't be solved by any general methodology. Although different numerical techniques provide approximate solutions assuming vital significance for nonlinear equation, there is no generalised benchmark. For example, the ordinary differential equation $\frac{dy}{dx} + y^3 = 0$ is satisfied by $y = \frac{1}{\sqrt{2(x+k)}}$, having k as arbitrary constant. But also note that $y = 0$ is another solution, which can be obtained without assigning any values of k.
- Interval of solution: Finite escape time is a vital issue in case of a nonlinear system. What does it mean? Let's consider the equation $\frac{d\alpha}{dt} = \alpha^2(t) + 1, \alpha(0) = 0, \forall t \geq 0$. After simplification, we get $\tan^{-1}\alpha = t + c$. Clearly the equation has an unique solution $\alpha(t) = \tan t$. But no continuously differentiable function $\alpha(t)$ defined over $[0,\ \infty)$ exists. Note that, as t approaches to $\frac{\pi}{2}$, function $\alpha(t)$ tends to ∞. So, this solution is unique but we can't consider it as global solution i.e. this initial value problem has no solution outside $(-\frac{\pi}{2},\ \frac{\pi}{2})$. Similarly, for the equation $\dot{y} = y^2$ having initial conditons $y(t_0) = y_0$, the solution can be written as $y = \frac{y_0}{1-y_0(t-t_0)}$. Note that if $y_0 = 0$, then solution is $y = 0$ for all t. But if $y_0 > 0$, then solution will not be good one as $t - t_0 \mapsto \frac{1}{y_0}$.
- Vector Space: Solutions of linear systems form a vector space while the solutions of nonlinear differential equations do not form a vector space.
- Superposition: Given homogeneous differential equation $L\alpha = 0$, we have $L(a\alpha_1 + b\alpha_2) = aL(\alpha_1) + bL(\alpha_2) = 0$ because L is linear operator. But what will happen for inhomogeneous differential equation $L\alpha = f(x)$? Before answering this question, note that general solution of linear inhomogeneous equation can be written as additive form of the complimentary function (which is linear superposition of linearly independent solution of homogeneous equation $L\alpha = 0$) and the particular integral. Therefore nth order linear homogeneous differential equation

$$\frac{d^n\alpha}{dx^n} + p_1(x)\frac{d^{n-1}\alpha}{dx^{n-1}} + \cdots + p_n(x)\alpha = 0 \tag{14}$$

(where $p_i(x)$ are the functions of x; $i = 1, 2, \ldots, n$) has the general solution as linear superposition of n linearly independent solutions $\alpha_i(x)$ i.e. $\alpha(x) = \sum_{i=1}^{n} k_i\alpha_i(x)$, where k_i's are arbitrary constants. Hence, we can write $L \equiv \frac{d^n}{dx^n} + p_1(x)\frac{d^{n-1}}{dx^{n-1}} + \cdots + p_n(x)$ Now if L is written in the differential operator form, then the above inhomogeneous differential equation $L\alpha = f(x)$ can have general solution as follows: $\alpha(x) = \sum_{i=1}^{n} \beta_i(x)\alpha_i(x)$ where unknown functions $\beta_i(x)$ can be computed through integration using n linearly independent functions $\alpha_i(x)$. But the idea of superposition of solutions is not applicable for nonlinear differential equations.

3 Chaos and Attractors

Chaotic analysis is an important segment of nonlinear dynamics and useful in the field of mathematical model. Issue of predictability comes into the picture. The term signifies that degree of a system's state prediction in qualitative or quantitative approach. Determinism is another key point of chaos theory which means the change of the system from given initial condition to a single state. In 1770, French philosopher d' Holbach first raised the issue of universal determinism in his book Le système de la nature. Then renowned mathematician Pierre-Simon Laplace explained the detailed scenario of universality of determinism, way out to compute the net effects in 1778. In 1848, Neptune was discovered by Joseph Le Verrier without using astronomical calculations, he further extended Laplace's ideas. James Maxwell and Ludwig Boltzmann imposed the existence of an isolated system in equilibria in all microstates having identical probability for statistical mechanical theory. Henri Poincare published his fantastic works during 1880–1899 on local and global stability of fixed points, notion of periodic solution. He emphasized on phase space. He introduced the idea of sensitivity to initial conditions of the systems in his thesis on n-body problem. G. D. Birkhoff proved Poincare's theorem in 1913 as a special case of three-body problem. In 1927, he published a book on dynamic systems. Jacques Hadamard came with the idea of geodesic flow on a surface, having nonpositive curvature around 1898. 'Hadamard billiards' is an interesting concept in chaotic analysis. G. D. Birkhoff, Andrey Kolmogorov, John Littlewood, Cartwright and many more had immense contributions for chaos theory in different perspective. Meteorologist Edward Lorenz first observed the scenario of sensitivity to initial conditions via a system of three coupled differential equations in 1963 [3]. He coined the term 'butterfly effect'. His title talk of 1972 conference is very popular till date, it was

> Predictability: does the flap of a butterfly's wing in Brazil set off tornedo in Texas?

He interpreted the strange attractors in his works, although Belgian physicist David Ruelle first coined this term in 1971. Mitchell Feigenbaum proposed the notion of period doubling incorporating connection between steady dynamics and chaos [10]. In 1975, James A. Yorke first used the term 'chaos' in mathematical field.

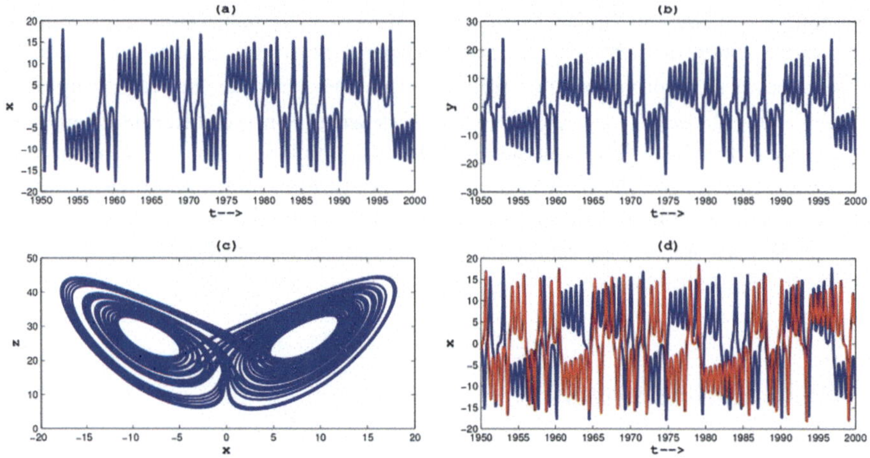

Fig. 11 **a** Time evolution of x-component, **b** y-component of Lorentz system, **c** Phase diagram in xz plane, **d** Sensitive dependence on initial conditions

Now what does it mean by attractor? Basically all neighboured trajectories converge to the attractor set A. Any trajectory starting in A remains within A throughout the course of time. Attractors should have some special features as follows: firstly, all trajectories starting close to it are attracted by set A,the set A is the minimal set having this feature. Limit cycles, stable fixed points are examples of attractors. Now the strange attractors have unique feature of sensitive dependence to initial conditions (Fig. 11).

4 Chaos Control and Synhronization

4.1 Chaos Control

Controlling chaos is the process of conversion from chaotic motion to stable periodicity. Small modification of parameters or inclusion of weak perturbation can eliminate random, irregular features of state values in a chaotic system. E. Ott, C. Grebogi, J.A. Yorke first proposed the control technique using the presence of unstable periodic orbits. Basic informations are collected through initial run of the system in this OGY method. If the trajectory appears near the unstable periodic orbit (embedded ergodically), small perturbation can be applied to get stability of that orbit. Ditto et al. applied OGY method to figure out the stabilization of oscillations of a magnetoelastic ribbon. Peng et al. introduced idea of occassional proportional feedback for controlling chaos in chemical models. For Belousov-Zhabotinsky reaction, perturbation in the control signal can be measured in terms of potential in Bromide electrode and fixed point evaluation by return map. The estimated value provides the scenario of stability. Pyragas and Tamasevicius experimentally analysed the continuous time delayed feedback signal for an externally driven nonlinear oscillator. Comparing delayed signal and output signal corresponding to the specific orbit, researchers observed stability of orbits using tunnel diode with different frequencies. As a proportional difference between actual state values and delayed value for time-lag, negative feedback link can be constructed to control chaos. Different works have been done in enzymatic reactions using Pyragas method. In the field of laser and nonlinear optics, Roy et al. and others proposed proportional feedback method for using perturbation to the driver of intracavity doubled Nd: YAG diodes.

4.2 Synchronization

Now we will look into the matter of chaotic synchronzation. When two or more chaotic systems are coupled, synchronization takes place. Due to coupling, two systems showcase common characteristics. According to Greek etymology, synchronization means to share the common time. Eminent physicist C. Huygens presented the striking synchronized phenomenon of two coupled pendulum clocks in 1665. Later B. Van Der Pol studied the synchronization in radio communication system. Andronov, Mayer, Bremsen and Feinberg also studied synchronization of relaxation oscillators extensively. Fujisaka and Yamada studied chaotic synchronization in the year 1983. Pecora et al. published few exciting works on nonlinear aspects of spin waves in 1988. Initially they used modified version of logistic map for coupling between transmitter and receiver in the network system. Their proposed 'master-slave' system contains two chaotic systems governed by same set of differential equations having same parametric values [7]. The output of coupled equations of one system should be incorporated to another one and as a result, synchronization will take place. So, one chaotic system will drive another. They tried to reproduce incoming signal for drive in case of synchronization of drive and response. They pointed the occurrence of synchronization for all negative values of Lyapunov exponents for response subsystem. In 1989, Volkavskii and others showcased stochastic locking in coupled systems. Around 1995, Gade et al. showed that coupled chaotic maps on a regular lattice are less robust in terms of synchronization than that on tree-structured networks. Different types of synchronization have been studied in recent times. Pecora-Carroll (PC) method was the part of complete synchronization whereas the lag synchronization is the intermediate class between phase and complete synchronization. Here two outputs lock phases in presence of time lag. There are various papers on synchronising technique used for electronics, encryption, chemistry.

5 Research Applications

5.1 Initial Description on Fractals

Fractals are basically irregular geometric patterns having a special virtue of self-similarity. What does it mean? Self-similarity is a special feature of fractal objects after rescaling. If any fractal (derived from Latin *fractus*, means broken or fractured) is divided into several identical parts, each one looks like a replica of the original one. The paper entitled 'How Long Is the Coast of Britain?' by an IBM researcher Benoit Mandelbrot made a vibration in the field of fractal geometry around 1960. Even he first coined the term 'fractal' around 1975. But the development of fractal geometry comprised of a long, rich history. The idea of recursive self-similarity was first proposed by Leibnitz around 17th century. Then Karl Weierstrass published an interesting example of everywhere continuous function which is nowhere differentiable (1872). Gradually different researches on Cantor Set (1883), Koch Snowflake (1904), Sierpinski Triangle (1915) have enriched the branch of fractals. The independent work of Fatou and Julia (1918) is another breakthrough in this field. But here the notion of fractal dimension is very important. Fractal dimension provides qualitative comparison of fractals. There are various approaches to compute noninteger dimension of fractals. But we should recall our elementary knowledge. For the one-dimensional line or two-dimensional square, we need the idea of covering objects. Let A be the subset of $\mathbf{R^d}$ and $N(\omega)$ is the minimum number of d-dimensional cubes having side ω to cover A. Then think about the relation between v, d-dimensional volume of the ball B_r with the radius r. For $d = 1$, $v \propto r$. For $d = 3$, $v \propto r^3$. Analogously, for a plane region having area S bounded by a smooth curve, $N(\omega) \propto \frac{S}{\omega^2}$. Similarly, for higher dimension, we can observe the power law as follows: $N(\omega) \propto \frac{1}{\omega^d}$. This gives the box dimension of A where the dimension is expressed by $d = -\lim_{\omega \to 0} \frac{lnN(\omega)}{ln\omega}$. In terms of metric space, we can say that d provides the estimate of number of smallest closed balls having radius ω to cover the set A where A is a subset of some metric space X. For example, box counting dimension of Von Koch curve is $\frac{ln4}{ln3}$.

Pointwise and correlation dimensions are another useful techniques for estimation. Procaccia and Grassberger proposed the technique to observe the frequency of a trajectory's appearance on specific neighborhood of a point on the attraction. Considering δ, a positive quantity and x as a point on the attractor, we get the relation $N_x(\delta) \propto \delta^d$ where d is pointwise dimension with respect to $N_x(\delta)$, the number of points inside a ball having radius δ centered at x. The summative form gives the correlation dimension.

5.2 Application to Ecology

Due to complexities in interaction for different ecological systems, primarily ecologists emphasize on empirical studies. But mathematical tools are being used to analyze ecological theories for a long time. Mathematical modeling depicts the complex dynamic characteristic of the systems in ecology as well as population dynamics. The study of the control of chaos in biological systems is very important. Chaotic dynamics of the systems have been studied in different context using multiple ecological factors, namely, Allee effect, fear effect, refugia, cannibalism etc. On the other hand, pattern formation is vital feature to relate population and space-time structure. Nowadays researchers are mainly looking for spatio-temporal pattern analysis in population dynamics for insect, plankton, vertebrate etc. Nonlinear ordinary and partial differential equations are respectively used for temporal and spatial systems. Lotka-Volterra model, commonly known as prey-predator model is a famous model in the dynamics of ecological systems. In 1920, American mathematician Alfred James Lotka published his works on interaction between prey and predator. Interestingly, another multi-talented mathematician Vito Volterra presented same kind of works with small difference in 1926. Various experiments are being performed till date on population dynamics using this model. But let's discuss about ingredients, assumptions of Lotka-Volterra model. Firstly, we have to consider unlimited resources and hunger capacity for prey and predator respectively. Secondly, prey population will rise exponentially in absence of predator barrier. Similarly, predator population will be diminished in absence of prey population. Next, the interaction between prey and predator will come into the picture. Finally, the model can be expressed as $\frac{dx}{dt} = ax - bxy$, $\frac{dy}{dt} = -cy + dxy$ where $x(t)$, $y(t)$ represent population densities of prey and predator at time instance t, a measures growth rate of prey in absence of predator, c signifies the loss of predator due to death or immigration, b and d are factors for interaction (Fig. 1). Logistic model is also a simple model of population dynamics, proposed by Robert May, we have already discussed about its structure. Another common model using Holling type-II functional response can be written as

$$\frac{dx}{dt} = gx\left(1 - \frac{x}{k}\right) - \frac{\alpha xy}{1 + \alpha hx}, \frac{dy}{dt} = \frac{\alpha\beta xy}{1 + \alpha hx} - dy \tag{15}$$

where k is carrying capacity of prey, g measures prey's growth, d denotes predator's mortality rate, α is intake rate [14]. Allee effect, an important notion of biology captures the correlation between growth rate and population density. In 1931, W.C. Allee first introduced this phenomenon. The simple equation using multiplicative factor can be written as

$$\frac{dx}{dt} = r\left(1 - \frac{x}{K}\right)(x - m)x \tag{16}$$

where m is multiplicative factor, r is growth rate and K is the carrying capacity.

5.3 Application to Climate Dynamics

Depending on three important factors (pressure, temperature and precipitation) of climate dynamics, different research works have been already performed using climate predictability indices from Indian perspective. Hurst exponent (H) computation of a stochastic stationary process is the fundamental goal in these kind of works. What is Hurst exponent? Why is it important in nonlinear dynamical analysis? Around 1951, renowned hydrologist Harold Edwin Hurst proposed a technique to optimize the storage capacity of a reservoir for the Nile river. Basically he tried to figure out a specific pattern in different natural phenomena which characterized by a random biased process. In other words, Hurst exponent H (nonlinear analytical tool named after him) is used to measure long-term memory processing in time series. Following conclusions can be drawn according to the range of the exponent H.

The value of H in the range 0–0.5 implies anti-persistence in the time series. So, the system is characterized by frequent switching tendency between low and high values. $H = 0.5$ is the value corresponding to an uncorrelated series, i.e. auto-correlations that may generate a completely uncorrelated series in small time delays but the absolute values decrease exponentially. The value of H in range 0.5–1 implies persistent process and it is characterized by long-term memory. Among different methods, R/S method is most popular technique to compute Hurst exponent. Mandelbrot and Wallis proposed the method in 1969. The scheme of this method is described as follows.

Firstly we need to divide the time series of length l into d non-overlapping blocks of length m. The sample mean and the standard deviation (σ_m) are calculated for each block, $m = 1, 2, \ldots d$. Normalized data is calculated by subtracting mean from each data for $i = 1, 2, \ldots, n$. Then cumulative time series need to be computed. Difference of maximum and minimum values of data values will be denoted by $range(n)$. Then the $range(n)$ need to be re-scaled dividing by (σ_m). The exponent H is defined in terms of statistically asymptotic behavior of the range as a function of time span. For the limiting case $n \to \infty$, we can write $E(\frac{range(n)}{\sigma(n)}) = kn^H$ where $\sigma(n)$ denotes sum of first n standard deviations in n number of data points, E denotes the expectation of data points and k is arbitrary constant. So, we can write the following relation $(\frac{R}{S})_n \approx kn^H$. Now fractal dimension D can be measured using the relation $D = 2 - H$.

To predict the spatio-temporal evolution of a multi-scale chaotic system, Lorenz 96 model is favourite to most of the researchers. Thornes et al. proposed this extended version of Lorenz, having a set of coupled nonlinear ordinary different equations [9]. Thornes used it to investigate chaotic fluctuations in climate system. The following system of equations represent Lorenz 96 model:

$$
\begin{aligned}
\frac{dX_k}{dt} &= X_{k-1}(X_{k+1} - X_{k-2}) + F - \frac{hc}{b}\sum_j Y_{j,k} \\
\frac{dY_{j,k}}{dt} &= -cbY_{j+1,k}(Y_{j+2,k} - Y_{j-1,k}) - cY_{j,k} + \frac{hc}{b}X_k - \frac{he}{d}\sum_i Z_{i,j,k} \\
\frac{dZ_{i,j,k}}{dt} &= edZ_{i-1,j,k}(Z_{i+1,j,k} - Z - i - 2, j, k) - geZ_{i,j,k} + \frac{he}{d}Y_{j,k}
\end{aligned}
\tag{17}
$$

where i, j, k varies from 1 to 8. $b = c = d = e = g = 10, h = 1$ are key factors to capture perfect scenario of proper temporal evolution and $F = 20$ is high scale force factor for exhibiting chaotic behaviour. Depending on the index modification, X, Y, Z show low to high frequency and amplitude variability.

5.4 Application to Economics

Chaotic behaviour of economic models have been studied for a long time from different perspective. Richard Godwin had presented extensive utility of chaos theory for economical theory and policy. R. Solow proposed descriptive growth model using neoclassical production function as follows:

$$(1 + p)c_{t+1} = sf(c_t) \tag{18}$$

where the capital stock exists for one period as per assumption of this first-order discrete system. Here f is production function, c denotes the capital for individual worker, p and s are population growth and savings rates respectively. In 1981, Benhabib and Day showcased the possibility of erratic behaviour generating from rational choice, based on previous experience and preference. We can obtain logistic model by modification of experience dependence parameter and Cobb-Douglas utility function. Around 1988, Barnett and Chen published works in support of chotic behaviour using monetary aggregates. Still there are lots of controversies for the evidence of chaos. Using BDS test and bispectral linearity, Copeland, Wong and Abhyankar observed nonlinearity in important stock market index like Financial Times Stock Exchange (FTSE 100) U.K., but chaotic presence was not observed.

5.5 Application to Cosmology

Einstein's remarkable paper on cosmological model in 1917 is considered as first recorded application of general relativity to the universe [17]. His idea about static universe, large-scale homogeneity, curved spatial descriptions were highly interesting and unique at all. In 1922, C. Charlier challenged Einstein's approach by introducing notion of fractal universe, unending pattern of clusters. A. Friedmann presented a work on a class of universes having homogeneity, isotropy, nonzero curvature etc. G. Lemaitre had incorporated the redshift phenomena with Friedmann's model. He had shown instability of Einstein's static universe, evolution of discontinuities over time. But in this regard, we can't ignore the importance of phenomenal work by E. Lifshitz on cosmological perturbation in 1946. There is a long historical journey of mathematical cosmology through the works of Cartan, Brand, Dicke. After a long development and qualitative research on cosmology using mathematical tools, chaotic inspection was performed by Misner [19]. For the field equations of general relativity, he used Hamiltonian notion. According to his idea, Mixmaster universe is a chaotic system. As per that interpretation, if the universe was Mixmaster at initialized state, having independence with initial conditions, an isotropic universe will be explored in futuristic evolution. Nowadays different qualitative research works are being performed to investigate improvisation from orderly initialization to chaotic behaviour.

5.6 Application to Epidemics

Mathematical modelling provide powerful insight of the field epidemiology. Using notion of susceptibility, exposure, infected, recovery, different models have been proposed for various infectious diseases throughout the globe. In epidemics, the fundamental objective of modelling lies on the observation of infectious agents transmission through different compartments. Sir Ronald Ross published a collection of research works using mathematical interpretation of the life cycle of the malaria parasite during 1890–1897. Due to limitations of simple Ross model, researchers have extended it further incorporating the factors of anti-parasite immunity, age-dependent susceptibility, latent period etc. Ross Model can be written as

$$\frac{dI_h}{dt} = abmI_m(1 - I_h) - rI_h, \frac{dI_m}{dt} = acI_h(1 - I_m) - \mu_2 I_m \tag{19}$$

where a, b are the rate of bites per day and rate of bites producing infection, respectively. Here r denotes the recovery rate of humans, μ_2 is the mortality rate of mosquitos per capita, m is the ratio between male and female mosquitos. The breakthrough work of Kermack and Mckendrick on SIR model (1927) is important from every perspective [23]. They tried to observe the dynamics of the Great Plague in London (1666). The model was expressed as

$$\dot{S} = -\beta SI, \dot{I} = (\beta S - \alpha)I \tag{20}$$

Here, infectious agents leave the infected persons at rate α per unit time and β is the rate in which infection is transmitted through other population. Anderson and May proposed modified version of Ross model including age factor and time for population. The model can be written as follows:

$$\begin{aligned} \frac{\partial I_h(\alpha,t)}{\partial t} + \frac{\partial I_h(\alpha,t)}{\partial \alpha} &= abmI_m[N - I_h(\alpha, t)] - (r + \mu_1)I_h(\alpha, t), \\ \frac{dI_m}{dt} &= acI_h[N' - I_m] - \mu_2 I_m \end{aligned} \tag{21}$$

where N is total human population with respect to age α. He et al. presented an interesting SEIR model on COVID-19 introducing quarantined, hospitalized category:

$$\begin{aligned}
\dot{S} &= -\tfrac{S}{N}(\beta_1 I_1 + \beta_2 I_2 + \chi E) + \rho_1 Q - \rho_2 S + \alpha R,\\
\dot{E} &= \tfrac{S}{N}(\beta_1 I_1 + \beta_2 I_2 + \chi E) - \theta_1 E - \theta_2 E,\\
\dot{I}_1 &= \theta_1 E - \gamma_1 I_1,\\
\dot{I}_2 &= \theta_2 E - \gamma_2 I_2 - \psi I_2 + \lambda(\Lambda + Q),\\
\dot{R} &= \gamma_1 I_1 + \gamma_2 I_2 + \phi H - \alpha R,\\
\dot{H} &= \psi I_2 - \phi H,\\
\dot{Q} &= \Lambda + \rho_2 S - \lambda(\Lambda + Q) - \rho_1 Q
\end{aligned} \tag{22}$$

where α, β_i, θ_i, γ_i, ρ_i represent temporary immunity rate, contact and infection rate through infected category, transition rate of exposed to infected, recovery rate of symptomtic respectively. Different models have been proposed to observe dynamics of Ebola, novel corona virus and many more using SIS, SEIR, SIR. Still there is a scope to explore dynamics of the epidemic model through modification of previously used parameters, to observe chaotic features, if any.

5.7 Application to Cognitive Science

Next, we will discuss about research works on cognitive science. As an interdisciplinary area, cognitive science is the branch dealing with the linkage between human response and psychological constraints. Complexity of human brain is a great mystery to the researchers. Different studies have explored temporal patterns for several signals (heartbeat, brain activity) over various time scales. Nowadays, contemporary cognitive study becomes computational due to large advancement of neuroscience, machine learning. Functional complexities of processes generating time signals are captured by fractal dimension where a signal is said to follow fractal nature if it possesses a scale-free behaviour. Nonlinear dynamic approach is being used in different neurophysiological applications based on Functional Magnetic Resolution Imaging (fMRI), Magnetoencephalogram (MEG), Electroencephalogram (EEG), Electromyogram (EMG). EEG is one of the common techniques. What is EEG study? Basically understanding complex psychological dynamics in EEG signal based study is nonlinear dynamical analysis of time series of the electrical activity of the brain. Different activities are performed by different sectors of the brain. Frontal lobe controls emotion, problem solving while speech, memory or even recognition is inter-connected with the the temporal lobe. Different lobes of the brain produce non-stationary and complex EEG signals. Electrodes are placed at the earlobe and with respect to these electrodes, signals are simply electrical potentials. In general, 128 is the maximum number of electrodes required for study. Frequency bands can be classified into 4 classes—delta ($<$3.5 Hz), theta, alpha, beta ($<$30 Hz). EEG signal studies capture neuronal impulses in the scalp which is the assimilated result of action potentials orginated from interaction of neurons. There is a long history on the attempt of brain activity using electrical devices. Around 1672, Otto von

Guericke, German physicist took an attempt for animal experiment through the discovery of electrostatic device. Nobili made a static galvanometer in 1825, which was redesigned by William Thomson in 1858. We should mention the contribution of G. Fritsch and Edvard Hitzig for electric simulation experiment (1870) in cerebrum of a dog. Richard Caton, an English physiologist observed the fluctuations of brain activities of rabbits and monkeys using a sensitive galvanometer in 1875. Around 1890, Beck published his similar research works entitled "Investigations into the Physiology of the Brain". Pavel Kaufman and Vladimirovich Neminsky, renowned physiologists from Russia studied first ever EEG for animals in 1912. Later, Hans Berger came into the picture for his work on neural activity on human brain in 1924. Then the works of Charles Sherrington (1932), Bishop(1932) enriched this domain of cognitive science. Around 1985, Babloyantz et al. published a revolutionary work on EEG study based on human sleep. On the same time, Rapp et al. presented an interesting issue on a monkey's neurological activities in the motor cortex. Sterman remarked that the underlying process of human behaviour is not stable as compared to the laws of physics. So, here nonlinear techniques are effective to capture these complexities. Different researchers have studied EEG signals using Hurst exponent, which we have already discussed in the above. Multifractal Detrended Fluctuation Analysis (MFDFA) is very common technique to observe long and small term fluctuations in EEG signals. Peng et al. first published the idea of DFA technique in 1994 for correlational aspects of DNA sequence. Then as a generalized version, Kantelhardt et al. proposed MFDFA technique for nonstationary time series in 2002. MFDFA tool captured self-similarity of a stochastic process, so it is employed for neurocognitive emotional effects of music, for diagnosis of rare neurological disorders like epilepsy. But, what is the procedure for MFDFA? Basic methodologies have been pointed out as follows:

- Computation of mean value of EEG signals : $\overline{x} = \frac{\sum x(i)}{N}$
- Computation of integrated time series: $X(i) = \sum [x(i) - \overline{x}]$
- Calculation of RMS trend for different non-overlapping blocks: $f^2(n, v) = \frac{1}{n} \sum_{i=1}^{n} \{X[(v-1)n + i] - x_v(i)\}^2$
- Computation of qth order fluctuation function : $f_q(n) = \{\frac{1}{nN} \sum [f^2(n, v)]^{\frac{q}{2}}\}^{\frac{1}{q}}$
- Checking the variation of fluctuation function $f_q(n)$ with respect to sample size n: $f_q(n) \sim n^H$, H is generalized Hurst exponent.

5.8 Applicaton to Neural Network

Human brain, consisted of approximately hundred billion neurons is very complex in nature. Neurons transfer information among them to control human behaviour, cognitive reactions. Complex neuronal network is interesting domain of research for a long time. Mathematical modelling quantitatively describe anatomy and physiological aspects of neuron systems. Neurons collect input from synapses and the output stimulate other neurons. Initially neuron's synaptic weight is multiplied to activated input neuron function and nonlinearty comes into the frame through the threshold function. Suitable chaotic dynamics provides better application of neural network models. Lapicque presented mathematical model of neuron in 1907. Hodgkin and Huxley presented a remarkable idea on nonlinear feature of membrane of a neuron. For their significant contribution, they awarded Nobel prize in 1969. FitzHugh-Nagumo model (FHN) is another famous model of neuronal dynamics. Richard FitzHugh and Jinichi Nagumo presented this simplified two-dimensional case of Hodgkin-Huxley model. Initially they proposed the name as Bonhoeffer-Van der Pol oscillator because this model reduces to Van der pol oscillator form as $a = b = 0$. As the external input crosses the threshold quantity, the system exhibits special feature in phase space. The model can be expressed as

$$\dot{v} = v - \frac{v^3}{3} - w + RI_{ext}, \tau\dot{w} = v + a - bw \tag{23}$$

where the variable v denotes the membrane voltage while the variable w measures the activation and deactivation of ingredients (like sodium, potassium) through the neural channel. In recent time, Cui et al. presented a chaotic neural network model using internal and external membrane conductance via Hopfield model as follows:

$\dot{x_i} = \sum_{j=1}^{3} S_{ij}x_i^T + \sum_{j=1}^{3} W_{ij}V_j + I_i$ where x_i represents the voltage on respective capacitor C_i, I_i is nonlinear input current with the matrix S_{ij} representing conductance of membranes in between neurons [27]. W_{ij} is another matrix representing synaptic weight of the neuron. Considering $n = 3$, $C_i = 1$, $I_i = \begin{pmatrix} 0 \\ 0 \\ i_3 \end{pmatrix}$, $i_3 = A\sin(x_1)$, $A = 20$ and the activation function $V_j = \tanh(x_i)$, they presented chaotic behaviour through the model. The system can be written as follows:

$$\begin{aligned} \dot{x_1} &= 2x_2 + 5\tanh(x_1) - \tanh(x_3), \\ \dot{x_2} &= x_3 - 6\tanh(x_1) + 2\tanh(x_2) + 4\tanh(x_3), \\ \dot{x_3} &= -3x_2 - 5x_3 - \tanh(x_1) - 3\tanh(x_2) + 10\tanh(x_3) + 20\sin(x_1) \end{aligned} \tag{24}$$

Nowadays different deep learning tools are being employed in this domain and gradually it is becoming a powerful field. Many useful research works can be explored using neural model and chaotic systems to observe different features of human brain, different features of mental trauma, mental disabilities.

5.9 Application to Cryptography

Cryptology is the study of secured communication in presence of unauthorised entities while chaos measures uncertainty of nonlinear dynamical systems. The term

'Crytography' was derived from Greek *kryptos logia*, whose meaning is the study of secrecy. With the large improvement of technology, we are too much dependent on insecure channels for communications. Using finite set of integers, different mathematical operations are performed to ensure confidentiality, integrity of transmitted data. To secure the transmission between the sender and receiver from the unauthorised eavesdropper, encryption and decryption techniques are performed. In first case, original message (or even image, audio, video) is converted into some ambiguous, complex, mystic codes via sender's private key. Then that is transmitted over the insecure platform. After receiving this mystic code, the receiver use his/her personal key to decipher the complex code. This technique of ciphering is not new at all. There are lots of evidential proofs on use of cipher. Hieroglyphs of Egypt, Scytale cipher from Greece are few common examples. Roman emperor Julius Caesar used an easy technique for ciphering while the Enigma machine used by German militaries during 2nd World War was too hard to break. Nowadays, the level of complexity in cryptographic algorithms is naturally too high. On the other hand, Randomness, sensitivity to initial conditions are important features of chaotic maps. So, nonlinear chaotic maps are employed for different encryption schemes. Confusion and diffusion are two stages in chaos-based image encryption. Firstly chaotic maps play the role of pseudo-random number generator. Initially, permutation operation is performed on pixel values of the original image. Then modifying the control parameter and performing several iterations, the original is converted into unreadable format. During diffusion stage, pixel values are arranged in sequential manner and finally the original image can be obtained from the encrypted image. In practice, it is difficult to predict future behaviour using initial iterations managing limited keyspace and time limit. In 1949, Shannon proposed idea of chaotic cryptography. For symmetric key encryption, he proposed the usefulness of Baker's map. L. Koracev pointed out the usefulness of chaotic cryptography in his famous paper 'Chaos-based cryptography: a brief overview'. There are enough literatures on this domain of chaotic cryptography. Pareek et al. implemented the technique of cross-coupling using two skew tent maps. On the other hand, the works of Huang et al. was very interesting, for the pixel shuffling via multi chaotic systems. Multi chaotic systems are used nowadays for remote sensing. Once Guan et al. used Arnold Cat map for spatial domain to encrypt and used Chen system to output signal. In recent times, different experiments have been performed using Henon and logistic maps for tele-medicine issues. Faragallah et al. nicely used Baker map, logistic and Arnold Cat map to observe efficiency of chaotic encryption in Fractional Fourier Transform domain. Still there is a scope to present efficient algorithms for image, video encryption through different combination of chaotic maps with necessary large key space (Figs. 12 and 13).

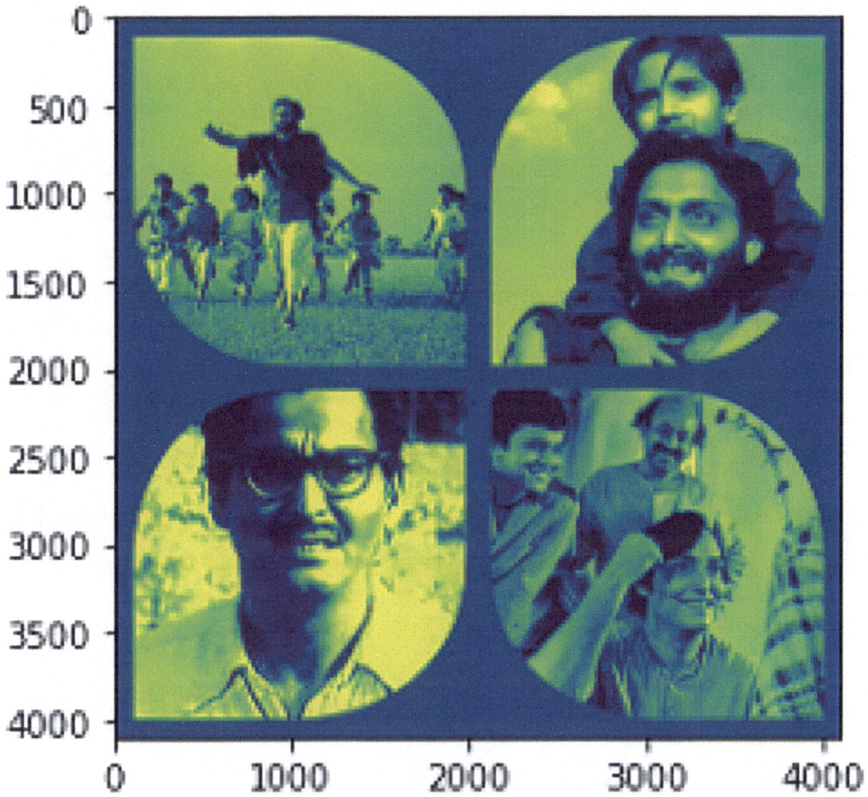

Fig. 12 Partial modification due to chaotic map

6 Conclusion

Nonlinear systems are widely used in various fields including engineering, ecological control, economic fluctuations, visual encryption etc. Self-similarity notion of fractals, chaotic behaviours are prominent in different aspects of our daily life. In terms of security and stability, complexity of the nonlinear system is an important part. After development in terms of continuous as well as discrete models, researchers are trying to explore other unknown characteristic of nonlinear systems, trying to control chaos and implementation in different exciting, real-life problems. In this regard, basic idea of nonlinear dynamical systems have been discussed in this chapter with relevant examples. Typical features of nonlinear system, different from linear model have been pointed out. And finally, applications of nonlinear systems in various fields have been discussed shortly. Still there is a long way to go for further exploration.

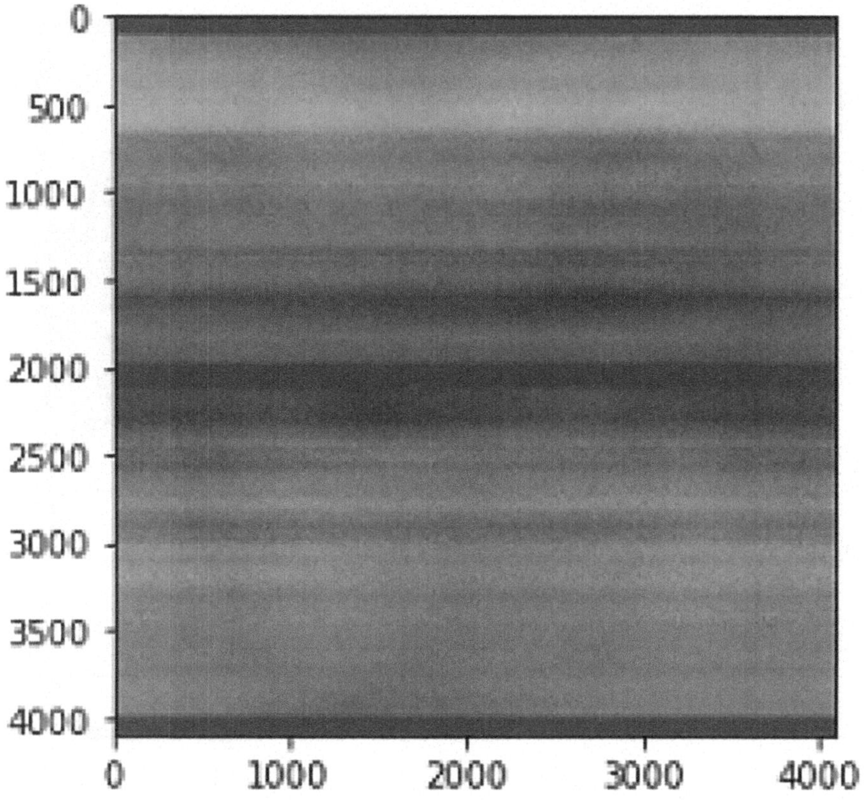

Fig. 13 Encrypted image

Competing Interests The authors have not received funding, research grant from any government or private institute, agency or anywhere else. The authors have no conflicts of interest to declare that are relevant to the content of this chapter.

Ethics Approval This chapter does not include any study of humans or other class of animals. No legal consent or ethics approval is required.

Acknowledgements The authors convey their heartiest gratitudes to the anonymous reviewers for their valuable comments, suggestions, rectifications to improve the content.

References

1. E.N. Lorenz, *The Essence of Chaos* (University of Washington Press, Seattle, WA, 1993)
2. S.H. Strogatz, *Nonlinear Dynamics and Chaos: With Application to Physics, Biology, Chemistry and Engineering*, 1st edn. (CRC Press, 2000). ISBN-13: 9780738204536
3. M. Lakshmanan, S. Rajasekar, *Nonlinear Dynamics: Integrability, Chaos and Patterns* (Springer, Berlin, Heidelberg, 2003)
4. M. Henon, A two-dimensional mapping with a strange attractor. Commun. Math. Phys. **50**(1), 69–77 (1976)
5. H. Poincaré, *Les Méthodes Nouvelles de la Mécanique Céleste*, vol. 1 (Gauthier-Villars, Paris, 1892)
6. B.B. Mandelbrot, *The Fractal Geometry of Nature* (Freeman, N.Y., 1977)
7. M.F. Barnsley, *Fractals Everywhere*, 2nd edn. (Morgan Kaufmann, USA, 1993)
8. A.A. Blumberg, Logistic growth rate functions. J. Theor. Biol. **21**, 42–44 (1968)
9. T. Thornes, P. Düben, P. Palmer, On the use of scale-dependent precision in Earth system modelling. Q. J. R. Meteorol. Soc. **143**(703), 897–908 (2017)
10. M.J. Feigenbaum, Quantitative universality for a class of nonlinear transformations. J. Stat. Phys. **19**(1), 25–52 (1978)
11. G. Sugihara, B. Grenfell, R.M. May, P. Chesson, H. Platt, M. Williamson, Distinguishing error from chaos in ecological time series [and discussion]. Philos. Trans. R. Soc. Lond. B Biol. Sci. **330**(1257), 235–251 (1990)
12. R.M. May, Simple mathematical models with very complicated dynamics. Nature **261**, 459–467 (1976)
13. L. Alsedà, J.T. Lazaró, R. Solé, B. Vidiella, J. Sardanyeś, Dynamics in a Time-Discrete Food-Chain Model with Strong Pressure on Preys. arXiv 1909.12501v1 (2019)
14. C.S. Holling, The functional response of predators to prey density and its role in mimicry and population regulation. Mem. Entomol. Soc. Can. **97**, 5–60 (1965)
15. A. Das, G.P. Samanta, Modeling the fear effect on a stochastic prey-predator system with additional food for the predator. J. Phys. A Math. Theor. **51**, 465601 (2018)
16. S. Mondal, G. Samanta, A comparison study of predator-prey system in deterministic and stochastic environments influenced by fear and its carry-over effects. Eur. Phys. J. Plus **137**, 70 (2022)
17. A. Einstein, Cosmological considerations in the general theory of relativity, Sitzungsber. Preuss. A Kad. Wiss, Berlin, Math. Phys. 142–152 (1917)
18. A. Einstein, W. de Sitter, On the relation between the expansion and the mean density of the universe. Proc. Natl. Acad. Sci. USA **18**, 213–214 (1932)
19. C.W. Misner, Mixmaster universe. Phys. Rev. Lett. **22**, 1071–1074 (1969)
20. J.D. Barrow, Chaotic behaviour in general relativity. Phys. Rep. **85**, 1–49 (1982)
21. F.S. Labini, M. Montuori, L. Pietronero, Scale invariance of galaxy clustering. Phys. Rep. **293** (1998)
22. R. Ross, H. Hudson, An application of the theory of probabilities to the study of a priori pathometry—Part II. Proc. R. Soc. Lond. A. **93**, 212–225 (1917)
23. W. Kermack, A.G. McKendrick, A contribution to the mathematical theory of epidemics. Proc. R. Soc. Lond. A. **115**, 700–721 (1927)
24. R.K. Upadhyay, A.K. Pal, S. Kumari, P. Roy, Dynamics of an SEIR epidemic model with nonlinear incidence and treatment rates. Nonlinear Dyn. **96**(4), 2351–2368 (2019)
25. S. He, Y. Peng, K. Sun, SEIR modeling of the COVID-19 and its dynamics. Nonlinear Dyn. **101**, 1667–1680 (2020)
26. S. Mandal, R.P. Sarkar, S. Sinha, Mathematical models of malaria—a review. Malaria J. **10**(202), 1–19 (2011)
27. L. Cui, C. Chen, J. Jin, F. Yu, Dynamic analysis and FPGA implementation of new chaotic neural network and optimization of traveling salesman problem. Complexity **2021**, Article ID 5521192 (2021). https://doi.org/10.1155/2021/5521192

28. M. Paterson, D. Stinson, *Cryptography: Theory and Practice*, 4th edn. (CRC Press, 2018)
29. G. Chen, Y. Mao, C.K. Chui, A symmetric image encryption scheme based on 3d chaotic cat maps. Chaos Solitons Fractals **21**(3), 749–761 (2004)
30. T. Gao, Z. Chen, A new image encryption algorithm based on hyper-chaos. Phys. Lett. A **372**(4), 394–400 (2008)
31. X. Wang, L. Liu, Y. Zhang, A novel chaotic block image encryption algorithm based on dynamic random growth technique. Opt. Lasers Eng. **66**, 10–18 (2015)
32. Y.Q. Zhang, X.Y. Wang, A new image encryption algorithm based on non-adjacent coupled map lattices. Appl. Soft Comput. **26**, 10–20 (2015)
33. P. Ping, F. Xu, Y. Mao, Z. Wang, Designing permutation—substitution image encryption networks with Henon map. Neurocomputing **283**, 53–63 (2018)
34. B. Lu, F. Liu, X. Ge, Z. Li, Cryptanalysis and improvement of a chaotic map-control-based and the plain image-related cryptosystem. Comput. Mater. Continua **61**(2), 687–699 (2019)
35. S. Banerjee, A.P. Misra, N.A. Fataf, A. Roy, H. Natiq, Enhancing chaos in multistability regions of Duffing map for an asymmetric image encryption algorithm. arxiv (2020)
36. N. Mukherjee, S. Poria, Preliminary concepts of dynamical systems. Int. J. Appl. Math. Res. **1**(4), 751–770 (2012)

Sasaki Gradient Tensor and Para-Complex Structures

Lakehal Belarbi and Hichem Elhendi

Abstract In this chapter we will talk about geometry on the tangent bundle $\mathcal{TM}$ equipped with Sasaki gradient metric and a paracomplex structure we study the integrability of the paracomplex and we calculate the tensor almost product connection which allows us to calculate the connection and torsion of almost product connection.

1 Introduction

The theory of the bundle Riemannian manifolds $\mathcal{TM}$ was first studied by Sasaki S. in his paper published in 1958 [22], where he has been using the connection on the manifold to define lifting noted by $\mathcal{F}_H$ and $\mathcal{G}_V$ by splitting the tangent of bundle $\mathcal{TM}$ between two parties horizontal and vertical, he also defined locally what was later generalized and we now know as the Sasaki metric G^S on $(x, u) \in \mathcal{TM}$:

$$G^S_{(x,u)}(\mathcal{F}_\mathcal{H}, \mathcal{G}_\mathcal{H}) = G_x(\mathcal{F}, \mathcal{G}) \circ \pi,$$
$$G^S_{(x,u)}(\mathcal{F}_\mathcal{V}, \mathcal{G}_\mathcal{V}) = G_x(\mathcal{F}, \mathcal{G}) \circ \pi,$$
$$G^S_{(x,u)}(\mathcal{F}_\mathcal{V}, \mathcal{G}_\mathcal{H}) = 0.$$

In 1962 Dombrowski [17], gave a definition of the horizontal and vertical part of $T\mathcal{TM}$.

Starting from there in 1966 Yano K., Kobayashi S. and Ishihara S. started developing the theory of vertical, complete and horizontal lift [24].

L. Belarbi (✉)
Department of Mathematics, Laboratory of Pure and Applied Mathematics, University of Mostaganem, Mostaganem, Algeria
e-mail: lakehalbelarbi@gmail.com

H. Elhendi
Department of Mathematics, University of Bechar, PO Box 417, 08000 Bechar, Algeria
e-mail: elhendihichem@univ-bechar.dz

M. Kharrat et al. (eds.), *Modeling of Discrete and Continuous Systems*, Forum for Interdisciplinary Mathematics, https://doi.org/10.1007/978-981-97-8715-9_15

In 1995 Cruceanu et al. [16], gave some properties on paracomplex geometry on a differential manifold.

2 Sasaki Gradient Tensor and Para-Complex Structures

2.1 Sasaki Gradient Tensor

Definition 1 Let h be a strictly smooth function, on the bundle $\mathcal{TM}$ we define a Sasaki gradient tensor by

1. $G_h(\mathcal{F}_\mathcal{H}, \mathcal{G}_\mathcal{H})_{(x,u)} = G_x(\mathcal{F}, \mathcal{G})$,
2. $G_h(\mathcal{F}_\mathcal{H}, \mathcal{G}_\mathcal{V})_{(x,u)} = 0$,
3. $G_h(\mathcal{F}_\mathcal{V}, \mathcal{G}_\mathcal{V})_{(x,u)} = G_x(\mathcal{F}, \mathcal{G}) + \mathcal{F}_x(h)\mathcal{G}_x(h)$,

where $\mathcal{F}, \mathcal{G} \in \Gamma(\mathcal{TM})$, $(x, u) \in \mathcal{TM}$.

2.2 Levi-Civita Connection

Theorem 1 *[8] The tensor connection of bundle $\mathcal{TM}$ equipped with the Sasaki gradient metric G_h is given by*

$$
\begin{aligned}
(\nabla^h_{\mathcal{F}_\mathcal{H}} \mathcal{G}_\mathcal{H})_x &= ((\nabla_\mathcal{F}\mathcal{G})_x)_\mathcal{H} - \frac{1}{2}((\mathcal{R}(\mathcal{F}, \mathcal{G})u)_x)_\mathcal{V}, \\
(\nabla^h_{\mathcal{F}_\mathcal{H}} \mathcal{G}_\mathcal{V})_x &= \frac{1}{2}((\mathcal{R}(u, \mathcal{G})\mathcal{F})_x)_\mathcal{H} + \frac{1}{2}\mathcal{G}(h)(\mathcal{R}(u, \nabla h)\mathcal{F})_\mathcal{H} + (\nabla_\mathcal{F}\mathcal{G})_V \\
&\quad + \frac{1}{2}\mathcal{G}(h)(\nabla_\mathcal{F}\nabla h)^V + \frac{1}{2\alpha}\Big[g(\mathcal{G}, \nabla_\mathcal{F}\nabla h) - \frac{1}{2}\mathcal{F}(\alpha)\mathcal{G}(h)\Big](\nabla h)_\mathcal{V}, \\
(\nabla^h_{\mathcal{F}_\mathcal{V}} \mathcal{G}_\mathcal{H})_x &= \frac{1}{2}((\mathcal{R}(u, \mathcal{F})\mathcal{G})_x)_\mathcal{H} + \frac{1}{2}\mathcal{F}(h)(\mathcal{R}(u, \nabla h)\mathcal{G})_\mathcal{H} + \frac{1}{2}\mathcal{F}(h)(\nabla_\mathcal{G}\nabla h)_\mathcal{V} \\
&\quad + \frac{1}{2\alpha}\Big[G(\mathcal{F}, \nabla_\mathcal{G}\nabla h) - \frac{1}{2}\mathcal{G}(\alpha)\mathcal{F}(h)\Big](\nabla h)_\mathcal{V}, \\
(\nabla^h_{\mathcal{F}_\mathcal{V}} \mathcal{G}_\mathcal{V})_x &= -\frac{1}{2}\mathcal{F}(h)(\nabla_\mathcal{G}\nabla h)_\mathcal{H} - \frac{1}{2}(\mathcal{G}(h)(\nabla_\mathcal{F}\nabla h)_\mathcal{H},
\end{aligned}
$$

$\forall \mathcal{F}, \mathcal{G} \in \Gamma(\mathcal{TM})$, where $\alpha = 1 + G(\nabla h, \nabla h)$.

2.3 Para-Kähler-Norden Structures of Sasaki Gradient Metric of Tangent Bundle

On the bundle $\mathcal{TM}$, we defined the almost para-complex structure by

$$\begin{cases} J\mathcal{F}_{\mathcal{H}} = -\mathcal{F}_{\mathcal{H}}, \\ J\mathcal{F}_{\mathcal{V}} = \mathcal{F}_{\mathcal{V}}, \end{cases} \tag{1}$$

$\forall \mathcal{F} \in \Gamma(\mathcal{TM})$.

Lemma 1 *Let J the almost para-complex structure defined by (1), then we have*

$$G_h(J\mathcal{F}_k, \mathcal{G}_h) = G_h(\mathcal{F}_k, J\mathcal{G}_h).$$

Proof

$$\begin{aligned} G_h(J\mathcal{F}_{\mathcal{H}}, \mathcal{G}_{\mathcal{H}}) &= G_h(-\mathcal{F}_{\mathcal{H}}, \mathcal{G}_{\mathcal{H}}) = G_h(\mathcal{F}_{\mathcal{H}}, -\mathcal{G}_{\mathcal{H}}) = G_h(\mathcal{F}_{\mathcal{H}}, J\mathcal{G}_{\mathcal{H}}), \\ G_h(J\mathcal{F}_{\mathcal{H}}, \mathcal{G}_{\mathcal{V}}) &= G_h(-\mathcal{F}_{\mathcal{H}}, \mathcal{G}_{\mathcal{V}}) = 0 = G_h(\mathcal{F}_{\mathcal{H}}, \mathcal{G}_{\mathcal{V}}) = G_h(\mathcal{F}_{\mathcal{H}}, J\mathcal{G}_{\mathcal{V}}), \\ G_h(J\mathcal{F}_{\mathcal{V}}, \mathcal{G}_{\mathcal{H}}) &= G_h(\mathcal{F}_{\mathcal{V}}, \mathcal{G}_{\mathcal{H}}) = 0 = G_h(\mathcal{F}_{\mathcal{V}}, -\mathcal{G}_{\mathcal{H}}) = G_h(\mathcal{F}_{\mathcal{V}}, J\mathcal{G}_{\mathcal{H}}), \\ G_h(J\mathcal{F}_{\mathcal{V}}, \mathcal{G}_{\mathcal{V}}) &= G_h(\mathcal{F}_{\mathcal{V}}, \mathcal{G}_{\mathcal{V}}) = G_h(\mathcal{F}_{\mathcal{V}}, J\mathcal{G}_{\mathcal{V}}). \end{aligned}$$

□

From Lemma 1, we have

Theorem 2 *The triple $(\mathcal{TM}, J, G_h)$ is an almost para-complex Norden manifold.*

Proposition 1 *For all $\mathcal{F}, \mathcal{G}, \mathcal{Z} \in \Gamma(\mathcal{TM})$, we have*

1. $(\phi_J G_h)(\mathcal{F}_{\mathcal{H}}, \mathcal{G}_{\mathcal{H}}, \mathcal{Z}_{\mathcal{H}}) = 0$,
2. $(\phi_J G_h)(\mathcal{F}_{\mathcal{V}}, \mathcal{G}_{\mathcal{H}}, \mathcal{Z}_{\mathcal{H}}) = 0$,
3. $(\phi_J G_h)(\mathcal{F}_{\mathcal{H}}, \mathcal{G}_{\mathcal{V}}, \mathcal{Z}_{\mathcal{H}}) = 2G(R(\mathcal{Z}, \mathcal{F})u, \mathcal{G}) + 2\mathcal{G}(h)g(\mathcal{R}(\mathcal{Z}, \mathcal{F})u, \nabla h)$,
4. $(\phi_J G_h)(\mathcal{F}_{\mathcal{H}}, \mathcal{G}_{\mathcal{H}}, \mathcal{Z}_{\mathcal{V}}) = 2G(\mathcal{R}(\mathcal{G}, \mathcal{F})u, \mathcal{Z}) + 2\mathcal{Z}(f)G(\mathcal{R}(\mathcal{G}, \mathcal{F})u, \nabla h)$,
5. $(\phi_J G_h)(\mathcal{F}_{\mathcal{V}}, \mathcal{G}_{\mathcal{V}}, \mathcal{Z}_{\mathcal{H}}) = 0$,
6. $(\phi_J G_h)(\mathcal{F}_{\mathcal{V}}, \mathcal{G}_{\mathcal{H}}, \mathcal{Z}_{\mathcal{V}}) = 0$,
7. $(\phi_J G_h)(\mathcal{F}_{\mathcal{H}}, \mathcal{G}_{\mathcal{V}}, \mathcal{Z}_{\mathcal{V}}) = -2\mathcal{Z}(f)G(\mathcal{G}, \nabla_{\mathcal{F}}\nabla h) - 2\mathcal{G}(h)G(\mathcal{Z}, \nabla_{\mathcal{F}}\nabla h)$,
8. $(\phi_J G_h)(\mathcal{F}_{\mathcal{V}}, \mathcal{G}_{\mathcal{V}}, \mathcal{Z}_{\mathcal{V}}) = 0$.

Proof

$$\begin{aligned} 1.\ \phi_J G_h(\mathcal{F}_{\mathcal{H}}, \mathcal{G}_{\mathcal{H}}, \mathcal{H}_{\mathcal{H}}) &= (J\mathcal{F}_{\mathcal{H}})G_h(\mathcal{G}_{\mathcal{H}}, \mathcal{H}_{\mathcal{H}}) - \mathcal{F}_{\mathcal{H}}G_h(J\mathcal{G}_{\mathcal{H}}, \mathcal{H}_{\mathcal{H}}) \\ &\quad + G_h\big((L_{\mathcal{G}_{\mathcal{H}}}J)\mathcal{F}_{\mathcal{H}}, \mathcal{H}_{\mathcal{H}}\big) + G_h\big(\mathcal{G}_{\mathcal{H}}, (L_{\mathcal{H}_{\mathcal{H}}}J)\mathcal{F}_{\mathcal{H}}\big) \\ &= (-\mathcal{F}_{\mathcal{H}})G_h(\mathcal{G}_{\mathcal{H}}, \mathcal{H}_{\mathcal{H}}) - \mathcal{F}_{\mathcal{H}}G_h(-\mathcal{G}_{\mathcal{H}}, \mathcal{H}_{\mathcal{H}}) \\ &\quad + G_h\Big(L_{\mathcal{G}_{\mathcal{H}}}(J\mathcal{F}_{\mathcal{H}}) - J(L_{\mathcal{G}_{\mathcal{H}}}\mathcal{F}_{\mathcal{H}}), \mathcal{H}_{\mathcal{H}}\Big) \end{aligned}$$

$$
\begin{aligned}
&+G_h\Big(\mathcal{G}_{\mathcal{H}}, (L_{\mathcal{H}_{\mathcal{H}}} J\mathcal{F}_{\mathcal{H}}) - J(L_{\mathcal{H}_{\mathcal{H}}}(\mathcal{F}_{\mathcal{H}}))\Big)\\
&= G_h\Big([\mathcal{G}_{\mathcal{H}}, -\mathcal{F}_{\mathcal{H}}] - J([\mathcal{G}_{\mathcal{H}}, \mathcal{F}_{\mathcal{H}}]), \mathcal{H}_{\mathcal{H}}\Big)\\
&+G_h\Big(\mathcal{G}_{\mathcal{H}}, [\mathcal{H}_{\mathcal{H}}, -\mathcal{F}_{\mathcal{H}}] - J([\mathcal{H}_{\mathcal{H}}, \mathcal{F}_{\mathcal{H}}])\Big)\\
&= G\Big(-[\mathcal{G}, \mathcal{F}]_{\mathcal{H}} + (R(\mathcal{G}, \mathcal{F})u)_{\mathcal{V}}, \mathcal{H}_{\mathcal{V}}\Big)\\
&-G\Big(J([\mathcal{G}, \mathcal{F}]_{\mathcal{H}} - (R(\mathcal{G}, \mathcal{F})u)_{\mathcal{V}}), \mathcal{H}_{\mathcal{V}}\Big)\\
&+G\Big(-[\mathcal{H}, \mathcal{F}]_{\mathcal{H}} + (R(\mathcal{H}, \mathcal{F})u)_{\mathcal{V}}, \mathcal{G}_{\mathcal{V}}\Big)\\
&-G\Big(J([\mathcal{H}, \mathcal{F}]_{\mathcal{H}} - (R(\mathcal{H}, \mathcal{F})u)_{\mathcal{V}}), \mathcal{G}_{\mathcal{V}}\Big)\\
&= 0,
\end{aligned}
$$

$$
\begin{aligned}
2.\ \phi_J G_h(\mathcal{F}_{\mathcal{V}}, \mathcal{G}_{\mathcal{H}}, \mathcal{H}_{\mathcal{H}}) &= (J\mathcal{F}_{\mathcal{V}})G_h(\mathcal{G}_{\mathcal{H}}, \mathcal{H}_{\mathcal{H}}) - \mathcal{F}_{\mathcal{V}}G_h(J\mathcal{G}_{\mathcal{H}}, \mathcal{H}_{\mathcal{H}})\\
&\quad + G_h\big((L_{\mathcal{G}_{\mathcal{H}}} J)\mathcal{F}_{\mathcal{V}}, \mathcal{H}_{\mathcal{H}}\big) + G_h\big(\mathcal{G}_{\mathcal{H}}, (L_{\mathcal{H}_{\mathcal{H}}} J)\mathcal{F}_{\mathcal{V}}\big)\\
&= (\mathcal{F}_{\mathcal{V}})G_h(\mathcal{G}_{\mathcal{H}}, \mathcal{H}_{\mathcal{H}}) - \mathcal{F}_{\mathcal{V}}G_h(-\mathcal{G}_{\mathcal{H}}, \mathcal{H}_{\mathcal{H}})\\
&\quad + G_h\big((L_{\mathcal{G}_{\mathcal{H}}} J\mathcal{F}_{\mathcal{V}}) - J(L_{\mathcal{G}_{\mathcal{H}}}\mathcal{F}_{\mathcal{V}}), \mathcal{H}_{\mathcal{H}}\big)\\
&\quad + G_h\big(\mathcal{G}_{\mathcal{H}}, (L_{\mathcal{H}_{\mathcal{H}}} J\mathcal{F}_{\mathcal{V}}) - J(L_{\mathcal{H}_{\mathcal{H}}}\mathcal{F}_{\mathcal{V}})\big)\\
&= (\mathcal{F}_{\mathcal{V}})G_h(\mathcal{G}, \mathcal{H}) - \mathcal{F}_{\mathcal{V}}G_h(-\mathcal{G}, \mathcal{H})\\
&\quad + G_h\big((\nabla_{\mathcal{G}}\mathcal{F})_{\mathcal{V}} - J((\nabla_{\mathcal{G}}\mathcal{F})_{\mathcal{V}}), \mathcal{H}_{\mathcal{H}}\big)\\
&\quad + G_h\big(\mathcal{G}_{\mathcal{H}}, (\nabla_{\mathcal{H}}\mathcal{F})_{\mathcal{V}}) - J((\nabla_{\mathcal{H}}\mathcal{F})_{\mathcal{V}})\big) = 0,\\
3.\ \phi_J G_h(\mathcal{F}_{\mathcal{H}}, \mathcal{G}_{\mathcal{V}}, \mathcal{H}_{\mathcal{H}}) &= G_h([\mathcal{G}_{\mathcal{V}}, -\mathcal{F}_{\mathcal{H}}], \mathcal{H}_{\mathcal{H}}) - G_h(J([\mathcal{G}_{\mathcal{V}}, \mathcal{F}_{\mathcal{H}}]), \mathcal{H}_{\mathcal{H}})\\
&\quad G_h([\mathcal{H}_{\mathcal{H}}, -\mathcal{F}_{\mathcal{H}}], \mathcal{G}_{\mathcal{V}}) - G_h(J([\mathcal{H}_{\mathcal{H}}, \mathcal{F}_{\mathcal{H}}]), \mathcal{G}_{\mathcal{V}})\\
&= G_h((\nabla_{\mathcal{F}}\mathcal{G})_{\mathcal{V}}, \mathcal{H}_{\mathcal{H}}) - G_h(J(-\nabla_{\mathcal{F}}\mathcal{G})_{\mathcal{V}}, \mathcal{H}_{\mathcal{H}})\\
&\quad + G_h(-[\mathcal{H}, \mathcal{F}]_{\mathcal{H}} + (\mathcal{R}(\mathcal{H}, \mathcal{F})u)_{\mathcal{V}}, \mathcal{G}_{\mathcal{V}})\\
&\quad - G_h(J[\mathcal{H}, \mathcal{F}]_{\mathcal{H}} + J(-\mathcal{R}(\mathcal{H}, \mathcal{F})u)_{\mathcal{V}}, \mathcal{G}_{\mathcal{V}})\\
&= 2G_h((R(\mathcal{H}, \mathcal{F})u)_{\mathcal{V}}, \mathcal{G}_{\mathcal{V}})\\
&= 2G((R(\mathcal{H}, \mathcal{F})u), \mathcal{G}) + 2\mathcal{G}(h)G(\mathcal{R}(\mathcal{H}, \mathcal{F})u, \nabla h),\\
4.\ \phi_J G_h(\mathcal{F}_{\mathcal{H}}, \mathcal{G}_{\mathcal{H}}, \mathcal{H}^{\mathcal{V}}) &= G_h([\mathcal{G}_{\mathcal{H}}, -\mathcal{F}_{\mathcal{H}}], \mathcal{H}_{\mathcal{V}}) - G_h(J([\mathcal{G}_{\mathcal{H}}, \mathcal{F}_{\mathcal{H}}]), \mathcal{H}_{\mathcal{V}})\\
&= 2G((\mathcal{R}(\mathcal{G}, \mathcal{F})u), \mathcal{H}) + 2\mathcal{H}(h)G(R(\mathcal{G}, \mathcal{F})u, \nabla h),\\
5.\ \phi_J G_h(\mathcal{F}_{\mathcal{V}}, \mathcal{G}_{\mathcal{V}}, \mathcal{H}_{\mathcal{H}}) &= G([\mathcal{G}_{\mathcal{V}}, \mathcal{F}_{\mathcal{V}}], \mathcal{H}_{\mathcal{H}}) - G_h(J([\mathcal{G}_{\mathcal{V}}, \mathcal{F}_{\mathcal{V}}]), \mathcal{H}_{\mathcal{H}})\\
&\quad + G_h([\mathcal{H}_{\mathcal{H}}, \mathcal{F}_{\mathcal{V}}], \mathcal{G}_{\mathcal{V}}) - G_h(J([\mathcal{H}_{\mathcal{H}}, \mathcal{F}_{\mathcal{V}}]), \mathcal{G}_{\mathcal{V}})\\
&= G_h(\nabla_{\mathcal{H}}\mathcal{F}, \mathcal{G}) - G_h(\nabla_{\mathcal{H}}\mathcal{F}, \mathcal{G}) = 0,\\
6.\ \phi_J G_h(\mathcal{F}_{\mathcal{V}}, \mathcal{G}_{\mathcal{H}}, \mathcal{H}_{\mathcal{V}}) &= G_h([\mathcal{G}_{\mathcal{H}}, \mathcal{F}_{\mathcal{V}}], \mathcal{H}_{\mathcal{V}}) - G_h(J([\mathcal{G}_{\mathcal{H}}, \mathcal{F}_{\mathcal{V}}]), \mathcal{H}_{\mathcal{V}}) = 0,\\
7.\ \phi_J G_h(\mathcal{F}_{\mathcal{V}}, \mathcal{G}_{\mathcal{V}}, \mathcal{H}_{\mathcal{V}}) &= \mathcal{F}_{\mathcal{V}}G_h(\mathcal{G}_{\mathcal{V}}, \mathcal{H}_{\mathcal{V}}) - \mathcal{F}_{\mathcal{V}}G_h(\mathcal{G}_{\mathcal{V}}, \mathcal{H}_{\mathcal{V}}) + G_h([\mathcal{G}_{\mathcal{V}}, \mathcal{F}_{\mathcal{V}}], \mathcal{H}_{\mathcal{V}})\\
&\quad - G_h(J[\mathcal{G}_{\mathcal{V}}, \mathcal{F}_{\mathcal{V}}], \mathcal{H}_{\mathcal{V}}) + G_h([\mathcal{H}_{\mathcal{V}}, \mathcal{F}_{\mathcal{V}}], \mathcal{G}_{\mathcal{V}})\\
&\quad - G(J[\mathcal{H}_{\mathcal{V}}, \mathcal{F}_{\mathcal{V}}], \mathcal{G}_{\mathcal{V}}) = 0,
\end{aligned}
$$

$$
\begin{aligned}
8.\ \phi_J G_h(\mathcal{F}_{\mathcal{H}}, \mathcal{G}_{\mathcal{V}}, \mathcal{H}_{\mathcal{V}}) &= -\mathcal{F}_{\mathcal{H}} G_h(\mathcal{G}_{\mathcal{V}}, \mathcal{H}_{\mathcal{V}}) - \mathcal{F}_{\mathcal{H}} G_h(\mathcal{G}_{\mathcal{V}}, \mathcal{H}_{\mathcal{V}}) \\
&+ G_h([\mathcal{G}_{\mathcal{V}}, -\mathcal{F}_{\mathcal{H}}], \mathcal{H}_{\mathcal{V}}) - G_h(J[\mathcal{G}_{\mathcal{V}}, \mathcal{F}_{\mathcal{H}}], \mathcal{H}_{\mathcal{V}}) \\
&+ G_h([\mathcal{H}_{\mathcal{V}}, -\mathcal{F}_{\mathcal{H}}], \mathcal{G}_{\mathcal{V}}) \\
&- G(J[\mathcal{H}_{\mathcal{V}}, \mathcal{F}_{\mathcal{H}}], \mathcal{G}_{\mathcal{V}}) \\
&= -2\mathcal{F}_{\mathcal{H}}\Big(g(\mathcal{G}, \mathcal{H}) + \mathcal{G}(h)\mathcal{H}(h)\Big) + 2G((\nabla_{\mathcal{F}}\mathcal{G})_{\mathcal{V}}, \mathcal{H}_{\mathcal{V}}) \\
&+ 2G((\nabla_{\mathcal{F}}\mathcal{H})_{\mathcal{V}}, \mathcal{G}_{\mathcal{V}}) \\
&= -2\Big[G(\nabla_X Y, \mathcal{H}) + G(\mathcal{G}, \nabla_X \mathcal{H}) + \mathcal{H}(h)\big(G(\nabla_{\mathcal{F}}\mathcal{G}, \nabla h) \\
&+ G(\mathcal{G}, \nabla_{\mathcal{F}}\nabla h)\big) + \mathcal{G}(h)\big(G(\nabla_{\mathcal{F}}\mathcal{H}, \nabla h) + G(\mathcal{H}, \nabla_{\mathcal{F}}\nabla h)\big)\Big] \\
&+ 2\Big[G((\nabla_{\mathcal{F}}\mathcal{G}), \mathcal{H}) + (\nabla_X Y)(h)\mathcal{H}(h) \\
&+ G(\nabla_{\mathcal{F}}\mathcal{H}, \mathcal{G}) + (\nabla_X \mathcal{H})(h)\mathcal{G}(h)\Big] \\
&= -2G(\mathcal{G}, \nabla_{\mathcal{F}}\nabla h) - 2g(\mathcal{H}, \nabla_{\mathcal{F}}\nabla h),
\end{aligned}
$$

□

2.4 Almost Product Connection Symmetric of Sasaki Gradient Metric

The almost product connection symmetric of bundle $\mathcal{TM}$ given by,

$$S^h(\overline{\mathcal{F}}, \overline{\mathcal{G}}) = \frac{1}{2}\Big[(\nabla^h_{J\overline{\mathcal{G}}} J)\overline{\mathcal{F}} + J(\nabla^h_{\overline{\mathcal{G}}} J)\overline{\mathcal{F}} - J(\nabla^h_{\overline{\mathcal{F}}} J)\overline{\mathcal{G}}\Big], \tag{2}$$

$$\overline{\nabla}_{\overline{\mathcal{F}}}\overline{\mathcal{G}} = \nabla^h_{\overline{\mathcal{F}}}\overline{\mathcal{G}} - S(\overline{\mathcal{F}}, \overline{\mathcal{G}}). \tag{3}$$

where $\overline{\nabla}$ is an almost product connection on $\mathcal{TM}$,

Then, we have

Lemma 2 *The tensor almost product connection of bundle $\mathcal{TM}$ is given by*

$$
\begin{aligned}
S^h(\mathcal{F}_{\mathcal{H}}, \mathcal{G}_{\mathcal{H}}) &= -\frac{1}{2}(\mathcal{R}(\mathcal{F}, \mathcal{G})u)_{\mathcal{V}}, \\
S^h(\mathcal{F}_{\mathcal{H}}, \mathcal{G}_{\mathcal{V}}) &= -\mathcal{G}(f)(\nabla_X \nabla h)_{\mathcal{V}} - \frac{1}{\alpha}\Big[g(\mathcal{G}, \nabla_X \nabla h) \\
&- \frac{1}{2}\mathcal{F}(\alpha)\mathcal{G}(f))\Big](\nabla h)_{\mathcal{V}} + \frac{1}{2}(R(u, \mathcal{G})\mathcal{F})_{\mathcal{H}} + \frac{1}{2}\mathcal{G}(f)(R(u, \nabla h)\mathcal{F})_{\mathcal{H}}, \\
S^h(\mathcal{F}_{\mathcal{V}}, \mathcal{G}_{\mathcal{H}}) &= -((R(u, \mathcal{F})\mathcal{G})_x)_{\mathcal{H}} - \mathcal{F}(f)(R(u, \nabla h)\mathcal{G})_{\mathcal{H}} + \frac{1}{2}\mathcal{F}(f)(\nabla_{\mathcal{G}}\nabla h)_{\mathcal{V}}
\end{aligned}
$$

$$+\frac{1}{2\alpha}\big[g(\mathcal{F},\nabla_{\mathcal{G}}\nabla h)-\frac{1}{2}\mathcal{G}(\alpha)\mathcal{F}(f)\big](\nabla h)_{\mathcal{V}},$$
$$S^h(\mathcal{F}_{\mathcal{V}},\mathcal{G}_{\mathcal{V}})=-\frac{1}{2}\mathcal{F}(f)(\nabla_{\mathcal{G}}\nabla h)_{\mathcal{H}}-\frac{1}{2}(\mathcal{G}(f)(\nabla_{\mathcal{F}}\nabla h)_{\mathcal{H}}.$$

For all $\mathcal{F},\mathcal{G}\in\Gamma(\mathcal{TM})$.

Proof By using the Theorem 1, we have :

$$\begin{aligned}
S^h(\mathcal{F}_{\mathcal{V}},\mathcal{G}_{\mathcal{V}})&=\frac{1}{2}\Big[(\nabla^h{}_{J\mathcal{G}_{\mathcal{V}}}J)\mathcal{F}_{\mathcal{V}}+J(\nabla^h_{\mathcal{G}_{\mathcal{V}}}J)\mathcal{F}_{\mathcal{V}}-J(\nabla^h_{\mathcal{F}_{\mathcal{V}}}J)\mathcal{G}_{\mathcal{V}}\Big]\\
&=\frac{1}{2}\Big[\nabla^h_{J\mathcal{G}_{\mathcal{V}}}J\mathcal{F}_{\mathcal{V}}-J(\nabla^h_{J\mathcal{G}_{\mathcal{V}}}\mathcal{F}_{\mathcal{V}})+J(\nabla^h_{\mathcal{G}_{\mathcal{V}}}J\mathcal{F}_{\mathcal{V}}-J(\nabla^h_{\mathcal{G}_{\mathcal{V}}}\mathcal{F}_{\mathcal{V}}))\\
&\quad-J(\nabla^h_{\mathcal{F}_{\mathcal{V}}}J\mathcal{G}_{\mathcal{V}}-J(\nabla^h_{\mathcal{F}_{\mathcal{V}}}\mathcal{G}_{\mathcal{V}}))\Big],\\
S^h(\mathcal{F}_{\mathcal{V}},\mathcal{G}_{\mathcal{V}})&=\frac{1}{2}\Big[\nabla^h_{\mathcal{G}_{\mathcal{V}}}\mathcal{F}_{\mathcal{V}}-J(\nabla^h_{\mathcal{G}_{\mathcal{V}}}\mathcal{F}_{\mathcal{V}})+J(\nabla^h_{\mathcal{G}_{\mathcal{V}}}J\mathcal{F}_{\mathcal{V}})-(\nabla^h_{\mathcal{G}_{\mathcal{V}}}\mathcal{F}_{\mathcal{V}})\\
&\quad-J(\nabla^h_{\mathcal{F}_{\mathcal{V}}}\mathcal{G}_{\mathcal{V}})+(\nabla^h_{\mathcal{F}_{\mathcal{V}}}\mathcal{G}_{\mathcal{V}})\Big],
\end{aligned}$$

from Theorem 1 we have that $\nabla^h_{\mathcal{F}_{\mathcal{V}}}\mathcal{G}_{\mathcal{V}}$ is horizontal, then we have $J(\nabla^h_{\mathcal{F}_{\mathcal{V}}}\mathcal{G}_{\mathcal{V}})=-\nabla^h_{\mathcal{F}_{\mathcal{V}}}\mathcal{G}_{\mathcal{V}}$ and we get

$$\begin{aligned}
S^h(\mathcal{F}_{\mathcal{V}},\mathcal{G}_{\mathcal{V}})&=\frac{1}{2}\Big[\nabla^h_{\mathcal{G}_{\mathcal{V}}}\mathcal{F}_{\mathcal{V}}+(\nabla^h_{\mathcal{G}_{\mathcal{V}}}\mathcal{F}_{\mathcal{V}})-(\nabla^h_{\mathcal{G}_{\mathcal{V}}}J\mathcal{F}_{\mathcal{V}})-(\nabla^h_{\mathcal{G}_{\mathcal{V}}}\mathcal{F}_{\mathcal{V}})\\
&\quad+(\nabla^h_{\mathcal{F}_{\mathcal{V}}}\mathcal{G}_{\mathcal{V}})+(\nabla^h_{\mathcal{F}_{\mathcal{V}}}\mathcal{G}_{\mathcal{V}})\Big]=\Big[(\nabla^h_{\mathcal{F}_{\mathcal{V}}}\mathcal{G}_{\mathcal{V}})\Big]\\
&=-\frac{1}{2}\mathcal{F}(h)(\nabla_{\mathcal{G}}\nabla h)_{\mathcal{H}}-\frac{1}{2}(\mathcal{G}(h)(\nabla_{\mathcal{F}}\nabla h)_{\mathcal{H}},\\
S^h(\mathcal{F}_{\mathcal{V}},\mathcal{G}_{\mathcal{H}})&=\frac{1}{2}\Big[(\nabla^h{}_{J\mathcal{G}_{\mathcal{H}}}J)\mathcal{F}_{\mathcal{V}}+J(\nabla^h_{\mathcal{G}_{\mathcal{H}}}J)\mathcal{F}_{\mathcal{V}}-J(\nabla^h_{\mathcal{F}_{\mathcal{V}}}J)\mathcal{G}_{\mathcal{H}}\Big]\\
&=\frac{1}{2}\Big[\nabla^h_{J\mathcal{G}_{\mathcal{H}}}J\mathcal{F}_{\mathcal{V}}-J(\nabla^h_{J\mathcal{G}_{\mathcal{H}}}\mathcal{F}_{\mathcal{V}})+J(\nabla^h_{\mathcal{G}_{\mathcal{H}}}J\mathcal{F}_{\mathcal{V}})-JJ(\nabla^h_{\mathcal{G}_{\mathcal{H}}}\mathcal{F}_{\mathcal{V}}))\\
&\quad-J(\nabla^h_{\mathcal{F}_{\mathcal{V}}}J\mathcal{G}_{\mathcal{H}})+JJ(\nabla^h_{\mathcal{F}_{\mathcal{V}}}\mathcal{G}_{\mathcal{H}})\Big]\\
&=\frac{1}{2}\Big[-\nabla^h_{\mathcal{G}_{\mathcal{H}}}\mathcal{F}_{\mathcal{V}}+J(\nabla^h_{\mathcal{G}_{\mathcal{H}}}\mathcal{F}_{\mathcal{V}})+J(\nabla^h_{\mathcal{G}_{\mathcal{H}}}\mathcal{F}_{\mathcal{V}})-(\nabla^h_{\mathcal{G}_{\mathcal{H}}}\mathcal{F}_{\mathcal{V}}))\\
&\quad+J(\nabla^h_{\mathcal{F}_{\mathcal{V}}}\mathcal{G}_{\mathcal{H}})+(\nabla^h_{\mathcal{F}_{\mathcal{V}}}\mathcal{G}_{\mathcal{H}})\Big]\\
&=\frac{1}{2}\Big[-2\nabla^h_{\mathcal{G}_{\mathcal{H}}}\mathcal{F}_{\mathcal{V}}+2J(\nabla^h_{\mathcal{G}_{\mathcal{H}}}\mathcal{F}_{\mathcal{V}})+J(\nabla^h_{\mathcal{F}_{\mathcal{V}}}\mathcal{G}_{\mathcal{H}})+(\nabla^h_{\mathcal{F}_{\mathcal{V}}}\mathcal{G}_{\mathcal{H}})\Big]\\
&=\frac{1}{2}\Big[-2((\mathcal{R}(u,\mathcal{F})\mathcal{G})_x)_{\mathcal{H}}-2\mathcal{F}(f)(\mathcal{R}(u,\nabla h)\mathcal{G})_{\mathcal{H}}\\
&\quad+\mathcal{F}(h)(\nabla_{\mathcal{G}}\nabla h)_{\mathcal{V}}+\frac{1}{\alpha}\big[g(\mathcal{F},\nabla_{\mathcal{G}}\nabla h)-\frac{1}{2}\mathcal{G}(\alpha)\mathcal{F}(h)\big](\nabla h)_{\mathcal{V}}\Big]
\end{aligned}$$

$$
\begin{aligned}
&= -((\mathcal{R}(u,\mathcal{F})\mathcal{G})_x)_{\mathcal{H}} - \mathcal{F}(h)(\mathcal{R}(u,\nabla h)\mathcal{G})_{\mathcal{H}} \\
&\quad + \frac{1}{2}\mathcal{F}(h)(\nabla_{\mathcal{G}}\nabla h)_{\mathcal{V}} + \frac{1}{2\alpha}\Big[G(\mathcal{F},\nabla_{\mathcal{G}}\nabla h) - \frac{1}{2}\mathcal{G}(\alpha)\mathcal{F}(h)\Big](\nabla h)_{\mathcal{V}}, \\
S^h(\mathcal{F}_{\mathcal{H}},\mathcal{G}_{\mathcal{H}}) &= \frac{1}{2}[\nabla^h_{-\mathcal{G}_{\mathcal{H}}}(-\mathcal{F}_{\mathcal{H}}) - J(\nabla^h_{-\mathcal{G}_{\mathcal{H}}}\mathcal{F}_{\mathcal{H}}) + J(\nabla^h_{\mathcal{G}_{\mathcal{H}}} - \mathcal{F}_{\mathcal{H}}) - \nabla^h_{\mathcal{G}_{\mathcal{H}}}\mathcal{F}_{\mathcal{H}} \\
&\quad - J(\nabla^h_{\mathcal{F}_{\mathcal{H}}} - \mathcal{G}_{\mathcal{H}}) + \nabla^h_{\mathcal{F}_{\mathcal{H}}}\mathcal{G}_{\mathcal{H}}] \\
&= \frac{1}{2}[J(\nabla^h_{\mathcal{F}_{\mathcal{H}}}\mathcal{G}_{\mathcal{H}}) + \nabla^h_{\mathcal{F}_{\mathcal{H}}}\mathcal{G}_{\mathcal{H}}] = -\frac{1}{2}(R(\mathcal{F},\mathcal{G})u)_{\mathcal{V}}, \\
S^h(\mathcal{F}_{\mathcal{H}},\mathcal{G}_{\mathcal{V}}) &= \frac{1}{2}\Big[(\nabla^h{}_{J\mathcal{G}_{\mathcal{V}}}J)\mathcal{F}_{\mathcal{H}} + J(\nabla^h_{\mathcal{G}_{\mathcal{V}}}J)\mathcal{F}_{\mathcal{H}} - J(\nabla^h_{\mathcal{F}_{\mathcal{H}}}J)\mathcal{G}_{\mathcal{V}}\Big] \\
&= \frac{1}{2}\Big[\nabla^h{}_{\mathcal{G}_{\mathcal{V}}}(-\mathcal{F}_{\mathcal{H}}) - J(\nabla^h{}_{\mathcal{G}_{\mathcal{V}}}(\mathcal{F}_{\mathcal{H}})) + J(\nabla^h_{\mathcal{G}_{\mathcal{V}}}(-\mathcal{F}_{\mathcal{H}})) \\
&\quad - JJ(\nabla^h_{\mathcal{G}_{\mathcal{V}}}(\mathcal{F}_{\mathcal{H}})) - J(\nabla^h_{\mathcal{F}_{\mathcal{H}}}\mathcal{G}_{\mathcal{V}}) + JJ(\nabla^h_{\mathcal{F}_{\mathcal{H}}}\mathcal{G}_{\mathcal{V}})\Big] \\
&= \frac{1}{2}\Big[-2\nabla^h{}_{\mathcal{G}_{\mathcal{V}}}\mathcal{F}_{\mathcal{H}} - 2J(\nabla^h{}_{\mathcal{G}_{\mathcal{V}}}\mathcal{F}_{\mathcal{H}}) - J(\nabla^h_{\mathcal{F}_{\mathcal{H}}}\mathcal{G}_{\mathcal{V}}) + (\nabla^h_{\mathcal{F}_{\mathcal{H}}}\mathcal{G}_{\mathcal{V}})\Big] \\
&= -\nabla^h{}_{\mathcal{G}_{\mathcal{V}}}\mathcal{F}_{\mathcal{H}} - J(\nabla^h{}_{\mathcal{G}_{\mathcal{V}}}\mathcal{F}_{\mathcal{H}}) - \frac{1}{2}J(\nabla^h_{\mathcal{F}_{\mathcal{H}}}\mathcal{G}_{\mathcal{V}}) + \frac{1}{2}(\nabla^h_{\mathcal{F}_{\mathcal{H}}}\mathcal{G}_{\mathcal{V}}) \\
&= -\mathcal{G}(h)(\nabla_X\nabla h)_{\mathcal{V}} - \frac{1}{\alpha}\Big[g(\mathcal{G},\nabla_X\nabla h) - \frac{1}{2}\mathcal{F}(\alpha)\mathcal{G}(h))\Big](\nabla h)_{\mathcal{V}} \\
&\quad + \frac{1}{2}(\mathcal{R}(u,\mathcal{G})\mathcal{F})_{\mathcal{H}} + \frac{1}{2}\mathcal{G}(h)(\mathcal{R}(u,\nabla h)\mathcal{F})_{\mathcal{H}}.
\end{aligned}
$$

□

Theorem 3 *The almost product connection* $\overline{\nabla}$ *defined by (3) is given by*

$$
\begin{aligned}
\overline{\nabla}_{\mathcal{F}_{\mathcal{H}}}\mathcal{G}_{\mathcal{H}} &= (\nabla_X Y)_{\mathcal{H}}, \\
\overline{\nabla}_{\mathcal{F}_{\mathcal{H}}}\mathcal{G}_{\mathcal{V}} &= (\nabla_{\mathcal{F}}\mathcal{G})_{\mathcal{V}} + \frac{3}{2}\mathcal{G}(h)(\nabla_{\mathcal{F}}\nabla h)_{\mathcal{V}} \\
&\qquad + \frac{3}{2\alpha}\Big[g(\mathcal{G},\nabla_{\mathcal{F}}\nabla h) - \frac{1}{2}\mathcal{F}(\alpha)\mathcal{G}(f)\Big](\nabla h)_{\mathcal{V}}, \\
\overline{\nabla}_{\mathcal{F}_{\mathcal{V}}}\mathcal{G}_{\mathcal{H}} &= \frac{3}{2}(\mathcal{R}(u,\mathcal{F})\mathcal{G})_{\mathcal{H}} + \frac{3}{2}\mathcal{F}(h)(\mathcal{R}(u,\nabla h)\mathcal{G})_{\mathcal{V}}, \\
\overline{\nabla}_{\mathcal{F}_{\mathcal{V}}}\mathcal{G}_{\mathcal{V}} &= 0.
\end{aligned}
$$

Lemma 3 *The torsion tensor* $\overline{T}$ *of almost product connection, is given by*

- $\overline{T}(\mathcal{F}_{\mathcal{H}},\mathcal{G}_{\mathcal{H}}) = (R(\mathcal{F},\mathcal{G})u)^{\mathcal{V}}$,
- $\overline{T}(\mathcal{F}_{\mathcal{H}},\mathcal{G}_{\mathcal{V}}) = \frac{3}{2}\Big[-(R(u,\mathcal{G})\mathcal{F})_{\mathcal{H}} - \mathcal{G}(h)(\mathcal{R}(u,\nabla h)\mathcal{F})_{\mathcal{H}} + \mathcal{G}(h)(\nabla_X\nabla h)_{\mathcal{V}} + \frac{3}{2\alpha}\Big(G(\mathcal{G},\nabla_X\nabla h) - \frac{1}{2}\mathcal{F}(\alpha)\mathcal{G}(h)\Big)(\nabla h)_{\mathcal{V}}\Big]$,

- $\overline{\mathcal{T}}(\mathcal{F}_{\mathcal{V}}, \mathcal{G}_{\mathcal{H}}) = -\frac{3}{2}\Big(- (\mathcal{R}(u, \mathcal{F})\mathcal{G})_{\mathcal{H}} - \mathcal{F}(h)(\mathcal{R}(u, \nabla h)\mathcal{G})_{\mathcal{H}} + \mathcal{F}(h)(\nabla_Y \nabla h)_{\mathcal{V}} + \frac{3}{2\alpha}\Big(G(\mathcal{F}, \nabla_Y \nabla h) - \frac{1}{2}\mathcal{G}(\alpha)\mathcal{F}(f)\Big)(\nabla h)_{\mathcal{V}}\Big),$
- $\overline{\mathcal{T}}(\mathcal{F}_{\mathcal{V}}, \mathcal{G}_{\mathcal{V}}) = 0,$

for all $\mathcal{F}, \mathcal{G} \in \Gamma(\mathcal{T}\mathcal{M})$.

References

1. M.T.K. Abbassi, M. Sarih, Killing vector fields on tangent bundles with Cheeger Gromoll metric. Tsukuba J. Math. 295–306 (2003)
2. M. Altunbas, R. Simsek, A. Gezer, A study concerning Berger type deformed Sasaki metric on the tangent bundle. J. Math. Phys. Anal. Geom. 435–447 (2003)
3. M. Altunbas, R. Simsek, A. Gezer, Some harmonic problems on the tangent bundle with a Berger-type deformed Sasaki metric. U.P.B. Sci. Bull. Ser. A (2020)
4. Y. Bahous, L. Belarbi, M. Belarbi, H. Elhendi, f-Biharmonic curves in three dimensional generalized symmetric spaces. J. Appl. Math. Inform. **42**(1), 85–92 (2024)
5. L. Belarbi, H. Elhendi, Harmonic vector fields on gradient Sasaki metric. Afr. Mat. 1–13 (2023). https://doi.org/10.1007/s13370-023-01052-w
6. M. Belarbi, H. Elhendi, L. Belarbi, Biharmonic curves in three-dimensional generalized symmetric spaces. J. Indian Math. Soc. 263–277 (2022)
7. L. Belarbi, M. Belarbi, H. Elhendi, Legendre curves on Lorentzian Heisenberg Space. Bull. Transilv. Univ. Brasov SER. **III**, 41–50 (2020)
8. L. Belarbi, H. Elhendi, On the geometry of the tangent bundle with gradient Sasaki metric. Arab J. Math. Sci. (2021). https://doi.org/10.1108/AJMS-11-2020-0125
9. L. Belarbi, H. Elhendi, F. Latti, On the geometry of the tangent bundle with vertical rescaled generalized Cheeger-Gromoll metric. Bull. Transilv. Univ. Brasov SER. **III**, 247–264 (2019)
10. L. Belarbi, H. Elhendi, Geometry of twisted Sasaki metric. J. Geom. Symmetry Phys. 1–19 (2019)
11. L. Belarbi, H. Elhendi, Harmonic and biharmonic maps between tangent bundles. Acta Math. Univ. Comenianae 187–199 (2019)
12. H. Elhendi, L. Belarbi, On paraquaternionic submersions of tangent bundle of order two. Nonlinear Stud. 653–664 (2018)
13. H. Elhendi, L. Belarbi, Deformed diagonal metrics on tangent bundle of order two and harmonicity. Panamer. Math. J. 90–106 (2017)
14. H. Elhendi, L. Belarbi, Deformed diagonal metrics on tangent bundle of order two and harmonicity. Panamer. Math. J. 90–106 (2018)
15. H. Mazouzi, H. Elhendi, L. Belarbi, On the generalized Bi-f-harmonic map equations on singly warped product manifolds. Commun. Appl. Nonlinear Anal. 52–76 (2018)
16. V. Cruceanu, P. Fortuny, P.M. Gadea, A survey on paracomplex geometry. Rocky Mount. J. Math. 83–115 (1995)
17. P. Dombrowski, On the geometry of the tangent bundle. J. für die reine und angewandte Mathematik 73–88 (1962)
18. S. Gudmundsson, E. Kappos, On the geometry of the tangent bundles. Expo. Math. 1–41 (2002)
19. E. Kappos, Natural metrics on tangent bundles. Master thesis, 2001
20. E. Musso, F. Tricerri, Riemannian metrics on tangent bundles. Ann. Mat. Pura Appl. 1–20 (1988)
21. A. Salimov, A. Gezer, M. Iscan, On para-Kähler-Norden structures on the tangent bundles. Annales Polonici Mathematici 247–261 (2012)

22. S. Sasaki, On the differential geometry of tangent bundle of Riemannian manifolds. Töhoku Math. J. 338–354 (1958)
23. S. Tachibana, M. Okumura, On the almost-complex structure of tangent bundles of Riemannian spaces. Tohoku Math. J. 156–161 (1962)
24. K. Yano, S. Ishihara, *Tangent and Cotangent Bundles* (Marcel Dekker, Inc., New York, 1973)

New Several Types of Difference Equations for Degenerate Quantum Bernoulli Polynomials Using (q, h)-Derivative

Waseem Ahmad Khan

Abstract The classical Bernoulli polynomials dealt within this paper contains many application in various areas. We construct new types of Bernoulli polynomials using degenerate quantum exponential functions, find some properties and several difference equations of these polynomials.

1 Introduction

For $\psi, q \in \mathbb{R}$ with $q \neq 1$, the quantum number (q-number) defined by Jackson is

$$[\psi]_q = \frac{1 - q^{\psi}}{1 - q}.$$

We note a relation as $\lim_{q \to 1} [\psi]_q = \psi$. Also, we call $[\psi]_q$ the q-integer for $k \in \mathbb{Z}$, (see [9, 10, 13, 14]).

With the advent of quantum numbers, new research in many areas of mathematics such as series, differential equations, and calculus has exploded, (see [1, 5, 11, 18]). For example, different types of trigonometric functions and hyperbolic functions defined by Ryoo and Kang [19, 20] bring generalized properties of different types of trigonometric functions and hyperbolic functions. In [1], Bangerezako combined q-number in an optimal control problem, making q-Euler Lagrange equation, q-optimal control and the q- Hamilton system q-Hamilton-Pontriaguine system. These studies have resulted in several researches combined with quantum numbers in application field.

Let ψ, υ be non-negative integer with $\upsilon \leq \psi$. Then, the Gaussian binomial coefficients is

$$\binom{\psi}{\upsilon}_q = \frac{[\psi]_q!}{[\psi - \upsilon]_q! [\upsilon]_q!}.$$

W. A. Khan (✉)
Department of Mathematics and Natural Sciences, Prince Mohammad Bin Fahd University, P.O. Box: 1664, Al Khobar 31952, Saudi Arabia
e-mail: wkhan1@pmu.edu.sa

M. Kharrat et al. (eds.), *Modeling of Discrete and Continuous Systems*, Forum for Interdisciplinary Mathematics, https://doi.org/10.1007/978-981-97-8715-9_16

We note $[0]_q! = 1$, $[\psi]_q! = [\psi]_q[\psi-1]_q \dots [2]_q[1]_q$. The q-Gaussian binomial coefficients that are appeared in the q-Taylor formula of ξ^ψ with $\xi = 1$ becomes the ordinary binomials coefficients when $q \to 1$. We can check some properties of these coefficients in Sect. 5, 6 of [15].

The following two quantum derivatives

$$D_q f(\xi) = \frac{d_q f(\xi)}{d_q \xi} = \frac{f(q\xi) - f(\xi)}{(q-1)\xi}, \quad D_h f(\xi) = \frac{d_h f(\xi)}{d_h \xi} = \frac{f(\xi + h) - f(\xi)}{h},$$

are called the q-derivative D_q and h-derivative D_h of the function $f(\xi)$, respectively. We note $\lim_{q\to1} D_q f(\xi) = \lim_{h\to0} D_h f(\xi) = \frac{df(\xi)}{d\xi}$ if $f(\xi)$ is differentiable, (see [15]).

Since 2010, mathematicians have tried to introduce a new concept involving the properties of two kinds of quantum numbers. In Ref. [2], Benaoum found (q, h)-Newton's binomial formula and its properties. Also, Cermak and Nechvatal studied a (q, h)-version of the fractional calculus in [6]. In 2019, the generalization of quantum Taylor formula and quantum binomial is made by Silindir and Yantir [22]. We are well aware that one of the ways available for solving linear differential equations with specific integral equations is Laplace transformation. In Ref. [21], we can find the (q, h)-Laplace transformation which is made by Rahmat is a generalized type for classical Laplace transformation.

A two-parameter time scale $\mathbb{T}_{q,h}$ was introduced as follows:

$$\mathbb{T}_{q,h} = \{q^\psi \xi + [\psi]_q h | \xi \in \mathbb{R},\ \psi \in \mathbb{Z},\ q \in \mathbb{R}^+,\ q \neq 1\} \cup \left\{\frac{h}{1-q}\right\}. \tag{1.1}$$

Definition 1.1 Let $f : \mathbb{T}_{q,h} \to \mathbb{R}$ be any function. Thus, the delta (q, h)-derivative of $D_{q,h}(f)$ is defined by (see [6, 22]):

$$D_q f(\xi) := \frac{f(q\xi + h) - f(\xi)}{(q-1)\xi + h}.$$

Definition 1.2 The generalized quantum binomial $(\xi - \xi_0)^\psi_{q,h}$ is defined by

$$(\xi - \xi_0)^\psi_{q,h} = \begin{cases} 1, & \text{if } \psi = 0, \\ \prod_{i=1}^{\psi}(\psi - (q^{i-1}\xi_0 + [i-1]_q h)), & \text{if } \psi \geq 0, \end{cases}.$$

where $\xi_0 \in \mathbb{R}$.

Definition 1.3 The generalized quantum exponential function is defined by (see [22])

$$\exp_{q,h}(\alpha\xi) = \sum_{i=0}^{\infty} \frac{\alpha^i (\xi - 0)^i_{q,h}}{[i]_q!}, \tag{1.2}$$

where α is arbitrary nonzero constant.

Clearly $\exp_{q,h}(0) = 1$. As $h \to 0$ with $\alpha = 1$, the generalized quantum exponential function (1.2) becomes the so called q-exponential function $e_q(\xi)$ (see [7, 8, 12]). Likewise, as $q \to 1$ with $\alpha = 1$, the generalized quantum exponential function $\exp_{q,h}(\alpha\xi)$ reduces to the so-called h-exponential function $\exp_{1,h}(\xi) = (1+h)^{\frac{\xi}{h}}$.

The definition of q-Bernoulli polynomials are determined by (see [16–18])

$$\sum_{\psi=0}^{\infty} \mathbb{B}_{\psi,q}(\xi)\frac{\varpi^{\psi}}{[\psi]_q!} = \frac{\varpi}{e_q(\varpi)-1}e_q(\xi\varpi). \tag{1.3}$$

Also the corresponding q-Bernoulli numbers are determined by $\mathbb{B}_{\psi,q} = \mathbb{B}_{\psi,q}(0)$.

For $q \to 1$, we note that q-Bernoulli numbers and polynomials become Bernoulli numbers and polynomials become Bernoulli numbers and polynomials, respectively.

The definition of degenerate Bernoulli polynomials are determined by (see [2–4])

$$\frac{\varpi}{(1+ht)^{\frac{1}{h}}+1}(1+h\varpi)^{\frac{\xi}{h}} = \sum_{\psi=0}^{\infty} \mathbb{B}_{\psi}(\psi : h)\frac{\varpi^{\psi}}{\psi!}. \tag{1.4}$$

Also the corresponding degenerate Bernoulli numbers are determined by $\mathbb{B}_{\psi}(h) = \mathbb{B}_{\psi}(0 : h)$.

Definition 1.4 Let $| q |< 1$ and h be non-negative integer, the degenerate q-tangent polynomials $\mathbb{T}_{\psi,q}(\xi : h)$ by using degenerate quantum exponential function as (see [12])

$$\frac{2}{q_{q,h}(2:\varpi)}e_{q,h}(\xi : \varpi) = \sum_{\psi=0}^{\infty} \mathbb{T}_{\psi,q}(\xi : h)\frac{\varpi^{\psi}}{[\psi]_q!}, \tag{1.5}$$

so that

$$\mathbb{T}_{\psi,q}(\xi : h) = \sum_{\upsilon=0}^{\psi} \binom{\psi}{\upsilon}_q (\xi)_{q,h}^{\psi-\upsilon}\mathbb{T}_{\upsilon,q}(h).$$

Also, the corresponding degenerate q-tangent numbers are determined by $\mathbb{T}_{\psi,q}(h) = \mathbb{T}_{\psi,q}(0 : h)$.

The application of Bernoulli numbers and polynomials as above motivated us to do new research. When the degenerate polynomials defined by L. Carlitz are combined with quantum numbers, many mathematicians wondered how Appell numbers and polynomials are defined and what properties are associated with them. An important objective of this paper is to construct a new type of Bernoulli polynomials containing the properties of quantum Bernoulli and degenerate Bernoulli polynomials, and to find the difference equations related to them.

In this paper, we construct a new degenerate quantum Bernoulli polynomials (degenerate quantum Bernoulli polynomials), obtain several types of difference equations for these polynomials by using (q, h)-derivative. Also, we shows expanded difference equations using classical Bernoulli, quantum Bernoulli (quantum Bernoulli), and degenerate Bernoulli polynomials.

2 Difference Equations for (q, h)-Analogue of Bernoulli Polynomials

In this section, we define (q, h)-analogue Bernoulli polynomials using generalized quantum exponential functions. Using the (q, h)-derivatives, we obtain several differential equations related to (q, h)-analogue of Bernoulli polynomials. Moreover, we determine relations among q-Bernoulli polynomials, degenerate Bernoulli polynomials and (q, h)-Bernoulli polynomials.

The degenerate quantum exponential function $e_{q,h}(\xi : \varpi)$ is defined by

$$e_{q,h}(\xi : \varpi) = \sum_{\psi=0}^{\infty} (\xi)_{q,h}^{\psi} \frac{\varpi^{\psi}}{[\psi]_q!}, \tag{2.1}$$

where $(\xi)_{q,h}^{\psi} = \xi(\xi - h)(\xi - [2]_q h) \cdots (\xi - [\upsilon - 1]_q h)$.

Setting $q \to 1$ in (2.1), we have

$$e_{1,h}(\xi : \varpi) = (1 + h\varpi)^{\frac{\xi}{h}} = \sum_{\psi=0}^{\infty} (\xi)_h^{\psi} \frac{\varpi^{\psi}}{\psi!},$$

where $e_h(\xi : \varpi)$ is the h-exponential function (or degenerate exponential function).

From the Eq. (2.1), we define a new type of degenerate quantum Bernoulli polynomials as follows:

Definition 2.1 Let $| q |< 1$, and h be non-negative integer, the (q, h)-analogue of Bernoulli polynomials $\mathbb{B}_{\psi,q}(\xi : h)$ are defined by

$$\frac{\varpi}{e_{q,h}(1 : \varpi) - 1} e_{q,h}(\xi : \varpi) = \sum_{\psi=0}^{\infty} \mathbb{B}_{\psi,q}(\xi : h) \frac{\varpi^{\psi}}{[\psi]_q!}. \tag{2.2}$$

Replacing $\xi = 0$ in Definition 2.1, we have

$$\sum_{\psi=0}^{\infty} \mathbb{B}_{\psi,q}(0 : h) \frac{\varpi^{\psi}}{[\psi]_q!} = \sum_{\psi=0}^{\infty} \mathbb{B}_{\psi,q}(h) \frac{\varpi^{\psi}}{[\psi]_q!} = \frac{\varpi}{e_{q,h}(1 : \varpi) - 1}, \tag{2.3}$$

where $\mathbb{B}_{\psi,q}(h)$ are called the (q,h)-Bernoulli numbers. From (2.1), we can see certain relations between the Bernoulli, degenerate Bernoulli and q-Bernoulli polynomials. Setting $h \to 0$ in (2.2) and (2.3), we can derive the q-Bernoulli numbers and polynomials as follows:

$$\sum_{\psi=0}^{\infty} \mathbb{B}_{\psi,q} \frac{\varpi^{\psi}}{[\psi]_q!} = \frac{\varpi}{e_q(1:\varpi)-1}, \quad \sum_{\psi=0}^{\infty} \mathbb{B}_{\psi,q}(\xi) \frac{\varpi^{\psi}}{[\psi]_q!} = \frac{\varpi}{e_q(1:\varpi)-1} e_q^{\xi\varpi}.$$

Also, the corresponding Bernoulli numbers and polynomials when $h \to 0$ and $q \to 1$ in (2.2) and (2.3) as

$$\sum_{\psi=0}^{\infty} \mathbb{B}_{\psi} \frac{\varpi^{\psi}}{\psi!} = \frac{\varpi}{e^{\varpi}-1}, \quad \sum_{\psi=0}^{\infty} \mathbb{B}_{\psi}(\xi) \frac{\varpi^{\psi}}{\psi!} = \frac{\varpi}{e^{\varpi}-1} e^{\xi\varpi}.$$

Also, the corresponding degenerate Bernoulli numbers and polynomials when $q \to 1$ in (2.2) and (2.3) as follows:

$$\sum_{\psi=0}^{\infty} \mathbb{B}_{\psi}(h) \frac{\varpi^{\psi}}{\psi!} = \frac{\varpi}{(1+h\varpi)^{\frac{1}{h}}-1}, \quad \sum_{\psi=0}^{\infty} \mathbb{B}_{\psi}(\xi; h) \frac{\varpi^{\psi}}{\psi!} = \frac{\varpi}{(1+h\varpi)^{\frac{1}{h}}-1} (1+h\varpi)^{\frac{\xi}{h}}.$$

Theorem 2.1 *Let $h \in \mathbb{N}$ with $\mid q \mid < 1$. Then, we obtain*

$$\mathbb{B}_{\psi,q}(\xi:h) = \sum_{\upsilon=0}^{\psi} \binom{\psi}{\upsilon}_q \mathbb{B}_{\upsilon,q}(h)(\xi)_{q,h}^{\psi-\upsilon}. \tag{2.4}$$

Proof One easily notices from (2.1) that

$$\sum_{\psi=0}^{\infty} \mathbb{B}_{\psi,q}(\xi:h) \frac{\varpi^{\psi}}{[\psi]_q!} = \frac{\varpi}{e_{q,h}(1:\varpi)-1} e_{q,h}(\xi:\varpi)$$

$$= \sum_{\psi=0}^{\infty} \mathbb{B}_{\psi,q}(h) \frac{\varpi^{\psi}}{[\psi]_q!} \sum_{\upsilon=0}^{\infty} (\xi)_{q,h}^{\upsilon} \frac{\varpi^{\upsilon}}{[\upsilon]_q!}$$

$$= \sum_{\psi=0}^{\infty} \left(\sum_{\upsilon=0}^{\psi} \binom{\psi}{\upsilon}_q \mathbb{B}_{\upsilon,q}(h)(\xi)_{q,h}^{\psi-\upsilon} \right) \frac{\varpi^{\psi}}{[\psi]_q!}. \tag{2.5}$$

We achieve the result (2.4) accordance with (2.1) and (2.5). □

Corollary 2.1 *On letting $q \to 1$ in Theorem 2.1, then*

$$\mathbb{B}_{\psi}(\xi : h) = \sum_{\upsilon=0}^{\psi} \binom{\psi}{\upsilon} \mathbb{B}_{\upsilon}(h)(\xi)_h^{\psi-\upsilon}.$$

Corollary 2.2 *On adjusting $h \to 0$ in Theorem 2.1, then*

$$\mathbb{B}_{\psi,q}(\xi) = \sum_{\upsilon=0}^{\psi} \binom{\psi}{\upsilon}_q \mathbb{B}_{\upsilon,q}\xi^{\psi-\upsilon}.$$

Theorem 2.2 *Let $h \in \mathbb{N}$ with $| q |< 1$. Then, we obtain*

$$D_{q,h,\xi}^{(1)} \mathbb{B}_{\psi,q}(\xi : h) = [\psi]_q \mathbb{B}_{\psi-1,q}(\xi : h). \tag{2.6}$$

Proof One easily notices from (2.1) that

$$\sum_{\psi=0}^{\infty} \mathbb{B}_{\psi,q}(\xi : h)\frac{\varpi^{\psi}}{[\psi]_q!} - \sum_{\psi=0}^{\infty} \mathbb{B}_{\psi,q}(h)\frac{\varpi^{\psi}}{[\psi]_q!} \sum_{\psi=0}^{\infty} (\xi)_{q,h}^{\psi} \frac{\varpi^{\psi}}{[\psi]_q!}$$

$$= \sum_{\psi=0}^{\infty} \left(\sum_{\upsilon=0}^{\psi} \binom{\psi}{\upsilon}_q \mathbb{B}_{\upsilon,q}(h)(\xi)_{q,h}^{\psi-\upsilon} \right) \frac{\varpi^{\psi}}{[\psi]_q!}. \tag{2.7}$$

From this equation, we have

$$\mathbb{B}_{\psi,q}(\xi : h) = \sum_{\upsilon=0}^{\psi} \binom{\psi}{\upsilon}_q \mathbb{B}_{\upsilon,q}(h)(\xi)_{q,h}^{\psi-\upsilon}. \tag{2.8}$$

Using the (q, h)-derivative in (2.8), we have

$$D_{q,h,\xi}^{(1)} \mathbb{B}_{\psi,q}(\xi : h) = \sum_{\upsilon=0}^{\psi} \binom{\psi}{\upsilon}_q [\psi - \upsilon]_q \mathbb{B}_{\upsilon,q}(h)(\xi)_{q,h}^{\psi-\upsilon} = [\psi]_q \mathbb{B}_{\psi-1,q}(\xi : h).$$

Hence, we find the desired result at once. □

Corollary 2.3 *Let h be a non-negative integer with $h \neq 0$ and $|q| < 1$. Then, we obtain*

$$\mathbb{B}_{\psi-\upsilon,q}(\xi : h) = \frac{[\psi - \upsilon]_q!}{[\psi]_q!} D_{q,h,\xi}^{(\upsilon)} \mathbb{B}_{\psi,q}(\xi : h).$$

Corollary 2.4 *Adjusting $q \to 1$ in Theorem 2.2, we have*

$$D^{(1)}_{h,\xi}\mathbb{B}_{\psi}(\xi : h) = \psi\mathbb{B}_{\psi-1}(\xi : h), \quad \mathbb{B}_{\psi-\upsilon}(\xi : h) = \frac{(\psi-\upsilon)!}{\psi!}D^{(\upsilon)}_{h,\xi}\mathbb{B}_{\psi}(\xi : h).$$

Corollary 2.5 *Letting $h \to 0$ in Theorem 2.2, we have*

$$D^{(1)}_{q,\xi}\mathbb{B}_{\psi,q}(\xi) = [\psi]_q\mathbb{B}_{\psi-1,q}(\xi), \quad \mathbb{B}_{\psi-\upsilon,q}(\xi) = \frac{[\psi-\upsilon]_q!}{[\psi]_q!}D^{(\upsilon)}_{q,\xi}\mathbb{B}_{\psi,q}(\xi).$$

Theorem 2.3 *A solutions of the following differential equation*

$$\frac{(1)^{\psi}_{q,h}}{[\psi]_q!}D^{(\psi)}_{q,h,\xi}\mathbb{B}_{\psi,q}(\xi : h) + \frac{(1)^{\psi-1}_{q,h}}{[\psi-1]_q!}D^{(\psi-1)}_{q,h,\xi}\mathbb{B}_{\psi,q}(\xi : h) + \frac{(1)^{\psi-2}_{q,h}}{[\psi-2]_q!}D^{(\psi-2)}_{q,h,\xi}\mathbb{B}_{\psi,q}(\xi : h)$$

$$+ \cdots + \frac{(1)^{2}_{q,h}}{[2]_q!}D^{(2)}_{q,h,\xi}\mathbb{B}_{\psi,q}(\xi : h) + (1)^{1}_{q,h}D^{(1)}_{q,h,\xi}\mathbb{B}_{\psi,q}(\xi : h) - \mathbb{B}_{\psi,q}(\xi : h) - [\psi]_q(\xi)^{\psi-1}_{q,h} = 0. \tag{2.9}$$

Proof One easily notices from (2.1) that

$$te_{q,h}(x : t) = \sum_{\psi=0}^{\infty}\mathbb{B}_{\psi,q}(\xi : h)\frac{\varpi^{\psi}}{[\psi]_q!}(e_{q,h}(1 : \varpi) - 1)$$

$$= \sum_{\psi=0}^{\infty}\mathbb{B}_{\psi,q}(\xi : h)\frac{\varpi^{\psi}}{[\psi]_q!}\left(\sum_{\upsilon=0}^{\infty}(1)^{\upsilon}_{q,h}\frac{\varpi^{\upsilon}}{[\upsilon]_q!} - 1\right)$$

$$= \sum_{\psi=0}^{\infty}\left(\sum_{\upsilon=0}^{\psi}\binom{\psi}{\upsilon}_q(1)^{\upsilon}_{q,h}\mathbb{B}_{\psi-\upsilon,q}(\xi : h) - \mathbb{B}_{\psi,q}(\xi : h)\right)\frac{\varpi^{\psi}}{[\psi]_q!}. \tag{2.10}$$

Other hand,

$$\varpi e_{q,h}(\xi : \varpi) = \sum_{\psi=0}^{\infty}[\psi]_q(\xi)^{\psi-1}_{q,h}\frac{\varpi^{\psi}}{[\psi]_q!}. \tag{2.11}$$

Therefore, by (2.10) and (2.11), we get

$$\sum_{\upsilon=0}^{\psi}\binom{\psi}{\upsilon}_q(1)^{\upsilon}_{q,h}\mathbb{B}_{\psi-\upsilon,q}(\xi : h) - \mathbb{B}_{\psi,q}(\xi : h) - [\psi]_q(\xi)^{\psi-1}_{q,h} = 0. \tag{2.12}$$

Taking the υ-th derivative of above equation, we obtain

$$\sum_{\upsilon=0}^{\psi}\frac{(1)_{q,h}^{\upsilon}}{[\upsilon]_q!}D_{q,h,\xi}^{(\upsilon)}\mathbb{B}_{\psi,q}(\xi:h)-\mathbb{B}_{\psi,q}(\xi:h)-[\psi]_q(\xi)_{q,h}^{\psi-1}=0.$$

Hence, we find the desired result at once. □

Corollary 2.6 *Letting $q\to 1$ in (2.9):*

$$\frac{(1)_{1,h}^{\psi}}{\psi!}D_{h,\xi}^{(\psi)}\mathbb{B}_{\psi}(\xi:h)+\frac{(1)_{1,h}^{\psi-1}}{(\psi-1)!}D_{h,\xi}^{(\psi-1)}\mathbb{B}_{\psi}(\xi:h)+\frac{(1)_{1,h}^{\psi-2}}{(\psi-2)!}D_{h,\xi}^{(\psi-2)}\mathbb{B}_{\psi}(\xi:h)$$

$$+\cdots+\frac{(1)_{1,h}^{2}}{(2)!}D_{h,\xi}^{(2)}\mathbb{B}_{\psi}(\xi:h)+(1)_h^1D_{h,\xi}^{(1)}\mathbb{B}_{\psi}(\xi:h)-\mathbb{B}_{\psi}(\xi:h)-\psi(x)_h^{\psi-1}=0.$$

Corollary 2.7 *Adjusting $h\to 0$ in (2.9):*

$$\frac{1}{[\psi]_q!}D_{q,\xi}^{(\psi)}\mathbb{B}_{\psi,q}(\xi)+\frac{1}{[\psi-1]_q!}D_{q,\xi}^{(\psi-1)}\mathbb{B}_{\psi,q}(\xi)+\frac{1}{[\psi-2]_q!}D_{q,\xi}^{(\psi-2)}\mathbb{B}_{\psi,q}(\xi)$$

$$+\cdots+\frac{1}{[2]_q!}D_{q,\xi}^{(2)}\mathbb{B}_{\psi,q}(\xi)+D_{q,\xi}^{(1)}\mathbb{B}_{\psi,q}(\xi)-\mathbb{B}_{\psi,q}(\xi)-[\psi]_q\xi^{\psi-1}=0.$$

Theorem 2.4 *A solutions of the following differential equation*

$$\frac{\mathbb{B}_{\psi,q}(1:h)-\mathbb{B}_{\psi,q}(h)}{[\psi]_q!}D_{q,h,\xi}^{(\psi)}\mathbb{B}_{\psi,q}(\xi:h)+\frac{\mathbb{B}_{\psi-1,q}(1:h)-\mathbb{B}_{\psi-1,q}(h)}{[\psi-1]_q!}D_{q,h,\xi}^{(\psi-1)}\mathbb{B}_{\psi,q}(\xi:h)+$$

$$\cdots+\frac{\mathbb{B}_{2,q}(1:h)-\mathbb{B}_{2,q}(h)}{[2]_q!}D_{q,h,\xi}^{(2)}\mathbb{B}_{\psi,q}(\xi:h)+(\mathbb{B}_{1,q}(1:h)-\mathbb{B}_{1,q}(h))D_{q,h,\xi}^{(1)}\mathbb{B}_{\psi,q}(\xi:h)$$

$$+\left(\mathbb{B}_{0,q}(1:h)-\mathbb{B}_{0,q}(h)\right)\mathbb{B}_{\psi,q}(\xi:h)-[\psi]_q\mathbb{B}_{\psi-1,q}(\psi:h)=0. \tag{2.13}$$

Proof One easily notices from (2.1) that

$$\sum_{\psi=0}^{\infty}\mathbb{B}_{\psi,q}(\xi:h)\frac{\varpi^{\psi}}{[\psi]_q!}=\frac{\varpi}{e_{q,h}(1:\varpi)-1}e_{q,h}(\xi:\varpi)$$

$$=\frac{1}{\varpi}\left(\frac{\varpi}{e_{q,h}(1:\varpi)-1}e_{q,h}(1:\varpi)-\frac{\varpi}{e_{q,h}(1:\varpi)-1}\right)\frac{\varpi}{e_{q,h}(1:\varpi)-1}e_{q,h}(\xi:\varpi).$$

Using the generating function of (q,h)-Bernoulli polynomials, we have

$$\sum_{\psi=0}^{\infty}[\psi]_qB_{\psi-1,q}(\xi:h)\frac{\varpi^{\psi}}{[\psi]_q!}=\sum_{\psi=0}^{\infty}\left(\sum_{\upsilon=0}^{\upsilon}\binom{\psi}{\upsilon}_q\left(\mathbb{B}_{\upsilon,q}(1:h)-\mathbb{B}_{\upsilon,q}(h)\right)\mathbb{B}_{\psi-\upsilon,q}(\xi:h)\right)\frac{\varpi^{\psi}}{[\psi]_q!}.$$

Comparing the coefficients of both sides, we have

$$\sum_{\upsilon=0}^{\psi}\binom{\psi}{\upsilon}_q\big(\mathbb{B}_{\upsilon,q}(1:h)-\mathbb{B}_{\upsilon,q}(h)\big)\,\mathbb{B}_{\psi-\upsilon,q}(\xi:h)-[\psi]_q\mathbb{B}_{\psi-1,q}(\xi:h)=0. \tag{2.14}$$

Replacing $\mathbb{B}_{\psi-\upsilon,q}(\xi:h)$ with $D^{(\upsilon)}_{q,h,\xi}\mathbb{B}_{\psi,q}(\xi:h)$ in Eq. (2.14), we have

$$\sum_{\upsilon=0}^{\psi}\frac{\big(\mathbb{B}_{\upsilon,q}(1:h)-\mathbb{B}_{\upsilon,q}(h)\big)}{[\upsilon]_q!}D^{(\upsilon)}_{q,h,\xi}\mathbb{B}_{\psi,q}(\xi:h)-[\psi]_q\mathbb{B}_{\psi-1,q}(\xi:h)=0.$$

The above equation allows us to complete the proof. □

Corollary 2.8 *On putting $h\to 0$ in (2.13):*

$$\frac{\mathbb{B}_{\psi,q}-\mathbb{B}_{\psi,q}}{[\psi]_q!}D^{(\psi)}_{q,\xi}\mathbb{B}_{\psi,q}+\frac{\mathbb{B}_{\psi-1,q}-\mathbb{B}_{\psi-1,q}}{[\psi-1]_q!}D^{(\psi-1)}_{q,\xi}\mathbb{B}_{\psi,q}(\psi)+$$

$$\cdots+\frac{\mathbb{B}_{2,q}-\mathbb{B}_{2,q}}{[2]_q!}D^{(2)}_{q,\xi}\mathbb{B}_{\psi,q}(\xi)+(\mathbb{B}_{1,q}-\mathbb{B}_{1,q})D^{(1)}_{q,\xi}\mathbb{B}_{\psi,q}(\xi)$$

$$+\big(\mathbb{B}_{0,q}-\mathbb{B}_{0,q}\big)\mathbb{B}_{\psi,q}(\xi)-[\psi]_q\mathbb{B}_{\psi-1,q}(\xi)=0.$$

Corollary 2.9 *On adjusting $q\to 1$ in (2.13):*

$$\frac{\mathbb{B}_{\psi}(1:h)-\mathbb{B}_{\psi}(h)}{[\psi]_q!}D^{(\psi)}_{h,\xi}\mathbb{B}_{\psi}(\xi:h)+\frac{\mathbb{B}_{\psi-1}(1:h)-\mathbb{B}_{\psi-1}(h)}{(\psi-1)!}D^{(\psi-1)}_{h,\xi}\mathbb{B}_{\psi}(\xi:h)+$$

$$\cdots+\frac{\mathbb{B}_{2}(1:h)-\mathbb{B}_{2}(h)}{[2]_q!}D^{(2)}_{h,\xi}\mathbb{B}_{\psi}(\xi:h)+(\mathbb{B}_{1}(1:h)-\mathbb{B}_{1}(h))D^{(1)}_{h,\xi}\mathbb{B}_{\psi}(\xi:h)$$

$$+(\mathbb{B}_{0}(1:h)-\mathbb{B}_{0}(h))\mathbb{B}_{\psi}(\xi:h)-\psi\mathbb{B}_{\psi-1}(\xi:h)=0.$$

Theorem 2.5 *A solutions of the following differential equation*

$$\frac{q^{\psi}\left(\mathbb{B}_{\psi,q}(1:q^{-1}h)-\mathbb{B}_{\psi,q}(q^{-1}h)\right)}{[\psi]_q!}D^{(\psi)}_{q,h,\xi}\mathbb{B}_{\psi,q}(q\xi:h)$$

$$+\frac{q^{\psi-1}\left(\mathbb{B}_{\psi-1,q}(1:q^{-1}h)-\mathbb{B}_{\psi-1,q}(q^{-1}h)\right)}{[\psi-1]_q!}D^{(\psi-1)}_{q,h,\xi}\mathbb{B}_{\psi-1,q}(q\xi:h)$$

$$+\cdots+\frac{q^2\left(\mathbb{B}_{2,q}(1:q^{-1}h)-\mathbb{B}_{2,q}(q^{-1}h)\right)}{[2]_q!}D^{(2)}_{q,h,\xi}\mathbb{B}_{\psi,q}(q\xi:h)$$

$$+q\left(\mathbb{B}_{1,q}(1:q^{-1}h)-\mathbb{B}_{1,q}(q^{-1}h)\right)D^{(1)}_{q,h,\xi}\mathbb{B}_{\psi,q}(q\xi:h)$$

$$+\left(\mathbb{B}_{0,q}(1:q^{-1}h)-\mathbb{B}_{0,q}(q^{-1}h)\right)\mathbb{B}_{\psi,q}(q\xi:h)-[\psi]_q\mathbb{B}_{\psi-1,q}(q\xi:h)=0. \tag{2.15}$$

Proof One easily notices from (2.1) that

$$\sum_{\psi=0}^{\infty}\mathbb{B}_{\psi,q}(\xi:h)\frac{\varpi^{\psi}}{[\psi]_q!}=\frac{\varpi}{e_{q,h}(1:\varpi)-1}e_{q,h}(q\xi:\varpi)$$

$$=\frac{1}{\varpi}\left(\frac{\varpi}{e_{q,q^{-1}h}(1:q\varpi)-1}e_{q,q^{-1}h}(1:q\varpi)-\frac{\varpi}{e_{q,q^{-1}h}(1:\varpi)-1}\right)\times\frac{\varpi}{e_{q,h}(1:\varpi)-1}e_{q,h}(q\xi:\varpi).$$

$$\varpi\sum_{\psi=0}^{\infty}\mathbb{B}_{\psi,q}(\xi:h)\frac{\varpi^{\psi}}{[\psi]_q!}=\sum_{\psi=0}^{\infty}\left(\sum_{\upsilon=0}^{\psi}\binom{\psi}{\upsilon}_q q^{\psi}\left(\mathbb{B}_{\psi,q}(1:q^{-1}h)-\mathbb{B}_{\psi,q}(q^{-1}h)\right)\mathbb{B}_{\psi-\upsilon,q}(q\xi:h)\right)\frac{\varpi^{\psi}}{[\psi]_q!}. \tag{2.16}$$

From the above equation (2.16), we get

$$\sum_{\upsilon=0}^{\psi}\binom{\psi}{\upsilon}_q q^{\upsilon}\left(\mathbb{B}_{\upsilon,q}(1:q^{-1}h)-\mathbb{B}_{\upsilon,q}(q^{-1}h)\right)\mathbb{B}_{\psi-\upsilon,q}(q\xi:h)-[\psi]_q\mathbb{B}_{\psi-1,q}(\xi:h)=0. \tag{2.17}$$

Substituting $q\xi$ for ξ in Corollary 2.3, we get

$$\mathbb{B}_{\psi-\upsilon,q}(q\xi:h)=\frac{[\psi-\upsilon]_q!}{[\psi]_q!}D^{(\upsilon)}_{q,h,\xi}\mathbb{B}_{\psi,q}(q\xi:h). \tag{2.18}$$

Applying equations (2.17) and (2.18), we obtain

$$\sum_{\upsilon=0}^{\psi}\frac{q^{\upsilon}\left(\mathbb{B}_{\upsilon,q}(1:q^{-1}h)-\mathbb{B}_{\upsilon,q}(q^{-1}h)\right)}{[\upsilon]_q!}D^{(\upsilon)}_{q,h,\xi}\mathbb{B}_{\psi,q}(q\xi:h)-[\psi]_q\mathbb{B}_{\psi-1,q}(q\xi:h)=0.$$

Therefore, we acquire at the desired result. □

Corollary 2.10 *Adjusting $h\to 0$ in (2.15):*

$$\frac{q^{\psi}\left(\mathbb{B}_{\psi,q}(1)-\mathbb{B}_{\psi,q}\right)}{[\psi]_q!}D^{(\psi)}_{q,\xi}\mathbb{B}_{\psi,q}(q\xi)+\frac{q^{\psi-1}\left(\mathbb{B}_{\psi-1,q}(1)-\mathbb{B}_{\psi-1,q}\right)}{[\psi-1]_q!}D^{(\psi-1)}_{q,\xi}\mathbb{B}_{\psi-1,q}(q\xi)$$

$$+\cdots+\frac{q^2\left(\mathbb{B}_{2,q}(1)-\mathbb{B}_{2,q}\right)}{[2]_q!}D^{(2)}_{q,\xi}\mathbb{B}_{\psi,q}(q\xi)$$

$$+q\left(\mathbb{B}_{1,q}(1)-\mathbb{B}_{1,q}\right)D^{(1)}_{q,\xi}\mathbb{B}_{\psi,q}(q\xi)+\left(\mathbb{B}_{0,q}(1)-\mathbb{B}_{0,q}\right)\mathbb{B}_{\psi,q}(q\xi)-[\psi]_q\mathbb{B}_{\psi-1,q}(q\xi)=0.$$

Theorem 2.6 *For* $|q|\geq 1$ *with* $a,b\neq 0$, *we derive a basic symmetry relation for difference equation as*

$$\frac{b^{\psi}\mathbb{B}_{\psi,q}(a\eta:b^{-1}h)}{[\psi]_q!}D^{(\psi)}_{q,h}\mathbb{B}_{\psi,q}(b\xi:a^{-1}h)+\frac{b^{\psi-1}a\mathbb{B}_{\psi-1,q}(a\eta:b^{-1}h)}{[\psi-1]_q!}D^{(\psi-1)}_{q,h}\mathbb{B}_{\psi,q}(b\xi:a^{-1}h)+$$

$$\cdots+\frac{b^2a^{\psi-2}\mathbb{B}_{2,q}(a\eta:b^{-1}h)}{[2]_q!}D^{(2)}_{q,h}\mathbb{B}_{\psi,q}(b\xi:a^{-1}h)+ba^{\psi-1}\mathbb{B}_{1,q}(a\eta:b^{-1}h)D^{(1)}_{q,h}\mathbb{B}_{\psi,q}(b\xi:a^{-1}h)$$

$$+a^{\psi}\mathbb{B}_{0,q}(a\eta:b^{-1}h)\mathbb{B}_{\psi,q}(b\xi:a^{-1}h)$$

$$=\frac{a^{\psi}\mathbb{B}_{\psi,q}(b\eta:a^{-1}h)}{[\psi]_q!}D^{(\psi)}_{q,h}\mathbb{B}_{\psi,q}(a\xi:b^{-1}h)+\frac{a^{\psi-1}b\mathbb{B}_{\psi-1,q}(b\eta:a^{-1}h)}{[\psi-1]_q!}D^{(\psi-1)}_{q,h}\mathbb{B}_{\psi,q}(a\xi:b^{-1}h)+$$

$$\cdots+\frac{a^2b^{\psi-2}\mathbb{B}_{2,q}(b\eta:a^{-1}h)}{[2]_q!}D^{(2)}_{q,h}\mathbb{B}_{\psi,q}(a\xi:b^{-1}h)+ab^{\psi-1}\mathbb{B}_{1,q}(b\eta:a^{-1}h)D^{(1)}_{q,h}\mathbb{B}_{\psi,q}(a\xi:b^{-1}h)$$

$$+b^{\psi}\mathbb{B}_{0,q}(b\eta:a^{-1}h)\mathbb{B}_{\psi,q}(a\xi:b^{-1}h). \tag{2.19}$$

Proof From (2.1), we find a relation

$$e_{q,h}(ab\xi:\varpi)=\sum_{\psi=0}^{\infty}a^{\psi}(b\xi)(b\xi-a^{-1}h)(b\xi-[2]_qa^{-1}h)\cdots(b\xi-[\psi-1]_qa^{-1}h)\frac{\varpi^{\psi}}{[\psi]_q!}$$

$$=e_{q,a^{-1}h}(b\xi:a\varpi).$$

If $e_{q,h}(ab\xi:\varpi)=e_{q,a^{-1}h}(b\xi:a\varpi)$, we can take

$$A(\varpi)=\frac{ab\varpi^2e_{q,h}(ab\xi:\varpi)e_{q,h}(ab\eta:\varpi)}{(e_{q,a^{-1}h}(1:a\varpi)-1)(e_{q,b^{-1}h}(1:b\varpi)-1)}$$

$$=\frac{a\varpi}{e_{q,a^{-1}h}(1:a\varpi)-1}e_{q,h}(ab\xi:\varpi)\frac{b\varpi}{e_{q,b^{-1}h}(1:b\varpi)-1}e_{q,h}(ab\eta:\varpi)$$

$$=\frac{a\varpi}{e_{q,a^{-1}h}(1:a\varpi)-1}e_{q,a^{-1}h}(b\xi:a\varpi)\frac{b\varpi}{e_{q,b^{-1}h}(1:b\varpi)-1}e_{q,b^{-1}h}(a\eta:b\varpi)$$

$$=\sum_{\psi=0}^{\infty} a^{\psi}\mathbb{B}_{\psi,q}(b\xi : a^{-1}h)\frac{\varpi^{\psi}}{[\psi]_q!}\sum_{\upsilon=0}^{\infty} b^{\upsilon}\mathbb{B}_{\upsilon,q}(a\eta : b^{-1}h)\frac{\varpi^{\upsilon}}{[\upsilon]_q!}$$

$$=\sum_{\psi=0}^{\infty}\left(\sum_{\upsilon=0}^{\psi}\binom{\psi}{\upsilon}_q a^{\psi-\upsilon}b^{\upsilon}\mathbb{B}_{\psi-\upsilon,q}(b\xi : a^{-1}h)\mathbb{B}_{\psi,q}(a\eta : b^{-1}h)\right)\frac{\varpi^{\psi}}{[\psi]_q!}. \tag{2.20}$$

Similarly, we obtain

$$A(\psi)=\sum_{\psi=0}^{\infty}\left(\sum_{\upsilon=0}^{\psi}\binom{\psi}{\upsilon}_q b^{\psi-\upsilon}a^{\upsilon}\mathbb{B}_{\psi-\upsilon,q}(a\xi : b^{-1}h)\mathbb{B}_{\psi,q}(b\eta : a^{-1}h)\right)\frac{\varpi^{\psi}}{[\psi]_q!}. \tag{2.21}$$

Comparing the coefficients of both sides in (2.20) and (2.21), we obtain

$$\sum_{\upsilon=0}^{\psi}\binom{\psi}{\upsilon}_q a^{\psi-\upsilon}b^{\upsilon}\mathbb{B}_{\psi-\upsilon,q}(b\xi : a^{-1}h)\mathbb{B}_{\psi,q}(a\eta : b^{-1}h)$$

$$=\sum_{\upsilon=0}^{\psi}\binom{\psi}{\upsilon}_q b^{\psi-\upsilon}a^{\upsilon}\mathbb{B}_{\psi-\upsilon,q}(a\xi : b^{-1}h)\mathbb{B}_{\psi,q}(b\eta : a^{-1}h). \tag{2.22}$$

From Corollary 2.3, we have

$$\mathbb{B}_{\psi-\upsilon,q}(b\xi, a^{-1}h)=\frac{[\psi-\upsilon]_q!}{[\psi]_q!}D^{(k)}_{q,h,\xi}\mathbb{B}_{\psi,q}(b\xi, a^{-1}h),$$

$$\mathbb{B}_{\psi-\upsilon,q}(a\xi, b^{-1}h)=\frac{[\psi-\upsilon]_q!}{[\psi]_q!}D^{(k)}_{q,h,\xi}\mathbb{B}_{\psi,q}(a\xi, b^{-1}h). \tag{2.23}$$

Replacing (2.22) with equation (2.23), we have

$$\sum_{\upsilon=0}^{\psi}\frac{a^{\psi-\upsilon}b^{\upsilon}\mathbb{B}_{\upsilon,q}(a\eta : b^{-1}h)}{[\upsilon]_q!}D^{(\upsilon)}_{q,h,\xi}\mathbb{B}_{\psi,q}(b\xi, a^{-1}h)$$

$$=\sum_{\upsilon=0}^{\psi}\frac{a^{\upsilon}b^{\psi-\upsilon}\mathbb{B}_{\upsilon,q}(b\eta : a^{-1}h)}{[\upsilon]_q!}D^{(\upsilon)}_{q,h,\xi}\mathbb{B}_{\psi,q}(a\xi, b^{-1}h). \tag{2.24}$$

From (2.24), we complete the proof of Theorem 2.6

Corollary 2.11 *On setting $q \to 1$ in (2.19), we have*

$$\frac{b^{\psi}\mathbb{B}_{\psi}(a\eta : b^{-1}h)}{\psi!}D^{(\psi)}_{h}\mathbb{B}_{\psi}(b\xi : a^{-1}h)+\frac{b^{\psi-1}a\mathbb{B}_{\psi-1}(a\eta : b^{-1}h)}{(\psi-1)!}D^{(\psi-1)}_{h}\mathbb{B}_{\psi}(b\xi : a^{-1}h)+$$

$$\cdots + \frac{b^2 a^{\psi-2}\mathbb{B}_2(a\eta : b^{-1}h)}{2!} D_h^{(2)}\mathbb{B}_\psi(b\xi : a^{-1}h) + ba^{\psi-1}\mathbb{B}_1(a\eta : b^{-1}h)D_h^{(1)}\mathbb{B}_\psi(b\xi : a^{-1}h)$$

$$+a^\psi \mathbb{B}_0(a\eta : b^{-1}h)\mathbb{B}_\psi(b\xi : a^{-1}h)$$

$$= \frac{a^\psi \mathbb{B}_\psi(b\eta : a^{-1}h)}{\psi!} D_h^{(\psi)}\mathbb{B}_\psi(a\xi : b^{-1}h) + \frac{a^{\psi-1}b\mathbb{B}_{\psi-1}(b\eta : a^{-1}h)}{(\psi-1)!} D_h^{(\psi-1)}\mathbb{B}_\psi(a\xi : b^{-1}h) +$$

$$\cdots + \frac{a^2 b^{\psi-2}\mathbb{B}_2(b\eta : a^{-1}h)}{2!} D_h^{(2)}\mathbb{B}_\psi(a\xi : b^{-1}h) + ab^{\psi-1}\mathbb{B}_1(b\eta : a^{-1}h)D_h^{(1)}\mathbb{B}_\psi(a\xi : b^{-1}h)$$

$$+b^\psi \mathbb{B}_0(b\eta : a^{-1}h)\mathbb{B}_\psi(a\xi : b^{-1}h).$$

Corollary 2.12 *Setting $h \to 0$ in (2.19):*

$$\frac{b^\psi \mathbb{B}_{\psi,q}(a\eta)}{[\psi]_q!} D_q^{(\psi)}\mathbb{B}_{\psi,q}(b\xi) + \frac{b^{\psi-1}a\mathbb{B}_{\psi-1,q}(a\eta)}{[\psi-1]_q!} D_q^{(\psi-1)}\mathbb{B}_{\psi,q}(b\xi) +$$

$$\cdots + \frac{b^2 a^{\psi-2}\mathbb{B}_{2,q}(a\eta)}{[2]_q!} D_q^{(2)}\mathbb{B}_{\psi,q}(b\xi) + ba^{\psi-1}\mathbb{B}_{1,q}(a\eta)D_q^{(1)}\mathbb{B}_{\psi,q}(bx)$$

$$+a^\psi \mathbb{B}_{0,q}(a\eta)\mathbb{B}_{\psi,q}(b\xi)$$

$$= \frac{a^\psi \mathbb{B}_{\psi,q}(b\eta)}{[\psi]_q!} D_q^{(\psi)}\mathbb{B}_{\psi,q}(a\xi) + \frac{a^{\psi-1}b\mathbb{B}_{\psi-1,q}(b\eta)}{[\psi-1]_q!} D_q^{(\psi-1)}\mathbb{B}_{\psi,q}(a\xi) +$$

$$\cdots + \frac{a^2 b^{\psi-2}\mathbb{B}_{2,q}(b\eta)}{[2]_q!} D_q^{(2)}\mathbb{B}_{\psi,q}(a\xi) + ab^{\psi-1}\mathbb{B}_{1,q}(b\eta)D_q^{(1)}\mathbb{B}_{\psi,q}(a\xi)$$

$$+b^\psi \mathbb{B}_{0,q}(b\eta)\mathbb{B}_{\psi,q}(a\xi).$$

Theorem 2.7 *For $\mid q \mid \geq 1$ with $\alpha, \beta \neq 0$, we derive a basic symmetry relation for difference equation as*

$$\frac{\beta^\psi \mathbb{B}_{\psi,q}(\beta^{-1}h)}{[\psi]_q!} D_{q,h}^{(\psi)}\mathbb{B}_{\psi,q}(\beta\xi : \alpha^{-1}h) + \frac{\beta^{\psi-1}\alpha\mathbb{B}_{\psi-1,q}(\beta^{-1}h)}{[\psi-1]_q!} D_{q,h}^{(\psi-1)}\mathbb{B}_{\psi,q}(\beta\xi : \alpha^{-1}h)$$

$$+\cdots + \frac{\beta^2 \alpha^{\psi-2}\mathbb{B}_{2,q}(\beta^{-1}h)}{[2]_q!} D_{q,h}^{(2)}\mathbb{B}_{\psi,q}(\beta\xi : \alpha^{-1}h) + \beta\alpha^{\psi-1}\mathbb{B}_{1,q}(\beta^{-1}h)D_{q,h}^{(1)}\mathbb{B}_{\psi,q}(\beta\xi : \alpha^{-1}h)$$

$$+\alpha^\psi \mathbb{B}_{0,q}(\beta^{-1}h)\mathbb{B}_{\psi,q}(\beta\xi : \alpha^{-1}h)$$

$$= \frac{\alpha^\psi \mathbb{B}_{\psi,q}(\alpha^{-1}h)}{[\psi]_q!} D_{q,h}^{(\psi)}\mathbb{B}_{\psi\psi,q}(\alpha\xi : \beta^{-1}h) + \frac{\alpha^{\psi-1}\beta\mathbb{B}_{\psi-1,q}(\alpha^{-1}h)}{[\psi-1]_q!} D_{q,h}^{(\psi-1)}\mathbb{B}_{\psi,q}(\alpha\xi : \beta^{-1}h)$$

$$+\cdots+\frac{\alpha^{2}\beta^{\psi-2}\mathbb{B}_{2,q}(\alpha^{-1}h)}{[\psi]_q!}D^{(2)}_{q,h}\mathbb{B}_{\psi,q}(\alpha\xi:\beta^{-1}h)+\alpha\beta^{\psi-1}\mathbb{B}_{1,q}(\alpha^{-1}h)D^{(1)}_{q,h}\mathbb{B}_{\psi,q}(\alpha\xi:\beta^{-1}h)$$

$$+\beta^{\psi}\mathbb{B}_{0,q}(\alpha^{-1}h)\mathbb{B}_{\psi,q}(\alpha\xi:\beta^{-1}h). \tag{2.25}$$

Proof To obtain another symmetric difference equation which is related to the degenerate quantum Bernoulli polynomials, we have

$$B(\varpi)=\frac{\alpha\beta\varpi^{2}e_{q,h}(\alpha\beta\xi:\varpi)}{(e_{q,\alpha^{-1}h}(1:\alpha\varpi)-1)(e_{q,\beta^{-1}h}(1:\beta\varpi)-1)}.$$

From the same way as proving Theorem 2.6, we find

$$\sum_{\upsilon=0}^{\psi}\binom{\psi}{\upsilon}_q\beta^{\upsilon}\alpha^{\psi-\upsilon}\mathbb{B}_{\upsilon,q}(\beta^{-1}h)\mathbb{B}_{\psi-\upsilon,q}(\beta\xi:\alpha^{-1}h)$$

$$=\sum_{\upsilon=0}^{\psi}\binom{\psi}{\upsilon}_q\alpha^{\upsilon}\beta^{\psi-\upsilon}\mathbb{B}_{\upsilon,q}(\alpha^{-1}h)\mathbb{B}_{\psi-\upsilon,q}(\alpha\xi:\beta^{-1}h). \tag{2.26}$$

Using Corollary 2.3 in (2.26), we obtain

$$\sum_{\upsilon=0}^{\psi}\frac{\binom{\psi}{\upsilon}_q\beta^{\upsilon}\alpha^{\psi-\upsilon}\mathbb{B}_{\upsilon,q}(\beta^{-1}h)}{[\upsilon]_q!}D^{(\upsilon)}_{q,h}\mathbb{B}_{\psi,q}(\beta\xi:\alpha^{-1}h)$$

$$=\sum_{\upsilon=0}^{\psi}\frac{\alpha^{\upsilon}\beta^{\psi-\upsilon}\mathbb{B}_{\upsilon,q}(\alpha^{-1}h)}{[\upsilon]_q!}D^{(\upsilon)}_{q,h}\mathbb{B}_{\psi,q}(\alpha\xi:\beta^{-1}h). \tag{2.27}$$

From (2.27), we complete the proof of Theorem 2.7 □

Theorem 2.8 *The degenerate quantum Bernoulli polynomials are solutions of the following higher-order differential equation combined with the degenerate quantum tangent numbers and polynomials*

$$\frac{\left(\mathbb{T}_{\psi,q}(2:h)+\mathbb{T}_{\psi,q}(h)\right)}{[\psi]_q!}D^{(\psi)}_{q,h,\xi}\mathbb{B}_{\psi,q}(\xi:h)+\frac{\left(\mathbb{T}_{\psi-1,q}(2:h)+\mathbb{T}_{\psi-1,q}(h)\right)}{[\psi-1]_q!}D^{(\psi-1)}_{q,h,\xi}\mathbb{B}_{\psi,q}(\xi:h)$$

$$+\cdots+\frac{\left(\mathbb{T}_{2,q}(2:h)+\mathbb{T}_{2,q}(h)\right)}{[2]_q!}D^{(2)}_{q,h,\xi}\mathbb{B}_{\psi,q}(\xi:h)+\left(\mathbb{T}_{1,q}(2:h)+\mathbb{T}_{1,q}(h)\right)D^{(1)}_{q,h,\xi}\mathbb{B}_{\psi,q}(\xi:h)$$

$$+\left(\mathbb{T}_{0,q}(2:h)+\mathbb{T}_{0,q}(h)-2\right)\mathbb{B}_{\psi,q}(\xi:h)=0. \tag{2.28}$$

Proof Using (1.5) and (2.1), we note that

$$\sum_{\psi=0}^{\infty} \mathbb{B}_{\psi,q}(\xi : h)\frac{\varpi^{\psi}}{[\psi]_q!} = \frac{\varpi}{e_{q,h}(1:\varpi)-1}e_{q,h}(\xi:\varpi)$$

$$= \frac{1}{2}\left(\frac{2}{e_{q,h}(2:\varpi)+1}e_{q,h}(2:\varpi) + \frac{2}{e_{q,h}(2:\varpi)+1}\right)\frac{\varpi}{e_{q,h}(1:\varpi)-1}e_{q,h}(\xi:\varpi)$$

$$= \frac{1}{2}\sum_{\psi=0}^{\infty}\left(\sum_{\upsilon=0}^{\psi}\binom{\psi}{\upsilon}_q \left(\mathbb{T}_{\upsilon,q}(2:h)+\mathbb{T}_{\upsilon,q}(h)\right)\mathbb{B}_{\psi-\upsilon,q}(\xi:h)\right)\frac{\varpi^{\psi}}{[\psi]_q!}. \tag{2.29}$$

Comparing the coefficients on both sides of equation (2.29), we have

$$2\mathbb{B}_{\psi,q}(\xi : h) = \sum_{\upsilon=0}^{\psi}\binom{\psi}{\upsilon}_q \left(\mathbb{T}_{\upsilon,q}(2:h)+\mathbb{T}_{\upsilon,q}(h)\right)\mathbb{B}_{\psi-\upsilon,q}(\xi:h). \tag{2.30}$$

Using the relationship of the (q, h)-Bernoulli polynomials to the υ-times (q, h)-derivative in (2.30), we get

$$\sum_{\upsilon=0}^{\psi}\frac{\left(\mathbb{T}_{\upsilon,q}(2:h)+\mathbb{T}_{\upsilon,q}(h)\right)}{[\upsilon]_q!}D_{q,h,\xi}^{(\upsilon)}\mathbb{B}_{\psi,q}(\xi:h) - 2\mathbb{B}_{\psi,q}(\xi:h) = 0.$$

The above equation allows us to complete the proof. □

Corollary 2.13 *Adjusting $h \to 0$ in (2.28):*

$$\frac{\left(\mathbb{T}_{\psi,q}(2)+\mathbb{T}_{\psi,q}\right)}{[\psi]_q!}D_{q,\xi}^{(\psi)}\mathbb{B}_{\psi,q}(\xi) + \frac{\left(\mathbb{T}_{\psi-1,q}(2)+\mathbb{T}_{\psi-1,q}\right)}{[\psi-1]_q!}D_{q,\xi}^{(\psi-1)}\mathbb{B}_{\psi,q}(\xi)$$

$$+\cdots+\frac{\left(\mathbb{T}_{2,q}(2)+\mathbb{T}_{2,q}\right)}{[2]_q!}D_{q,\xi}^{(2)}\mathbb{B}_{\psi,q}(\xi) + \left(\mathbb{T}_{1,q}(2)+\mathbb{T}_{1,q}\right)D_{q,\xi}^{(1)}\mathbb{B}_{\psi,q}(\xi)$$

$$+\left(\mathbb{T}_{0,q}(2)+\mathbb{T}_{0,q}-2\right)\mathbb{B}_{\psi,q}(\xi) = 0.$$

Corollary 2.14 *Setting $q \to 1$ in (2.28):*

$$\frac{\left(\mathbb{T}_{\psi}(2:h)+\mathbb{T}_{\psi}(h)\right)}{\psi!}D_{h,\xi}^{(\psi)}\mathbb{B}_{\psi}(\xi:h) + \frac{\left(\mathbb{T}_{\psi-1}(2:h)+\mathbb{T}_{\psi-1}(h)\right)}{[\psi-1]!}D_{h,\xi}^{(\psi-1)}\mathbb{B}_{\psi}(\xi:h)$$

$$+\cdots+\frac{(\mathbb{T}_{2}(2:h)+\mathbb{T}_{2}(h))}{2!}D_{h,\xi}^{(2)}\mathbb{B}_{\psi}(\xi:h) + (\mathbb{T}_{1}(2:h)+\mathbb{T}_{1}(h))\,D_{h,\xi}^{(1)}\mathbb{B}_{\psi}(\xi:h)$$

$$+(\mathbb{T}_{0}(2:h)+\mathbb{T}_{0}(h)-2)\,\mathbb{B}_{\psi}(\xi:h) = 0.$$

3 Conclusion

We constructed (q, h)-analogue of Bernoulli polynomials and numbers and found several differential equations with these polynomials as solutions. We also found differential equations combining with q-Bernoulli and q-tangent polynomials. We established some properties of (q, h)-analogue of Bernoulli polynomials and numbers. The results deduced in this paper are general in advantage and likely to discover certain applications in the theory of special functions. On the other hand, the main results have importance as from which we have deduced some important integral formulas for special values of the parameters which are potentially useful for the laser specialists.

Competing Interests The authors have no conflicts of interest to declare that are relevant to the content of this chapter.

References

1. G. Bangerezako, Variational q-calculus. J. Math. Anal. Appl. **289**, 650–665 (2004)
2. H.B. Benaoum, (q, h)-analogue of Newton's binomial formula. J. Phys. A Math. Gen. **32**, 2037–2040 (1999)
3. L. Carlitz, Degenerate Stirling, Bernoulli and Eulerian numbers. Utilitas Math. **15**, 51–88 (1979)
4. L. Carlitz, A degenerate Staudt-Clausen theorem. Arch. Math. (Basel) **7**, 28–33 (1956)
5. R.D. Carmichael, The general theory of linear q-difference equations. Am. J. Math. **34**, 147–168 (1912)
6. J. Cermak, L. Nechvatal, On (q, h)-analogue of fractional calculus. J. Nonlinear Math. Phys. **17**, 51–68 (2010)
7. Z. Chen, N. Omur, S. Koparal, W.A. Khan, Some identities with multi generalized q-hyper harmonic numbers of order r. Symmetry **15**, 917 (2023). https://doi.org/10.3390/sym15040917
8. H. Guan, N. Omur, S. Koparal, W.A. Khan, q-analogs of $H^r_{n,m}$ (σ) and their applications. Mathematics **11**, 4159 (2023). https://doi.org/10.3390/math11194159
9. H.F. Jackson, q-Difference equations. Am. J. Math. **32**, 305–314 (1910)
10. H.F. Jackson, On q-functions and a certain difference operator. Trans. R. Soc. Edinb. **46**, 253–281 (2013)
11. J.Y. Kang, W.A. Khan, A new class of q-Hermite based Apostol type Frobenius Genocchi polynomials. Commun. Korean Math. Soc. **35**(3), 759–771 (2020)
12. J.Y. Kang, Properties of differential equations related to degenerate q-tangent numbers and polynomials. Symmetry **15**, 874 (2023)
13. W.A. Khan, A note on q-analogues of degenerate Catalan-Daehee numbers and polynomials. J. Math. **2022**, Article ID 9486880, 9 p. https://doi.org/10.1155/2022/9486880
14. Khan, W.A. A note on q-analogue of degenerate Catalan numbers associated p-adic integral on $\mathbb{Z}_p$. Symmetry **2022**, 14(119). https://doi.org/10.3390/sym14061119
15. V. Kac, P. Cheung, *Quantum Calculus; Part of the Universitext Book Series (UTX)* (Springer Nature, Switzerland, 2002); ISBN 978-0-387-95341-0
16. N.I. Mahmudov, q-analogues of the Bernoulli and Genocchi polynomials and the Srivastava-Pinter addition theorems. Discrete Dyn. Nat. Soc. Article number 169348 (2012). https://doi.org/10.1155/2012/169348

17. N.I. Mahmudov, On a class of q-Bernoulli and q-Euler polynomials. Adv. Differ. Equ. (2013). https://doi.org/10.1186/1687-1847-2013-1
18. K.S. Nisar, W.A. Khan, Notes on q-Hermite based unified Apostol type polynomials. J. Interdiscip. Math. **22**(7), 1185–1203 (2019)
19. C.S. Ryoo, J.Y. Kang, Properties of q differential equations of higher order and visualization of fractal using q-Bernoulli polynomials. Fract. Fract. **6**, 296 (2022)
20. C.S. Ryoo, J.Y. Kang, Various types of q-differential equations of higher-order for q-Euler and q-Genocchi polynomials. Mathematics **10**, 1181 (2022)
21. M.R.S. Rahmat, The (q, h)-Laplace transform on discrete time scales. Comput. Math. Appl. **62**, 272–281 (2011)
22. B. Silindir, A. Yantir, Generalized quantum exponential function and applications. Filomat **33**, 4907–4922 (2019)

Optimal Control Problems Governed by a Class Nonlinear Dynamical System

Maawiya Ould Sidi, Abdulmajeed Aljadi Alshammari, and Rabie Zine

Abstract This chapter proposes two optimal control problems that are governed by a class of non-linear dynamical systems. The first problem is constrained by a convective-diffusive model, and we study the existence of an optimal control solution for the considered problem and characterize it using the adjoint method. The second problem investigates a multiplicative distributive control for a vibrating string; we show that there exists an optimal control solution of the problem considered. The optimality system offers a standard for defining the unique optimal control. The chapter ends with some conclusions.

1 Introduction

The term "parabolic distributed systems" refers to a particular class of partial differential equations (PDEs) that describes how a function changes over time and space, with the value of the function and its partial derivatives at nearby points influencing the value of the function at any given point in space and time.

The presence of a second-order positive partial differential equation that changes sign as a function of space and time and does not abide by the maximum principle distinguishes the parabolic distributed systems from other types of systems. In other words, it is impossible to identify the solutions of parabolic PDEs from the values of the function at any other location in time or space without first providing the initial conditions and boundary conditions [10].

Examples of parabolic PDEs include the Black-Scholes equation, which represents the evolution of the price of a financial derivative, and the heat equation (sometimes referred to as the diffusion equation), which describes the diffusion of heat in

M. O. Sidi (✉) · A. A. Alshammari
Mathematics Department, College of Science, Jouf University, P.O. Box: 2014, Sakaka, Saudi Arabia
e-mail: msidi@ju.edu.sa

R. Zine
School of Science and Engineering, Al Akhawayn University in Ifrane, Ifrane, Morocco

M. Kharrat et al. (eds.), *Modeling of Discrete and Continuous Systems*, Forum for Interdisciplinary Mathematics, https://doi.org/10.1007/978-981-97-8715-9_17

a medium. Physics, chemistry, biology, finance, and engineering are just a few of the scientific and engineering disciplines that use the solutions to these equations [3].

The term hyperbolic distributed systems refers to a particular kind of partial differential equation (PDE) that models how a function changes over time and space, the value of the function at any given point in time and space being determined by the values of the function and its partial derivatives at other nearby points [20].

The hyperbolic distributed systems are characterized by the fact that they contain a second-order partial differential equation that is negative (meaning it changes sign as a function of space and time) and that satisfies the wave equation. In other words, solutions to hyperbolic PDEs can be determined by specifying the initial conditions and boundary conditions, and also by knowing the values of the function at other points in space and time.

The wave equation, which describes how waves travel across a medium, and the Euler equations, which describe how an ideal fluid flows, are two examples of hyperbolic PDEs. Numerous scientific and engineering disciplines, including physics, engineering, and geophysics, use the solutions to these equations [15].

Finding control inputs for a system that minimize a cost function or maximize a performance criterion while taking into account input and system dynamics is the focus of the field of control theory known as optimum control.

The goal of optimal control is to find the best possible control actions that will bring a system from an initial state to a desired final state in the most efficient way possible, while also taking into account any constraints on the system or the control inputs.

The use of optimal control is widespread within areas such as robotics, aeronautical engineering, economics, and finance, among others. Pontryagin's maximum principle, calculus of variations, dynamic programming, and gradient descent algorithms are a few of the methods frequently employed in optimal control [1].

Distributed bilinear systems are mathematical models of systems that combine bilinear processes and distributed processes. A bilinear process is a process that is linear in two variables, while a distributed process is a process that is defined over a spatial domain in which the state variables have continuous values that vary depending on their location in the domain.

Distributed bilinear systems are employed to simulate a variety of physical, biological, and engineering systems, including chemical reactors, ecological systems, and control systems [31].

The dynamics of distributed bilinear systems are typically explained by partial differential equations (PDEs). Numerical simulation methods can be used to study the system's behavior over time even though these PDEs are usually nonlinear and difficult to comprehend.

The general form of bilinear distributed system defined by the following evolution equation:

$$\begin{aligned} \dot{z}(m) &= Az(m) + f(x,m)B(z(m)), \\ z(0) &= z_0 \end{aligned} \tag{1}$$

where A generates a C^0 semigroup of bounded linear operators in a Banach space X, $B : H \rightarrow H$ is a map C^1 and $f \in H_0^1([0, M]; I\!R)$ is a control defined in a specified interval [0, M]. Usually we assume that B is linear, so that (1) is bilinear in the pair (f, z); note that even in this case, the solution z of (1) is a nonlinear function of f. The existence and uniqueness of the solution are given by Ball et Slemrod in [25].

One of the main challenges in analyzing distributed bilinear systems is to characterize their stability properties. Stability analysis relies on determining the behavior of the system under various perturbations and disturbances. Different stability criteria have been developed, such as Lyapunov stability, exponential stability, and input-to-state stability.

Recent advances in control theory have led to the development of control strategies for distributed bilinear systems. These strategies typically rely on feedback control methods, such as decentralized control and distributed control. These control methods can be used to stabilize the system and enhance its performance [4].

Quadratic functional cost and optimal control problems are a class of optimization problems that involve minimizing a quadratic cost function subject to constraints on the system dynamics. These problems are commonly encountered in control theory, where the goal is to find an input signal that minimizes a cost function while satisfying the dynamics of the system [27].

The general form of a quadratic functional cost can be expressed as

$$C(f) = \mu \int_0^M \int_0^I z(x, m) dx dm + \frac{\gamma}{2} \int_0^M \int_0^I \left[f_m^2(x, m) + f_x^2(x, m) \right] dx dm,$$

where $f(x, m)$ is the control input and $z(m)$ is the state of the system.

The optimal control problem is to find the input signal $f(x, m)$ that minimizes the cost function $C(f)$ subject to the dynamics of the system (1).

Gradient control problems are a subset of optimization problems where the objective function is a scalar function that depends on the gradient of another function. Such issues typically take the following form: minimize

$$C(f) = \mu \int_D \nabla z dx dm + \frac{\gamma}{2} \int_D \left[f_m^2(x, m) + f_x^2(x, m) \right] dx dm \tag{2}$$

subject to (1)
where $C(f)$ is the objective function, (1) is a constraint system, and ∇z is the gradient of $z(x)$ [22].

These issues arise in a variety of areas such as engineering, economics, and finance. Examples include parameter estimation for partial differential equations, portfolio optimization in finance, and design optimization in engineering [30].

Gradient control problems present a number of difficulties, one of which is that the resulting optimization problem is frequently non-convex and difficult to resolve.

Iterative techniques, numerical optimization algorithms, and simulated annealing are a few of the numerical approaches that have been developed to address these issues. Another strategy is to reformulate the issue as a partial differential equation (PDE) and resolve it through numerical techniques [22].

For background, a number of authors, including Bradly and Lenhart in [16], Bradly et al. in [7], and Lenhart in [6], have used quadratic costs, such as (2). They control the coefficient term in a variety of one-dimensional bilinear equations while minimizing a spatiotemporal velocity. In [12], Joshi examines the management of velocity in a class of bilinear systems. In [5], Boulaaras discusses a parabolic quasi-variational inequality. Zakeri and Abchouyeh investigate a quadratic optimum control problem constrained by elliptic systems in their article [21]. The results of the controllability of two-dimensional bilinear systems are provided by Ayalaa et al. in [2]. The parameterization method is used by Kafash and Delavarkhalafi in their paper [13] for problems involving optimum control. Quadratic control problems are used by Zerrik and Ould Sidi in their publications [23, 28, 29] to guide the location of infinite-dimensional bilinear systems to the desired state on a particular subregion. [32] to find articles by Zine and Ould Sidi that discuss quadratic control problems in the context of hyperbolic bi-linear systems.

In this chapter, we investigate a class of distributed bilinear systems in infinite-dimensional spaces, and we offer a solution to various optimum control problems that are limited by a genuine model of bilinear systems. The easiest technique to control the first spatial derivative's coefficient in a bilinear equation with a parabolic function is covered in the first section. Our convective velocity coefficient $z(x, m)$ is considered in the situation of one spatial dimension where the control is space- and time-dependent. Controlling such a coefficient demands more regularity on the control set than the management of a coefficient of a zeroth-order term. The existence of an optimal control is proven, and the optimality system is derived. We define an optimal control through the application of an elliptic partial differential equation. The optimality system is then composed of two parabolic equations and an elliptic variational inequality.

The second section investigates a multiplicative distributive control problem for a vibrating string. The multiplicative control coefficient $z(x, m)$ of the velocity term. We use multiplicative control with a quadratic cost of control to get the state solution to the necessary profile. We show that there is an ideal control. The optimality system offers a standard for defining the unique optimal control.

2 Optimal Control Problem Constrained by Convective Velocity

2.1 Preliminary

We explore the optimal approach to deal with the coefficient of the first spatial derivative in a parabolic equation. We deal with this issue when there is only one spatial dimension and management requires consideration of both space and time. Because controlling such a coefficient requires more regularity on the control set than controlling a coefficient of a zeroth-order term, as opposed to the conventional explicit description, an optimal control is described by a partial differential equation (PDE) instead.

We consider the parabolic equation.

$$\begin{cases} z_m(x,m) = bz_{xx} - f(x,m)z_x + g \text{ in } & D = (0,I) \times (0,M). \\ z(x,0) = z_0(x) & 0 < x < I, m = 0 \end{cases} \tag{3}$$

with the boundary conditions
$z(0, m) = 0$ for all $m \in (0, M)$
$z_x(I, m) = 0$ for all $m \in (0, M)$

Control is assumed to be the flow velocity coefficient, or $f(x, m)$. we define the objective functional;

$$C(f) = \mu \int_0^M \int_0^I z(x,m)dxdm + \frac{\gamma}{2} \int_0^M \int_0^I \left[f_m^2(x,m) + f_x^2(x,m) \right] dxdm$$

where μ and γ are weighting functions, which are positive numbers. Our aim is to reduce the functional purpose:

$$C(f^*) = \min_{f \in Z} C(f)$$

where the control set Z is defined by $Z = \{f \in H_0^1(D) \| f(x,m)\| \leq F\}$.

The purpose of the example is to more generally apply two recent flow applications while also demonstrating this approach with more routine controls. The treatment of objective functionals with a broader, differentiable reliance on z is also possible. Handagama and Lenhart looked at the best way to manage the time-dependent flow rate into the bioreactor when a pollutant is broken down by the metabolism of bacteria in a gas-phase bioreactor [17]. Lenhart [16] studied optimal control of a parabolic differential equation to describe a one-dimensional fluid traveling through a soil-packed tube in which a contaminant is initially distributed. Only as a function of time, she considers the flow velocity coefficient. See Chawla and Lenhart [8] for a similar illustration employing flow in a bioreactor.

2.2 Existence of an Optimal Control

The state equation's solution is a weak one in

$$S = L^2(0, M; P)$$

where $P = H_0^1(0, I)$. (The functions disappear when $x = 0$ when a subscript of 0 is present.)

Definition 1 A function $z \in S$ with $z_m \in L^2(0, M; P^*)$ is a solution of

$$\begin{cases} z_m(x, m) = bz_{xx} - f(x, m)z_x + g \text{ in } & D = (0, I) \times (0, M) \\ z(0, m) = 0 & \text{on } (0, M), x = 0 \\ z_x(I, m) = 0 & \text{on } (0, M), x = I \\ z(x, 0) = z_0(x) & \text{on } (0, I), m = 0 \end{cases} \tag{4}$$

if

$$\int_0^M \left[\langle \acute{z}(m), p\rangle + \int_0^I (bz_x p_x + f z_x p - gp)dx\right]dm = 0 \tag{5}$$

for all $p \in P$ and a.e. time $0 \leq m \leq M$, and $z(x, 0)$, where the brackets $\langle , \rangle$ in the integrand denote the duality between P and P^*.

The following assumptions are made:

(1) b, μ, γ are positive constants,
(2) $z_0 \in H^1(0, I)$ with compact support,
(3) $g \in L^\infty(D)$.

For $f \in Z$, a classic result in [11] shows that there must be a single solution $z = z(f)$ in S satisfying (4) in the sense (5).

Theorem 1 *[12] There is an optimal control $f^* \in Z$ that reduces the objective functional $C(f)$.*

Proof Choose a minimizing sequence $f^l, l = 1, 2, \ldots,$ such that

$$\lim_{l\to\infty} C(f^l) = \inf_{f\in Z} C(f)$$

Let $z^l = z(f^l)$ be the appropriate state responses to (4). We need a change in variable to get the required a priori estimates s on our minimizing sequence and related states. $s = e^{-\lambda m}$, where λ is to select a positive constant. Therefore, Eq. (3) becomes

$$s_m + \lambda s = bs_{xx} - f(x, m) + e^{-\lambda m} g.$$

Utilizing the aforementioned equation's weak formulation and time integration, we can obtain

$$\int_0^M \langle s_m^l, s^l \rangle dm + \int_0^M \int_0^I (bs_x^l s_x^l + f^l s_x^l s^l + \lambda (s^l)^2 - e^{-\lambda m} s^l g) dx dm = 0$$

and then

$$\frac{1}{2} \int_0^I (s^l)^{(2)}(x, M) dx + b \int_0^M \int_0^I (s_x^l)^2 dx dm + \lambda \int_0^M \int_0^I (s^l)^2 dx dm \quad (6)$$
$$= \frac{1}{2} \int_0^I z_0^2 dx - \int_0^M \int_0^I f^l s_x^l s^l dx dm + \int_0^M \int_0^I e^{-\lambda m} s^l dx dm.$$

Note: $\|f^l\|_{H^1(D)} \leq C$ through the use of the objective functional.
We now employ Cauchy's inequality on the right-hand side in order to achieve.

$$\frac{1}{2} \int_0^I (s^l)^2(x, M) dx + b \int_0^M \int_0^I (s_x^l)^2 dx dm + \lambda \int_0^M \int_0^I (s^l)^2 dx dm$$
$$\leq \frac{1}{2} \int_0^I z_0^2 dx + \int_0^M \int_0^I g^2 dx dm \quad (7)$$

According to (7), we can say that $\|s^l\|_S$ are independently limited and uniform l. Thus $\|z^l\|_S$ are independently limited uniformly l. Using (3) and the aforementioned limits, we also have uniform bounds on $\|z_m^l\|_{L^2(0,M;P^*)}$.

Using the aforementioned constraints, we can find a subsequence that satisfies the following convergence conditions:

$$\begin{cases} z^l \longrightarrow z^* & \text{weakly in} \quad S, \\ z_m^l \longrightarrow z_m^* & \text{weakly in} \quad L^2(0, M; P^*), \\ f^l \longrightarrow f^* & \text{strongly in} \quad H^1(D), \\ f_x^l \longrightarrow f_x^* & \text{weakly in} \quad L^2(D), \\ f_m^l \longrightarrow f_m^* & \text{weakly in} \quad L^2(D), \end{cases} \quad (8)$$

The strong convergence of f^l in L^2 makes sense as $H^1(D) \subseteq L^2(D)$ [18].

Now, we must demonstrate $z^* = z(f^*)$, i.e. z^* is the state solution associated with f^*.

Take into account the PDE's weak form, which is integrated over time and fulfilled by z^l (for $p \in P$):

$$\int_0^M \langle z^l_m, p\rangle dm + b\int_0^M\int_0^I z^l_x p_x dxdm + \int_0^M\int_0^I f^l z^l_x pdxdm - \int_0^M\int_0^I gpdxdm = 0 \tag{9}$$

The third term's convergence makes use of the notion that

$$\begin{aligned}
&|\int_0^M\int_0^I f^l z^l_x p - f z_x pdxdm| \\
&\leq \int_0^M\int_0^I |(f^l - f)||z^l_x|dxdm + |\int_0^M\int_0^I fp(z^l_x - z_x)dxdm| \\
&\leq (\int_0^M\int_0^I |(f^l - f)p|^2 dxdm)^{\frac{1}{2}}(\int_0^M\int_0^I |z^l_x|dxdm)^{\frac{1}{2}} \\
&+|\int_0^M\int_0^I fp(z^l_x - z_x)dxdm| \\
&\leq J_1(\int_0^M\int_0^I |f^l - f|^2 dxdm)^{\frac{1}{2}}(\int_0^M\int_0^I |z^l_x|^2 dxdm)^{\frac{1}{2}} \\
&+|\int_0^M\int_0^I fp(z^l_x - z_x)dxdm| \\
&\to 0, as, l \to \infty
\end{aligned} \tag{10}$$

where the J_1 is reliant on the L^∞ limit on p. The L^2 strong convergence of the f^l sequence and the L^2 uniform limit on the z^l_x sequence are both necessary for the first term to converge. Since fp is in $L^2(D)$, the second term converges via weak L^2 convergence of the z^l_x series. Passing to the limit in the PDE satisfied by z^l, we receive $z^* = z(f^*)$.

We then confirm that f^* is the best control. Utilizing the goal functional's lower semicontinuity with regard to these weak convergence, we derive

$$\begin{aligned}
&C(f^*) \leq \inf_l \mu \int_0^M\int_0^I z^l(x, m)dx + \inf_l \frac{\gamma}{2}\int_0^M\int_0^I [(f^l_m)^2 + (h^l_x)^2]dxdm \\
&= \lim_{l\to\infty} C(f^l) = \inf_l C(f)
\end{aligned} \tag{11}$$

Therefore, f^* is the best control. □

Remark 1 The exact rationale for having a H^1 control set is the convergence of the term $f^l z_x^l$ with $f^l = f^l(x, m)$). The L^∞ controls and methods of [8, 16] would be effective if the controls were only spatial functions.

2.3 *Characterization of an Optimal Control*

By distributing the objective functional $C(f)$ with respect to f, we may derive the optimality system.

An elliptic differential equation will be used to characterize an optimum control. So, two parabola equations and a variational inequality will make up the optimality system [14].

Lemma 1 *[12] The solution map*

$$f \in Z \to z = (f) \in S$$

is differentiable in the following sense: $\dfrac{z(f + \varepsilon\ell) - z(f)}{\varepsilon} \rightharpoonup \phi$ *weakly in* S *as* $\varepsilon \to 0$, *where* $f + \varepsilon\ell \in Z$, $\ell \in L^\infty(D)$. *Moreover the sensitivity function* $\phi = \phi(f; \ell)$ *satisfies*

$$\begin{cases} \phi_m(x, m) = b\phi_{xx} - f\phi_x - \ell z_x \ in & D = (0, I) \times (0, M), \\ \phi(x, 0) \quad = 0 & when \ \ m = 0, \\ \phi(x, m) \quad = 0 & when \ \ x = 0, \\ \phi_x(I, m) \quad = 0 & when \ \ x = I, \\ where \ z \quad = z(f) \end{cases} \tag{12}$$

Proof To generate a priori estimations of the difference quotients, the variables are shifted after that

$$s = e^{-\lambda m} z, s^\varepsilon = e^{-\lambda m} z^3,$$

where $z = 4(f)$ and $z^\varepsilon = z(f + \varepsilon\ell)$. When we take into account the PDE that is solved by $s^\varepsilon - s$, we get (multiplying by test function $(s^\varepsilon - s)$)

$$\frac{1}{2}\int_0^I (s^\varepsilon - s)^2(x, M)dx + b\int_0^M\int_0^I ((s^\varepsilon - s)_x)^2 dxdm + \lambda\int_0^M\int_0^I (s^\varepsilon - s)^2 dxdm$$
$$= -\int_0^M\int_0^I f(s^\varepsilon - s)_x(s^\varepsilon - s)dxdm - \varepsilon\int_0^M\int_0^I \ell s_x^\varepsilon(s^\varepsilon - s)dxdm$$
$$\leq \frac{b}{2}\int_0^M\int_0^I ((s^\varepsilon - s)_x)^2 dxdm + \frac{1}{2b}\int_0^M\int_0^I (f(s^\varepsilon - s))^2 dxdm$$
$$+\frac{\varepsilon^2}{4}\int_0^M\int_0^I (\ell s_x)^2 dxdm + \int_0^M\int_0^I (s^\varepsilon - s)^2 dxdm$$
$$+\frac{\varepsilon^2}{4}\int_0^M\int_0^I (\ell s_x)^2 dxdm + \int_0^M\int_0^I (s^\varepsilon - s)^2 dxdm + Y_1\varepsilon^2, \tag{13}$$

denoting the L^∞ bound on f by F. Using the facts that $\|s_x^\varepsilon\|_{L^2}$ is uniformly bounded from (7), we obtain

$$\frac{b}{2}\int_0^M\int_0^I ((s^\varepsilon - s)_x)^2 dxdm + (\lambda - \frac{F^2}{2b} - 1)\int_0^M\int_0^I (s^\varepsilon - s)^2 dxdm \leq Y_1\varepsilon^2,$$

where Y_1 depends on $\|\ell\|_\infty$ and $\|s_x^\varepsilon\|_{L^2}$. For large λ, we have

$$\|\frac{s^\varepsilon - s}{\varepsilon}_S \leq Y_2$$

or

$$\|\frac{z^\varepsilon - z}{\varepsilon}\|_S \leq Y_3$$

The result is a mild convergence to ϕ. Additionally, we

$$(\frac{z^\varepsilon - z}{\varepsilon})_m \rightharpoonup \phi_m$$

weakly in $L^2(0, M; P^*)$ and

$$\frac{z^\varepsilon - z}{\varepsilon} \rightharpoonup \phi$$

firmly in $L^2(D)$ as in Theorem 1 As a result, we draw the conclusion that ϕ is satisfied (12).

Now that $C(f)$ has been differentiated with regard to f at an optimal control, we can derive a variational inequality that describes an optimal control. □

Theorem 2 *[12] The adjoint issue can be resolved using the provided optimal controls, f in Z and v in S.*

$$\begin{cases} -v_m = bv_{xx} + (fv)_x + 1 \;\; in \;\; D, \\ v(x, M) = 0 \qquad on \;\; (0, I), m = M, \\ v(0, m) = 0 \qquad on \;\; (0, M), x = 0, \\ bv_x(I, m) + fv(I, m) = 0 \; on \;\; (0, M), x = I. \end{cases} \tag{14}$$

The following variational condition is satisfied by $f(x, m)$ as well:

$$\min(\max(-(f_{mm} + f_{xx} + \tfrac{\mu}{\gamma} v z_x), f - F), f + F) = 0 \tag{15}$$

where $z = z(f)$.

Remark 2 For information on what constitutes a weak answer to such a variational inequality, see ([9, 14]).

Proof Assume that $f(x, m)$ is a perfect control. Let $\ell \in H_0^1(D)$ s.t $f + \varepsilon\ell \in Z$ for $\varepsilon > 0$. The derivative of $C(f)$with respect to f in the direction of l fulfills the condition.

$$\begin{aligned} 0 &\le \lim_{\varepsilon \to 0^+} \frac{C(f + \varepsilon\ell) - C(f)}{\varepsilon} \\ &= \lim_{\varepsilon \to 0^+} [\mu \int_0^M \int_0^I \frac{z^\varepsilon - z}{\varepsilon}(x, m) dx dm \\ &+ \frac{\gamma}{2} \int_0^M \int_0^I (f_m + \varepsilon\ell_m)^2 - f_m^2 + (f_x + \varepsilon\ell_x)^2 - f_x^2(x, m) dx dm] \\ &= \mu \int_0^M \int_0^I \phi(x, m) dx dm + \gamma \int_0^M \int_0^I [f_m \ell_m + f_x \ell_x] dx dm \end{aligned} \tag{16}$$

We have this directional derivative being zero because our control set is not subject to any restrictions on point-wise inequality. The limited convergence of Lemma 1 was examined using $(z^\varepsilon - z)/\varepsilon$ When the weak version of the adjoint problem (14) is substituted in (16), given that v is the answer to the adjoint problem (14) [19],

$$0 \le \mu \int_0^M \int_0^I (-v_m\phi + bv_x\phi_x - (fv)_x\phi)(x,m)dxdm + \mu \int_0^M fv\phi(I,m)dm$$
$$+\gamma \int_0^M \int_0^I [f_m\ell_m + f_x\ell_x](x,m)dxdm \tag{17}$$

Adding components together produces

$$0 \le \mu\{\int_0^I -[\phi v]_0^M dx - \int_0^M [\phi f v]_0^I dm \int_0^M (fv\phi)(I,m)dm$$
$$+\int_0^M \int_0^I (v\phi_m + fv\phi_x + bv_x\phi_x)dxdm + \gamma \int_0^M \int_0^I [f_m\ell_m + f_x\ell_x](x,m)dxdm\}$$
$$= \mu\{\int_0^I [-\phi v(x,M) + \phi v(x,0)]dx - \int_0^M [fv\phi(I,m) - fv\phi(0,m)]dm\}$$
$$\mu \int_0^M \int_0^I (v\phi_m + bv_x\phi_x + fv\phi_x)dxdm + \gamma \int_0^M \int_0^I [f_m\ell_m + f_x\ell_x](x,m)dxdm$$
$$= \int_0^M \int_0^I (-\mu v\ell z_x + \gamma[f_m\ell_m + f_x\ell_x](x,m)dxdm \text{ (using 12 and using 16).} \tag{18}$$

To achieve the final equality, we substituted the inhomogeneous term $-\ell z_x$ from the sensitivity (12). The final equation provides the variational inequality (15) that a perfect control f must meet. □

Remark 3 Our optimality system (15) consists of two parabolic PDEs and an elliptic variational inequality, denoted as (3) and (14).

3 Optimal Control Problem Constrained by a Vibrating String

3.1 Preliminary

It takes into account the multiplicative distributive control problem for a vibrating string. The velocity term z_m coefficient, or $z(x,m)$, is multiplicative control. To get the desired profile with a quadratic cost, we use multiplicative control to acquire the state solution.

We demonstrate the existence of ideal control. An optimality system is used to offer criteria for establishing the unique optimal control.

Suppose that $I = (0, \ell)$ and $M > 0$. For the following hyperbolic equation, consider the initial boundary value issue:

$$\begin{cases} z_{mm} = z_{xx} - f z_m + g z + n, & \text{in} \quad D = I \times (0, M), \\ z = 0 & x = 0, \quad m \in (0, M), \\ z = 0 & x = \ell, \quad m \in (0, M), \\ z = z_0(x), & m = 0, \quad x \in I, \\ z_m = z_1(x), & m = 0, \quad x \in I, \end{cases} \tag{19}$$

where $z_0 \in H_0^1(I)$, $z_1 \in L^2(I)$, $n \in L^2(D)$, $g \in L^\infty(D)$. $f \in H_0^1(D)$ is a proportional control that multiplicates, or is bilinear. We presum that the control functions are a part of the permitted controls:

$$U = \{f \in H_0^1(D) | 0 \leq f \leq F\} \tag{20}$$

where F is a positive constant. Consider the objective function:

$$C(f) = \frac{\mu}{2} \int_D (z - u)^2 dx dm + \frac{\gamma}{2} \int_D [(f_x^2 + (f_m)^2] dx dm \tag{21}$$

where the designated target function is $z \in L^2(D)$. The cost of control is combined with a measurement of how closely the desired profile z resembles the actual profile in this objective function. We try to characterize f^* in a way that minimizes the function over the range of legal controls U.

$$C(f^*) = \min_{f \in U} C(f) \tag{22}$$

We'll make the claim that optimal control functions exist and back it up with examples. The control $f(x, m)$, which is a coefficient of a velocity term for the provided $f(x, m)$, will then be described. We show that there exist ideal control strategies that reduce the cost function. The optimality system is derived by differentiating the cost functional with regard to the control function f, and an optimal control is then described using an elliptic variational inequality.

3.2 Existence of an Optimal Control

The following definition identifies the weak solution of the problem (19).

Definition 2 Given $f \in U$, we say that a function $z \in C([0, M]; H_0^1(I))$, with $z_m \in C([0.M]; L^2(I))$, $z_{mm} \in C([0, M]; H^{-1}(I))$ is a weak solution of the problem (19) if

- For any $\phi \in H_0^1$, and a,e. $0 \leq m \leq M$,

$$\langle z_{mm}, \phi \rangle + B[z, \phi] + \int_I f z_m \phi \mathrm{dx} = \int_I n \phi dx$$

- $z(x.0) = z_0(x)$, $z_m(x, 0) = z_1(x)$; where $\langle ., . \rangle$ denotes the duality pairing of $H^{-1}(I)$, $H_0^1(I)$, and $B[z, \phi] = \int_I z_x \phi_x dx - \int_I gz\phi dx$ is bilinear from.

Set the notational conventions as follows:

$$\mathcal{H} = H_0^1(I) \times L^2(I)$$

$$\tilde{z} = (z, z_m).$$

For weak solutions to equations (19), the existence and uniqueness are taken into account by [26]. Consider the following theorem as we look into the possibility of optimal control issue solutions (22).

Theorem 3 *[26] (Presence of optimal control). The objective functional $C(f)$ for $f \in U$ can be minimized by an optimum control with the function $f^* \in U$.*

Proof Suppose $\{f^l\} \in U$ is a minimizing sequence that has the property

$$\lim_{l \longrightarrow \infty} C(f^l) = \inf_{f \in U} C(f)$$

Let z^l be a solution of Eq. (19) with control f^l. Denote $z^l = z(f^l)$. Using [26], we have

$$\begin{aligned} &\sup_{0 \leq m \leq M} (\|z^l(m)\|_{H_0^1(I)} + \|(z^l)m\|_{L^2(I)}) \\ &+\|(z^l)_{mm}(m)\|_{L^2([0,M];H^{-1}(I))} \\ &\leq C(\|n\|_{L^2([0,M];L^2(I))} | \|z_0\|_{H_0^1(I)} + \|z_1\|_{L^2(I)}) \end{aligned} \tag{23}$$

where a constant is C.

By weak compactness, a subsequence on which z^* exists in $C([0, M]; H^1(I))$ has the property that

$$\begin{aligned} z^l &\rightarrow z^* \text{ weakly in } L^\infty([0, M]; H_0^1(I)), \\ (z^l)_m &\rightarrow (z^*)_m \text{ weakly in } L^\infty([0, M]; L^2(I)), \\ (z^l)_{mm} &\rightarrow (z^*)_{mm} \text{ weakly in } L^\infty([0, M]; H^{-1}(I)), \\ f^l &\rightarrow f^* \text{ weakly in } H^1(D) \\ (f^l)_x &\rightarrow (f^*)_x \text{ weakly in } L^2(D) \\ (f^l)_m &\rightarrow (f^*)_m \text{ weakly in } L^2(D) \end{aligned} \tag{24}$$

Using the Simon [24] compactness result, we have $z^l \to z^*$ firmly in$L^\infty([0, M]; L^2(I))$, as defined by weak approach, we have

$$\int_0^M \langle (z^l)_{mm}, \phi \rangle dm$$

$$= -\int_D [(z^l)_x \phi_x + f^l((z^l)_m)\phi - z^l \phi - n\phi] dxdm$$

We have,

$$
\begin{aligned}
&|\int_0^M \int_0^\ell (f^l(z^l)_m \phi - f^*(z^*)_m \phi) dxdm| \\
&\le \int_0^M \int_0^\ell |(f^l - f^*)\phi||(z^l)_m| dxdm \\
&+|\int_0^M \int_0^\ell f^* \phi((z^l)_m - (z^*)_m) dxdm| \\
&\le (\int_0^M \int_0^\ell |f^l - f^*|\phi|^2 dxdm)^{\frac{1}{2}} (\int_0^M \int_0^\ell |(z^l)_m|^2 dxdm)^{\frac{1}{2}} \qquad (25) \\
&+|\int_0^M \int_0^\ell f^* \phi((z^l)_m - (z^*)_m) dxdm| \\
&\le C_1 (\int_0^M \int_0^\ell |f^l - f^*|^2 dxdm)^{\frac{1}{2}} (\int_0^M \int_0^\ell |(z^l)_m|^2 dxdm)^{\frac{1}{2}} \\
&|\int_0^M \int_0^\ell f^* \phi((z^l)_m - (z^*)_m) dxdm| \to 0, as (l \to \infty)
\end{aligned}
$$

where C_1 depends on the L^∞ bound on ϕ.

The first term converges, since $f^l \to f^*$ weakly in $H_1(D)$ and $\{(z^l)_m\}$is uniformly bounded in $L^2(D)$. Also, $f^*\phi \in L^2(D)$ and$(z^l)_m \to (z^*)_m$ weakly in $L^\infty([0, M]; L^2(I))$,the second term converges. Thus, $f^l(z^l)_m \to f^*(z^*)_m$ weakly in $L^2(D)$.

Using the weak formulation of $z^l$and going to the limit as $l \to \infty$, we obtain

$$\int_0^M \langle (z^*)_{mm}, \phi \rangle dm$$

$$= -\int_D [(z^*)_x \phi_x + f^*((z^*)_m)\phi - z^*\phi - n\phi] dxdm$$

Therefore, the state equation (19) with controls f^* has the solution $z^* = z(f^*)$ Since

$$\begin{aligned} C(f^*) &= \frac{\mu}{2} \int_D (z^* - u)^2 dxdm \\ &+ \frac{\gamma}{2} \int_D [((f^*)_x)^2 + ((f^*)_m)^2] dxdm \end{aligned} \tag{26}$$

When weak convergence is taken into account, lower-semicontinuity of the L^2 norm allows us to

$$\begin{aligned} C(f^*) &\leq \lim_{l \longrightarrow \infty} \frac{\gamma}{2} \int_D [((f^l)_x)^2 + ((f^l)_m)^2] dxdm \\ &+ \lim_{l \longrightarrow \infty} \frac{\mu}{2} \int_D (z^l - u)^2 dxdm \\ &\leq \lim_{l \longrightarrow \infty} \inf C(f^l) \\ &= \inf_{f \in U} C(f). \end{aligned} \tag{27}$$

Since f^* is the best control, we draw this conclusion. □

3.3 *Characterization of an Optimal Control*

By distinguishing the goal functional function $C(f)$ from the control functional function f, the optimality system is obtained. In order to show that the transformation is sufficiently differentiable, we must first show that $z = z(f)$ is a factor in $C(f)$.

$$f \to z(f)$$

Lemma 2 *[26] The mapping*

$$f \in U \to \tilde{z}(f) = (z(f), z_m(f)) \in L^2([L^2([0, M]; \mathcal{H})])$$

is distinguishable in the sense that:

$$\frac{\tilde{z}(f + \epsilon h) - \tilde{z}(f)}{\epsilon} \rightharpoonup \Psi$$

weakly in $L^2([0, M]; \mathcal{H})$ *as* $\epsilon \rightharpoonup 0$ *for any* f *satisfying* $f + \epsilon h \in U$ *for small* ϵ *and* $h \in H_0^1(D)$.

Moreover $\Psi = (\Psi, \Psi_m) \in L^2([0, M]; \mathcal{H})$ *is a weak solution of the following problem:*

$$\begin{cases} \Psi_{mm} = \Psi_{xx} - f\Psi_m - g\Psi - hz_m & in\ D \\ \Psi \quad = 0 & x = 0, \quad m \in (0, M) \\ \Psi \quad = 0 & x = \ell, \quad m \in (0, M) \\ \Psi \quad = 0 & m = 0, \quad x \in I \\ \Psi_m \ = 0 & m = 0, \quad x \in I \end{cases} \tag{28}$$

where $z = z(f)$

Proof Let $z^\epsilon = z(f + \epsilon\ell)$ and $z = z(f)$, then $(z^\epsilon - z)/\epsilon$ is a weak solution of

$$\begin{cases} (\frac{z^\epsilon - z}{\epsilon})_{mm} = (\frac{z^\epsilon - z}{\epsilon})_{xx} + f(\frac{z^\epsilon - z}{\epsilon})_m + g(\frac{z^\epsilon - z}{\epsilon}) - hz^\epsilon_m, & \text{in D}, \\ \frac{z^\epsilon - z}{\epsilon} = 0, & x = 0, m \in (0, M), \\ \frac{z^\epsilon - z}{\epsilon} = 0, & x = \ell, m \in (0, M), \\ \frac{z^\epsilon - z}{\epsilon} = 0, & m = 0, x \in I, \\ (\frac{z^\epsilon - z}{\epsilon})_m = 0, & m = 0, x \in I, \end{cases} \tag{29}$$

Using [26], we have

$$\begin{aligned} &\sup_{0 \le m \le M} (\|\frac{(z^\epsilon - z)}{\epsilon}\|_{H_0^1(I)} + \|\frac{z^\epsilon - z}{\epsilon}\|_{L^2(I)}) \\ &+ \|\frac{z^\epsilon_{mm} - z_{mm}}{\epsilon}\|_{L^2([0,M];H^{-1}(I))} \\ &\le C\|hz^\epsilon_m\|_{L^2([0,M];L^2(M))} \\ &\le C_1\|h\|_{L^\infty(D)} \sup_{0 \le m \le M} \|z^\epsilon_m\|_{L^2(I)} \end{aligned} \tag{30}$$

where the constant C_3 is used. Consequently, via weak compactness on a subspace, we have

$\frac{z^\epsilon_{mm} - z_{mm}}{\epsilon} \to \Psi_{mm}$ weakly in $L^\infty([0, M]; H^{-1}(I))$,

$\frac{z^\epsilon_m - z_m}{\epsilon} \to \Psi_m$ weakly in $L^\infty([0, M]; L^2(I))$,

$\frac{z^\epsilon - z}{\epsilon} \to \Psi$ weakly in $L^\infty([0, M]; H_0^1(I))$,

According to the definition of a weak solution, we have

$$\langle \frac{z_{mm}^{\epsilon} - z_{mm}}{\epsilon}, \phi \rangle + \int_I (\frac{z^{\epsilon} - z}{\epsilon})_x \phi_x dx$$
$$= - \int_I f(\frac{z^{\epsilon} - z}{\epsilon})_m \phi dx + \int_I g(\frac{z^{\epsilon} - z}{\epsilon}) \phi dx - \int_I h z_m^{\epsilon} \phi dx \quad (31)$$

for any $\phi \in H_0^1(I)$, a.e. $0 \leq m \leq M$.

Letting $\epsilon \to 0$ and using $z^{\epsilon} \to z, z_m^{\epsilon} \to z_m$ weakly in $L^2(I)$, we obtain that Ψ is the weak solution of (28).

We now show what is needed for optimal control. □

Theorem 4 *[26] Considering optimal control f and matching*

$$\tilde{z} = \tilde{z}(f) = (z, z_m),$$

there is a dubious answer $\tilde{v} = (v, v_m) in \mathcal{H}$ to the adjoint problem:

$$\begin{cases} v_{mm} = v_{xx} + (fv)_m + gv + (z - u), & inD, \\ v = 0, & x = 0, \quad m \in (0, M), \\ v = 0, & x = \ell, \quad m \in (0, M), \\ v = 0, & m = M, \quad x \in I, \\ v_m = 0, & m = M, \quad x \in I, \end{cases} \quad (32)$$

Furthermore, f satisfies the following variational inequality

$$\min(\max(-(f_{mm} + f_{xx} + \tfrac{\mu}{\gamma} v z_m), f - F), f + F) = 0 \quad (33)$$

where $z = z(f)$.

Proof Let $f \in U$ be an optimal control and $z = z(f)$ be the corresponding optimal solution. Let $f + \epsilon h \in U$ for $\epsilon > 0$ and$z\epsilon = z(f + \epsilon h)$ be the corresponding weak solution of equation (19). We calculate the objective functional $C(f)$ directional derivative with respect to f in the direction of h. Since f is the intended target for C to reach its minimum, we obtain

$$
\begin{aligned}
0 \leq & \lim_{\epsilon \longrightarrow 0^+} \frac{C(f+\epsilon h)-C(f)}{\epsilon} \\
= & \lim_{\epsilon \longrightarrow 0^+} \{\frac{\mu}{2\epsilon} \int_D [(z^\epsilon - u)^2 - (z-u)^2] dx dm\} \\
+ & \lim_{\epsilon \longrightarrow 0^+} \{\frac{\gamma}{2\epsilon} \int_D [(f_m + \epsilon h_m)^2 - (f_m^2)] dx dm\} \\
+ & \lim_{\epsilon \longrightarrow 0^+} \{\frac{\gamma}{2\epsilon} \int_D [(f_x + \epsilon h_x)^2 - (f_x)^2] dx dm\} \\
= & \mu \int_D \Psi(z-u) dx dm + \gamma \int_D [f_x h_x + f_m h_m] dx dm
\end{aligned} \tag{34}
$$

where *Lemma* 2 has the definition of Ψ. We need to show that there is a solution to the adjoint problem (32). It does not follow that the state equation (19) has an existence and uniqueness of solution because the adjoint equation contains the term $(fv)_m$.

When stated in semigroup form, the adjoint equation looks like this:

$$
\frac{d}{dm}\tilde{v}(m) = A\tilde{v}(m) + \begin{pmatrix} 0 \\ (fv)_m + gv + z - u \end{pmatrix}
$$

where $A = \begin{pmatrix} 0 & I \\ \frac{\partial^2}{\partial x^2} & 0 \end{pmatrix}$ creates a unitary group on the matrix $\mathcal{H}$.

We anticipate a (32) solution that takes the following form

$$
\tilde{v}(m) = \int_m^M e^{A(m-\tau)} \begin{pmatrix} 0 \\ (fv)_m + gv + z - u \end{pmatrix} d\tau
$$

Integration from components results in,

$$
\begin{aligned}
\tilde{v}_m = & \int_m^M (-Ae^{A(m-\tau)} \begin{pmatrix} 0 \\ fv \end{pmatrix} d\tau \\
+ & \int_m^M (e^{A(m-\tau)} \begin{pmatrix} 0 \\ gv + z - u \end{pmatrix} + \begin{pmatrix} 0 \\ fv(m) \end{pmatrix}
\end{aligned} \tag{35}
$$

Now consider the map

$$
\begin{aligned}
F\tilde{v}(m) &= \int_m^M (-Ae^{A(m-\tau)} \begin{pmatrix} 0 \\ fv \end{pmatrix} d\tau \\
&+ \int_m^M (e^{A(m-\tau)} \begin{pmatrix} 0 \\ gv + z - u \end{pmatrix} d\tau \begin{pmatrix} 0 \\ fv(m) \end{pmatrix}
\end{aligned}
\tag{36}
$$

Sonawane et al. [26] show that the map F has a singular fixed point on $[0, M]$, establishing the existence and singularity of the weak solution of the adjoint equation.

Assume that v is an inadequate solution to the problem (32), satisfying just v.

$$
\langle v_{mm}, \phi\rangle(m) + \int_I v_x\phi_x(m)dx = \int_I (fv)_m\phi dx + \int_I gv\phi dx + \int_I (z-u)\phi(m)dx
$$

for any $\phi \in H_0^1(I)$, and $a.e.\ 0 \le m \le M$. Then

$$
\begin{aligned}
0 &\le \mu \int_D \Psi(z-u)dxdm + \gamma \int_D [f_x h_x + f_m h_m]dxdm \\
&= \mu \int_0^M \{\langle v_{mm}, \Psi\rangle(m) + \int_I v_x\Psi_x(m)dx - \int_I (fv)_m\Psi\}dxdm \\
&- \mu \int_0^M \{\int_I gv\Psi dx\}dm + \gamma \int_D [f_x h_x + f_m h_m]dxdm \\
&= \mu \int_0^M \{\int_I -hz_m v dx\}dm + \gamma \int_D [f_x h_x + f_m h_m]dxdm \\
&= \int_D \{-\mu h z_m v + \gamma[f_x h_x + f_m h_m]\}dxdm
\end{aligned}
\tag{37}
$$

The variational inequality is provided by the final equation (33).

Using (19, 32), and the variational inequality, we are able to create the optimality system (33). □

The best management of the state and gradient in distributed bilinear systems will be discussed in this section.

Competing Interests The authors declare that they have no conflicts of interest.

References

1. B.D. Anderson, J.B. Moore, Optimal Control: Linear Quadratic Methods. (Courier Corporation, 2007)
2. V. Ayala, E. Cruz, W. Kliemann, L.R. Laura-Guarachi, Controllability properties of bilinear systems in dimension 2. J. Math. Comput. Sci. **16**, 554–575 (2016)
3. H. Banks, P. Lamm, Estimation of variable coefficients in parabolic distributed systems. IEEE Trans. Autom. Control **30**(4), 386–398 (1985)
4. L. Berrahmoune, Stabilization and decay estimate for distributed bilinear systems. Syst. Control Lett. **36**(3), 167–171 (1999)
5. S. Boulaaras, Existence and uniqueness for solutions of parabolic quasi-variational inequalities with impulse control and nonlinear source terms. J. Nonlinear Sci. Appl. **9**, 568–583 (2016)
6. M.E. Bradley, S. Lenhart, Bilinear optimal control of a Kirchhoff plate. Syst. Control Lett. **22**(1), 27–38 (1994)
7. M.E. Bradley, S. Lenhart, J. Yong, Bilinear optimal control of the velocity term in a Kirchhoff plate equation. J. Math. Anal. Appl. **238**(2), 451–467 (1999)
8. S. Chawla, S.M. Lenhart, Application of optimal control theory to bioremediation. J. Comput. Appl. Math. **114**(1), 81–102 (2000)
9. M. Chipot, *Variational Inequalities and Flow in Porous Media*, vol. 52 (Springer Science and Business Media, 2012)
10. R.F. Curtain, Finite dimensional compensators for parabolic distributed systems with unbounded control and observation. SIAM J. Control Optim. **22**(2), 255–276 (1984)
11. L.C. Evans, *Partial Differential Equations*, vol. 19 (American Mathematical Society, 2022)
12. H.R. Joshi, Optimal control of the convective velocity coefficient in a parabolic problem. Nonlinear Anal. Theory Methods Appl. **63**(5–7), e1383–e1390 (2005)
13. B. Kafash, A. Delavarkhalafi, Restarted state parameterization method for optimal control problems. J. Math. Comput. Sci. **14**, 151–161 (2015)
14. D. Kinderlehrer, G. Stampacchia, *An introduction to variational inequalities and their applications* (Society for Industrial and Applied Mathematics, 2000)
15. S. Klainerman, S. Selberg, Bilinear estimates and applications to nonlinear wave equations. Commun. Contemp. Math. **4**(02), 223–295 (2002)
16. S. Lenhart, Optimal control of a convective-diffusive fluid problem. Math. Models Methods Appl. Sci. **5**(02), 225–237 (1995)
17. S. Lenhart, M. Liang, V. Protopopescu, Identification problem for a wave equation via optimal control, in *Control of Distributed Parameter and Stochastic Systems: Proceedings of the IFIP WG 7.2 International Conference*, June 19–22, 1998 Hangzhou, China (Springer, US, 1999), pp. 79–84
18. X. Li, J. Yong, *Optimal Control Theory for Infinite Dimensional Systems* (Springer Science and Business Media, 2012)
19. J.L. Lions, E. Magenes, *Non-homogeneous Boundary Value Problems and Applications: Vol. 1*, vol. 181 (Springer Science and Business Media, 2012)
20. N. Nguyen, M. Ardema, Predictive optimal control of a hyperbolic distributed model for a wind tunnel. J. Guidance Control Dyn. **29**(3), 626–634 (2006)
21. J.A. Rad, S. Kazem, K. Parand, Optimal control of a parabolic distributed parameter system via radial basis functions. Commun. Nonlinear Sci. Numer. Simul. **19**(8), 2559–2567 (2014)
22. M.O. Sidi, S.A. Beinane, Gradient optimal control problems for a class of infinite dimensional systems. Nonlinear Dyn. Syst. Theory **20**(3), 316–326 (2020)
23. M.O. Sidi, An output controllability of bilinear distributed system. Int. Rev. Autom. Control **3**(5), 466–473 (2010)
24. J. Simon, Compact sets in the space L p (O, T; B). Annali di Matematica pura ed applicata **146**, 65–96 (1986)
25. Slemrod, M.: Controllability of distributed bilinear systems, in *1981 20th IEEE Conference on Decision and Control including the Symposium on Adaptive Processes* (IEEE, 1981), pp. 240–240

26. R.B. Sonawane, A. Kumar, S.B. Nimse, Optimal control for a vibrating string with variable axial load and damping gain. IFAC Proc. Vol. **47**(1), 75–81 (2014)
27. J. Yong, Linear-quadratic optimal control problems for mean-field stochastic differential equations. SIAM J. Control Optim. **51**(4), 2809–2838 (2013)
28. El Hassan Zerrik, O. Castillo, *Stabilization of Infinite Dimensional Systems* (Springer International Publishing, 2021)
29. E.H. Zerrik, M.O. Sidi, Constrained regional control problem for distributed bilinear systems. IET Control Theory Appl. **7**(15), 1914–1921 (2013)
30. Y. Zhang, Y. Yin, H. Wu, D. Guo, Zhang dynamics and gradient dynamics with tracking-control application, in *2012 Fifth International Symposium on Computational Intelligence and Design*, vol. 1 (IEEE, 2012), pp. 235–238
31. W. Zheng, W. Wu, A. Gomez-Exposito, B. Zhang, Y. Guo, Distributed robust bilinear state estimation for power systems with nonlinear measurements. IEEE Trans. Power Syst. **32**(1), 499–509 (2016)
32. R. Zine, M. Ould Sidi, Regional optimal control problem with minimum energy for a class of bilinear distributed systems. IMA J. Math. Control Inf. **35**(4), 1187–1199 (2018)

The manufacturer's authorised representative in the EU is Springer Nature Customer Service Centre GmbH, Europaplatz 3, 69115 Heidelberg, Germany. If you have any concerns regarding our products, please contact ProductSafety@springernature.com

Printed and bound by CPI Group (UK) Ltd, Croydon, CR0 4YY
01/10/2026
02225941-0007